AA004059

International Workshop on Quantum Chromodynamics – Theory and Experiment (QCD@Work 2018)

EPJ Web of Conferences Volume 192 (2018)

Matera, Italy
25 - 28 June 2018

Editors:

Giuseppe E. Bruno
Gabriele Chiodini
Pietro Colangelo
Claudio Coriano
Donato M. Creanza
Fulvia De Fazio
Eugenio Nappi

ISBN: 978-1-5108-7478-7

Printed from e-media with permission by:

Curran Associates, Inc.
57 Morehouse Lane
Red Hook, NY 12571

Some format issues inherent in the e-media version may also appear in this print version.

This work is licensed under a Creative Commons Attribution 4.0 International Licence. Licence details: http://creativecommons.org/licenses/by/4.0/.

Printed by Curran Associates, Inc. (2019)

For additional information, please contact EDP Sciences – Web of Conferences at the address below.

EDP Sciences – Web of Conferences
17, Avenue du Hoggar
Parc d'Activité de Courtabœuf
BP 112
F-91944 Les Ulis Cedex A
France

Phone: +33 (0) 1 69 18 75 75
Fax: +33 (0) 1 69 28 84 91

contact-edps@webofconferences.org

Additional copies of this publication are available from:

Curran Associates, Inc.
57 Morehouse Lane
Red Hook, NY 12571 USA
Phone: 845-758-0400
Fax: 845-758-2633
Email: curran@proceedings.com
Web: www.proceedings.com

TABLE OF CONTENTS

Editors

Giuseppe E. Bruno
Dipartimento Interateneo di Fisica "M. Merlin", Università e Politecnico di Bari
Via Orabona 4, 70126 Bari, Italy
e.mail: giuseppe.bruno@ba.infn.it

Gabriele Chiodini
Istituto Nazionale di Fisica Nucleare – Sezione di Lecce
Via Arnesano, 73100 Lecce, Italy
e.mail: gabriele.chiodini@le.infn.it

Pietro Colangelo
Istituto Nazionale di Fisica Nucleare – Sezione di Bari
Via Orabona 4, 70126 Bari, Italy
e.mail: pietro.colangelo@ba.infn.it

Claudio Corianò
Dipartimento di Matematica e Fisica "E. De Giorgi", Università del Salento
Via Arnesano, 73100 Lecce, Italy
e.mail: claudio.coriano@le.infn.it

Donato M. Creanza
Dipartimento Interateneo di Fisica "M. Merlin", Università e Politecnico di Bari
Via Orabona 4, 70126 Bari, Italy
e.mail: donato.creanza@ba.infn.it

Fulvia De Fazio
Istituto Nazionale di Fisica Nucleare – Sezione di Bari
Via Orabona 4, 70126 Bari, Italy
e.mail: fulvia.defazio@ba.infn.it

Eugenio Nappi
Istituto Nazionale di Fisica Nucleare – Sezione di Bari
Via Orabona 4, 70126 Bari, Italy
e.mail: eugenio.nappi@ba.infn.it

Scientific Advisory Committee

Roberto Casalbuoni (Firenze University)
Fernando Ferroni (Roma Sapienza University and INFN)
Michelangelo Mangano (CERN Geneva)
Nello Paver (Trieste University)
Mikhail A. Shifman (Minnesota University)
George Sterman (SUNY Stony Brook)
Sheldon Stone (Syracuse University)

Local Organizing Committee

Leonardo Angelini (Bari University)
Giuseppe E. Bruno (Bari Polytechnic)
Gabriele Chiodini (INFN Lecce)
Pietro Colangelo (INFN Bari)
Claudio Corianò (Università del Salento)
Donato M. Creanza (Bari Polytechnic)
Fulvia De Fazio (INFN Bari)
Eugenio Nappi (INFN Bari)
Stefania Spagnolo (Università del Salento)

Scientific organization

Fabio Colamaria (INFN Bari)
Floriana Giannuzzi (Bari University)
Francesco Loparco (Bari University)
Stefano Nicotri (INFN Bari)

Participants

ABUKI, Hiroaki	Aichi University of Education, Kariya, JAPAN
ANGELINI, Leonardo	University of Bari and INFN - Sezione di Bari, ITALY
BARABANOV, Mikhail	JINR, Dubna, RUSSIA
BEDOLLA HERNANDEZ, Marco A.	INFN - Sezione di Genova, ITALY
BHATTACHARYA, Srimoy	IIT, Guwati, INDIA
BRUNO, Giuseppe Eugenio	Politecnico di Bari and INFN - Sezione di Bari, ITALY
BURAS, Andrzej	TUM- IAS, Munich, GERMANY
CAFFARRI, Davide	NIKHEF, Amsterdam, NETHERLANDS
CAMPANELLA, Stefano	University of Bari, ITALY
GARCIA, Josu	Oklahoma State University, Stillwater, USA
CASALBUONI, Roberto	University of Firenze, ITALY
CHIODINI, Gabriele	INFN - Sezione di Lecce, ITALY
CHIRILLI, Giovanni Antonio	University of Regensburg, Regensburg, GERMANY
COLAMARIA, Fabio	INFN - Sezione di Bari, ITALY
COLANGELO, Pietro	INFN - Sezione di Bari, ITALY
CORIANÓ, Claudio	University of Salento, Lecce, ITALY
COSTANTINI, Antonio	University of Salento and INFN - Sezione di Lecce, ITALY
CREANZA, Donato	Politecnico di Bari, ITALY
D'AMBROSIO, Giancarlo	INFN - Sezione di Napoli, ITALY
D'ANGELO, Annalisa	University of Tor Vergata and INFN - Roma 2, ITALY
DE FAZIO, Fulvia	INFN - Sezione di Bari, ITALY
DEL DUCA, Vittorio	ETH Zurich and INFN LNF, Zurich, SWITZERLAND
DELLE ROSE, Luigi	INFN - Sezione di Firenze, ITALY
DI GIACOMO, Adriano	University of Pisa, ITALY
FAJFER, Svjetlana	Ljubljana Univ. and J. Stefan Inst., Ljubljana, SLOVENIA
FANTECHI, Riccardo	INFN - Sezione di Pisa, ITALY
FAZIO, Angelo Raffaele	Universidad Nacional de Colombia, Bogotá, COLOMBIA
FINI, Rosa Anna	INFN - Sezione di Bari, ITALY
FRAMPTON, Paul	University of Salento, Lecce, ITALY
GASTALDI, Ugo	INFN - Sezione di Ferrara, ITALY
GIACHINO, Alessandro	University of Genova and INFN, ITALY
GIANNUZZI, Floriana	University of Bari, Bari, ITALY
GUEVARA, Adolfo	Madrid University, Madrid, SPAIN
GUIOT, Benjamin	Univ. Técnica Federico Santa Maria, Valparaiso, CHILE
GUZZI, Marco	Kennesaw State University, Kennesaw, USA
HAKOBYAN, Hayk	Univ. Técnica Federico Santa Maria, Valparaiso, CHILE
HUANG, Mei	IHEP, CAS, Beijing, CHINA
IVANOV, Mikhail	JINR, Dubna, RUSSIA
IYER, Abhishek	INFN - Sezione di Napoli, ITALY
KATRAMATOU, Mina	Kent State University, Kent, USA
KHACHATRYAN, Grigor	Yerevan Physics Institute, Yerevan, ARMENIA
KIDONAKIS, Nikolaos	Kennesaw State University, Kennesaw, USA
KOERNER, Juergen	University of Mainz, Mainz, GERMANY
KULCHITSKY, Yuri	JINR/ Belarus Academy, Dubna, RUSSIA

LOPARCO, Francesco	University of Bari, Bari, ITALY
LUCHA, Wolfgang	Austrian Academy of Sciences, Vienna, AUSTRIA
MAGLIO, Matteo Maria	INFN - Sezione di Lecce, ITALY
MANNARELLI, Massimo	LNGS INFN, ITALY
MARGETIS, Spyridon	Kent State University, Kent, USA
MARTINI, Alberto	INFN - Sezione di Roma 3, ITALY
MASTROSERIO, Annalisa	INFN - Sezione di Bari, ITALY
MEGIAS, Eugenio	Max Planck Institut fur Physik, Munich, GERMANY
MELIKHOV, Dmitri	Moscow State University, Moscow, RUSSIA
MUNIER, Stéphane	CPHT, Ecole Polytechnique, CNRS, Palaiseau, FRANCE
MURGIA, Francesco	INFN - Sezione di Cagliari, ITALY
MY, Salvatore	University of Bari and INFN, Bari, ITALY
NAPPI, Eugenio	INFN - Sezione di Bari, ITALY
NICCOLAI, Silvia	IPN Orsay, Orsay, FRANCE
NICOTRI, Stefano	INFN - Sezione di Bari, ITALY
OZPINECI, Altug	METU, Ankara, TURKEY
PALANO, Antimo	University of Bari and INFN, Bari, ITALY
PETRATOS, Gerassimos	Kent State University, Kent, USA
REDMER, Christoph Florian	University of Mainz, Mainz, GERMANY
REYES ROJAS, Edilson Alfonso	Universidad Nacional de Colombia, Bogotá, COLOMBIA
RUGGIERI, Marco	Lanzhou University, Lanzhou, CHINA
SANDACZ, Andrzej	National Centre for Nuclear Research, Warsaw, POLAND
SANTOPINTO, Elena	INFN - Sezione di Genova, ITALY
SANTORELLI, Pietro	University of Napoli and INFN, ITALY
SCRIMIERI, Egidio	University of Bari and INFN, ITALY
SOHAIL, Muhammad	INFN - Sezione di Roma 3, ITALY
STROLOGAS, John	University of Ioannina, Ioannina, GREECE
TRAN, Thang	Joint Institute for Nuclear Research, Dubna, RUSSIA
TRENTADUE, Luca	University of Parma and INFN, ITALY
TRIANTAFYLLOPOULOS, Dionysios	ECT*, Trento, ITALY
TROMBETTA, Giuseppe	INFN - Sezione di Bari, ITALY
VECCHI, Stefania	INFN - Sezione di Ferrara, ITALY
VENDITTI, ROSAMARIA	INFN - Sezione di Bari, ITALY
VON TOERNE, Eckhard	University of Bonn, Bonn, GERMANY
WATANABE, Akira	IHEP, CAS, Beijing, CHINA
WOLF, György	MTA Wigner RCP, Budapest, HUNGARY
ZAMIRALOV, Valeriy	Moscow State University, Moscow, RUSSIA

Preface

The 2018 edition of **QCD@Work - International Workshop on Quantum Chromodynamics - Theory and Experiment** was held in Matera, Italy, from 25 to 28 June 2018. The series started in 2001 in Martina Franca, and continued in 2003 and 2005 in Conversano, 2007 and 2010 in Martina Franca, 2012 in Lecce, 2014 in Giovinazzo, and 2016 again in Martina Franca, Italy. The aim of all the nine meetings was to bring together senior and young researchers, theorists and experimentalists, putting QCD to work in different frameworks, for a common discussion.

The topics debated during the workshop have been:

- Low energy physics and nonperturbative QCD
- Advances in perturbative QCD
- Heavy quarks
- QCD in extreme conditions of temperature and density
- Holographic methods for strong interactions.

About 80 researchers (with a large component of young scientists) lively participated in the workshop, that comprised oral and poster contributions. This volume registers the outcome of the meeting.

As organizers of the event, we wish to thank the institutions and individuals who have contributed to its realization. We acknowledge the financial support of Istituto Nazionale di Fisica Nucleare (INFN), Mathematics and Physics Department "Ennio De Giorgi" of Salento University in Lecce, Politecnico di Bari, Commissioni Scientifiche Nazionali INFN III and IV, Theory group of INFN-Sezione di Bari. We are grateful to Mariarosa Buonfiglio, INFN Sezione di Bari, and Loredana Napolitano, Physics Department "Michelangelo Merlin" of Bari University and Polytechnic, for advice in the administrative management, and to Fabio Colamaria, Floriana Giannuzzi, Francesco Loparco and Stefano Nicotri for support in the scientific organization.

We are grateful to Soprintendenza Archeologia, Belle Arti e Paesaggio della Basilicata, in particular to Architetto Biagio Lafratta, for the availability, as prestigious conference venue, of Chiesa del Cristo Flagellato in the ancient Ospedale di San Rocco in Matera, town belonging to the Unesco World Heritage and elected as 2019 European Capital of Culture. We also thank Donato Loparco and LAB43 - Officina delle idee, for the on site technological organization.

Finally, we are grateful to maestro Michele Volpicella, Bari, for having allowed the use of one of his paintings as logo of QCD@Work 2018.

Giuseppe E. Bruno, Gabriele Chiodini, Pietro Colangelo, Claudio Corianò,
Donato M. Creanza, Fulvia De Fazio, Eugenio Nappi

edp sciences

EPJ Web of Conferences

Statement of Peer review

In submitting conference proceedings to EPJ Web of Conferences, I certify to the Publisher that I adhere to the **Policy on Publishing Integrity** of the journal in order to safeguard good scientific practice in publishing.

1. All articles have been subjected to peer review administered by the proceedings editors.
2. Reviews have been conducted by expert referees, who have been requested to provide unbiased and constructive comments aimed, whenever possible, at improving the work.
3. Proceedings editors have taken all reasonable steps to ensure the quality of the materials they publish and their decision to accept or reject a paper for publication has been based only on the merits of the work and the relevance to the journal.

Title, date and place of the conference QCD@Work 2018

June 25-28, 2018, Matera (Italy)

Proceedings editor(s):

Giuseppe Eugenio Bruno, Gabriele Chiodini, Donato Maria Creanza, Pietro Colangelo
Fulvia De Fazio, Eugenio Nappi

Date and editor's signature Bari, 18 October 2018

17, avenue du Hoggar - PA de Courtabœuf – BP 112 - 91944 Les Ulis Cedex A (France)
Tél. : 33 (0)1 69 18 75 75 - Fax : 33(0)1 69 07 45 17 – www.edpsciences.org

Confinement, Monopoles and Gauge Invariance

Adriano Di Giacomo[1,*]

[1]Dipartimento di Fisica Universita' Pisa and INFN Sezione di Pisa , Largo Pontecorvo 56127 Pisa Italy

Abstract. The status is reviewed of the lattice studies of confinement by dual superconductivity of QCD vacuum. Progress is made by proving that the existence and the creation of a monopole are gauge invariant concepts. This proof fully legitimates an existing order parameter, and with it the evidence it provides of monopole condensation in the confining vacuum.

1 Introduction

Quarks are strictly confined in nature. No free quark has ever been observed neither in ordinary matter nor in the final state of high energy reactions. This results in upper limits on the abundance n_q of quarks and on the inclusive cross section σ_q for production of quarks in, say, proton-proton collisions

$$R_n \equiv \frac{n_q}{n_p} \leq 10^{-27} \qquad\qquad R_\sigma \equiv \frac{\sigma_q}{\sigma_{TOT}} \leq 10^{-15} \tag{1}$$

Here n_q is the abundance of protons, σ_{TOT} the total $p - p$ cross section. The expectations for these numbers would be, if quarks could exist as free particles[1],

$$R_n \approx 10^{-12} \qquad\qquad R_\sigma \approx 1 \tag{2}$$

The natural explanation of an inhibition factor smaller than 10^{-15} is that R_n and σ_q are strictly zero: confinement is an absolute property protected by some symmetry. As a consequence the deconfinement transition is a change of symmetry.

The symmetry involved can not be a subgroup of the gauge group, since the order parameter has to be gauge invariant [2]. A hint for the nature of this symmetry is provided by the following argument. Let $U_G(\vec{x})$ be a time-independent transformation of the gauge group on a system of fields. In general

$$U_G(\vec{x}) = U(\vec{x})U_B(\vec{x}) \tag{3}$$

where $U(\vec{x})$ is the usual gauge transformation acting on the bulk of the system, which is trivial [$U(\vec{x}) = 1$] on the surface S_2 at spatial infinity. $U_B(\vec{x})$ is instead trivial [$U_B(\vec{x}) = 1$] everywhere except on S_2.

If all the fields vanish on S_2 $U_B(\vec{x})$ is irrelevant and can be put equal to 1 and U_G reduces to $U(\vec{x})$, the usual gauge transformation. It is instead relevant if there are degrees of freedom living on S_2. These degrees of freedom can produce a gauge invariant change of symmetry. Living on a sphere they can have non trivial topology.

*e-mail: adriano.digiacomo@df.unipi.it

© The Authors, published by EDP Sciences. This is an open access article distributed under the terms of the Creative Commons Attribution License 4.0 (http://creativecommons.org/licenses/by/4.0/).

The natural topological excitations in 3 spatial dimensions are monopoles, defined by the mapping $S_2 \rightarrow SU(2)$. Vortices would be the natural excitations in 2 spatial dimensions with a mapping $S_1 \rightarrow U(1)$.

The above argument supports the idea that confinement be produced by dual superconductivity of the vacuum [3] [4]. Dual means that electric and magnetic are interchanged with respect to ordinary superconductors. Monopoles condense in the QCD vacuum in the same way as Cooper pairs do in ordinary superconductors, and quarks are confined by Abrikosov chromo-electric flux tubes in the same way as magnetic charges are confined by magnetic flux tubes in ordinary superconductors.

Lattice is a convenient non perturbative tool to explore large distance properties of QCD like confinement, for which the perturbative expansion fails. In particular monopoles have been extensively studied on lattice to verify the mechanism of confinement by dual superconductivity of the vacuum, proposed in [3] [4], and further developed in [5]. Two main strategies have been used to that purpose

1) Define an operator $\mu(\vec{x})$ which creates a monopole in the point $\vec{x}$ and use its vacuum expectation value $\langle\mu\rangle \equiv \langle 0|\mu|0\rangle$ as the order parameter for dual superconductivity of vacuum [6] [7][8][9][10][11].

2) Detect monopoles in lattice configurations and determine the constraint effective potential of their field. A minimum at non zero field signals dual superconductivity of the vacuum [12].

Both approaches look for a change of symmetry at the deconfinement transition. Both have to choose a $U(1)$ subgroup of the gauge group in which monopoles live, which is named Abelian Projection (AP) . A natural AP exists in the Higgs model in which monopoles were originally discovered [13][14], and is the unitary representation in which the Higgs field Φ is diagonal. Monopoles sit at the zeroes of Φ.[See Sect. 2 below]

In QCD there is no fundamental Higgs field. A possibility could be that any operator in the adjoint representation can act as effective Higgs field [5], and its zeroes as locations of the monopoles. Somehow physics should be independent on the specific choice of the operator [5].

Following this suggestion the operator $\mu(\vec{x})$ of strategy 1) was defined and developed in Ref [8][10] using the simplest choice for the colour orientation of the effective Higgs field, the nominal 3-direction used in the numerical updating procedure.

As for strategy 2) what one needs is a reliable way to detect monopoles in lattice configurations in order to determine numerically from them the effective potential. The way this is done is by measuring the magnetic flux of the abelian field in the colour direction of the effective Higgs field: excess of magnetic flux through a plaquette is interpreted as Dirac string of a monopole[15]. If one repeats this game in different abelian projections and compares, the result is rather disturbing: the number and the location of monopoles strongly depend on the abelian projection. The existence of a monopole seems to be a gauge dependent concept.

In this paper we discuss these issues and in particular we show that the existence or the creation of a monopole is a gauge invariant, AP independent concept [Sect. 2]. This legitimates the construction of $\langle\mu\rangle$ used to implement strategy 1) and the idea that all choices for effective Higgs field are equivalent [5].

As for strategy 2) we discuss in some detail the operators in the theory which could act as effective Higgs fields in the sense of Ref [5]. The result is that a key role is played by the temporal component $\vec{A}_4 = i\vec{A}_0$ of the gauge field in the gauge $\partial_4\vec{A}_4 = 0$, i.e. the AP in which the Polyakov line is diagonal.

An intrinsic difficulty in the method is to detect monopoles in the confined phase, in which the magnetic field is shielded, the magnetic photon being massive [Sect. 3].

Regarding the AP dependence of the monopoles observed in lattice configurations we argue that it is due to discretisation artefacts in the procedure used to detect monopoles on the lattice. This is discussed in Sect.3. Sect 4 contains some concluding remarks.

A copious literature exists on monopoles in QCD based on the concept of dominance, i.e. on the demonstration that removing somehow the monopoles from the lattice configurations removes confinement [16] [17]. In this optics some abelian projection (Maximal Abelian Gauge) looks better than others, with less lattice artefacts. We do not consider this approach relevant to understand the symmetry governing confinement.

2 Monopoles in Non-Abelian Gauge Theories

Monopoles were discovered as solitons in the broken phase of the $SU(2)$ Higgs model with the scalar field $\vec{\Phi}$ in the adjoint representation [13][14]. The Lagrangian is

$$L = -\frac{1}{4}\vec{G}_{\mu\nu}\vec{G}_{\mu\nu} + \frac{1}{2}D_\mu\vec{\Phi}D_\mu\vec{\Phi} - \frac{m^2}{2}\vec{\Phi}\vec{\Phi} - \frac{\lambda}{4}(\vec{\Phi}\vec{\Phi})^2 \tag{4}$$

The monopole solution has the form

$$\Phi^a = \rho\frac{x^a}{|\vec{x}|}H(\xi), \quad A_0^a = 0, \quad A_i^a = -\frac{1}{gx^2}\epsilon_{iab}x_b[1 - K(\xi)] \tag{5}$$

Here $\xi = g\rho|\vec{x}|$, $\quad \rho \equiv \sqrt{(-\frac{m^2}{\lambda})}$.

$H(\xi)$ and $K(\xi)$ can be determined numerically. In order to have a solution with finite energy (soliton) they must obey the boundary conditions

$$K(\infty) = 0, \quad K(0) = 1 \qquad H(\infty) = 1, \quad H(0) = 0 \tag{6}$$

$\vec{\Phi}$ has a zero at $\vec{x} = 0$, which is conventionally assumed to be the position of the monopole. At small $|\vec{x}|$

$$\Phi^a_{\vec{x}\to 0} \approx constant \times x^a \tag{7}$$

At large $|\vec{x}|$

$$\Phi^a{}_{|\vec{x}|\to\infty} = \rho\frac{x^a}{|\vec{x}|} \tag{8}$$

Eq(8) is a non trivial mapping of the sphere S_2 on $SU(2)/U(1)$: its winding number is the magnetic charge.

The unitary gauge, in which $\vec{\Phi}\vec{\sigma}$ is diagonal ($\vec{\Phi}\vec{\sigma} = |\vec{\Phi}|\sigma_3$) is by definition the AP identified by Φ. The abelian magnetic field in this gauge $\quad B_k^3 \equiv \frac{1}{2}\epsilon_{kij}F_{ij}^3$, ($F_{ij}^3 \equiv \partial_i A_j^3 - \partial_j A_i^3$), can be given a manifestly gauge invariant form in terms of the 'tHooft tensor F_{ij} [13] . If we denote by $\hat{\Phi} = \frac{\vec{\Phi}}{|\vec{\Phi}|}$ is the orientation of $\vec{\Phi}$ in colour space

$$F_{ij} \equiv \hat{\Phi}[\vec{G}_{ij} - \frac{1}{g}D_i\hat{\Phi} \wedge D_j\hat{\Phi}] \tag{9}$$

In F_{ij} the terms quadratic in the gauge field cancel by construction, linear terms are proportional to a spatial derivative $\partial_k\hat{\Phi}$ and vanish in the unitary gauge in which $\hat{\Phi}$ is constant, so that only the abelian part of $\vec{G}_{ij}$ is left , $F_{ij} = F^3_{ij}$ and for the magnetic field $B^3_k = B_k = \frac{1}{2}\epsilon_{kij}F_{ij}$. At distances $r > \frac{1}{\rho}$ the field $\vec{B}$ is that of an abelian magnetic monopole

$$\vec{B} \approx_{r\to\infty} \frac{1}{g}\frac{\vec{r}}{r^3} + Dirac - string \tag{10}$$

We shall choose to put the Dirac string along the positive z-axis $z+$. This can be done since the magnetic charge obeys the Dirac quantisation condition and any string crossing space from side to side is invisible. The magnetic charge M is equal to the magnetic flux through the Dirac string and is given by

$$M = \oint_c \hat{\Phi}\vec{A}_i dx_i \tag{11}$$

with c a closed path on S_2 encircling $z+$. Indeed, by use of Stokes theorem, in the unitary gauge where $\vec{\nabla}\hat{\Phi}^a = 0$, $M = \int_S d\vec{\sigma}\hat{\Phi}^a\vec{\nabla} \wedge (\vec{A}^a) = \int_S d\vec{\sigma}\vec{B}$. S is a surface which has c as border, and is thus pierced by the Dirac string.

M is gauge invariant in the sense that $M' = \oint_c \hat{\Phi}'\vec{A}'_i dx_i = M$. Indeed if $\hat{\Phi}' = R\hat{\Phi}$ and $\vec{A}'_i = \vec{A}_i + \frac{i}{g}R\partial_i R^\dagger$

$$\oint_c \hat{\Phi}'\vec{A}'_i dx_i = \oint_c (\hat{\Phi}\vec{A}_i + \frac{i}{g}\hat{\Phi}\partial_i R^\dagger)dx_i \tag{12}$$

The second term in Eq(12) vanishes if R is a global transformation, but also if $\hat{\Phi}$ is constant i.e. in the unitary representation.

Suppose that two (or more) different fields of the theory both in the adjoint representation, $\vec{\Phi}$ and $\vec{\Phi}'$ can act as effective Higgs fields and define monopole configurations and let $\hat{\Phi}$ and $\hat{\Phi}'$ be the corresponding directions in colour space. A transformation R of the gauge group exists , with non trivial U_B Eq.(3), such that on S_2 $\hat{\Phi}' = R\hat{\Phi}$, and hence also a transformation R' which transforms the AP of Φ into the AP of Φ', which is a global transformation. As a consequence $M = M'$: the magnetic charge is AP invariant. Creating a monopole in one specific AP means creating it in all AP's.

The argument can be generalized to larger gauge groups G [21][22], where monopoles live in $SU(2)$ subgroups.

In QCD there is no fundamental Higgs field, but effective Higgs fields can exist allowing the existence of monopoles. One example is the temporal component $\vec{A}_4 = i\vec{A}_0$ in the Gauge $\partial_0\vec{A}_0 = 0$ [18] [19] [20]. We shall discuss this issue in Sect.3. Anyhow the statement that the magnetic charge is AP invariant is the same no matter if the Higgs field is fundamental or effective.

This rigorously justifies the simplification used in the strategy 1) by creating the monopole in the AP defined by the 3-axis used in numerical updates of the configurations [9][10]. However it makes hard to explain the empirical observation made on lattice in strategy 2) that monopoles defined by different abelian projections look different both in their number and in their locations.

3 Monopoles in QCD

Monopoles are needed in QCD to produce by condensation the dual superconductivity of the vacuum, but no fundamental Higgs field exists in the system to produce a non trivial mapping of S_2 on $SU(2)$ subgroups of the gauge group. A solution to this difficulty proposed in Ref[5] is that any operator $\vec{O}$ in the adjoint representation can act as an effective Higgs field. The corresponding AP is the representation in which $\vec{O}\vec{\sigma}$ is diagonal, and the zeroes of $\vec{O}$ are the locations of the monopoles. The argument is [5] that in that gauge near the zero

$$\langle \vec{O}\vec{\sigma} \rangle \approx costant \times |\vec{x}|\sigma_3 \tag{13}$$

at least if there is invariance under rotations and O is a scalar under space rotations. For a classical configuration $\vec{O} = \langle \vec{O} \rangle$ and the behaviour Eq(13) is the same as that of the Higgs field in the monopole solution Eq(7). Of course the behaviour Eq(7) is a necessary, but not generally sufficient condition for a monopole solution. The condition Eq(8) must also be satisfied which expresses the topological nature of monopoles. The effective Higgs field is a function of the classical gauge fields, transforming in the adjoint representation

$$\Phi_{eff} = \Phi(\vec{A}_\mu) \tag{14}$$

at least for pure gauge theory (no quarks).

If we insist in putting $\vec{A}_0 = 0$ as in the solution Eq(5) Φ_{eff} only depends on $\vec{A}_i$ which vanishe on S_2, and must have at the same time a constant length on S_2 to satisfy Eq.(8). This looks impossible.

If instead we allow non zero $\vec{A}_0$ the requirement that the solution be static implies

$$\partial_0 \vec{A}_0 = 0 \tag{15}$$

An obvious possibility for the effective Higgs field is the choice $\vec{\Phi}_{eff} = \vec{A}_4 = i\vec{A}_0$. To have a monopole the boundary conditions Eq(7) and Eq(8) must be satisfied by $\vec{A}_4$. This is the solution of Ref. [18] [19][20].

Of course $\vec{A}_4$ on S_2 is the same for all monopoles: this fixes the monopoles which can co-exist. A detailed discussion is presented in Ref.[22], where the construction is also extended to $SU(N)$ gauge groups with $N > 2$.

$\partial_0 \vec{A}_0 = 0$ is a gauge condition. In this gauge the Polyakov line $\Pi(\vec{x})$ has a simple expression

$$\Pi(\vec{x}) \equiv P\exp(ig\int_0^T \vec{A}_0(\vec{x},t)\frac{\vec{\sigma}}{2}) = \exp(i\frac{g}{T}\vec{A}_0(\vec{x})\frac{\vec{\sigma}}{2}) \tag{16}$$

T is the temperature, P indicates time ordered product, which simplifies to ordinary product if $\vec{A}_0$ is time independent.

The gauge in which $\vec{A}_0(\vec{x})\vec{\sigma}$ is diagonal coincides with that in which the Polyakov line is diagonal, and has been extensively studied on lattice in connection with confinement.

4 The order parameter

An operator $\mu(\vec{x}, t)$ which creates a monopole in the location $\vec{x}$ at time t is [8]

$$\mu(\vec{x}, t) = \exp\left(i\int d^3y \vec{E}^3_\perp(\vec{y}, t)\vec{A}^0_\perp(\vec{y} - \vec{x})\right) \tag{17}$$

$\vec{A}^0_\perp(\vec{y} - \vec{x})$ is the classical vector potential produced in the point $\vec{y}$ by a monopole sitting at the point $\vec{x}$, in the transverse gauge $\vec{\nabla}\vec{A}^0_\perp = 0$. It can either be a point magnetic charge with a Dirac string carrying the flux at infinity , or a Wu-Yang type configuration in which the flux is carried to infinity by a 2-dimensional surface of singularity. In the convolution in Eq(17) only the transverse components of the electric field $\vec{E}^3$ contribute. $\vec{E}^3$ is the (canonical) electric field if the theory has gauge group $U(1)$, or its component in the direction 3 of the effective Higgs field in a AP of non abelian theories.

In any case, no matter how the theory is quantized, $\vec{E}^3_\perp(\vec{y},t)$ is the canonically conjugate field to $\vec{A}^3_\perp(\vec{y},t)$. Therefore, in the same way as for a motion in one dimension

$$\exp(ipa)|x\rangle = |x + a\rangle \tag{18}$$

in the Schroedinger representation of the field theory

$$\mu(\vec{x},t)|\vec{A}^3_\perp(\vec{z},t)\rangle = |\vec{A}^3_\perp(\vec{z},t) + \vec{A}^0_\perp(\vec{z}-\vec{x})\rangle \tag{19}$$

The operator $\mu(\vec{x},t)$ adds a monopole to field configurations.

The peculiar dependence on the coupling constant allows to compute $\langle\mu\rangle$ in a convenient way [7] [9]. If we normalise the fields as done in lattice formulation the action depends on the coupling constant as βS , with $\beta \equiv \frac{2N}{g^2}$ for gauge group $SU(N)$ and S independent on β

In Eq(17) the electric field is the canonical field, i.e. $\frac{1}{g}$ times the field in the lattice normalisation. An extra factor $\frac{1}{g}$ comes from the magnetic charge in the monopole field $\vec{A}_0$, so that finally μ has the form

$$\mu(\vec{x}) = \exp(-\beta\Delta S)$$

with $\Delta S = \frac{g^2}{2N}(i\int d^3y \vec{E}^3_\perp(\vec{y},t)\vec{A}^0_\perp(\vec{y}-\vec{x}))$ independent on β.

In the language of Feynman integrals

$$\mu(\beta) \equiv \langle\mu(\beta)\rangle = \frac{\int dA \exp(-\beta(S+\Delta S))}{\int dA \exp(-\beta S)} = \frac{Z(S+\Delta S)}{Z(S)} \tag{20}$$

the ratio of two partition functions. The quantity $\rho(\beta) = \frac{d}{d\beta}\ln(\langle\mu(\beta)\rangle)$ is easily computed from the expression (20) giving

$$\rho(\beta) = \langle S\rangle_S - \langle(S+\Delta S)\rangle_{(S+\Delta S)} \tag{21}$$

The subscript indicates the action used to perform the average. Since $\mu(0) = 1$

$$\mu(\beta) = \exp\left(\int_0^\beta \rho(\beta')d\beta'\right) \tag{22}$$

Computing numerically ρ on using the expression (21) presents no difficulty. The expectation is, if confinement is produced by dual superconductivity, that in the confined phase ($\beta < \beta_c$) ρ stays finite in the infinite volume limit so that $\mu(\beta)$ is different from zero; instead ρ goes large and negative in the deconfined phase, $\beta > \beta_c$ so that $\mu(\beta)$ vanishes in the infinite volume limit. In pure gauge theory there is an independent order parameter for confinement, the string tension measured from correlators of Polyakov lines, which permits the check.

At the beginning this research program, which was already rigorously checked in the simple case of gauge group $U(1)$ [7], worked perfectly within the statistical errors for gauge groups $SU(2)$ and $SU(3)$ [9]. In trying to extend the analysis to G_2 gauge group, to test the general validity of the mechanism a divergence of ρ in the large volume limit showed up, which would give $\mu(\beta) = 0$ both below ad above the critical temperature β_c[23]. A reanalysis of the previous work showed that the same disease would appear at large volumes or with larger statistics ad smaller errors also for the case of $SU(2)$ and $SU(3)$ gauge groups.

The origin of the problem was understood and removed in Ref [10] by improving the order parameter

$$\langle 0|\mu|0\rangle \to \mu' = \frac{\langle 0|\mu|0\rangle}{\sqrt{(\langle 0|0\rangle)}\sqrt{(\langle\mu 0|\mu 0\rangle)}} \tag{23}$$

which acts as an order parameter even when $\langle \mu 0 | \mu 0 \rangle$ and $\langle 0 | \mu | 0 \rangle$ vanish in the infinite volume limit. In the case of gauge group $U(1)$ the problem does not exist and μ and μ' differ by a finite multiplicative constant.

For gauge group $SU(2)$ the improvement was numerically checked in all details in Ref [10], including the scaling behaviour at the deconfining phase transition. For $SU(3)$ and possibly G_2 we are still collecting data.

The last open question about the order parameter, the choice of the abelian projection along the 3-axis used in numerical updating, is given a positive answer by the arguments presented above. Any AP is legitimate, in particular the one used in [9] and [10]. A rigorous test thus exists of the confinement mechanism of Ref. [3] [4].

Of course strategy 2) should provide the same answer. However the last existing works on the subject, by the ITEP group and the Kanazawa Group in which the effective potential of the monopole field is computed by a sophisticated procedure from the monopoles observed in lattice configurations, do not provide stable and clear evidence for a minimum at non zero field [See Ref. [24] for a detailed discussion and an accurate bibliography]. One reason for that can be that a fraction of the monopoles used in that analysis are not AP invariant, they are discretisation artefacts, and can pollute the procedure employed to determine the potential. Maybe repeating the procedure at smaller lattice spacings thus reducing the probability of large fluctuations of the abelian magnetic flux through plaquettes would improve the result.

5 Discussion

The stringent experimental limits on the existence of free quarks indicate that there is a symmetry behind confinement, and that deconfinement is a change of symmetry. The only possible gauge invariant symmetry breaking in a gauge theory involves degrees of freedom living on the 2-dimensional surface at spatial infinity, and monopoles come out as natural candidates, and with them dual superconductivity of the vacuum [3] [4].

Monopoles are $U(1)$ configurations, and there exist infinitely many ways of choosing that $U(1)$ in a non abelian system. Each choice is called an Abelian Projection. For long time it has been believed that monopoles living in different AP 's can be different from each other. The number and the positions of the monopoles observed in lattice configurations look indeed generically different.

Monopoles, however, are defined as classical configurations with non trivial topology [13] [14], and topology can not be changed by continuous gauge transformations. What we show is that the magnetic charge of a configuration is gauge and AP invariant. Creating a monopole in any abelian projection is equivalent to create it in all AP's. This allows to define in a simple way the operator which creates a monopole and with it the order parameter, which is its vacuum expectation value [10]. Both in $U(1)$ [7] and in $SU(2)$ pure gauge [10] theories it has been demonstrated that confinement is due to condensation of monopoles in the vacuum. Work is in progress on $SU(3)$ and G_2 pure gauge systems.

An alternative technique to compute the order parameter consists in computing the effective potential of the monopole field and in showing that it has a minimum at non zero field. The effective potential is computed from the positions of the monopoles in lattice configurations. The fact that no convincing results are achieved [24] can depend on the fact that the input monopoles are strongly AP dependent, which means that many of them are probably lattice artefacts. The situation, however, can be improved e.g. by going at larger values of

$\beta = \frac{2N}{g^2}$ thus reducing fluctuation on the magnetic flux across plaquettes in terms of which monopole are usually defined on Lattice. Finally the two strategies must give the same result.

References

[1] L. B. Okun *Leptons and Quarks* (North Holland, Amsterdam, New York, Oxford, 1982) 260

[2] S. Elitzur, Phys.Rev.**D12** 3978 (1975)

[3] G. 't Hooft, in *High Energy Physics: Proceedings EPS International Conference, Palermo,Italy 23-28 june 1975, A. Zichichi ed.*(Ed. Compositori, Bologna, 1976) 1225

[4] S. Mandelstam, Phys. Rep. **23C**, 245 (1976).

[5] G. 't Hooft,Nucl. Phys. **B 190** 455 (1981) .

[6] A. Di Giacomo, Acta Phys. Polon. **B 25**, 215 (1994)

[7] A. Di Giacomo, G. Paffuti, Phys. Rev. **D56**, 6816 (1997)

[8] L. Del Debbio, A. Di Giacomo, G. Paffuti, P. Pieri, Phys.Lett. **B355**, 255 (1995)

[9] A. Di Giacomo, B. Lucini, L. Montesi, G. Paffuti. Phys. Rev. **D61**, 034503 and 034504 (2000)

[10] C. Bonati, G. Cossu, M. D'Elia, A. Di Giacomo, Phys. Rev. **D85** 065001 (2012)

[11] P. Cea, L. Cosmai, Phys. Rev. **D62** 094510 (2000)

[12] M. N. Chernodub, M. I. Polikarpov, A. I. Veselov, Phis.Lett. **B399**, 267 (1997)

[13] G. 't Hooft, Nucl. Phys. **B 79**, 276 (1974)

[14] A. M. Polyakov, JETP Lett. **20**, 194 (1974)

[15] T. Degrand, D. Toussaint, Phys.Rev. **D 22**, 2478 (1980)

[16] T. Suzuki, I. Yotsuyanagi, Phys. Rev. **D 42**, 4257 (1990).

[17] M. I. Polikarpov, Nucl. Phys. B (Proc. Suppl.) **53**, 134 (1997)

[18] B. Julia, A. Zee Phys. Rev **D11** 2227 (1975)

[19] E. B. Bogomonlyi, Sov. J. Nucl. Phys **24**, 449 (1976)

[20] M. K. Prasad, C. M. Sommerfeld, Phys. Rev. Lett **35**, 760 (1973)

[21] A. Di Giacomo, QCD Monopoles, abelian projections and gauge invariance, arXiv:**1707.07896** (2017)

[22] A. Di Giacomo, Nucl.Part.Phis.Proc. **270-272** 175 (2016)

[23] G. Cossu, M. D'Elia, A. Di Giacomo,B. Lucini, C. Pica, **PoS LATTICE2007** 296 (2007)

[24] M. I. Polikarpov, Nucl. Phys.**B (Proc. Suppl.) 53**, 134 (1997) [arXiv:hep-lat/9609020].

Probing QCD with the ATLAS Detector

Yuri Kulchitsky[1,2,*], on behalf of the ATLAS Collaboration

[1]B.I.Stepanov Institute of Physics, National Academy of Sciences of Belarus, Minsk, Belarus
[2]Joint Institute for Nuclear Research, Dubna, Russia

Abstract. Perturbative QCD calculations at next-to-next-to-leading-order are available for many processes since several years and can be rigorously tested with a large variety of final states. The latest results from the ATLAS Collaboration at the LHC (CERN) involving jets, dijets, photons in association with heavy flavors and vector bosons in association with jets, measured at center of mass energies of $\sqrt{s}$ = 8 and 13 TeV are presented. All measured cross-sections are compared to state-of-the art theory Monte Carlo predictions. The paper concludes with the results of jet-substructure studies at 13 TeV, in particular the measurement of the jet soft-drop mass.

Introduction

Precise measurements of jet cross-sections are crucial in understanding physics at hadron colliders. In quantum chromodynamics (QCD), jets are interpreted as resulting from the fragmentation of quarks and gluons produced in a short-distance scattering process. Jet cross-sections provide valuable information about the strong coupling constant, α_s, and the structure of the proton. Inclusive jet and dijet events represent a background to many other processes at hadron colliders. The predictive power of fixed-order QCD calculations is therefore relevant in many searches for new physics. The ATLAS collaboration[1] [1] at the LHC (CERN) [2] have measured inclusive jet cross-sections in pp collisions at $\sqrt{s}$ = 13 TeV [3] and 8 TeV [4]. Dijet production at the LHC has been measured at $\sqrt{s}$ = 13 TeV [3]. The production of isolated prompt photons in association with a heavy flavour (HF) jet containing a b or c hadron at $\sqrt{s}$ = 8 TeV [5] provides a testing ground for perturbative quantum chromodynamics (pQCD), the content of the proton and the treatment of heavy quarks (HQ) in matrix element (ME) and parton shower (PS) computations. Prompt photons, which refer to those not arising from hadron decays, are targeted by requiring that their signals are isolated. A measurement of the W boson production cross section and the W^+/W^- cross-section ratio, both in association with jets, in pp collisions at $\sqrt{s}$ = 8 TeV are presented in Ref. [6]. A measurement of the first jet substructure quantity was calculated at next-to-next-to-leading-logarithm (NNLL) accuracy are presented in Ref. [7].

Inclusive jet and dijet cross-sections at $\sqrt{s}$ = 13 TeV

This paper presents measurements of the inclusive jet and dijet cross-sections in pp collisions at $\sqrt{s}$ = 13 TeV corresponding to an integrated luminosity of 3.2 fb^{-1} [3]. The inclusive jet cross-sections are measured double-differentially as a function of the jet transverse momentum, p_T, and absolute jet rapidity, $|y|$. The double-differential dijet production

*e-mail: Iouri.Kultchitski@cern.ch

[1]Copyright [2018] CERN for the benefit of the [ATLAS Collaboration]. Reproduction of this article or parts of it is allowed as specified in the CC-BY-4.0 license.

© The Authors, published by EDP Sciences. This is an open access article distributed under the terms of the Creative Commons Attribution License 4.0 (http://creativecommons.org/licenses/by/4.0/).

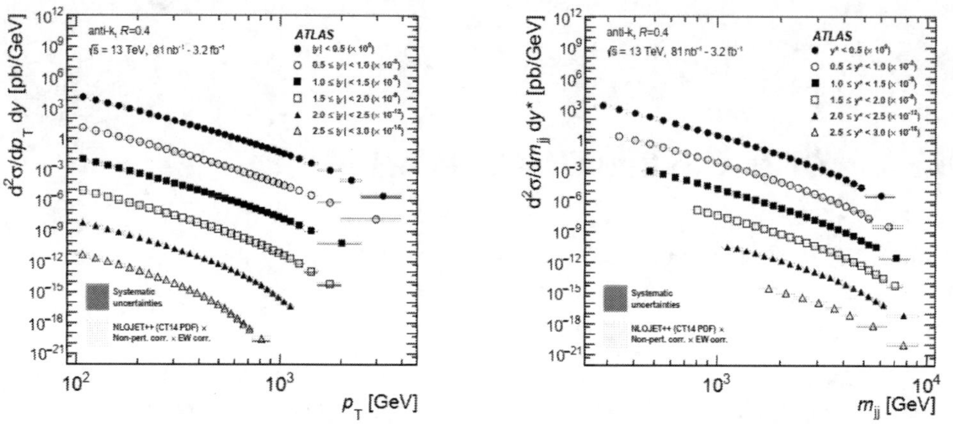

Figure 1. Inclusive (left) jet and (right) dijet cross-sections as a function of p_T for jet and m_{jj} for dijet and $|y|$, for anti-k_t jets with $R = 0.4$ in pp collisions at $\sqrt{s} = 13$ TeV. The statistical uncertainties are smaller than the size of the symbols used to plot the cross-section values. The dark gray shaded areas indicate the experimental systematic uncertainties. The data are compared to NLO pQCD predictions calculated using NLOJET++ with p_T^{max} for jet and $p_T^{max}\exp(0.3\,y^*)$ for dijet as the QCD scale and the CT14 NLO PDF set. The yellow shaded areas indicate the predictions with their uncertainties. At low and intermediate p_T bins the experimental systematic uncertainties are comparable to the theory uncertainties (drawn on top) and therefore are barely visible. In most m_{jj} bins the experimental systematic uncertainty is smaller than the theory uncertainties and is therefore not visible. Taken from Ref. [3].

cross-sections are presented as a function of the invariant mass of the dijet system, m_{jj}, and as a function of half the absolute rapidity separation between the two highest-p_T jets satisfying $|y| < 3$, denoted y^*. Jets are reconstructed using the anti-k_t jet clustering algorithm with a radius parameter value of $R = 0.4$. The measurements cover the kinematic region of $0.1 < p_T < 3.5$ TeV and $|y| < 3$ for the inclusive jet cross-section, and of $0.3 < m_{jj} < 9$ TeV and $y^* < 3$ for the dijet cross-section. NLO pQCD predictions calculated using several parton distribution function (PDF) sets, corrected for electroweak (EW) and non-perturbative (NP) effects, are quantitatively compared to the measurement results. In addition, the inclusive jet cross-sections are compared to the recently published complete next-to-next-to-leading-order (NNLO) pQCD calculation. The measured double-differential inclusive jet cross-sections are shown in Fig. 1(left) as a function of p_T for the 6 jet rapidity bins, and the measured double-differential dijet cross-sections are shown in Fig. 1(right) as a function of m_{jj} for the 6 y^* bins. The NLO pQCD predictions using the CT14 PDF set corrected for NP and EW effects are also shown in Fig. 1. The ratios of the NLO pQCD predictions to the measured inclusive jet cross-sections as a function of p_T in the 6 jet $|y|$ bins are shown in Fig. 2 for the CT14, MMHT2014, NNPDF3.0, HERAPDF2.0, ABMP16 PDF sets. No significant deviation of the data points from the predictions is seen; the NLO pQCD predictions and data agree within uncertainties. In the forward region ($|y| > 2$) there is a tendency for the NLO pQCD prediction using the CT14, MMHT2014 and NNPDF3.0 PDF sets to overestimate the measured cross-section in the high p_T range, although the difference from data does not exceed the range covered by the experimental and theoretical uncertainties. A quantitative comparison of the measurements to fixed-order NLO QCD calculations shows overall fair agreement when considering jet cross-sections in individual jet $|y|$ bins independently. In the inclusive jet measurement, a significant tension between data and theory is observed when considering data points from all jet p_T and $|y|$ regions. No significant differences between the inclusive jet cross-sections and the fixed-order NNLO QCD calculations corrected for NP and EW effects

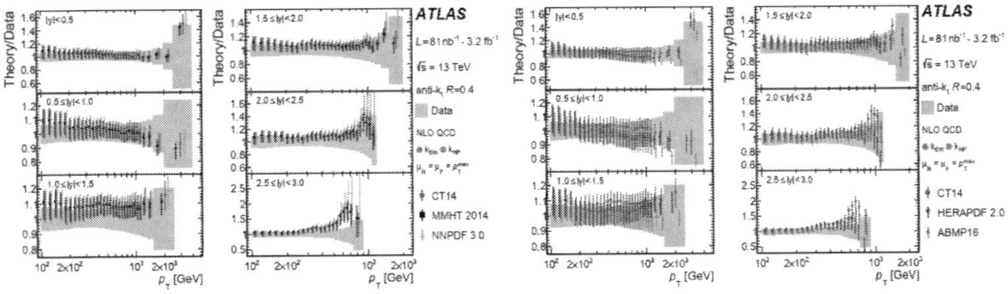

Figure 2. Comparison of the measured inclusive jet cross-sections and the NLO pQCD predictions shown as the ratios of predictions to the measured cross-sections in pp collisions at $\sqrt{s}$ = 13 TeV. The ratios are shown as a function of the jet p_{T} in 6 $|y|$ bins for anti-k_t jets with $R = 0.4$. The predictions are calculated using NLOJet++ with different PDF sets (left) [CT14, MMHT2014, NNPDF3.0] and (right) [CT14, HEPAPDF3.0, ABMP16]. The NP and EW corrections are applied. The gray bands show the total data uncertainty including both the systematic and statistical uncertainties. Taken from Ref. [3].

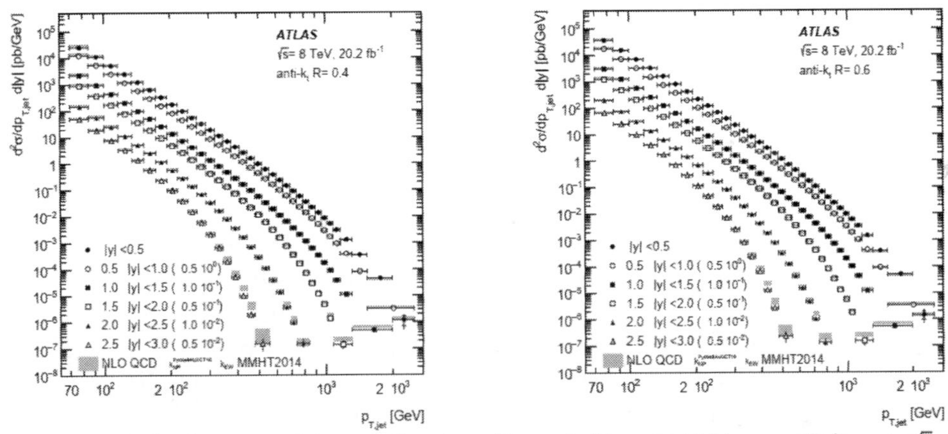

Figure 3. Inclusive jet cross-section as a function of jet p_{T} in bins of jet $|y|$ in pp collisions at $\sqrt{s}$ = 8 TeV. The results are shown for jets identified using the anti-k_t algorithm with (left) $R = 0.4$ and (right) $R = 0.6$. For better visibility the cross-sections are multiplied by the factors indicated in the legend. The data are compared to the NLO QCD prediction with the MMHT2014 PDF set. The error bars indicate the statistical and the systematic uncertainties in the measurement added in quadrature. The statistical uncertainty is shown separately by the inner vertical line. Taken from Ref. [4].

are observed when using $p_{\mathrm{T}}^{\mathrm{jet}}$ as the QCD scale. The NLO pQCD predictions using $p_{\mathrm{T}}^{\mathrm{max}}$ as the scale overestimate the measured inclusive jet cross-sections.

Inclusive jet cross-sections at $\sqrt{s} = 8$ TeV

The measured double-differential inclusive jet cross-sections at $\sqrt{s} = 8$ TeV are shown in Fig. 3 as a function of the jet p_{T} for anti-k_t jets with $R = 0.4$ and $R = 0.6$ for each jet $|y|$ bin [4]. The cross-section covers 11 orders of magnitude in the central $|y|$ region and 9 orders of magnitude in the forward $|y|$ region. Jet transverse momenta above $p_{\mathrm{T}} = 2$ TeV are observed. In the most forward region the jet p_{T} reaches about 0.5 TeV. The measurement is compared to an NLO QCD prediction using the MMHT2014 PDF set based on NLOJet++ corrected for NP and EW effects. This theory prediction describes the gross features in the data. The ratio of NLO QCD calculations to data corrected for NP and EW effects for various PDF sets is shown in Fig. 4(left) for anti-k_t jets with $R = 0.4$. [4]. At low p_{T} the level of agreement

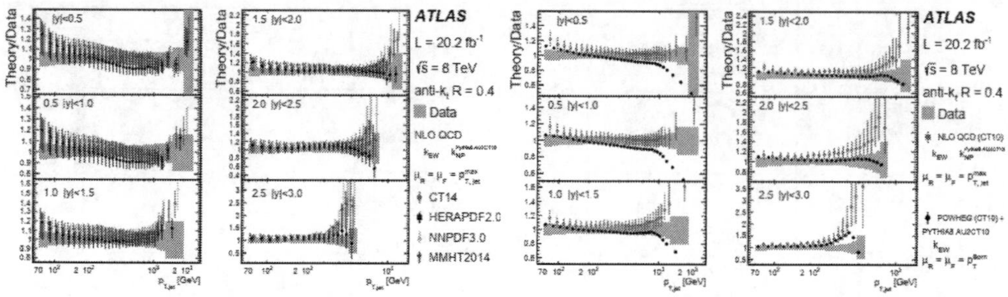

Figure 4. (Left) Ratio of the inclusive jet cross-section predicted by NLO QCD corrected for NP and EW effects to the cross-section in data as a function of the jet p_T in each jet $|y|$ bin in pp collisions at $\sqrt{s} = 8$ TeV. Shown are the predictions for various PDF sets for anti-k_t jets with $R = 0.4$. The points are offset in jet p_T for better visibility. The error bars indicate the total theory uncertainty. The gray band shows the total uncertainty in the measurement. (Right) Ratio of the inclusive jet cross-section predicted by the Powheg MC event generator with respect to the cross-section in data as a function of the jet p_T in each jet $|y|$ bin for anti-k_t jets with $R = 0.4$. Only the nominal values of this ratio are indicated. Also shown is the prediction by NLO QCD corrected for NP effects, where the error bars indicate the total theory uncertainty. EW corrections are applied for both theory predictions and the CT10 PDF set is used. Taken from Ref. [4].

is very sensitive to NP effects. When using Pythia8 as the nominal NP correction, the NLO QCD prediction is typically about 10–20% above the data at low p_T, whereas the NLO QCD prediction corrected with Herwig++ follows the data well for jets with $R = 0.4$, while it is 5–10% below the data for jets with $R = 0.6$. The comparisons of the Powheg predictions with the measurement for jets with $R = 0.4$ are shown in Fig. 4(right) as a function of the jet p_T in jet $|y|$ bins. The measurements are also compared to the NLO QCD prediction using the CT10 PDF set and corrected for NP effects with the same Monte Carlo (MC) generator configuration as was used for Powheg. Strong tension between data and theory is observed when considering data points from all jet p_T and $|y|$ regions, with a full treatment of the correlations. This tension can be reduced, but not completely resolved, using alternative correlation scenarios for the experimental and theoretical systematic uncertainties [4].

Differential cross-sections of γ with b or c jet at $\sqrt{s} = 8$ TeV

Prompt photons (γ) can be used as a colourless non-hadronizing probe of parton dynamics that yields a clean experimental signature. Processes containing final state b or c quarks play an important role in many LHC physics analyses and therefore the accuracy of the description of this HF content of the proton must be investigated. HF jets are defined as jets which contain either a b or c hadron. In Fig. 5 the values for the measured differential $\gamma + c$ cross sections are plotted and in Fig. 6 the ratios of $\gamma + b$ and $\gamma + c$ for central and forward regions are given, with the relevant theory predictions [5]. In general, considering the LO predictions in pQCD, those from Sherpa agree well with the measured values and provide a better description of the data than those from Pythia. Comparisons of the $\gamma + c$ measurement to NLO+PS predictions from MadGraph5_aMC@NLO in the $5F$ scheme using NNPDF3.1nlo with 0.26% of the intrinsic charm (IC) and CT14nnlo are shown in Fig. 5. The predictions are found to agree with the data within the uncertainties across the entire E_T^γ range. However, those using the BHPS or the fitted charm PDF sets predict higher cross-section values in the forward region at high E_T^γ, above 105 GeV, than those using the nominal PDF sets. As shown in Fig. 6(left), the $4F$ and $5F$ NLO predictions for the cross-section ratios consistently overestimate the data for $E_T^\gamma > 65$ GeV; the $5F$ predictions are at the edge of agreement with the measured values within uncertainties. Sherpa, which generates additional partons in the ME and uses a

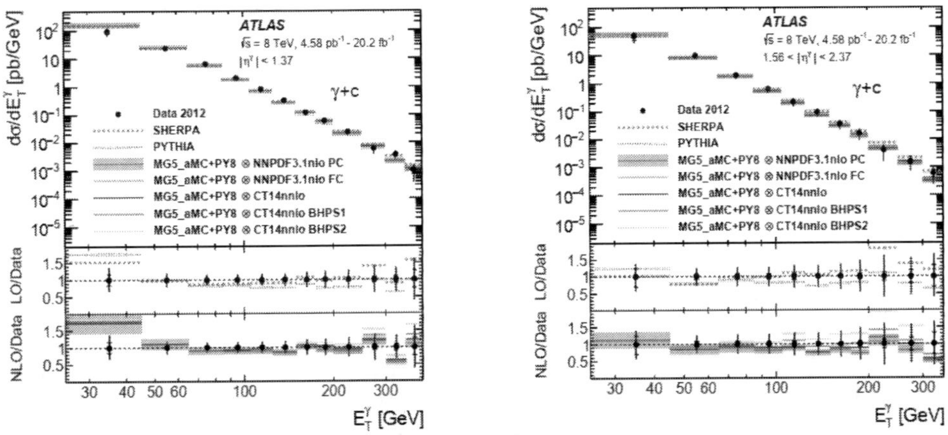

Figure 5. Differential cross sections as a function of the photon transverse energy, E_T^γ, for $\gamma + c$ in (left) the central and (right) the forward regions in pp interactions at $\sqrt{s}$ = 8 TeV. The statistical uncertainty is represented as horizontal marks on the error bars of the data points, while the total measurement uncertainty is represented by the complete length of the error bars. The MG5_aMC+PY8 label in the legend refers to the MadGraph5_aMC@NLO calculations interfaced to Pythia8. The $5F$ and $4F$ labels in the legend refer to PDF sets with five and four quark flavours, respectively. The PC and FC labels in the legend refer to perturbative charm and fitted charm PDF sets, respectively. All of the predictions for $\gamma + c$ use $5F$ PDF sets. The theoretical uncertainty in the MadGraph5_aMC@NLO predictions is displayed for a single PDF set since it is similar for each of the PDF sets. The Sherpa and Pythia cross sections are not normalized to data and no uncertainties are provided for them. Taken from Ref. [5].

massive $5F$ scheme, provides a better description of the measured cross sections and cross-section ratios than MadGraph5_aMC@NLO in either the $5F$ or $4F$ scheme. The predicted values for the cross-section ratios, shown in Fig. 6(right), are smaller for the predictions using the BHPS or the fitted charm sets than for those using the nominal sets. This is the expected behaviour of the IC contributions from these PDF sets in the theory predictions. The predictions with the BHPS2 PDF set (2.1% of the IC) deviate the most from those using the nominal PDF sets, by about a factor 1.5, while those using the BHPS1 (0.6% of the IC) and the fitted charm PDF sets give intermediate values. The precision of the data is comparable to the size of these deviations in the predictions. For the $\gamma + b$ final state, the best description of the data is provided by the Sherpa predictions, which include up to three additional partons and are computed in the massive $5F$ scheme. The NLO predictions underestimate the data in the highest E_T^γ intervals measured. The $5F$ scheme of the theoretical calculations provides a better description of the data than the $4F$ scheme. For the γ+c final state, which exhibits larger measurement uncertainties, all the predictions are in agreement with the data. Differences of about the size of the measurement uncertainties are seen between the predictions using PDF sets with IC contributions and those without. These measured cross sections provide a test of pQCD calculations with HQs and are sensitive to the b and c quark PDFs.

Differential cross-sections of W + jets and W⁺/W⁻ ratios at $\sqrt{s}$ = 8 TeV

Using data corresponding to an integrated luminosity of 20.2 fb^{-1} at $\sqrt{s}$ = 8 TeV, the results for W + jets production in final states containing one electron and missing transverse momentum, focusing on events with one or two additional jets are presented [6]. The data are measured for W production as well as for W^+ and W^- production and the cross-section ratio of W^+/W^- as a function of the number of jets (N_{jets}). For events with at least one jet, the differential cross sections are shown as a function of the scalar sum of the transverse

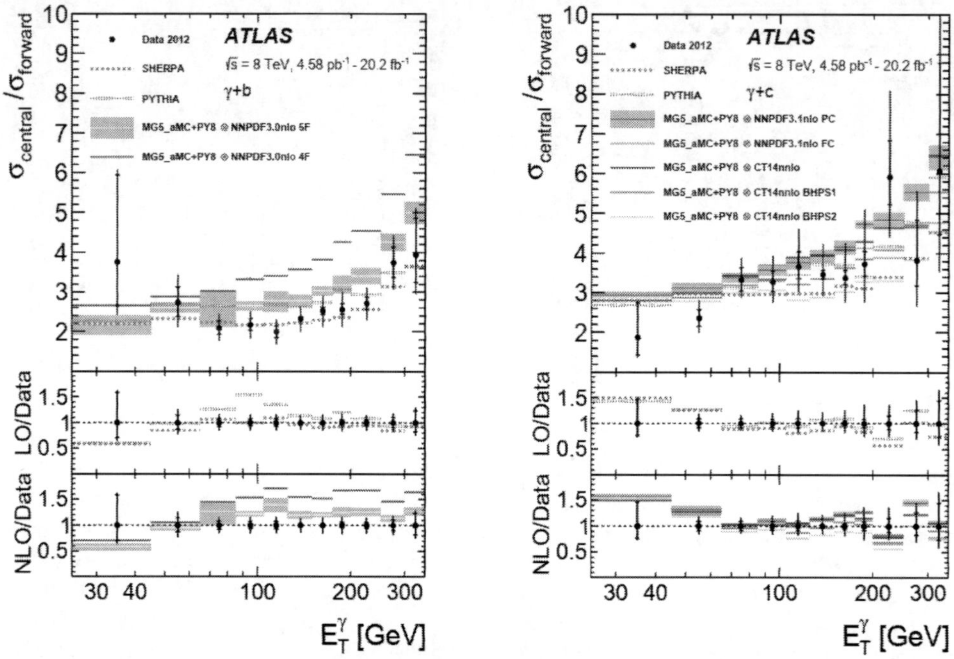

Figure 6. Cross-section ratios of the central region, $|\eta^{\gamma}| < 1.37$, to the forward region, $1.56 < |\eta^{\gamma}| < 2.37$, as a function of the photon transverse energy, E_{T}^{γ}, for (left) $\gamma + b$ and (right) $\gamma + c$. The statistical uncertainty is represented as horizontal marks on the error bars of the data points, while the total measurement uncertainty is represented by the complete length of the error bars. The MG5_aMC+PY8 label in the legend refers to the MadGraph5_aMC@NLO calculations interfaced to Pythia8. The $5F$ and $4F$ labels in the legend refer to PDF sets with five and four quark flavours, respectively. The PC and FC labels in the legend refer to perturbative charm and fitted charm PDF sets, respectively. All of the predictions for $\gamma + c$ use $5F$ PDF sets. The theoretical uncertainty in the MadGraph5_aMC@NLO predictions is displayed for a single PDF set since it is similar for each of the PDF sets, except for NNPDF3.1nlo FC for which the total uncertainty is similar to that in NNPDF3.1nlo PC at a value of 25 GeV in E_{T}^{γ}, but rises steadily relative to it to be a factor of three larger at 350 GeV. No uncertainties are provided for Sherpa and Pythia. Taken from Ref. [5].

momenta of electron, neutrino and jets (H_{T}), the p_{T} of the W boson, and the p_{T} and rapidity of the most energetic jet (leading jet). These observables are sensitive to higher-order terms in the prediction as well as the PDFs. For events with at least two jets, the differential cross sections are shown for W boson production only and include distributions as a function of the p_{T} and rapidity of the second leading jet, the dijet angular separation, and the dijet invariant mass. These observables are sensitive to hard parton radiation at large angles and different ME/PS matching schemes. The differential cross section for W production and the ratio of W^{+} and W^{-} as a function of H_{T} are shown in Fig. 7 for $N_{\mathrm{jets}} \geq 1$. The H_{T} distribution is a very important test of pQCD as the higher values are sensitive to higher jet multiplicities and topologies such as $qq \to qq'W$. The LO predictions of Sherpa and Alpgen, which both include multiple jets in the ME calculation describe the data best, although these predictions have large theoretical uncertainties. The BlackHat+Sherpa predictions underestimate the data at large values of H_{T}. This is expected since, at these large values of H_{T}, contributions from additional jets are important, which are only partially present in this calculation. The predictions from the BlackHat+Sherpa exclusive sums method and from the NNLO N_{jetti} calculation, which include an additional jet emission at NLO, provide better agreement with

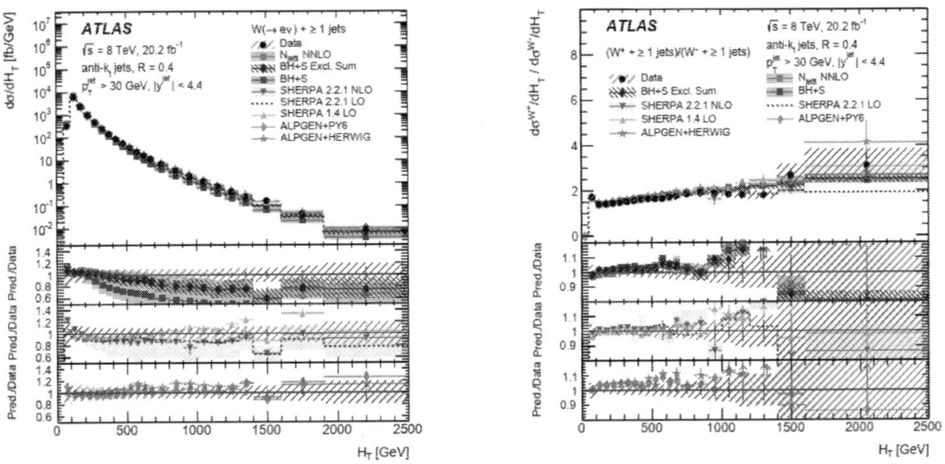

Figure 7. Differential cross sections for the production of (left) W bosons and (right) the W^+/W^- ratio as a function of H_T for events with $N_{\text{jets}} \geq 1$. The last bin in the left figure includes values beyond the shown range. For the data, the statistical uncertainties are indicated as vertical bars, and the combined statistical and systematic uncertainties are shown by the hatched bands. The uppermost panel in plot shows the differential cross sections. The lower panels show the ratios of the predictions to the data. The arrows on the lower panels indicate points that are outside the displayed range. Taken from Ref. [6].

the data. At the largest measured values of H_T, where the measured cross section is small, the total experimental uncertainty in the W^+/W^- cross-section ratio increases due to larger statistical uncertainties in the data and some systematic uncertainties that do not fully cancel out in the ratio. Overall, the measured distributions show that NNLO and NLO predictions are able to describe the data. However, at high transverse momenta, large jet rapidities, or large dijet angular separations, many of these predictions underestimate or overestimate the cross sections. In many places, multi-leg LO generators, such as ALPGEN and SHERPA, which consider a larger number of parton emissions from the ME calculation, model the data best, although with large theoretical uncertainties. The H_T, jet rapidity, and dijet invariant mass distributions are in general the least well described, suggesting that better modelling of events with energetic jets as well as jets with large rapidities is needed.

Measurement of the soft-drop jet mass at $\sqrt{s} = 13$ TeV

Calculations of jet substructure observables that are accurate beyond LL accuracy have recently become available. In the Ref. [7] the normalized, differential cross-section is measured as a function of $\log_{10}(\rho^2)$, where ρ is the ratio of the soft-drop mass to the ungroomed jet p_T. This quantity is measured in dijet events from 32.9 fb^{-1} of $\sqrt{s} = 13$ TeV pp collisions. The soft-drop procedure acts on the clustering history of a sequential recombination jet algorithm. In these algorithms, all inputs to jet-finding start as a proto-jet and are combined pairwise using a distance metric in $y - \phi$ space. When the smallest distance is above some threshold R, the algorithm terminates and the remaining proto-jets are the final jets. The clustering history is the sequence of pairwise combinations that lead to a particular jet. The soft-drop algorithm starts by re-clustering an anti-k_t jet's constituents. Next, the clustering tree is traversed from the latest branch to the earliest and at each node the following criterion is applied to proto-jets j_1 and j_2: $\frac{min(p_{T_{j_1}}, p_{T_{j_2}})}{p_{T_{j_1}} + p_{T_{j_2}}} > z_{\text{cut}}(\frac{\Delta R_{12}}{R})^\beta$, where p_T is the momentum of a jet transverse to the beam pipe, z_{cut} and β are parameters, and $\Delta R_{12} = \sqrt{\Delta y^2 + \Delta\phi^2}$ is the distance in $y - \phi$ between the proto-jets. The data are unfolded to correct for detector

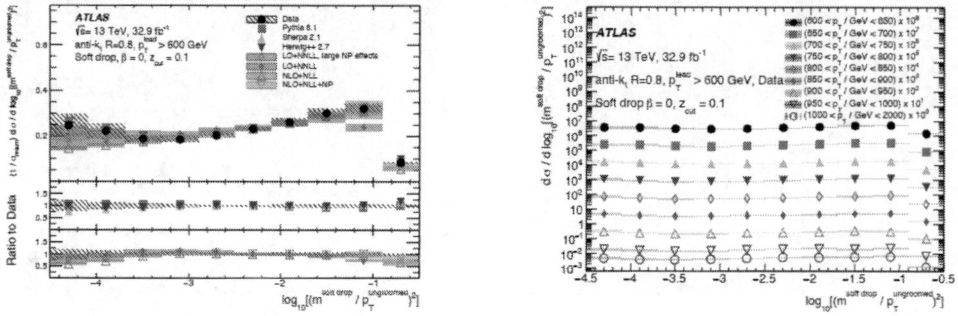

Figure 8. (Left) The unfolded $\log_{10}(\rho^2)$ distribution for anti-k_t $R = 0.8$ jets with $p_T^{\text{lead}} > 600$ GeV, after the soft drop algorithm is applied for $\beta = 0$, in data compared to Pythia, Sherpa, and Herwig++ particle-level, and NLO+NLL+NP and LO+NNLL theory predictions. The LO+NNLL calculation does not have NP corrections; the region where these are expected to be large is shown in a open marker, while regions where they are expected to be small are shown with a filled marker. The distributions are normalized to the integrated cross section, σ_{resum}, measured in the resummation region, $-3.7 < \log_{10}(\rho^2) < -1.7$. The NLO+NLL+NP cross-section in this resummation regime is 0.14 nb for $\beta = 0$. (Right) The unfolded $\log_{10}(\rho^2)$ distribution in data for each of the p_T bins in the analysis, overlaid with a multiplicative factor indicated in the legend. Taken from Ref. [7].

effects and compared to precise QCD calculations and LL particle-level MC simulations and are shown in Fig. 8(left). They are compared to the predictions of the Pythia, Sherpa, and Herwig++ generators, as well as the NLO+NLL and the LO+NNLL. The NLO+NLL calculation includes NP corrections based on the average of various MC models with NP effects turned on and off; the envelope of predictions is added as an uncertainty. The LO+NNLL predictions do not contain NP effects, but the open makers in Fig. 8(left) indicate where NP are expected to be large. The $\log_{10}(\rho^2)$ distribution in data for $\beta = 0$ and each p_T bin is shown in Fig. 8(right). There is no strong dependence of the shape on p_T.

Summary

The latest results from the ATLAS collaboration at the LHC (CERN) involving jets, dijets, photons in association with HF jets and vector bosons in association with jets, measured at $\sqrt{s} = 8$ and 13 TeV are obtained. The first measurement of γ + HF jet at the LHC are presented. All measured cross-sections are compared to state-of-the art theory predictions.

References

[1] ATLAS Collaboration, JINST **3**, S08003 (2008).
[2] L. Evans and P. Bryant (editors), JINST **3**, S08001 (2008).
[3] ATLAS Collaboration, JHEP **1805**, 195 (2018)
[4] ATLAS Collaboration, JHEP **1709**, 020 (2017)
[5] ATLAS Collaboration, Phys. Lett. B **776**, 295 (2018)
[6] ATLAS Collaboration, JHEP **1805**, 077 (2018)
[7] ATLAS Collaboration, Phys. Rev. Lett. **121**, 092001 (2018)

CTEQ-TEA parton distribution functions with intrinsic charm

Marco Guzzi[1,*], *Tie-Jiun* Hou[2,6], *Sayipjamal* Dulat[3], *Jun* Gao[4], *Joey* Huston[5], *Pavel* Nadolsky[6], *Carl* Schmidt[5], *Jan* Winter[5], *Keping* Xie[6], and *C.-P.* Yuan[5]

[1]Department of Physics, Kennesaw State University, 370 Paulding Ave., 30144 Kennesaw, GA, U.S.A.
[2]National Tsing Hua University, Hsinchu City, Taiwan.
[3]School of Physics Science and Technology & Center for Theoretical Physics, Xinjiang University, Urumqi, Xinjiang 830046 China.
[4]INPAC, Shanghai Key Laboratory for Particle Physics and Cosmology & School of Physics and Astronomy, Shanghai Jiao Tong University, Shanghai 200240, China.
[5]Department of Physics and Astronomy, Michigan State University, East Lansing, MI 48824 U.S.A.
[6]Department of Physics, Southern Methodist University, Dallas, TX 75275-0181, U.S.A.

Abstract. We present a study in which the possibility of a (sizable) nonperturbative contribution to the charm parton distribution function (PDF) in a nucleon is investigated together with theoretical issues arising in its interpretation. The separation of the universal component of the nonperturbative charm from the rest of the radiative contributions is also discussed. We illustrate the potential impact of a nonperturbative charm PDF on LHC scattering processes. An estimate of nonperturbative charm magnitude in the CT14 and CT14HERA2 global QCD analyses at the next-to-next-to leading order (NNLO) in the QCD coupling strength is given by including the latest experimental data from HERA and the Large Hadron Collider. We show a comparison between different models of intrinsic charm and illustrate prospects for standard candle observables at the LHC.

1 Introduction

The principle of the global analysis is to use QCD theory to analyze a broad range of experimental data, including precision data from HERA, the Tevatron, and the Large Hadron Collider (LHC). In particular, theoretical predictions for short-distance scattering processes allow the measurement, within some approximations, of universal parton distribution functions (PDFs) for the proton. As new experimental measurements from the LHC grow increasingly precise, novel challenges arise in extracting accurate predictions for the parton content of the proton in global QCD analysis of PDFs needed for advanced tests of the Standard Model and possible physics beyond the Standard Model. A recently published CTEQ-TEA (CT) global analysis of experimental data [1] produced the CT14NNLO PDFs, referred to as the CT14 PDFs. The analysis is based on the next-to-next-to-leading order (NNLO) approximation for perturbative QCD.

*e-mail: mguzzi@kennesaw.edu

© The Authors, published by EDP Sciences. This is an open access article distributed under the terms of the Creative Commons Attribution License 4.0 (http://creativecommons.org/licenses/by/4.0/).

1.1 The perturbative charm PDF

In global PDF analyses, all QCD parameters, such as α_s and the quark masses, are correlated with the PDFs. The determination of the PDFs depends not only on the data sample included in the fits, but also on the specific theory assumptions and underlying physics models. As one such choice made in the standard CT PDF sets, the charm quark and antiquark PDFs are taken to be zero below a low energy scale $Q_c = Q_0$ of order of the charm mass. In the CT14 analysis, the charm quark and antiquark PDFs were turned on at the scale $Q_c = Q_0 = m_c =$ 1.3 GeV, with an initial $O(\alpha_s^2)$ distribution consistent with NNLO matching [2, 3] to the three-flavor result. At higher Q, most of the charm PDF is generated from the DGLAP evolution that proceeds through perturbative splittings of gluons and light-flavor quarks. Hence, the charm PDF from a standard global analysis is called "perturbative" and it is obtained by perturbatively evolving the PDFs from the initial scale Q_c to the experimental data scale Q.

1.2 A nonperturbative component for the charm PDF

In addition to the perturbative charm production mechanism, it is believed that "intrinsic charm quarks" may emerge from the nonperturbative structure of the hadronic bound state. The plausibility of the intrinsic charm (IC) component, its dynamical origin, and its actual magnitude have been a subject of a long-standing debate. Indeed, QCD theory rigorously predicts existence of power-suppressed (higher-twist) channels for charm quark production that are independent of the leading-power (twist-2, or perturbative) production of charm quarks. The IC quarks have been associated with the excited $|uudc\bar{c}\rangle$ Fock state of the proton wave function [4–9] and predicted by meson-baryon models [10–13]. The range of validity of the PDF models with nonperturbative charm has been appraised in a recent CT analysis published in Ref. [14] and in other recent studies [15–19].

1.3 Fitted charm and nonperturbative charm parametrizations

Starting from the factorization theorem for DIS cross sections with massive fermions, that is a fundamental QCD result, one can draw a consequential distinction between the "fitted" charm PDF parametrization and the nonperturbative charm PDF. The fitted charm PDF accounts for the nonperturbative charm plus other (possibly not universal) higher $O(\alpha_s)$ higher power suppressed terms. Since the perturbative charm PDF component cancels near the threshold up to a higher order, the fitted charm component may approximate for a missing higher-order term or a power-suppressed nonperturbative component. The genuine nonperturbative charm PDF instead, is defined by the means of power counting of radiative contributions to DIS. Assuming that this additional nonperturbative charm component can be factorized like the perturbative charm component, one is able to examine how it differs from the perturbative charm, and how it depends on theoretical inputs in a global QCD analysis of PDFs. In principle, the intrinsic charm content would be suppressed by powers of $(\Lambda^2_{\rm QCD}/m_c^2)$, but, since this ratio is not very small, it may be relevant in some processes such as precise DIS. For a more detailed description of QCD factorization with power suppressed charm contributions, we refer the reader to the recent CTEQ-TEA analysis of Ref. [14].

1.4 Valence-like and sea-like models for the charm PDF parametrization

Various model estimates suggest a power-suppressed charm cross section of a modest size: of order of a fraction of the α_s^2 component in DIS charm production, carrying less than about a percent of the proton's momentum. We examine a more extensive list of nonperturbative models, fit the most complete set of DIS data from HERA as well as the data from the LHC and (optionally) the EMC experiment [20], and utilize a PDF parametrization that results in a more physical behavior. Given that several mechanisms may give rise to the fitted charm, we will parametrize it by two generic shapes, a "*valence-like*" and a "'*sea-like*" shape. The two shapes arise in a variety of dynamical models.

A valence-like shape has a local maximum at x above 0.1 and satisfies $f_{q/p}(x, Q_c) \sim x^{-a_1}$ with $a_1 \lesssim 1/2$ for $x \to 0$ and $f_{q/p}(x, Q_c) \sim (1-x)^{a_2}$ with $a_2 \gtrsim 3$ for $x \to 1$. The distributions for valence u and d quarks fall into this broad category, as well as the "intrinsic" sea-quark distributions that can be naturally generated in several ways [6], e.g., for all flavors, nonperturbatively from a $|uudQ\overline{Q}\rangle$ Fock state in light-cone [4, 5, 7–9] and meson-baryon models [10–13]; for $\bar{u}$ and $\bar{d}$, from connected diagrams in lattice QCD [21].[1] The approximate Brodsky-Hoyer-Peterson-Sakai (BHPS) model [4, 5] parametrizes the charm PDF at Q_0 by a valence-like nonperturbative function

$$\widehat{c}(x) = \frac{1}{2} A\, x^2 \left[\frac{1}{3}(1-x)(1+10x+x^2) - 2x(1+x)\ln(1/x) \right]. \tag{1}$$

This function is obtained from a light-cone momentum distribution by taking the charm mass to be much heavier than the masses of the proton and light quarks: $m_c \gg M_p, m_u, m_d$. A is the normalization factor that is to be determined from the fit. The BHPS1 and BHPS2 global fits are obtained with this parametrization choice and are illustrated in Sec. 2 below.

Instead of approximating the probability integral as in the original BHPS model, the $\widehat{c}(x)$ can also be obtained by solving the BHPS model for the $|uudc\bar{c}\rangle$ Fock state numerically and keeping the exact dependence on M_p, m_u, and m_d. In this BHPS model, the intrinsic quark distributions are determined by starting from a $|uudq\bar{q}\rangle$ proton Fock state, where the probability differential for a quark i to carry a momentum fraction x_i is given by

$$d\mathcal{P}(x_1, \ldots, x_5) = A\, dx_1 \ldots dx_5\, \delta(1 - \sum_{i=1}^{5} x_i) \frac{1}{\left[M_p^2 - \sum_{i=1}^{5} \frac{m_i^2}{x_i} \right]^2}. \tag{2}$$

This generalized BHPS model, used in the context of the CT14HERA fit with IC, is named BHPS3. We do not, however, include the intrinsic contribution to the s quark PDF, because it is overwhelmed by the very large strange PDF uncertainty. The presence of an intrinsic component for the strange quark does not affect our conclusions about the nonperturbative charm.

A sea-like component is usually monotonic in x and satisfies $f_{q/p}(x, Q_c) \sim x^{-a_1}$ for $x \to 0$ and $f_{q/p}(x, Q_c) \sim (1-x)^{a_2}$ for $x \to 1$, with a_1 slightly above 1, and $a_2 \gtrsim 5$. This behavior is typical for the leading-power, or "extrinsic" production. For example, an (anti)quark PDF with this behavior originates from $g \to q\bar{q}$ splittings in perturbative QCD, or from disconnected diagrams in lattice QCD (see Ref. [21] for details). Even a missing next-to-next-to-next-to-leading order (N3LO) leading-power correction may produce a sea-like contribution at $x \ll 0.1$, where the valence-like components are suppressed. In the SEA model, the charm PDF is parametrized by a "sea-like" nonperturbative function that is proportional to the light quark distributions:

$$\widehat{c}(x) = A\left(\bar{d}(x, Q_0) + \bar{u}(x, Q_0)\right). \tag{3}$$

This model is assumed with the SEA1 and SEA2 PDF sets which are illustrated in Sec. 2 below. Finally, the normalization coefficient A in the models described above can be derived

[1] In contrast to the light flavors, in lattice QCD a charm PDF arises exclusively from disconnected diagrams [22]. This suggests that c and $\bar{c}$ contributions in DIS are connected to the hadron target by gluon insertions.

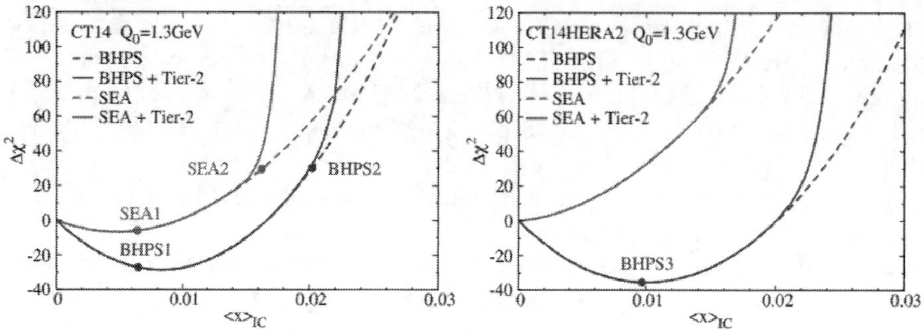

Figure 1. The change $\Delta\chi^2$ in the goodness of fit to the CT14 (left) and CT14HERA2 (right) data sets as a function of the charm momentum fraction $\langle x\rangle_{\rm IC}$ for the BHPS (blue) and SEA (red) models. Solid (dashed) lines represent the total χ^2 and the partial χ^2_{global}, as defined in Sec. 2.

from the charm momentum fraction (first moment) at scale Q:

$$\langle x\rangle_{\rm IC} = \int_0^1 x\,[c(x,Q_0) + \bar{c}(x,Q_0)]\,dx. \tag{4}$$

By its definition, $\langle x\rangle_{\rm IC}$ is evaluated at the initial scale Q_0. It is to be distinguished from the full charm momentum fraction $\langle x\rangle_{c+\bar{c}}(Q)$ at $Q > Q_c$, which rapidly increases with Q because of the admixture of the twist-2 charm component.

2 Results of the global analysis

In this section we illustrate the main findings of our global analysis in which we assume that the additional nonperturbative charm component can be factorized in a similar fashion to the perturbative one. To gauge the preference of the global QCD data to a specific $\langle x\rangle_{\rm IC}$, we examine the goodness-of-fit function $\chi^2 \equiv \chi^2_{global} + P$, constructed in the CT14 method from the global χ^2_{global} and a "tier-2" statistical penalty P [1]. It is convenient to compare each fit with an $\langle x\rangle_{\rm IC} \neq 0$ to the "null-hypothesis" fit obtained assuming $\langle x\rangle_{\rm IC} = 0$. Thus, we start by computing $\Delta\chi^2 \equiv \chi^2 - \chi^2_0$, where χ^2 and χ^2_0 are given for $\langle x\rangle_{\rm IC} \neq 0$ and $\langle x\rangle_{\rm IC} = 0$, respectively, at 50 values of $\langle x\rangle_{\rm IC}$ and default $Q_0 = m_c^{pole} = 1.3$ GeV. We plot the resulting $\Delta\chi^2$ behavior in Fig. 1. The CT14 (CT14HERA2) data sets are compared against the approximate (exact) solution of the BHPS model, respectively. The SEA charm parametrizations are constructed as in Eq. (3) in terms of the respective CT14 or CT14HERA2 light-antiquark parametrizations. We see from Fig. 1 that large amounts of intrinsic charm are disfavored for all models under scrutiny. A mild reduction in χ^2, however, is observed for the BHPS fits, roughly at $\langle x\rangle_{\rm IC} = 1\%$, both in the CT14 and CT14HERA2 frameworks. The significance of this reduction and the upper limit on $\langle x\rangle_{\rm IC}$ depends on the assumed criterion. In CTEQ practice, a set of PDFs with $\Delta\chi^2$ smaller (larger) than 100 units is deemed to be accepted (disfavored) at about 90% C.L. Thus, a reduction of χ^2 by less than forty units for the BHPS curves has significance roughly of order one standard deviation. The new upper limits on $\langle x\rangle_{\rm IC}$ in the CT14 and CT14HERA2 analyses at the 90% C.L.: $\langle x\rangle_{\rm IC} \lesssim 0.021$ for CT14 BHPS; $\langle x\rangle_{\rm IC} \lesssim 0.024$ for CT14HERA2 BHPS; $\langle x\rangle_{\rm IC} \lesssim 0.016$ for CT14 and CT14HERA2 SEA.

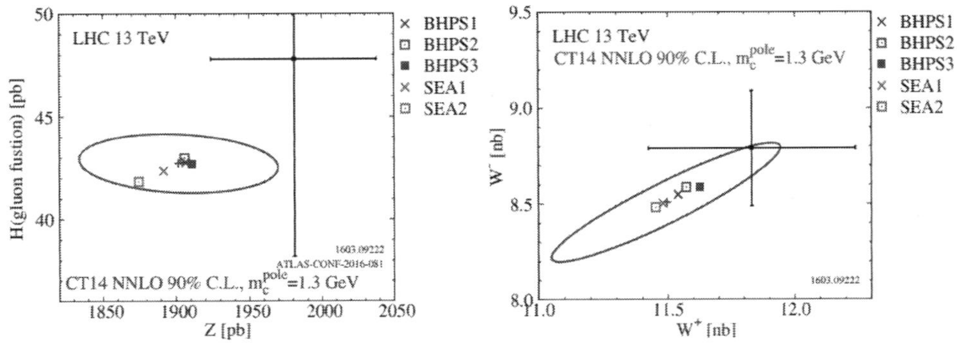

Figure 2. CT14 NNLO H (gluon-gluon fusion), Z, $W+$, and $W-$ production cross sections with an IC PDF component at the LHC $\sqrt{s}$ = 13 TeV, with pole mass m_c^{pole} = 1.3 GeV. The 90% C.L. uncertainty regions for CT14 at NNLO and experimental points from ATLAS [23, 24] are also shown.

2.1 Impact of IC on the electroweak Z and H boson production cross sections at the LHC

Figure 2 illustrates predictions of the total cross sections for inclusive production of electroweak bosons $W^{\pm}$, Z^0, and H (via gluon-gluon fusion) for the BHPS and SEA models at the LHC $\sqrt{s}$ = 13 TeV with charm quark mass m_c = 1.3 GeV. To provide a visual measure of the CT14NNLO uncertainty, each figure shows an error ellipse corresponding to CT14 NNLO at the 90% C.L. and experimental data points relative to measurements from the ATLAS collaboration [23, 24]. The W and Z inclusive cross sections (multiplied by branching ratios for the decay into one charged lepton flavor), are calculated by using the VRAP v0.9 program [25, 26] at NNLO in QCD, with the renormalization and factorization (μ_R and μ_F) scales set equal to the invariant mass of the vector boson. The Higgs boson cross sections via gluon-gluon fusion are calculated at NNLO in QCD by using the iHixs v1.3 program [27], in the heavy-quark effective theory (HQET) with finite top quark mass correction, and with the QCD scales set equal to the invariant mass of the Higgs boson. The central value predictions for the BHPS and SEA models are all within the CT14 NNLO uncertainties, with BHPS very close to the CT14 nominal fit. The impact of IC on these key LHC observables is mild.

3 Conclusions

We explored the possibility of sizeable nonperturbative contribution to charm PDF assuming that factorization for such contributions exists. We have determined the magnitude of the IC component of the proton that is consistent with the CT14 global QCD analysis of hard scattering data: $\langle x \rangle < 2\%$ for BHPS IC and $\langle x \rangle < 1.6\%$ for SEA IC at 90% C.L.. As of today, the experimental confirmation of the IC component in the proton is still missing, and data from far more sensitive measurements are required.

Acknowledgments. This work was supported in part by the U.S. Department of Energy under Grant No. DESC0010129; by the U.S. National Science Foundation under Grant No. PHY-1417326; by the National Natural Science Foundation of China under the Grant No. 11465018; and by the Lancaster-Manchester-Sheffield Consortium for Fundamental Physics under STFC Grant No. ST/L000520/1. The work of J.G. is sponsored by Shanghai Pujiang Program.

References

[1] S. Dulat et al., Phys. Rev. **D93** (2016) 033006, 1506.07443.
[2] M. Buza et al., Nucl. Phys. **B472** (1996) 611, hep-ph/9601302.
[3] M. Buza et al., Eur. Phys. J. **C1** (1998) 301, hep-ph/9612398.
[4] S.J. Brodsky et al., Phys. Lett. **B93** (1980) 451.
[5] S.J. Brodsky, C. Peterson and N. Sakai, Phys. Rev. **D23** (1981) 2745.
[6] J. Pumplin, Phys. Rev. **D73** (2006) 114015, hep-ph/0508184.
[7] W.C. Chang and J.C. Peng, Phys. Rev. Lett. **106** (2011) 252002, 1102.5631.
[8] J. Bluemlein, Phys. Lett. **B753** (2016) 619, 1511.00229.
[9] S.J. Brodsky et al., Adv. High Energy Phys. **2015 231547**, (2015) 1504.06287.
[10] F.S. Navarra et al., Phys. Rev. **D54** (1996) 842, hep-ph/9504388.
[11] S. Paiva et al., Mod. Phys. Lett. **A13** (1998) 2715, hep-ph/9610310.
[12] F.M. Steffens, W. Melnitchouk and A.W. Thomas, Eur. Phys. J. **C11** (1999) 673, hep-ph/9903441.
[13] T.J. Hobbs, J.T. Londergan and W. Melnitchouk, Phys. Rev. **D89** (2014) 074008, 1311.1578.
[14] T.J. Hou, S. Dulat, J. Gao, M. Guzzi, J. Huston, P. Nadolsky, C. Schmidt, J. Winter, K. Xie, and C. -P. Yuan JHEP **1802** (2018) 059, 1707.00657.
[15] P. Jimenez-Delgado et al., Phys. Rev. Lett. **114** (2015) 082002, 1408.1708.
[16] P. Jimenez-Delgado et al., Phys. Rev. Lett. **116** (2016) 019102, 1504.06304.
[17] F. Lyonnet et al., JHEP 07 (2015) **141**, 1504.05156.
[18] NNPDF, R.D. Ball et al., Eur. Phys. J. **C76** (2016) 647, 1605.06515.
[19] NNPDF, R.D. Ball et al., (2017), Eur.Phys.J. **C77** (2017) no.10, 663 1706.00428.
[20] European Muon, J.J. Aubert et al., Nucl. Phys. **B213** (1983) 31.
[21] K.F. Liu et al., Phys. Rev. Lett. **109** (2012) 252002, 1206.4339.
[22] K.F. Liu, private communication.
[23] ATLAS, G. Aad et al., Phys. Lett.**B759** (2016) 601, 1603.09222.
[24] ATLAS Collaboration, ATLAS-CONF-2016-081.
[25] C. Anastasiou et al., Phys. Rev. **D69** (2004) 094008, hep-ph/0312266.
[26] C. Anastasiou et al., Phys. Rev. Lett. **91** (2003) 182002, hep-ph/0306192.
[27] C. Anastasiou et al., JHEP **12** (2011) 058, 1107.0683.

Single-top and top-antitop cross sections

Nikolaos Kidonakis[1,*]

[1]Department of Physics, Kennesaw State University, Kennesaw, GA 30144, USA

Abstract. I present high-order calculations, including soft-gluon corrections, for single-top and top-antitop production cross sections and differential distributions. For single-top production, results are presented for the three different channels in the Standard Model, for associated production with a charged Higgs, and for processes involving anomalous couplings. For top-antitop pair production, total cross sections and top-quark transverse-momentum and rapidity distributions are presented for various LHC energies.

1 Introduction

Higher-order soft-gluon corrections have been calculated through N^{3}LO for top-quark production via various processes; see Ref. [1] for a review. Here I present the latest results for t-channel and s-channel single-top production, tW and tH^- production, tZ production via anomalous couplings, and $t\bar{t}$ production. For all these processes QCD corrections are very significant and are dominated by soft-gluon corrections.

I calculate and resum these soft corrections at next-to-next-to-leading logarithm (NNLL) accuracy for the double-differential cross section, which is then used to calculate top-quark transverse-momentum and rapidity distributions and total cross sections. Finite-order expansions at approximate NNLO (aNNLO) and approximate N^{3}LO (aN^{3}LO), matched to exact results, provide the best predictions for these quantities.

We calculate soft-gluon corrections for partonic processes of the form

$$f_1(p_1) + f_2(p_2) \to t(p_t) + X$$

and we define $s = (p_1 + p_2)^2$, $t = (p_1 - p_t)^2$, $u = (p_2 - p_t)^2$, and $s_4 = s + t + u - \sum m^2$. At partonic threshold $s_4 \to 0$. The nth-order soft-gluon corrections appear as $\left[\ln^k(s_4/m_t^2)/s_4\right]_+$ where $k \le 2n - 1$.

Moments of the partonic cross section, $\hat{\sigma}(N) = \int (ds_4/s)\, e^{-Ns_4/s}\hat{\sigma}(s_4)$, can be written in factorized form

$$\sigma^{f_1 f_2 \to tX}(N,\epsilon) = H_{IL}^{f_1 f_2 \to tX}(\alpha_s(\mu_R))\, S_{LI}^{f_1 f_2 \to tX}\left(\frac{m_t}{N\mu_F}, \alpha_s(\mu_R)\right) \prod J_{\rm in}(N,\mu_F,\epsilon) \prod J_{\rm out}(N,\mu_F,\epsilon)$$

in $4-\epsilon$ dimensions, where $H_{IL}^{f_1 f_2 \to tX}$ is a hard function and $S_{LI}^{f_1 f_2 \to tX}$ is a soft-gluon function. $S_{LI}^{f_1 f_2 \to tX}$ satisfies the renormalization group equation

$$\left(\mu\frac{\partial}{\partial\mu} + \beta(g_s)\frac{\partial}{\partial g_s}\right) S_{LI}^{f_1 f_2 \to tX} = -(\Gamma_S^\dagger)_{LK}^{f_1 f_2 \to tX} S_{KI}^{f_1 f_2 \to tX} - S_{LK}^{f_1 f_2 \to tX}(\Gamma_S)_{KI}^{f_1 f_2 \to tX}$$

*e-mail: nkidonak@kennesaw.edu

© The Authors, published by EDP Sciences. This is an open access article distributed under the terms of the Creative Commons Attribution License 4.0 (http://creativecommons.org/licenses/by/4.0/).

where the soft anomalous dimension $\Gamma_S^{f_1 f_2 \to tX}$ controls the evolution of the soft function, giving the exponentiation of logarithms of N. To achieve NNLL accuracy we need to calculate the relevant soft anomalous dimensions at two loops.

2 Single-top production

We now provide results for single-top production in the t-channel, s-channel, and via tW production. Fixed-order results for these processes are known at NNLO for the t-channel [2–4] and s-channel [5], and at NLO for tW production [6]. Here we provide results with soft-gluon corrections at aNNLO for t- and s-channel production [7], and at aN^{3}LO for tW production [8]; these results are obtained from NNLL resummation [7–9]. We use MMHT2014 NNLO pdf [10].

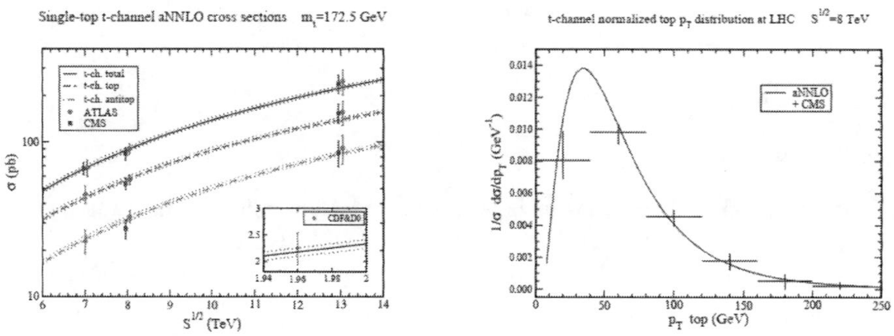

Figure 1. (Left) Single-top t-channel aNNLO cross sections compared with CMS and ATLAS data at 7 TeV [12, 13], 8 TeV [14, 15], and 13 TeV [16, 17], and with CDF and D0 combined data at 1.96 TeV [18]. (Right) Top-quark aNNLO normalized p_T distributions in t-channel production at 8 TeV compared to CMS [19] data.

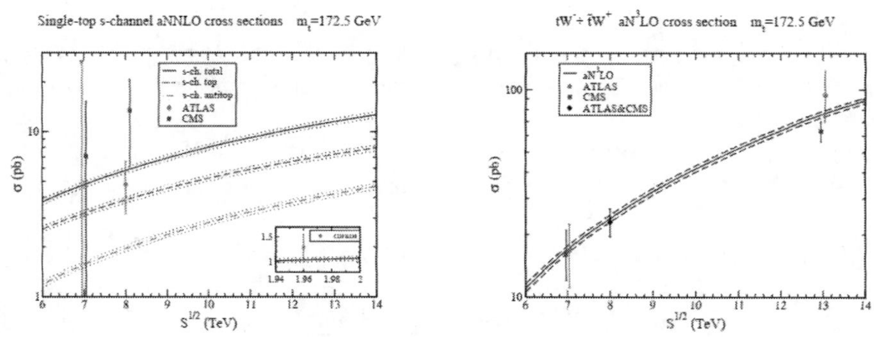

Figure 2. (Left) Single-top s-channel aNNLO cross sections compared to ATLAS and CMS data at 7 TeV [20, 21] and 8 TeV [21, 22], and to CDF and D0 combined data [23] at 1.96 TeV. (Right) aN^{3}LO cross sections for tW production compared to ATLAS and CMS data at 7 TeV [24, 25], 8 TeV [26], and 13 TeV [27, 28].

We begin with t-channel production at aNNLO. In the left plot of Fig. 1 we display total t-channel cross sections as functions of collider energy at the LHC and (inset) at the Tevatron.

Results are given for the single-top cross section, the single-antitop cross section, and their sum. We observe very good agreement of the aNNLO theory curves [7, 11] with all available data from the LHC and the Tevatron at various energies. The right plot of Fig. 1 shows the aNNLO normalized top-quark p_T distributions at 8 TeV LHC energy, which describe the corresponding data from CMS [19] quite well.

We continue with s-channel production at aNNLO. In the left plot of Fig. 2 we show cross sections for single-top and single-antitop s-channel production, and their sum, as functions of energy at the LHC and (inset) at the Tevatron. We again observe very good agreement of the aNNLO theory curves [7, 11] with available data at LHC and Tevatron energies.

We next discuss tW production at aN^{3}LO. In the right plot of Fig. 2 we show the total $tW^-+\bar{t}W^+$ cross section as a function of LHC energy. Very good agreement is observed between the aN^{3}LO results [8, 11] and the data at 7, 8, and 13 TeV energies.

3 tH^- production

We continue with tH^- production in the MSSM or other two-Higgs-doublet models [29]. We use MMHT2014 NNLO pdf [10] for our numerical results.

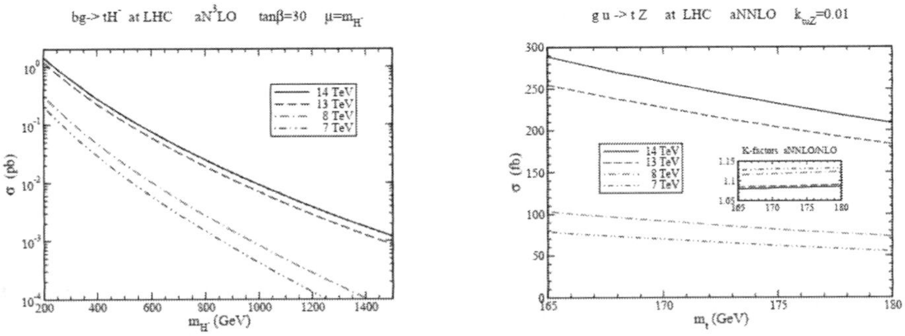

Figure 3. (Left) aN^{3}LO cross sections for tH^- production. (Right) aNNLO cross sections for tZ production via anomalous couplings.

In the left plot of Fig. 3 we show the aN^{3}LO total cross section for tH^- production [11] as a function of charged-Higgs mass at LHC energies of 7, 8, 13, and 14 TeV. We use $\tan\beta = 30$. The soft-gluon corrections are large for this process. Top-quark p_T and rapidity distributions in this process have also been presented in [29].

4 tZ production via anomalous couplings

Next, we discuss soft-gluon corrections in tZ production in models with anomalous t-q-Z couplings [30, 31]. The NLO corrections for this process were calculated in Ref. [32]. The complete NLO corrections are very well approximated by the soft-gluon corrections at that order.

We use CT14 pdf [33] for our numerical results in this process. In the right plot of Fig. 3 we plot the aNNLO total cross section for tZ production as a function of top-quark mass at LHC energies of 7, 8, 13, and 14 TeV.

The K-factors shown in the inset plot show that the aNNLO corrections are large and significantly enhance the NLO cross section, especially at lower energies. This is important in providing theoretical input to experimental limits on the couplings [34, 35]. Top-quark differential distributions in this process have been presented in Ref. [31].

Similar results have more recently been presented for $t\gamma$ production via anomalous couplings in Ref. [36].

5 Top-antitop pair production

Finally, we discuss top-antitop pair production [37, 38]. The soft anomalous dimensions are 2×2 matrices for the $q\bar{q} \to t\bar{t}$ channel, and 3×3 matrices for the $gg \to t\bar{t}$ channel.

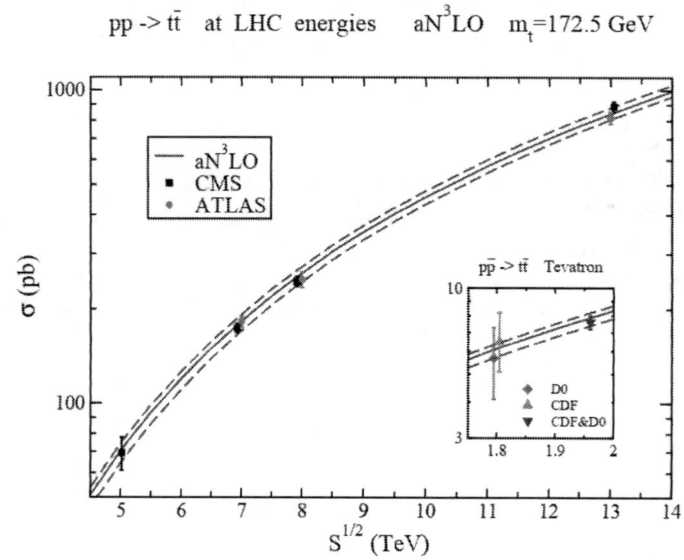

Figure 4. Top-antitop aN³LO cross sections compared with CMS data at 5.02 TeV [41] and with ATLAS and CMS data at 7 TeV [42, 43], 8 TeV [43, 44], and 13 TeV [45, 46] LHC energies. The inset shows the aN³LO cross section compared with CDF [47] and D0 [48] data at 1.8 TeV, and CDF&D0 combined data [49] at 1.96 TeV Tevatron energy.

At one loop for $q\bar{q} \to t\bar{t}$, the elements of the $\Gamma_S^{q\bar{q}\to t\bar{t}}$ matrix in an s-channel singlet-octet color basis are

$$\Gamma_{11}^{q\bar{q}\to t\bar{t}\,(1)} = \Gamma_{\rm cusp}^{(1)}\,, \quad \Gamma_{12}^{q\bar{q}\to t\bar{t}\,(1)} = \frac{C_F}{C_A}\ln\left(\frac{t_1}{u_1}\right), \quad \Gamma_{21}^{q\bar{q}\to t\bar{t}\,(1)} = 2\ln\left(\frac{t_1}{u_1}\right),$$

$$\Gamma_{22}^{q\bar{q}\to t\bar{t}\,(1)} = \left(1-\frac{C_A}{2C_F}\right)\Gamma_{\rm cusp}^{(1)} + 4C_F\ln\left(\frac{t_1}{u_1}\right) - \frac{C_A}{2}\left[1+\ln\left(\frac{s m_t^2 t_1^2}{u_1^4}\right)\right].$$

At two loops for $q\bar{q} \to t\bar{t}$, the elements of the $\Gamma_S^{q\bar{q}\to t\bar{t}}$ matrix are

$$\Gamma_{11}^{q\bar{q}\to t\bar{t}\,(2)} = \Gamma_{\rm cusp}^{(2)}\,, \quad \Gamma_{12}^{q\bar{q}\to t\bar{t}\,(2)} = \left(\frac{K}{2} - \frac{C_A}{2}N_{2l}\right)\Gamma_{12}^{q\bar{q}\to t\bar{t}\,(1)},$$

$$\Gamma_{21}^{q\bar{q}\to t\bar{t}\,(2)} = \left(\frac{K}{2} + \frac{C_A}{2}N_{2l}\right)\Gamma_{21}^{q\bar{q}\to t\bar{t}\,(1)},$$

$$\Gamma_{22}^{q\bar{q}\to t\bar{t}\,(2)} = \frac{K}{2}\Gamma_{22}^{q\bar{q}\to t\bar{t}\,(1)} + \left(1-\frac{C_A}{2C_F}\right)\left(\Gamma_{\rm cusp}^{(2)} - \frac{K}{2}\Gamma_{\rm cusp}^{(1)}\right).$$

See also Ref. [1] for more details. Here $\Gamma_{\rm cusp}$ is the cusp anomalous dimension [39].

The soft-gluon corrections are large and they are excellent approximations to the complete corrections at both NLO and NNLO. The additional corrections at aN^{3}LO provide further significant enhancements and must be included for precision physics. Another approach for the aN^{3}LO corrections has appeared recently in Ref. [40].

In our numerical results below we use the MMHT2014 NNLO pdf [10]. The total top-antitop cross sections at aN^{3}LO [38] are shown in Fig. 4 and compared with data at Tevatron and LHC energies. We find remarkable agreement between theory and data at all energies.

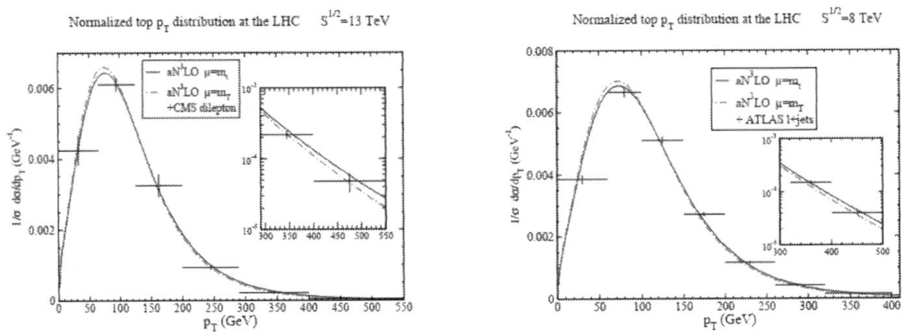

Figure 5. aN^{3}LO top-quark normalized p_T distributions (left) at 13 TeV energy compared with CMS [50] data and (right) at 8 TeV energy compared with ATLAS [51] data.

The aN^{3}LO top-quark normalized p_T distributions in $t\bar{t}$ production are shown in Fig. 5 at 13 TeV (left plot) and 8 TeV (right plot) and compared with CMS and ATLAS data respectively. We find excellent agreement of the theoretical predictions with the data.

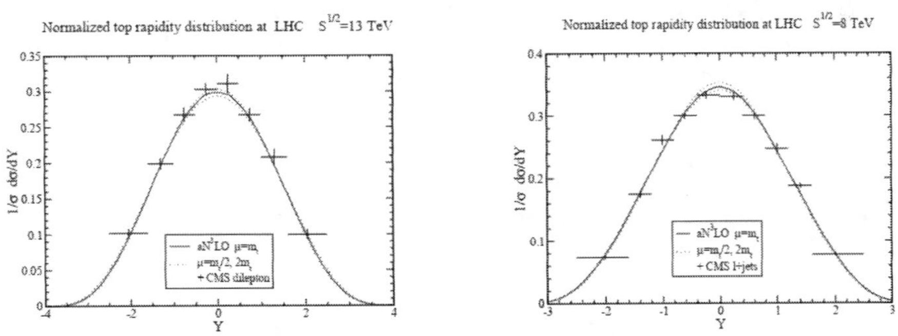

Figure 6. Top-quark aN^{3}LO normalized rapidity distribution at (left) 13 TeV energy compared with CMS data [50] and (right) 8 TeV energy compared with CMS data [52].

The aN^{3}LO top-quark normalized rapidity distributions in $t\bar{t}$ production are shown in Fig. 6 at 13 TeV (left plot) and 8 TeV (right plot) and compared with CMS data. Again, we find that the theory curves provide an excellent description of the data.

6 Summary

We have discussed cross sections and distributions for various top-quark production processes. Soft-gluon corrections are important in all cases. We have shown results for t-channel and s-channel single-top production at aNNLO, tW production at aN^{3}LO, tH^- production at aN^{3}LO, tZ production via anomalous couplings at aNNLO, and $t\bar{t}$ production at aN^{3}LO. We find excellent agreement with available collider data. The higher-order corrections are very significant and need to be included for better theoretical predictions.

Acknowledgments

This material is based upon work supported by the National Science Foundation under Grant No. PHY 1519606.

References

[1] N. Kidonakis, Int. J. Mod. Phys. A **33**, 1830021 (2018) [arXiv:1806.03336 [hep-ph]].
[2] M. Brucherseifer, F. Caola, and K. Melnikov, Phys. Lett. B **736**, 58 (2014) [arXiv:1404.7116 [hep-ph]].
[3] E.L. Berger, J. Gao, C.-P. Yuan, and H.X. Zhu, Phys. Rev. D **94**, 071501 (2016) [arXiv:1606.08463 [hep-ph]].
[4] E.L. Berger, J. Gao, and H.X. Zhu, JHEP **1711**, 158 (2017) [arXiv:1708.09405 [hep-ph]].
[5] Z. Liu and J. Gao, arXiv:1807.03835 [hep-ph].
[6] S.H. Zhu, Phys. Lett. B **524**, 283 (2002) [Erratum: *ibid.* **537**, 351 (2002)] [hep-ph/0109269].
[7] N. Kidonakis, Phys. Rev. D **81**, 054028 (2010) [arXiv:1001.5034 [hep-ph]]; **83**, 091503(R) (2011) [arXiv:1103.2792 [hep-ph]]; **88**, 031504(R) (2013) [arXiv:1306.3592 [hep-ph]]; **93**, 054022 (2016) [arXiv:1510.06361 [hep-ph]].
[8] N. Kidonakis, Phys. Rev. D **96**, 034014 (2017) [arXiv:1612.06426 [hep-ph]]; in *Proceedings of DPF 2017*, eConf C170731 [arXiv:1709.06975 [hep-ph]].
[9] N. Kidonakis, Phys. Rev. D **82**, 054018 (2010) [arXiv:1005.4451 [hep-ph]].
[10] L.A. Harland-Lang, A.D. Martin, P. Molytinski, and R.S. Thorne, Eur. Phys. J. C **75**, 204 (2015) [arXiv:1412.3989 [hep-ph]].
[11] N. Kidonakis, in *Proceedings of CIPANP2018* [arXiv:1808.02934 [hep-ph]].
[12] CMS Collab., JHEP **1212**, 035 (2012) [arXiv:1209.4533 [hep-ex]].
[13] ATLAS Collab., Phys. Rev. D **90**, 112006 (2014) [arXiv:1406.7844 [hep-ex]].
[14] CMS Collab., JHEP **1406**, 090 (2014) [arXiv:1403.7366 [hep-ex]].
[15] ATLAS Collab., Eur. Phys. J. C **77**, 531 (2017) [arXiv:1702.02859 [hep-ex]].
[16] ATLAS Collab., JHEP **1704**, 086 (2017) [arXiv:1609.03920 [hep-ex]].
[17] CMS Collab., Phys. Lett. B **772**, 752 (2017) [arXiv:1610.00678 [hep-ex]].
[18] CDF and D0 Collab., Phys. Rev. Lett. **115**, 152003 (2015) [arXiv:1503.05027 [hep-ex]].
[19] CMS Collab., CMS-PAS-TOP-14-004.
[20] ATLAS Collab., ATLAS-CONF-2011-118.
[21] CMS Collab., JHEP **1609**, 027 (2016) [arXiv:1603.02555 [hep-ex]].
[22] ATLAS Collab., Phys. Lett. B **756**, 228 (2016) [arXiv:1511.05980 [hep-ex]].
[23] CDF and D0 Collab., Phys. Rev. Lett. **112**, 231803 (2014) [arXiv:1402.5126 [hep-ex]].
[24] ATLAS Collab., Phys. Lett. B **716**, 142 (2012) [arXiv:1205.5764 [hep-ex]].
[25] CMS Collab., Phys. Rev. Lett. **110**, 022003 (2013) [arXiv:1209.3489 [hep-ex]].

[26] ATLAS and CMS Collab., ATLAS-CONF-2016-023, CMS-PAS-TOP-15-019.
[27] ATLAS Collab., JHEP **1801**, 063 (2018) [arXiv:1612.07231 [hep-ex]].
[28] CMS Collab., arXiv:1805.07399 [hep-ex].
[29] N. Kidonakis, JHEP **0505**, 011 (2005) [hep-ph/0412422]; Phys. Rev. D **94**, 014010 (2016) [arXiv:1605.00622 [hep-ph]].
[30] A. Belyaev and N. Kidonakis, Phys. Rev. D **65**, 037501 (2002) [hep-ph/0102072]; N. Kidonakis and A. Belyaev, JHEP **0312**, 004 (2003) [hep-ph/0310299].
[31] N. Kidonakis, Phys. Rev. D **97**, 034028 (2018) [arXiv:1712.01144 [hep-ph]].
[32] B.H. Li, Y. Zhang, C.S. Li, J. Gao, and H.X. Zhu, Phys. Rev. D **83**, 114049 (2011) [arXiv:1103.5122 [hep-ph]].
[33] S. Dulat, T.-J. Hou, J. Gao, M. Guzzi, J. Huston, P. Nadolsky, J. Pumplin, C. Schmidt, D. Stump, and C.-P. Yuan, Phys. Rev. D **93**, 033006 (2016) [arXiv:1506.07443 [hep-ph]].
[34] CMS Collaboration, JHEP **1707**, 003 (2017) [arXiv:1702.01404 [hep-ex]].
[35] ATLAS Collaboration, ATLAS-CONF-2017-070.
[36] M. Forslund and N. Kidonakis, arXiv:1808.09014 [hep-ph].
[37] N. Kidonakis, Phys. Rev. D **82**, 114030 (2010) [arXiv:1009.4935 [hep-ph]]; in *Physics of Heavy Quarks and Hadrons, HQ2013* [arXiv:1311.0283 [hep-ph]].
[38] N. Kidonakis, Phys. Rev. D **90**, 014006 (2014) [arXiv:1405.7046 [hep-ph]]; **91**, 031501(R) (2015) [arXiv:1411.2633 [hep-ph]]; **91**, 071502(R) (2015) [arXiv:1501.01581 [hep-ph]].
[39] N. Kidonakis, Phys. Rev. Lett. **102**, 232003 (2009) [arXiv:0903.2561 [hep-ph]]; Int. J. Mod. Phys. A **31**, 1650076 (2016) [arXiv:1601.01666 [hep-ph]].
[40] J. Piclum and C. Schwinn, JHEP **1803**, 164 (2018) [arXiv:1801.05788 [hep-ph]].
[41] CMS Collab., JHEP **1803**, 115 (2018) [arXiv:1711.03143 [hep-ex]].
[42] ATLAS Collab., Eur. Phys. J. C **74**, 3109 (2014) [Addendum: *ibid.* **76**, 642 (2016)] [arXiv:1406.5375 [hep-ex]].
[43] CMS Collab., JHEP **1608**, 029 (2016) [arXiv:1603.02303 [hep-ex]].
[44] ATLAS Collab., Eur. Phys. J. C **78**, 487 (2018) [arXiv:1712.06857 [hep-ex]].
[45] ATLAS Collab., Phys. Lett. B **761**, 136 (2016) [Erratum: *ibid.* **772**, 879 (2017)] [arXiv:1606.02699 [hep-ex]].
[46] CMS Collab., JHEP **1709**, 051 (2017) [arXiv:1701.06228 [hep-ex]].
[47] CDF Collab., Phys. Rev. D **64**, 032002 (2001) [Erratum: *ibid.* **67**, 119901 (2003)] [hep-ex/0101036].
[48] D0 Collab., Phys. Rev. D **67**, 012004 (2003) [hep-ex/0205019].
[49] CDF and D0 Collab., Phys. Rev. D **89**, 072001 (2014) [arXiv:1309.7570 [hep-ex]].
[50] CMS Collab., JHEP **1804**, 060 (2018) [arXiv:1708.07638 [hep-ex]].
[51] ATLAS Collab., Eur. Phys. J. C **76**, 538 (2016) [arXiv:1511.04716 [hep-ex]].
[52] CMS Collab., Eur. Phys. J. C **75**, 542 (2015) [arXiv:1505.04480 [hep-ex]].

EPJ Web of Conferences **192**, 00005 (2018) https://doi.org/10.1051/epjconf/201819200005
QCD@Work 2018

Experimental results from CMS

John Strologas[1](for the CMS collaboration)

[1]Department of Physics, University of Ioannina, Ioannina 45110, Greece

Abstract. We present some of the latest results of the CMS experiment, covering analyses in QCD multijet, top, bottom, Higgs, forward/small-x QCD, heavy-ion, exotic, and supersymmetry physics utilizing LHC integrated luminosity up to ~80 fb^{-1}.

1 Introduction

CMS, as a major general-purpose detector at the LHC, provides an abundance of high-quality data that enable the performance of analyses on a plethora of physics topics. We present some of the legacy and most recent results with some emphasis on Quantum Chromodynamics (QCD) relevance.

2 QCD multijet physics

Double-differential inclusive jet production cross sections have been measured with 8 TeV [1] and 13-TeV [2] data, corresponding to luminosity of 19.7 fb^{-1} and 71 pb^{-1} respectively. Particle-flow [3] jets are reconstructed with the anti-k_t algorithm [4] with R distance parameter of 0.4 and 0.7. Results demonstrate excellent understanding of QCD next-to-leading order (NLO) fixed-order calculations corrected for non-perturbative (NP) effects from parton showering, hadronization and multiparton interactions. As shown in Figs. 1 and 2, the Standard Model (SM) agrees with the observed distributions for a cross-section range of 18 orders of magnitude, up to 2 TeV in jet-transverse momentum p_T and in seven absolute rapidity y bins up to 4.7, for jets reconstructed with both R=0.4 and R=0.7.

Studies of azimuthal correlations between the leading two jets or any pair of jets [5] compare 35.9-fb^{-1}, 13-TeV data to event generators at LO (Pythia8, Herwig++, MadGraph+Pythia8) and NLO (Powheg, Herwig7). Levels of agreement depend on jet multiplicity.

The triple-differential dijet cross section, measured as a function of p_T, the difference, and sum of jet rapidities with 19.7-fb^{-1}, 8-TeV data [6], is not described sufficiently by NLO QCD calculations with NP and electroweak corrections. This fact is used for constraining the proton parton-distribution functions (PDF). The inclusion of these CMS results in the PDF fits changes the gluon PDF shape at lower x values and reduces the systematic uncertainty effect at higher x, compared to deep-inelastic-scattering-only fits, as shown in Fig. 3.

A measurement of the ratio of inclusive 3-jet to 2-jet cross section with R=0.7 jets with 19.7-fb^{-1} data from 8-TeV collisions provides an accurate measurement of the strong coupling constant a_s [7], due to canceling common systematic uncertainties on the numerator

© The Authors, published by EDP Sciences. This is an open access article distributed under the terms of the Creative Commons Attribution License 4.0 (http://creativecommons.org/licenses/by/4.0/).

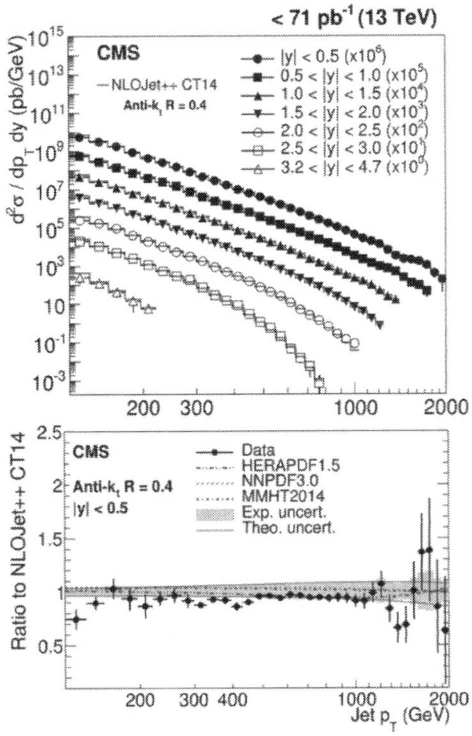

Figure 1. Upper plot: Double-differential inclusive jet cross section for R=0.4 jets at 13 TeV. Lower plot: comparison with NLO-in-QCD theoretical prediction, with NP and electroweak corrections, for $|y| < 0.5$.

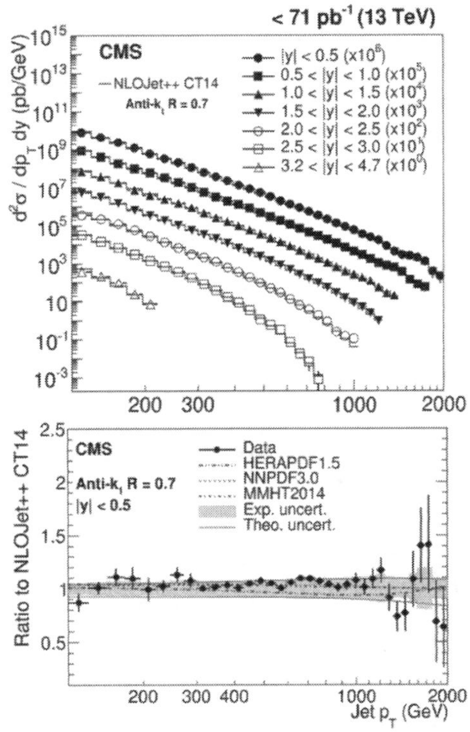

Figure 2. Upper plot: Double-differential inclusive jet cross section for R=0.7 jets at 13 TeV. Lower plot: comparison with NLO-in-QCD theoretical prediction, with NP and electroweak corrections, for $|y| < 0.5$.

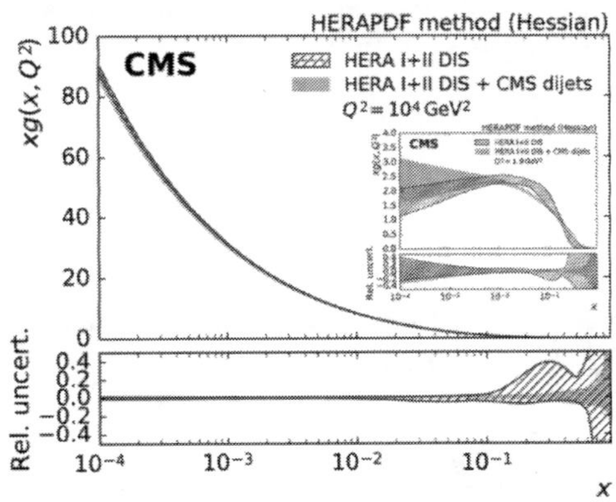

Figure 3. Extraction of the gluon PDF with inclusion of input from the CMS triple-differential inclusive jet cross section at $Q^2 = 10^4$ GeV^2 (insert: $Q^2 = 1.9$ GeV^2).

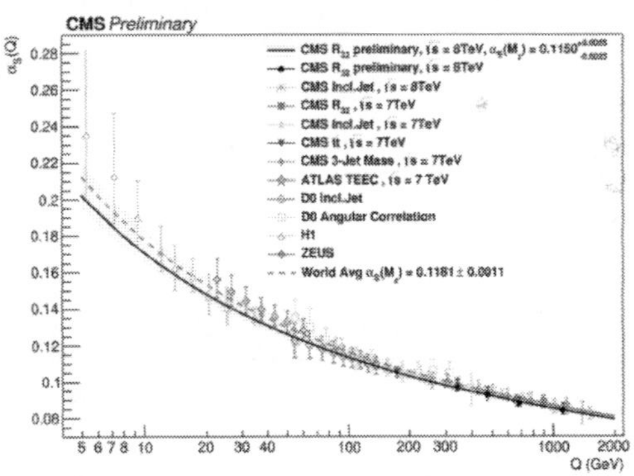

Figure 4. Extraction of the α_S from the CMS R_{32} analysis (black data points) and comparison with world average and other experimental measurements.

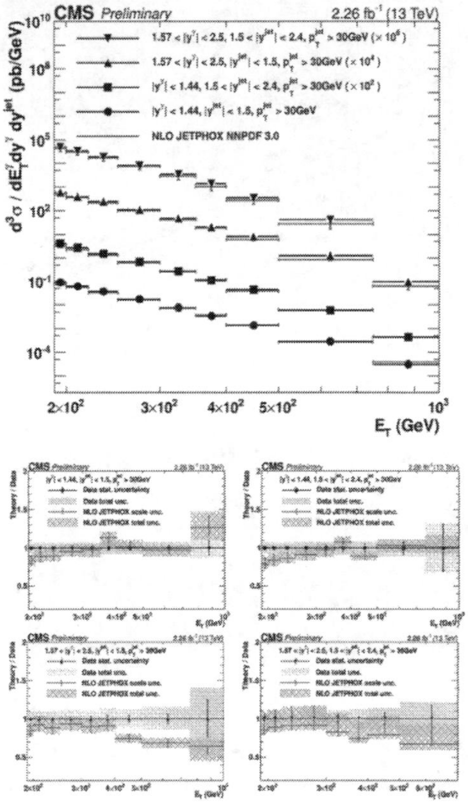

Figure 5. Upper plot: Triple-differential cross section of the production of photon in association with jets in four combinations of rapidities of photon and the highest-p_T jet, as a function of photon E_T. Lower plot: Theory/Data plots for these rapidity bins.

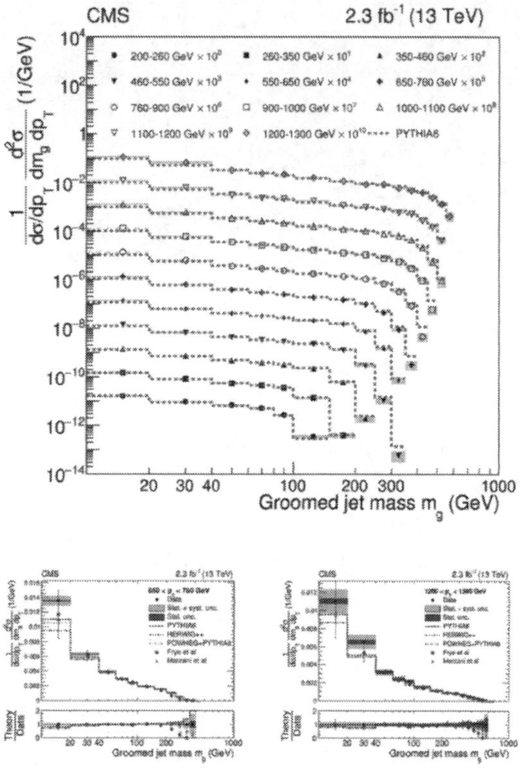

Figure 6. Upper plot: Normalized double-differential inclusive dijet cross section as a function of groomed jet mass for 11 p_T bins. Lower plot: Comparison of groomed jet mass with generators and theoretical calculation in two p_T bins.

and denominator quantities. The analysis is done in four average leading-two-jet p_T bins, resulting in measurement of a_S for four values of energy scale Q, as shown in Fig. 4.

The production of a photon in association with jets is an excellent test of QCD. Inclusive-photon double-differential and photon+jet triple-differential measurements (shown in Fig. 5) with 2.26-fb^{-1}, 13-TeV data agree well with the NLO-in-QCD JetPhox generator [8].

The same luminosity was used for the measurement of the inclusive dijet cross section as a function of groomed and ungroomed jet mass, for 11 p_T bins of R=0.8 jets, as a test of soft QCD effects [9]. Good agreement with Pythia8 is observed, especially for groomed masses below 200 GeV, and comparisons with theoretical calculations are made, as shown in Fig. 6.

The measurement of the jet charge, defined as the sum of jet-constituent charge with three different variations of momentum weights, has been measured in 8 TeV dijet data that correspond to luminosity of 19.7 fb^{-1} [10].

3 Top physics

A 13-TeV, 35.8-fb^{-1} analysis looks for $t\bar{t}$ events in final states with lepton+jets [11]. Along with the lepton, at least 4 jets are required, at least two of which should be b-tagged. The

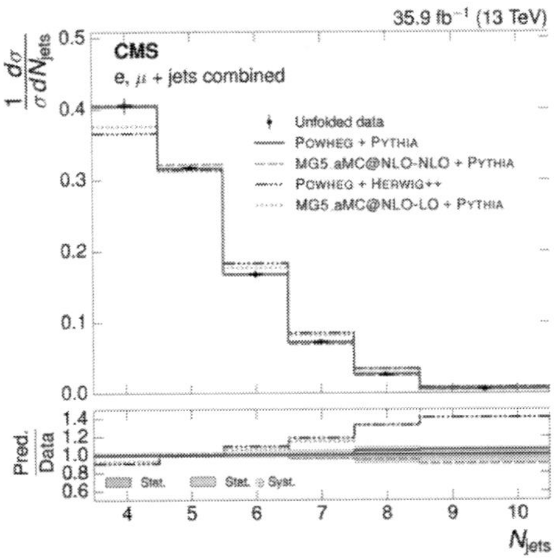

Figure 7. Jet multiplicity in top pair events in lepton+jets final states.

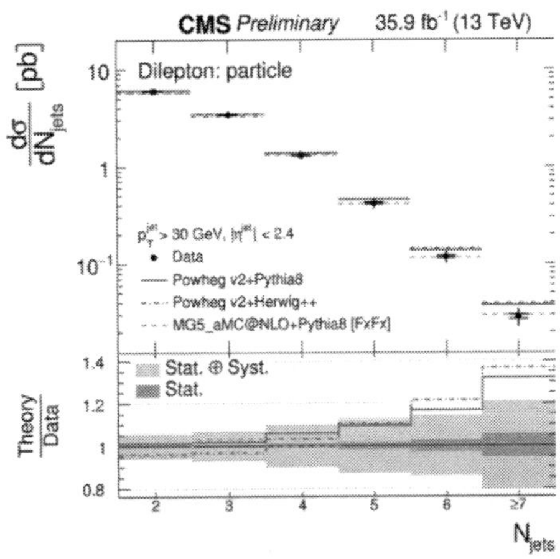

Figure 8. Jet multiplicity in top pair events in dilepton final states.

presence of extra jets is studied as a QCD test and for the determination of backgrounds in new-physics searches. Main background comes from QCD multijet events, bosons+jets and other $t\bar{t}$ channels. POWHEG, with PYTHIA8 showering and hadronization, describes the jet multiplicity, as shown in Fig. 7, but some kinematics are only consistent with prediction after the inclusion of theoretical uncertainties in the unfolding.

The corresponding decays of the top pair to dileptons were also recently analyzed with the same luminosity [12] by requiring two leptons and at least two jets, one of which b-tagged. The analysis was performed in separate lepton-flavor channels which were eventually combined. Main backgrounds come from top+W, W/Z+jet, diboson and $t\bar{t}$+W. Good agreement in jet-multiplicities compared to POWHEG+PYTHIA8 were observed, as shown in Fig. 8, and the top chromomagnetic dipole moment and its y-charge asymmetry were determined.

A recently released analysis measures the single-top production cross section in association with a photon, for muonic top decays [13], in 13-TeV, 35.9-fb^{-1} data. A muon, photon, missing transverse momentum and at least two jets are required. Main background is $t\bar{t}$+γ. The result is $\sigma(t\gamma + \text{jets}) \times \mathcal{B}(t \to \mu\nu b) = 115 \pm 17(\text{stat}) \pm 30(\text{syst})$ fb, whereas the SM prediction is 81 ± 4 fb, which translates to a 4.4-σ evidence.

4 Higgs physics

The current CMS combined RunII measured Higgs boson cross section divided by the SM prediction (signal strength μ) is 1.17 ± 0.1 [14]. Improvements in the precision of the ggH production rate of around 50%, compared with RunI, has been achieved due to higher luminosity and collider energy. Fig. 9 shows the good agreement of measured cross section with SM prediction for all Higgs production and decay channels.

The combination of 13-TeV data with the 7- and 8-TeV data allowed for the first observation of the Higgs boson production in association with a top-quark pair [15]. This is a 5.2-σ discovery (expected 4.2-σ). All Higgs-decay channels signal strength agree with expectation, as shown in Fig. 10.

Latest results include the decay $H \to ZZ \to \ell\ell\ell\ell$ [16]. The analysis was performed separately in the $4e$, 4μ, $ee\mu\mu$ channels and then combined. The data collected in 2017 were analyzed separately and combined with 2016 data for a total integrated luminosity of 77.4 fb^{-1}. The resulted signal strength is $\mu = 1.10^{+0.19}_{-0.17}$.

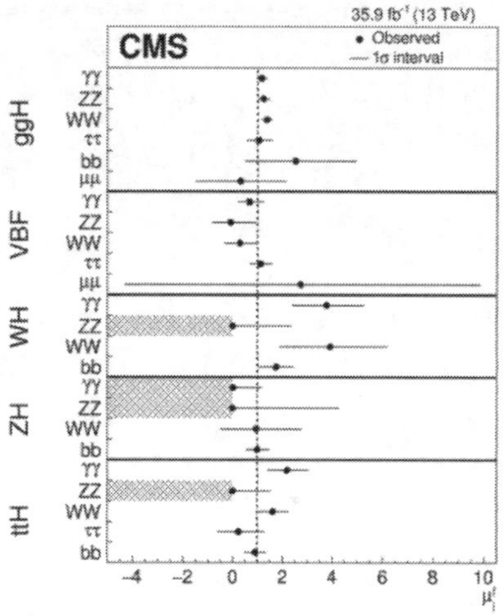

Figure 9. The current status of signal strength μ for all Higgs production and decay combinations at 13 TeV.

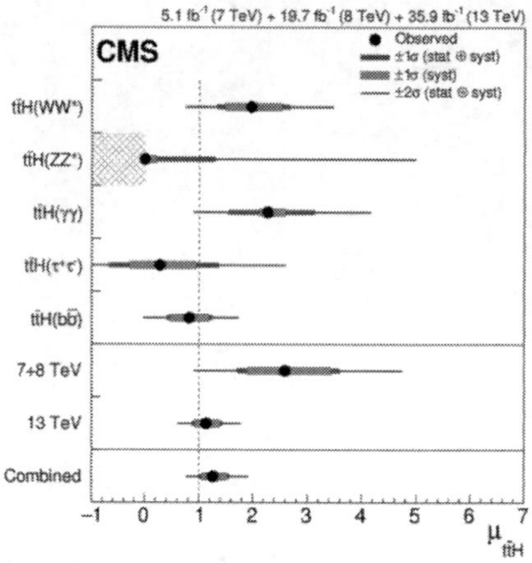

Figure 10. Signal strength for the different Higgs decay channels used in the $Ht\bar{t}$ discovery.

The production of Higgs in association with a vector boson and the subsequent Higgs decay to two tau leptons was recently completed, at 13 TeV with 35.9 fb^{-1} [17]. At least one of the tau hadrons is required to decay hadronically. A 2.3-sigma excess in the signal strength was observed ($\mu = 2.54^{+1.35}_{-1.26}$). The results are combined with earlier $H \to \tau\tau$ analyses.

5 Heavy-ion physics

The CMS maintains a very active and productive heavy-ion collisions program. One of the latest released results measures the center-of-mass pseudorapidity distribution of the muon from $W^{\pm}$ boson decays in 8.16-TeV pPb collisions [18]. This analysis probes the nuclear PDFs of u and d quarks. A shown in Fig. 11, modified PDFs are needed to describe the pseudorapidity distributions.

A recent study of anisotropic flow in XeXe collisions at 5.44 TeV and comparison with the respective PbPb results probes the system-size dependence and the role of initial state effects [19].

6 Forward and small-x QCD physics

Forward and small-x analyses offer valuable insight into a different range of parton dynamics. A 5-TeV PbPb search for "light-by-light" scatter [20] led to a 4.1-σ evidence with $\sigma(\gamma\gamma \to \gamma\gamma) = 120 \pm 46(\text{stat}) \pm 28(\text{syst}) \pm 4(\text{theory})$ nb. The powerful $\gamma\gamma$ acoplanarity variable was used for the separation of $\gamma\gamma \to \gamma\gamma$ signal from $\gamma\gamma \to ee$ and $gg \to \gamma\gamma$ events, as shown in Fig. 12.

7 B physics

Bottomonia help us study perturbation and lattice calculations of non-relativistic potentials. An analysis of 13-TeV data using 80 fb^{-1} of luminosity [21] led to the first observation of the resolved $\chi_{b1}(3P)$ and $\chi_{b2}(3P)$ resonances. Decays to $(Y(3S) \to \mu\mu)(\gamma \to ee)$ were

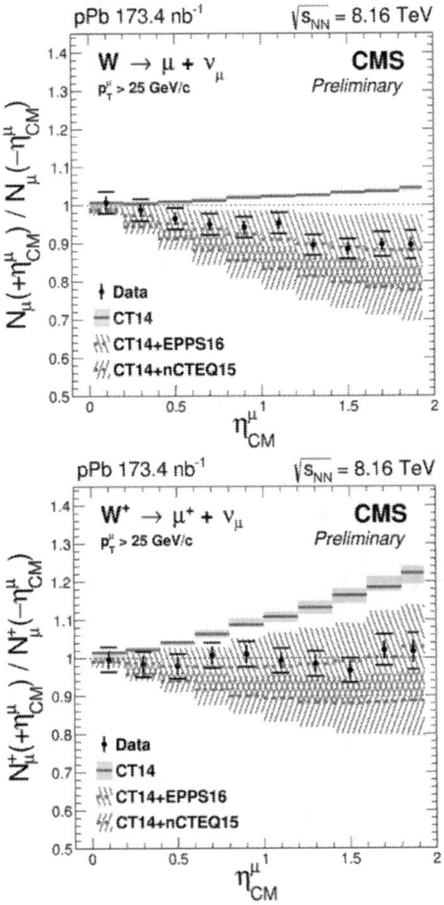

Figure 11. The forward-backward pseudorapidity ratios of the muons from the production of W^- (upper) and $W+$ (lower) bosons in pPb collisions can be explained only with nuclear modifications in the PDFs.

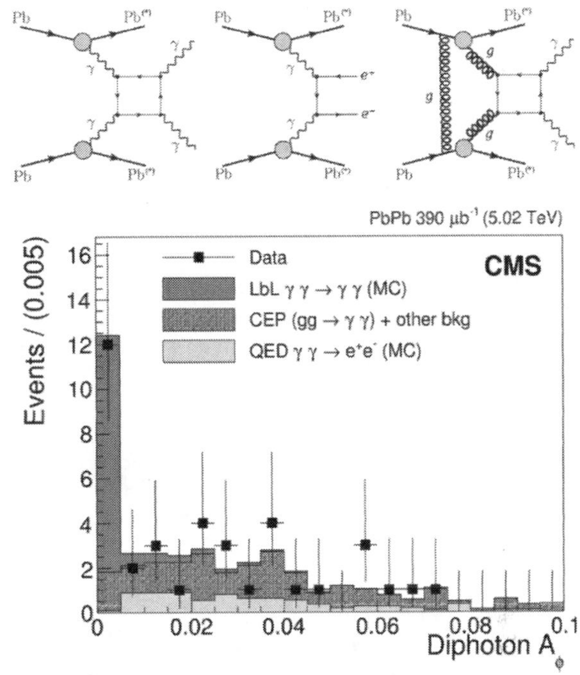

Figure 12. The diphoton system acoplanarity A_ϕ is a valuable quantity for separating light-by-light scatter from $\gamma\gamma \to ee$ and $gg \to \gamma\gamma$ background processes shown in the Feynman diagrams.

considered. The separation was achieved after a photon-energy scale calibration that utilizes $\chi_b(nP)$ decays to $Y(nS)\gamma$ leading to a measured $Y(3S)\gamma$ invariant mass resolution of 2.2 MeV, as shown in Fig. 13(a). The invariant mass is shown in Fig. 13(b).

The first observation of the rare decay $Z \to J/\psi + \ell^+\ell^-$ was performed with 35.9 fb^{-1} of 13-TeV data [22]. Both direct production of J/ψ and through $\psi(2S) \to J/\psi + X$ were considered. The signal is extracted from an unbinned log-likelihood fit, resulting to 13 ± 3.9(stat) $\mu^+\mu^-\mu^+\mu^-$ events and 11.2 ± 3.4(stat) $e^+e^-\mu^+\mu^-$ events leading to the measured ratio of branching fractions $\mathcal{B}(Z \to J/\psi\ell^+\ell^-)/\mathcal{B}(Z \to \mu^+\mu^-\mu^+\mu^-) = 0.67 \pm 0.18(\text{stat}) \pm 0.05(\text{syst})$.

8 Standard Model overview

An overview of the measurement of production cross sections of SM processes with a great variety of final-state objects [23] shows good agreement with predictions for 7-, 8-, and 13-TeV data.

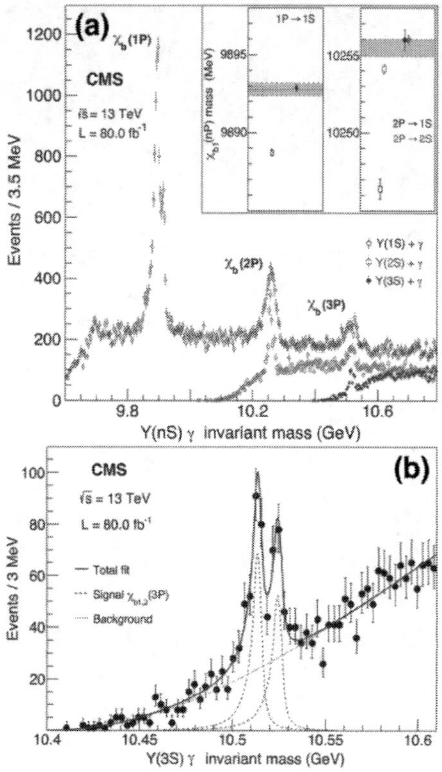

Figure 13. (a) The inset shows the $\chi_{b1}(1P)$ and $\chi_{b1}(1P)$ masses fitted before and after the photon energy calibration. (b) The resolved $\chi_{b1}(3P)$ and $\chi_{b2}(3P)$ resonances.

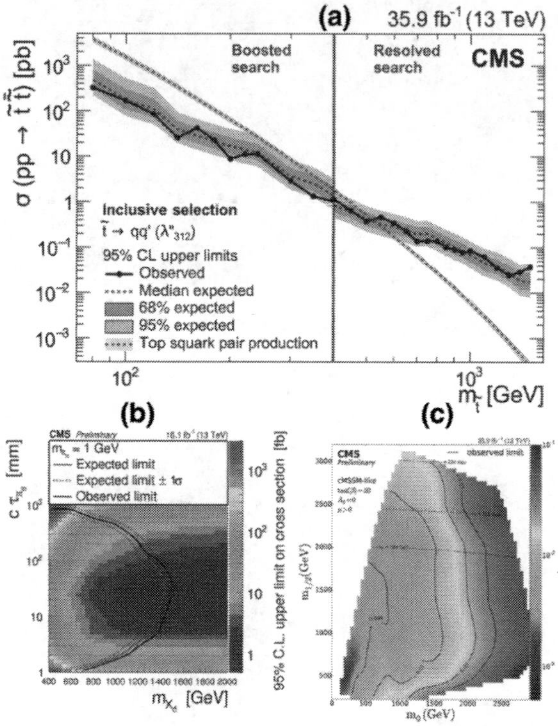

Figure 14. (a) The limits on the pair-produced scalar top decaying to diquark resonances. (b) The limit on dark quarks assuming dark pion mass of 1 GeV. (c) The $m_{1/2}$ vs m_0 limits of the same-sign dimuon analysis.

9 Exotic and supersymmetry physics

CMS runs a diverse program of analyses dedicated to direct discovery of new physics, either model-independent or described by existing supersymmetric (SUSY) or exotic models.

A 13-TeV, 35.9-fb^{-1} analysis looks for pair-produced resonances decaying to diquark pairs [24]. The four jets are either reconstructed as boosted dijets, with use of appropriate grooming, for resonances below 400 GeV or they are fully resolved above 400 GeV. A version of the analysis also requires two of the jets to be b-tagged. The results are consistent with expectation and limits are set on R-parity violating SUSY production pair of scalar top quarks. Fig. 14(a) shows the limit for the sub-analysis that does not require b-tagged jets.

Using multijet events in 13-TeV, 38.5-fb^{-1} data , an analysis searched for potential production and decay of long-lived particles through the detection of displaced vertices in the tracker [25]. The results are interpreted in R-parity violating SUSY as pair production of pairs of scalar-top quarks or gluinos or neutralinos. Cross section upper limits are set as a function of the mass and lifetime of the particles. For masses between 800 and 2600 GeV and mean decay lengths between 1 and 40 mm, cross sections above 0.3 fb are excluded at 95% confidence level.

A search for heavy, narrow resonances decaying to a Higgs boson and a photon has been completed with 13-TeV, 35.9-fb^{-1} data [26]. Investigating masses above 700 GeV leads to boosted Higgs bosons which decay to $b\bar{b}$ pairs reconstructed as single "fat" jet. Anti-k_t jets with R=0.8 are used with soft-drop grooming and dedicated b-tagging. The results agree with the SM and confidence limits are set.

A 13-TeV, 16.1-fb^{-1} analysis [27] looks for dark quarks, which are the decay products of heavy mediators in the BSSW dark-QCD model. The signature consists of 4 high-p_T jets two from ordinary quarks and two from dark quarks, which result in emerging jets with many displaced vertices from the decay of dark hadrons to the SM charged particles. Mediator particles with masses between 400 and 1250 GeV are excluded for dark hadron decay lengths between 5 and 225 mm. Fig. 14(b) shows exclusion limits for models with a 1-GeV dark pion.

A 13-TeV, 35.9-fb^{-1} analysis looks for same-sign dimuons with at least two jets [28]. Results are consistent with SM expectation and limits are set in R-parity-violating single chargino or neutralino production in association with a muon, as shown in Fig. 14(c).

10 Conclusions

An overview of some of the latest and legacy CMS physics results were presented. A complete collection of CMS publications and preliminary public results can be found in https://cms-results.web.cern.ch/cms-results/public-results/publications/.

References

[1] CMS Collaboration, J. High Energy Phys. 03 (2017) 156.
[2] CMS Collaboration, Eur. Phys. J. C **76** (2016) 451.
[3] CMS Collaboration, 2017 JINST **12** P10003.
[4] M. Cacciari, G. P. Salam, and G. Soyez, J. High Energy Phys. **04** (2008) 063.
[5] CMS Collaboration, Eur. Phys. J. C **78** (2018) 566.
[6] CMS Collaboration, Eur. Phys. J. C **77** (2017) 746.
[7] CMS Collaboration, CMS-PAS-SMP-16-008 (2017).
[8] CMS Collaboration, arXiv:1807.00782, submitted to Eur. Phys. J. C (2018).
[9] CMS Collaboration, arXiv:1807.05974, submitted to J. High Energy Phys. (2018).
[10] CMS Collaboration, J. High Energy Phys. **10** (2017) 131.
[11] CMS Collaboration, J. High Energy Phys. **06** (2018) 002.
[12] CMS Collaboration, CMS-PAS-TOP-17-014 (2018).
[13] CMS Collaboration, arXiv:1808.02913, submitted to Phys. Rev. Lett. (2018).
[14] CMS Collaboration, arXiv:1809.10733, submitted to Eur. Phys. J. C (2018).
[15] CMS Collaboration, Phys. Rev. Lett. **120** (2018) 231801.
[16] CMS Collaboration, CMS-PAS-HIG-18-001 (2018).
[17] CMS Collaboration, arXiv:1809.03590, submitted to J. High Energy Phys. (2018).
[18] CMS Collaboration, CMS-PAS-HIN-17-007 (2018).
[19] CMS Collaboration, CMS-PAS-HIN-18-001 (2018).
[20] CMS Collaboration, arXiv:1810.04602, submitted to Phys. Lett. B (2018).
[21] CMS Collaboration, Phys. Rev. Lett. **121**, 092002 (2018).
[22] CMS Collaboration, Phys. Rev. Lett. **121**, 141801 (2018).
[23] CMS Collaboration, https://twiki.cern.ch/twiki/bin/view/CMSPublic/PhysicsResultsCombined/.
[24] CMS Collaboration, arXiv:1808.03124, submitted to Phys. Rev. D (2018).
[25] CMS Collaboration, arXiv:1808.03078, accepted by Phys. Rev. D (2018).
[26] CMS Collaboration, CMS-EXO-17-019 (2018).
[27] CMS Collaboration, CMS-PAS-EXO-18-001 (2018).
[28] CMS Collaboration, CMS-PAS-SUS-17-008 (2018).

NLO and NNLO corrections to polarized top quark decays

Stefan Groote[1,*] and *Jürgen G.* Körner[2,**]

[1]Füüsika Instituut, Tartu Ülikool, W. Ostwaldi 1, EE-50411 Tartu, Estonia
[2]PRISMA Cluster of Excellence, Institut für Physik, Johannes-Gutenberg-Universität, D-55099 Mainz, Germany

Abstract. We present partial results on NLO and NNLO QCD, and NLO electroweak corrections to polarized top quark decays. In parallel we derive positivity bounds for the polarized structure functions in polarized top quark decays and check them against the perturbative corrections to the structure functions.

1 Introduction

In the limited space available to us in this write-up of a talk given at the International Workshop on QCD Theory and Experiment (QCD@Work 2018) in Matera, Italy, we cannot review the subject of polarized top quark decays in any depth. Instead we take the opportunity to report on results on radiative corrections to polarized top quark decays obtained by our group in the last few years. We will share our insights into the problem, why we did the calculations and how we did them. We also take the opportunity to specify which perturbative calculations have been done and which remain to be done.

The motivation for studying polarized top quark decays is provided by the huge sample of singly produced polarized top quarks at the LHC. The dominant source of polarized top quarks is from weak t–channel production with an average polarization of ≈ 90 % for the produced top quarks. Up to date the LHC detectors have seen $\sim 10^7$ singly produced top quarks. The projected overall luminosity of the future high luminosity HL-LHC is 3 ab^{-1} which corresponds to $\sim\ 10^9$ singly produced top quarks. Top quarks retain their polarization at birth when they decay since the life time of the top quark is so short.

2 The polarized top quark three-body decay $t(\uparrow) \to X_b + \ell^+ + \nu_\ell$

The full angular decay distribution for polarized top quark decay $t(\uparrow) \to X_b + \ell^+ + \nu_\ell$ in the top quark rest frame can be written in terms of the four structure functions A, B, C and D. The decay distributuion reads [1]

$$\begin{aligned}\frac{d\Gamma}{d\cos\theta d\phi} &= A + B\,P_t\cos\theta_P + C\,P_t\sin\theta_P\cos\phi + D\,P_t\sin\theta_P\sin\phi \\ &= A\Big(1 + \frac{B}{A}\,P_t\cos\theta_P + \frac{C}{A}P_t\sin\theta_P\cos\phi + \frac{D}{A}P_t\sin\theta_P\sin\phi\Big), \qquad (1)\end{aligned}$$

where the angles θ_p and ϕ are defined in Fig. 1. In the classification of Ref. [2] this is the helicity system Ib.

---

*e-mail: groote@ut.ee
**e-mail: jukoerne@uni-mainz.de

© The Authors, published by EDP Sciences. This is an open access article distributed under the terms of the Creative Commons Attribution License 4.0 (http://creativecommons.org/licenses/by/4.0/).

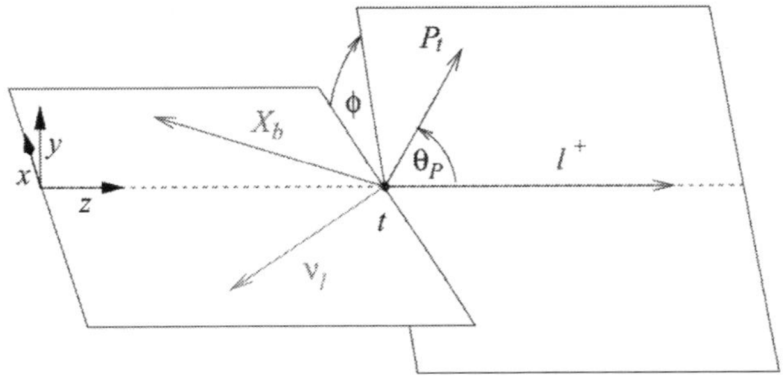

Figure 1. Definition of the polar angle θ_P and the azimuthal angle ϕ in the three-body decay of a polarized top quark.

In the usual classification the structure functions A, B and C are T–even structure functions and D is a T–odd structure function as can be seen by rewriting the angular factor multiplying D in the form

$$\sin\theta_P \sin\phi \propto \vec{p}_\nu \cdot (\vec{p}_\ell \times \vec{s}_t)\,. \tag{2}$$

The T–odd structure function D can be fed by final-state interactions (also called rescattering corrections) or by CP–violating interactions. We will present examples of both contributions further on.

It is clear that the angular decay distribution must remain positive definite over all of phase space. The positivity of the rate will be a recurring theme in this write-up.

2.1 The LO angular distribution

For the leading order (LO) Born term contribution one obtains the amplitude

$$\begin{aligned} M &= \bar{u}(b)\gamma^\mu(1-\gamma_5)u(t)\,\bar{u}(\ell)\gamma_\mu(1-\gamma_5)v(\nu) \\ &= 2\bar{u}(b)(1+\gamma_5)v(\nu)\,\bar{u}(\ell)(1-\gamma_5)u(t)\,, \end{aligned} \tag{3}$$

where we have used a Fierz transformation of the second kind to convert the $(V-A)^\mu(V-A)_\mu$ form to a $(S+P)(S-P)$ form (see e.g. [3]). The polarized angular decay distribution then reads (we set $P_t = |\vec{P}_t| = 1$)

$$\begin{aligned} W^P(\cos\theta_P) &= \sum_{spins} |M|^2 = 8\,tr\{\not{p}_b \not{p}_\nu\}\,tr\{(\not{p}_t + m_t)\tfrac{1}{2}(1+\gamma_5\not{s}_t)(1+\gamma_5)\not{p}_\ell\} \\ &= 16 m_t^4\, x_\ell(1-x_\ell)(1+\cos\theta_P)\,, \end{aligned} \tag{4}$$

where $x_\ell = 2E_\ell/m_t$. Quite naturally, the same result is obtained more tediously if one uses the $(V-A)^\mu(V-A)_\mu$ form of the amplitude (3). At LO there are no azimuthal correlations, i.e. $C = D = 0$! It is not difficult to see that the abscence of LO azimuthal correlations is in line with the postulate of positivity for the LO rate.

2.2 The NLO angular distribution

The NLO QCD contribution to the structure functions A, B and C have been calculated in [3]. We denote the NLO contribution by $A^{NLO} = A^{(0)} + A^{(1)}$ etc.. Setting $\phi = 0$ the angular decay distribution at NLO reads

$$W(\theta_P) = A^{(0)}\left(\left(1 + \frac{A^{(1)}}{A^{(0)}}\right) + \left(1 + \frac{B^{(1)}}{A^{(0)}}\right)\cos\theta_P + \frac{C^{(1)}}{A^{(0)}}\sin\theta_P\right) \tag{5}$$

where $A^{(1)}/A^{(0)} = -0.0846955$, $B^{(1)}/A^{(0)} = -0.0863048$ and $C^{(1)}/A^{(0)} = -0.0024$. The above values of the coefficient functions represent average values of the respective functions averaged over x_ℓ in the interval $[x, 1]$ where $x = m_W/m_t$. As concerns the $\cos\theta_P$ dependence, the NLO rate remains barely positive for $\cos\theta_P = -1$ and $P_t = 1$ as can be seen from

$$W(\theta_P) \sim 1 + \frac{1 + B^{(1)}/A^{(0)}}{1 + A^{(1)}/A^{(0)}} \cos\theta_P = 1 + 0.99824 \cos\theta_P \,. \tag{6}$$

Next we analyze positivity including the T–even azimuthal correlation proportional to the strucrture function C. We expand the angular rate around the point of risk $\theta_P = \pi$. One has $\cos(\pi - \delta) = -1 + \frac{1}{2}\delta$ and $\sin\delta = \delta$. We then obtain ($\Delta = (A^{NLO} - B^{NLO})/A^{NLO} = 0.001758$)

$$W(\theta_P) = \Delta - \frac{C^{(1)}}{A^{(0)}} \frac{1}{1 + A^{(1)}/A^{(0)}} \delta + \frac{(1-\Delta)}{2} \delta^2 \,. \tag{7}$$

A lower bound on the positivity of the rate is obtained when the discriminant of the quadratic equation vanishes. One obtains

$$\left|\frac{C^{(1)}}{A^{(0)}}\right| \leq \underbrace{\sqrt{2\Delta(1-\Delta)}(1 + \frac{A^{(1)}}{A^{(0)}})}_{0.05422} \,. \tag{8}$$

It is apparent that the NLO contributions listed above ($|C^{(1)}/A^{(0)}| = 0.0024$) easily satisfy the bound. The lower bound occurs at $\delta = -8.36 \times 10^{-4}\,\pi$ which shows that the small-δ expansion is well justified. This calculation provides the setting for deriving a bound on the size of the T–odd structure function D to be discussed in the next subsection.

2.3 NLO positivity bounds for the T–odd structure function D

There is a CP–conserving Standard Model contribution to the T–odd structure function D coming from the NLO electroweak rescattering correction as shown as absorptive parts in Fig. 2. One can check that there are no NLO absorptive QCD contributions.

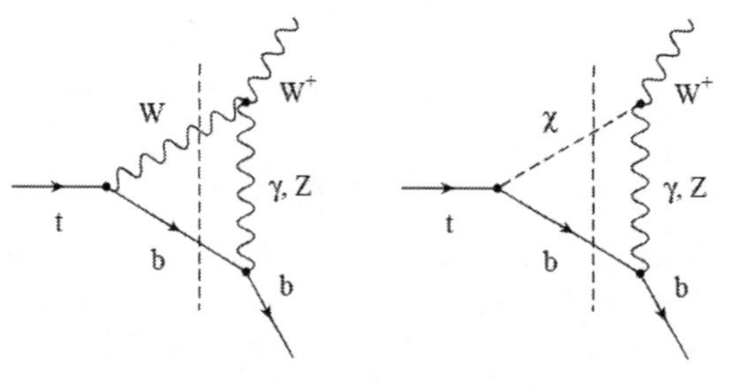

Figure 2. Electroweak absorptive parts of the four Feynman diagrams that contribute to T-odd correlations in polarized top quark decays

We present the result of calculating the absorptive parts of the two diagrams in terms of the imaginary part of the effective coupling constant g_R in the effective Lagrangian

$$\mathcal{J}^\mu_{\rm eff} = -\frac{g_w}{\sqrt{2}} \bar{b} \Big\{ \gamma_\mu (V_L P_L + V_R P_R) + \frac{i\sigma^{\mu\nu} q_\nu}{m_W} (g_L P_L + g_R P_R) \Big\} t \tag{9}$$

where $P_{L,R} = (1 \mp \gamma_5)/2$. The SM structure of the tbW^+ vertex is obtained by dropping all terms except for the contribution proportional to $V_{tb}^* \sim 1$. $\mathrm{Im}\, g_R$ is contributed to by CP–conserving rescattering effects and by CP–violating New Physics effects.

The electroweak rescattering contribution to $\mathrm{Im}\, g_R$ can be calculated to be [4]

$$\mathrm{Im}\, g_R = -2.175 \times 10^{-3}\,. \tag{10}$$

We agree with the result in [5] but disagree with [6].

A bound on $\mathrm{Im}\, g_R$ can be obtained by first calculating the contribution of $\mathrm{Im}\, g_R$ to the structure function D. The result is

$$\frac{D}{A^{(0)}} = -\frac{3\pi}{4}\frac{(1-x^2)}{(1+2x^2)}\,\mathrm{Im}\, g_R \tag{11}$$

A positivity bound on $\mathrm{Im}\, g_R$ can then be derived in analogy to the bound on the T–even structure function C in Eq. (8) but now setting $\sin\phi = 1$ and replacing C by D. One then obtains the $O(\alpha_s)$ bound [1]

$$-0.0420 \le \mathrm{Im}\, g_R \le 0.0420\,. \tag{12}$$

For once, the CP–conserving electroweak absorptive contribution to $\mathrm{Im}\, g_R = -2.175 \times 10^{-3}$ can be seen to easily satisfy the bound.

As concerns experiment the bound is tighter than the experimental bound obtained by the ATLAS collaboration [7]

$$-0.18 \le \mathrm{Im}\, g_R \le 0.06\,. \tag{13}$$

We mention that the NLO electroweak corrections to the two polarized T–even structure functions B and C have not been done.

3 The two-stage sequential polarized top quark decay $t(\uparrow) \to b + W^+(\to \ell^+ + \nu_\ell)$

The two-stage sequential two-body decay process $t(\uparrow) \to b + W^+(\to \ell^+ + \nu_\ell)$ is described by two polar angles θ and θ_P, and the azimuthal angle ϕ as defined in Fig. 3.

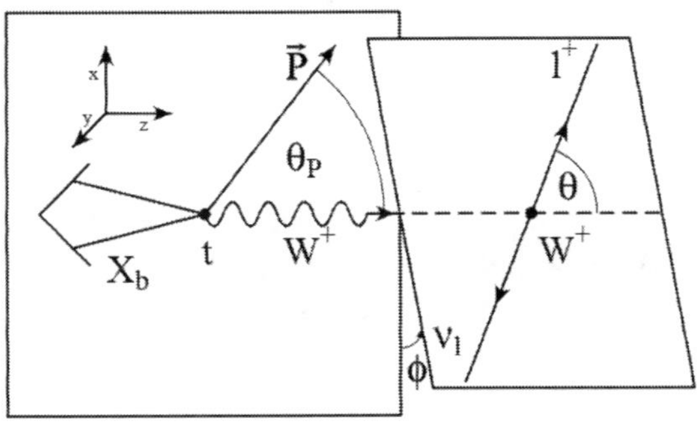

Figure 3. Definition of the polar angles θ and θ_P, and the azimuthal angle ϕ in the two-stage sequential two-body decay $t(\uparrow) \to X_b + W^+(\to \ell^+ + \nu_\ell)$.

The count of the independent structure functions is best done by considering the independent double spin density matrix elements $H^{\lambda_t\,\lambda_t'}_{\lambda_W\,\lambda_W'}$ of the W^+ which form a hermitian (3×3) matrix

$$\left(H^{\lambda_t\,\lambda_t'}_{\lambda_W\,\lambda_W'}\right)^\dagger = \left(H^{\lambda_t'\,\lambda_t}_{\lambda_W'\,\lambda_W}\right). \tag{14}$$

There are altogether ten independent double spin density matrix elements

$$H^{++}_{++},\ H^{--}_{++},\ H^{++}_{--},\ H^{--}_{--},\ H^{++}_{00},\ H^{--}_{00},\ \mathrm{Re}\,H^{+-}_{+0},\ \mathrm{Im}\,H^{+-}_{+0},\ \mathrm{Re}\,H^{-+}_{-0},\ \mathrm{Im}\,H^{-+}_{-0} \tag{15}$$

out of which eight are T-even and two are T-odd structure functions. Compare this to the three T-even and one T-odd structure functions that describe the direct three-body decays of polarized top quarks discussed in Sec. 2.

Let us concentrate on the polar angle distribution which is obtained by integrating over the azimuthal angle ϕ. One obtains

$$\begin{aligned} W(\theta,\theta_P) &= \frac{3}{8}\Big(1+\cos\theta\Big)^2\Big(T_+ + T^P_+ P_t\cos\theta_P\Big) + \frac{3}{8}\Big(1-\cos\theta\Big)^2\Big(T_- + T^P_- P_t\cos\theta_P\Big) \\ &+ \frac{3}{4}\sin^2\theta\Big(L + L^P\,P_t\cos\theta_P\Big) \end{aligned} \tag{16}$$

where $T_+, T^P_+, T_-, T^P_-, L, L^P$ are linear combinations of $H^{++}_{++},\ H^{--}_{++},\ H^{++}_{--},\ H^{--}_{--},\ H^{++}_{00},\ H^{--}_{00}$.

At LO and for $m_b = 0$ one has $T_- = -T^P_-$, $L = +L^P$, $T_+ = T^P_+ = 0$, i.e. the $\cos\theta_P$ dependence of the longitudinal and transverse-minus rates are given by $L : (1 + P_t\cos\theta_P)$ and $T_- : (1 - P_t\cos\theta_P)$. Similar to the discussion in Sec. 2.2 one is precariously close to a violation of positivity when $P_t = 1$. Using the $O(\alpha_s)$ results of [8, 9] we have checked that positivity is not spoiled at NLO QCD. In the same vein one can derive NLO bounds for the T–even and T–odd structure functions $\mathrm{Re}\,H^{+-}_{+0}$, $\mathrm{Im}\,H^{+-}_{+0}$, $\mathrm{Re}\,H^{-+}_{-0}$, $\mathrm{Im}\,H^{-+}_{-0}$ which again are satisfied by the NLO QCD and the NLO electroweak results.

We mention that we are in the process of calculating the NLO electroweak corrections to the eight T–even structure functions of sequential polarized top quark decay [10].

3.1 NNLO QCD corrections to sequential polarized top quark decay

As a last topic of this presentation we discuss the calculation of NNLO QCD corrections to the eight T–even structure functions in sequential polarized and unpolarized top quark decays for which we have obtained some partial results [11, 12]. The main idea behind our approach has been laid down in the NNLO calculation of the total rate in [13]. One converts a two-scale problem $\Gamma(m_t, m_W)$ to a one scale problem $\Gamma(m_t)$ bei expanding in the ratio $x = m_W/m_t$ such that

$$\Gamma(m_t,\,m_W) \to \Gamma(m_t,\,\sum a_i\,x^i) \tag{17}$$

In practice we terminate the expansion at $i = 10$. We found very satisfactory convergence of the expansion.

The NNLO results are obtained from the absorptive parts of 36 $O(\alpha_s^2)$ three-loop top quark self-energy diagrams which we denote by Σ. The unpolarized and polarized rates are then obtained from the trace

$$\Gamma + \Gamma^P = \frac{1}{m_t}\ \mathrm{Im}\,tr\Big\{(\not p_t + m_t)(1 + \gamma_5 \not s^\ell_t)\,\Sigma\Big\} \tag{18}$$

where s^ℓ_t is the longitudinal polarization four-vector of the top quark. One needs to avail of a covariant representation of s^ℓ_t which is given by

$$s^{l,\mu}_t = \frac{1}{|\vec q\,|}\Big(q^\mu - \frac{p_t\cdot q}{m_t^2}\,p^\mu_t\Big), \tag{19}$$

The unwieldy denominator factor $|\vec q\,|$ comes in through the normalization condition $s_t \cdot s_t = -1$. Express $|\vec q\,| = \sqrt{q_0^2 - q^2} = \sqrt{(pq/m_t)^2 - q^2}$ through the (unphysical) inverse propagator of the top quark $N = (p_t + q)^2 - m_t^2$. Then expand in terms of inverse powers of N up to the desired order

$$\frac{1}{|\vec{q}|} = \frac{2m_t}{N} \sum_{i=0}^{\infty} \binom{2i}{i} \left(\frac{2q^2 N - q^4 + 4m_t^2 q^2}{4N^2} \right)^i . \tag{20}$$

One can replace q^2 by m_W^2 everywhere since one is cutting through the W^+–line when taking the absorptive parts of the three-loop diagrams.

In this way we have calculated NNLO QCD results for the three helicity fractions $\mathcal{F}_{T_+}$, $\mathcal{F}_{T_-}$ and $\mathcal{F}_L$ [11] and the polarized rate $\Gamma^P_{T_+ + T_- + L}$ [12]. One finds that the perturbation series' are well-behaved. There is no real obstacle but hard work and the handling of huge computer codes to calculate the NNLO corrections to the remaining structure functions.

We have checked elements of our three-loop calculation by doing the corresponding NLO calculation involving the absorptive parts of four NLO two-loop top quark self energy diagrams. The results of the $x = m_W/m_t$ expansion agrees with the corresponding expansion of the known closed-form NLO results [8, 9].

For example, when one expands up to $O(x^{10}, x^{10}\ln x)$ the NLO expansion reads

$$\begin{aligned}
\hat{\Gamma}^{(1)}_{(T_+ + T_- + L)} &= C_F \Bigg[\frac{5}{4} + \frac{3}{2}x^2 - 6x^4 + \frac{46}{9}x^6 - \frac{7}{4}x^8 - \frac{49}{300}x^{10} + \\
&\quad - 2(1 - x^2)^2(1 + 2x^2)\zeta(2) + \left(3 - \frac{4}{3}x^2 + \frac{3}{2}x^4 + \frac{2}{5}x^6 \right) x^4 \ln x \Bigg], \\
\hat{\Gamma}^{(1)}_{(T_+ + T_- + L)^P} &= C_F \Bigg[-\frac{15}{4} - \frac{17}{8}x^4 - \frac{1324}{225}x^5 - \frac{31}{36}x^6 + \\
&\quad + \frac{48868}{11025}x^7 - \frac{23}{288}x^8 + \frac{884}{6615}x^9 - \frac{3}{100}x^{10} + (1 + 4x^2)\zeta(2) \Bigg]. \tag{21}
\end{aligned}$$

Note that the expansion of the parity-conserving (pc) rate $(T_+ + T_- + L)$ involves only even powers of x while the expansion of the parity-violating (pv) polarized rate $(T_+ + T_- + L)^P$ involves even and odd powers of x. This pattern holds true for all pc and pv $O(\alpha_s)$ and the known $O(\alpha_s^2)$ T-even structure functions for which we lack a deep understanding. Our hope is that some chance reader of this presentation can provide us with a solution to this empirical paradigma.

Acknowledgement

J, G, K. is grateful to the organizers of the International Workshop on QCD, theory and experiment (QCD@Work) for the invitation to the workshop and their hospitality in Matera. Special thanks go to Fulvia de Fazio and Pietro Colangelo. S. G. is supported by the Estonian Science Foundation under the grant No. IUT2-27. S. G. acknowledges the hospitality of the theory group THEP at the Institute of Physics at the University of Mainz where part of this work was done and the support of the Cluster of Excellence PRISMA at the University of Mainz.

References

[1] S. Groote and J. G. Körner, Phys. Rev. D **96** (2017) no.11, 111301 [arXiv:1711.06432 [hep-ph]].
[2] J. G. Körner and D. Pirjol, Phys. Rev. D **60** (1999) 014021 [hep-ph/9810511].

[3] S. Groote, W. S. Huo, A. Kadeer and J. G. Körner, Phys. Rev. D **76** (2007) 014012 [hep-ph/0602026].

[4] M. Fischer, S. Groote and J. G. Körner, Phys. Rev. D **97** (2018) no.9, 093001 [arXiv:1802.02492 [hep-ph]].

[5] A. Arhrib and A. Jueid, JHEP **1608** (2016) 082 [arXiv:1606.05270 [hep-ph]].

[6] G.A. Gonzalez-Sprinberg, R. Martinez and J. Vidal, JHEP **1107** (2011) 094; Erratum: [JHEP **1305** (2013) 117] [arXiv:1105.5601 [hep-ph]].

[7] M. Aaboud *et al.* [ATLAS Collaboration], JHEP **1704** (2017) 124 [arXiv:1702.08309 [hep-ex]].

[8] M. Fischer, S. Groote, J. G. Körner, M. C. Mauser and B. Lampe, Phys. Lett. B **451** (1999) 406 [hep-ph/9811482].

[9] M. Fischer, S. Groote, J. G. Körner and M. C. Mauser, Phys. Rev. D **65** (2002) 054036 [hep-ph/0101322].

[10] S. H. Do, M. Fischer, S. Groote and J. G. Körner, to be published

[11] A. Czarnecki, J. G. Körner and J. H. Piclum, Phys. Rev. D **81** (2010) 111503 [arXiv:1005.2625 [hep-ph]].

[12] A. Czarnecki, S. Groote, J. G. Körner and J. H. Piclum, Phys. Rev. D **97** (2018) no.9, 094008 [arXiv:1803.03658 [hep-ph]].

[13] I. R. Blokland, A. Czarnecki, M. Slusarczyk and F. Tkachov, Phys. Rev. Lett. **93** (2004) 062001 [hep-ph/0403221].

Jet Substructure Measurements Sensitive to Soft QCD effects with the ATLAS Detector

Gabriele Chiodini[1,*] On behalf of the ATLAS collaboration

[1]Istituto Nazionale di Fisica Nucleare, sezione di Lecce, Italia

Abstract. Jet substructure observables are powerful tools for the identification of boosted heavy particles and for probing QCD at different energy scales in proton-proton collisions at the LHC. In this work three recent ATLAS measurements sensitive to soft QCD effects and performed with the use of jet substructures are described. Their are k_t splitting scales, soft-drop jet masses and pull-jet angles. Some discrepancies between data and predictions are observed in the non-perturbative regime.

1 Introduction

The features of energy flow within the body of hadronic jets are very useful for probing perturbative and non-perturbative QCD at hadron colliders. Nowadays, their analysis is used in many studies.

Jet substructure observables are very relevant for an efficient identification of hadronically decaying boosted objects, such as W, Z, H and t, extending significantly the probed energy range and discovery capability of LHC experiments such as ATLAS [1]. They are measured in data, unfolded for detector effects using an iterative Bayesian technique [2], and compared to QCD calculations and particle-level MC simulations.

2 Sequential clustering algorithm

Jets are reconstructed from a set of particle momenta (reconstructed from calorimeter energy deposits, charged tracks, simulated particles) which are close in the rapidity-azimuthal y-ϕ plane according to a clustering algorithm.

At the LHC sequential clustering algorithms [3] are used to reconstruct jets. They have the nice property to be infrared- and collinear-safe, allowing a meaningful comparison between data and theoretical predictions.

The algorithms iteratively combine the nearest pair of jet constituent j_i and j_j, in the y-ϕ plane, according to a distance d_{ij} computed as the geometrical distance $\Delta R_{ij}^2 = \sqrt{\Delta y_{ij}^2 + \Delta\phi_{ij}^2}$ normalized to the square of the radius parameter R adopted and weighted by the minimum of a power n of the transverse momenta and defined by the formula:

$$d_{ij} = \min(p_{T,i}^n, p_{T,j}^n)\frac{\Delta R_{ij}^2}{R^2}, \qquad (1)$$

*e-mail: gabriele.chiodini@le.infn.it

© The Authors, published by EDP Sciences. This is an open access article distributed under the terms of the Creative Commons Attribution License 4.0 (http://creativecommons.org/licenses/by/4.0/).

where the value of R is usually chosen between 0.2 and 1. If the minimum distance at the k-th step $d_k = \min_{i,j}\{d_{ij}\}$ is larger than $d_{ib} = p^n_{T,i}$ then the jet constituent j_i is promoted to a jet and removed from the jet constituent list. The clustering algorithm is repeated again on the list until all jet constituents are clustered in a jet.

The value of the exponent n sets for specific clustering algorithm used:

- n=2 corresponds to the k_t algorithm where close and soft constituents pairs are clustered first, miming the infrared and collinear splittings.
- n=−2 corresponds to the anti-k_t algorithm where close and hard constituents are clustered first, producing regularly shaped jets in y-ϕ space.
- n=0 corresponds to the Cambridge-Aachen (CA) algorithm ($d_{ij} = \frac{\Delta R^2_{ij}}{R^2}$ and $d_{ib} = 0$) where the geometrically closest constituent pairs are clustered first, miming the angular-ordered parton shower (PS).

If one is interested in studying jets, the anti-k_t algorithm is the best choice because it is only weakly susceptible to the underlying event and pile-up. However, if one is interested in jet substructures, only the k_t and CA algorithms are appropriate. This is clearly seen at the second-last stage of the clustering sequence where the last two sub-jets can reveal a hard underlying structure for the k_t and CA algorithms but not for the anti-k_t algorithm. As shown in the next sections, jet substructure observables are built by following the steps of a sequential clustering algorithm in reverse order.

3 Measurements of k_t splitting scales

A measurement of the splitting scales of the k_t jet-clustering algorithm was done with the ATLAS detector, in p-p collisions at 8 TeV center-of-mass energy and 20.2 fb^{-1} integrated luminosity, for jets produced in association with a Z boson decaying leptonically (p+p→Z(ll)+jets) and reconstructed with charged tracks (track-jets) [4].

The k-th order splitting scale is defined by $\sqrt{d_k}$ for n=2 (see previous section) with the number k indexing the k_t clustering sequence in reverse order. The 0-th order splitting scale corresponds to the final jet transverse momentum $\sqrt{d_0} = p_T$, the 1-st order splitting scale corresponds to $\sqrt{d_1} = \min(p_{T,1}, p_{T,2})\frac{\Delta R_{12}}{R}$ of the penultimate stage of the clustering, and so on for all other orders.

Figure 1 shows two examples of splitting scale distribution measured for events with Z decaying in the muon and electron channel and jet radius 0.4 and 1, respectively. In reference [4] measurements up to 7-th order splitting scales are reported. Large discrepancy between data and predictions are observed at low values of the splitting scale. It is relevant to note that soft and collinear splitting have low values of $\sqrt{d_k}$, instead the hard splitting has high values.

4 Measurements of soft-drop jet mass

The calculation beyond leading log (LL) accuracy for observable related to jets is problematic due to non-global logarithms (NGLs) related to particles that radiate out of, and then radiate back into, a jet [5]. The so-called soft-drop grooming algorithm makes jet substructure insensitive to NGLs terms because it removes energy in jet related to soft QCD processes and pile-up [6]. The algorithm starts by re-clustering the jet's constituents with the CA algorithm and removes sub-jet pairs from the clustering history sequence if they do not satisfy the following criteria:

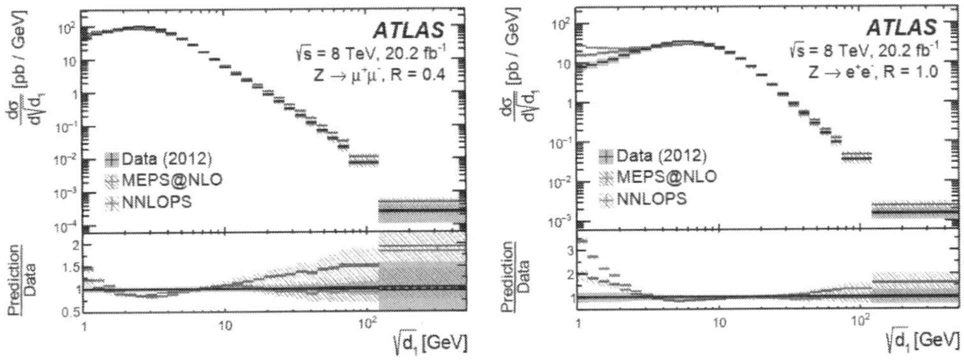

Figure 1. Left plot: charged (measured) distribution for the 1-st order splitting scales in the muon channel using the jet-radius parameter R = 0.4. Right plot: charged (measured) + neutral (extrapolated) distribution for the 1-st order splitting scales in the electron channel using the jet-radius parameter R = 1. The size of the error bars reflects the statistical uncertainty, while the combined statistical and systematic uncertainty is indicated by the grey band. Theoretical predictions from Sherpa with NLO multijet merging ("MEPS@NLO") and from POWHEG+PYTHIA 8 with NNLO matching ("NNLOPS") are displayed including error bands for the generator uncertainties [4].

$$\frac{\min(p_{T,1}, p_{T,2})}{p_{T,1} + p_{T,2}} > z_{cut}(\frac{\Delta R_{12}}{R})^{\beta} \tag{2}$$

where the parameter z_{cut} controls how much energy is removed from the jet and β tunes the sensitivity of the algorithm to wide-angle radiation.

A very important application of the soft-drop is the measurements of the jet mass which is a powerful tool for identifying boosted hadronically decaying massive particles [7]. In fact, the jet mass of soft-drop groomed jet is insensitive to NGLs terms and it has been calculated at both next-to-leading order (NLO) with next-to-leading-logarithm NLL and leading order (LO) with next-to-next-to-leading-logarithm (NNLL) accuracy.

The soft-drop jet mass $m^{\text{soft-drop}}$ of the two leading anti-k_t jets in di-jet events (p+p→jet+jet) was measured with the ATLAS detector, in p-p collisions at 13 TeV center-of-mass energy and 32.9 fb^{-1} integrated luminosity [8]. The soft-drop jet mass is measured relative to the jet transverse momentum before grooming in terms of the dimensionless ratio $\rho = m^{\text{soft-drop}} / p_T^{\text{ungroomed}}$.

The experimental distributions of $\log_{10}\rho^2$ is compared with several predictions and two of them are reported in Fig.2. The main focus is on the region $-3.7 < \log_{10}\rho^2 < -1.7$ of the recent precision calculations, where resummation dominates over non-perturbative $-3.7 < \log_{10}\rho^2$ or fixed-order $\log_{10}\rho^2 < -1.7$ regions. Studying the distribution in log-scale allows the re-summation region to be studied more closely. In regions where non-perturbative effects are expected to be significant, the calculations disagree with the data and a better description is provided by MC simulation.

5 Measurements of color flow

Despite the strong effort in the past to measure the potential effects of colour connections between partons sharing color flow, where more radiation is expected from colour coherence effects [9], it remains a poorly constrained effect of QCD and further experimental input are clearly welcome.

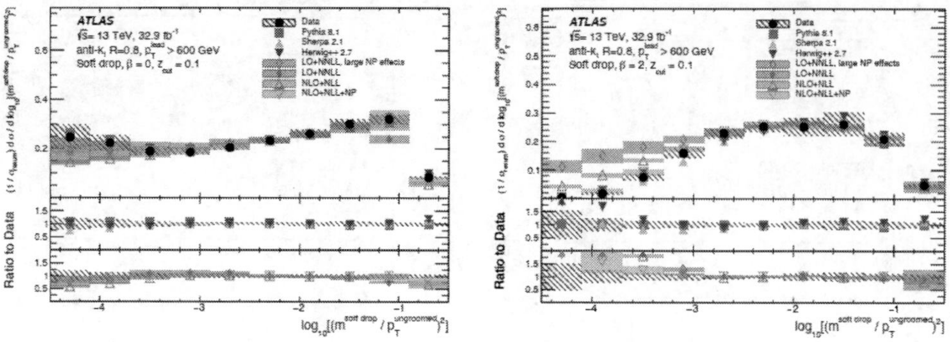

Figure 2. The unfolded $\log_{10}\rho^2$ distribution for anti-k_t jets of radius R = 0.8 with $p_T^{\text{leading}} > 600$ GeV, after the soft-drop algorithm is applied for β=0 (left plot) and β=2 (right plot) in data compared to Pythia, Sherpa, and Herwig++ particle-level, and NLO+NLL+NP and LO+NNLL theory predictions. The LO+NNLL calculation does not have non-perturbative (NP) corrections; the region where these are expected to be large is shown in a open marker, while regions where they are expected to be small are shown with a filled marker. All uncertainties described in the text are shown on the data; the uncertainties from the calculations are shown on each one. The distributions are normalized to the integrated cross section measured in the resummation region [8].

In the production of a $t\bar{t}$ pair four colour-charged final state particles can be identified: the two b-quarks produced directly by the decay of the top-quarks and the two quarks produced by the hadronically decaying W boson. As the W boson does not carry colour charge the two daughter quarks must share a colour flow. The two b-quarks carry the colour charge of their respective top-quark parents, and are thus not expected to be colour connected (see Fig.3 left).

The color connection between two jets j_1 and j_2 (see Fig.3 right) can be studied using the jet-pull angle $\theta_{\mathcal{P}}$ defined as the angle between the jet-pull vector $\mathcal{P}(\vec{j}_1)$ and the vector difference in the y-ϕ plane of j_1 and j_2 (jet-connection vector). The jet-pull vector of j_1 defined by the formula:

$$\mathcal{P}(\vec{j}_1) = \sum_{i \epsilon j_1} \frac{|\Delta\vec{r}_i| p_T^i}{p_T^{j_1}} \Delta\vec{r}_i \quad (3)$$

is a p_T-weighted radial moment of the jet constituents.

Figure 4 shows the jet-pull angle measured with the ATLAS detector, in p-p collisions at 13 TeV center-of-mass energy and 36.1 fb^{-1} integrated luminosity [10]. Jets are reconstructed using the anti-k_t algorithm with R = 0.4 starting from three-dimensional, massless, positive-energy topological clusters [11] reconstructed from energy deposited in the calorimeters. Semileptonic top pair events were selected in such a way to have a jet pair from the hadronic W decay (color connected) of a top decay and a b-jet pair from the two top decays (not-color connected) and to trigger on the high p_T lepton from the leptonic W decay of the other top decay.

As expected, the jet-pull angle shows a flat distribution for pairs of jets not sharing a color flow and a distribution accumulating at 0 in case of color connection. In Fig.4 it is evident that color connection looks stronger in predictions than in data.

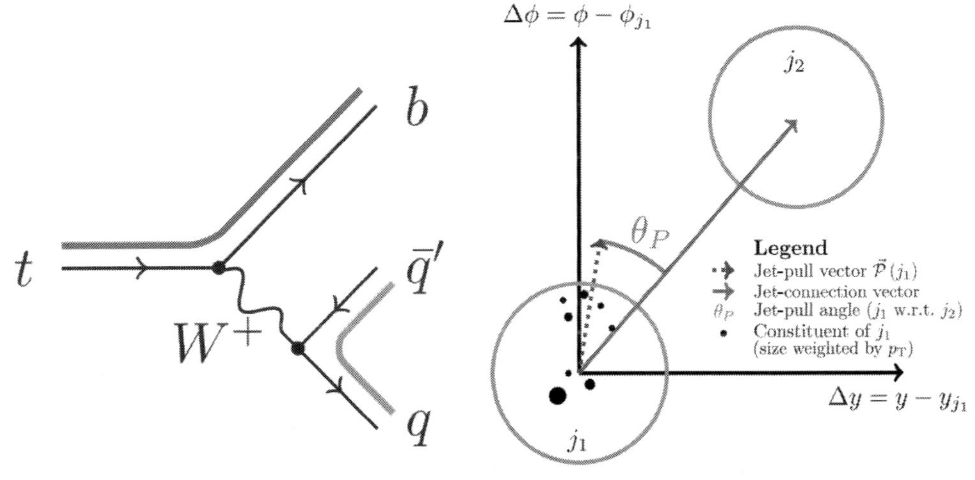

Figure 3. Left: Illustration of a hadronic top decay with typical colour connections (thick coloured lines). Right: illustration of jet-pull observables for a dijet system [10].

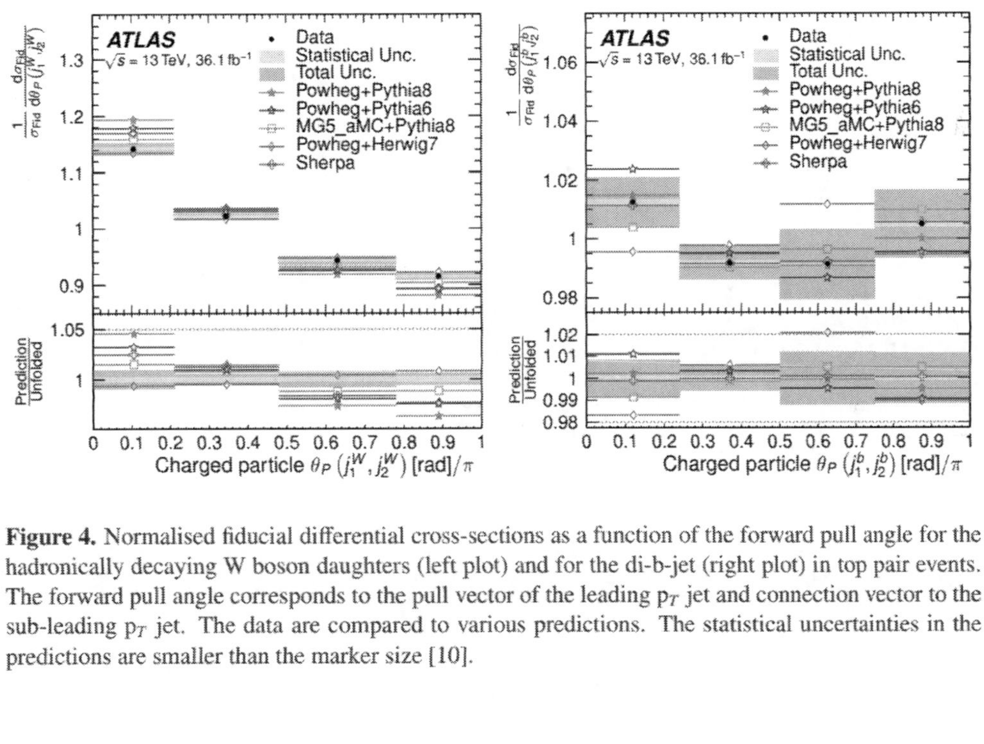

Figure 4. Normalised fiducial differential cross-sections as a function of the forward pull angle for the hadronically decaying W boson daughters (left plot) and for the di-b-jet (right plot) in top pair events. The forward pull angle corresponds to the pull vector of the leading p_T jet and connection vector to the sub-leading p_T jet. The data are compared to various predictions. The statistical uncertainties in the predictions are smaller than the marker size [10].

6 Conclusions

The reported ATLAS measurements of soft QCD effects at the LHC constrain analytical calculations in the perturbative regime. The discrepancy between predictions and data observed in the non-perturbative regime are helpful to tune MC in this regime, which is sensitive to pile-up and underlying event.

References

[1] ATLAS Collaboration, 2008 JINST 3 S08003.

[2] G. D'Agostini, Nucl. Instrum. Meth. A **362**, 487 (1995).

[3] G.P. Salam, Eur. Phys. J. C 67 **637**, (2010). arXiv:0906.1833 [hep-ex]. R. Atkin, J. Phys. Conf. Ser. **645** 012008 (2015).

[4] ATLAS Collaboration, JHEP08 **026** (2017). arXiv:1704.01530 [hep-ex].

[5] A. Banfi et al., JHEP 08 **064** (2010). arXiv: 1004.3483 [hep-ex].

[6] M. Rubin, JHEP 05 **005** (2010). arXiv: 1002.4557 [hep-ex].

[7] A. Abdesselam, et al., Eur. Phys. J. **C71**, 1661 (2011). arXiv:1012.5412 [hep-ex].

[8] ATLAS Collaboration, Phys. Rev. Lett. **121**, 092001 (2018). arXiv:1711.08341 [hep-ex].

[9] R. K. Ellis, W. J. Stirling and B. R. Webber, *QCD and Collider Physics*, Cambridge University Press, 2003.

[10] ATLAS Collaboration, CERN-EP-2018-041. http://cds.cern.ch/record/2316982. arXiv:1805.02935 [hep-ex]. Submitted to EPJC.

[11] ATLAS Collaboration, Eur. Phys. J. C 77 *490*, (2017). arXiv: 1603.02934 [hep-ex].

Diffractive patterns in deep-inelastic scattering and parton genealogy

Stéphane Munier[1,*]

[1]CPHT, École Polytechnique, CNRS, Université Paris-Saclay, Route de Saclay, 91128 Palaiseau, France

Abstract. We report on our recent observation that the occurrence of diffractive patterns in the scattering of electrons off nuclei obeys the same law as the fluctuations of the height of genealogical trees in branching diffusion processes.

1 Introduction

Consider electrons scattering off atomic nuclei in a high-energy particle collider, a process known as "deep-inelastic scattering". Diffractive events are characterized by the striking property that the nucleus is left intact in the process, and is surrounded by an angular region in which no particle is observed, called the "gap", the size of which, a quantity fluctuating from event-to-event, is used to classify the diffractive patterns (see Fig. 1).

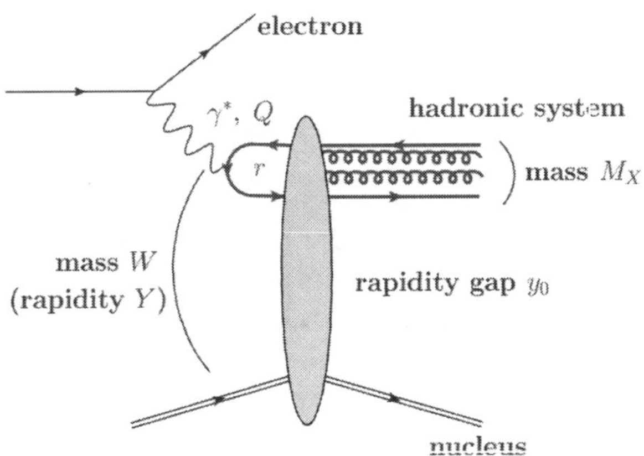

Figure 1. [Adapted from Ref. [1]] Graph contributing to the electron-nucleus diffractive scattering amplitude. The electron emits a virtual photon, which converts to a quark-antiquark pair. The latter scatters elastically, coherently, off the nucleus, through a particular quantum fluctuation (containing two gluons in addition to the quark and to the antiquark in this illustration). The final state exhibits a gap of size y_0 here, which is a signature of the elastic scattering process.

Diffraction in the context of high-energy hadronic and nuclear physics is a surprising phenomenon. Indeed, a nucleus being a lose boundstate of nucleons, themselves being rather fragile compounds of quarks (whenever involved in a reaction of center-of-mass energy larger than a few GeV), one may think that the probability that a nucleus "survives" a high-energy collision with a small (compared to nuclear scales) object is negligible. The observation at the DESY-HERA collider of a significant fraction (about 10% overall) of diffractive events came as a surprise for many physicists (for a review, see Ref. [2]). It boosted the study of quantum chromodynamics (QCD) in the dense, semi-classical, regime, since the most elegant

*e-mail: stephane.munier@polytechnique.edu

© The Authors, published by EDP Sciences. This is an open access article distributed under the terms of the Creative Commons Attribution License 4.0 (http://creativecommons.org/licenses/by/4.0/).

explanation for the important fraction of these events and for their properties came from so-called "saturation models", which assumed that at HERA, the proton appeared as a dense system of gluons at momentum scales as high as 1 GeV [3].

Now consider an apparently completely different problem: a particle evolves stochastically in time by diffusing on a line and randomly splitting, at some fixed rate, into two particles, which subsequently follow independently the same rules (see Fig. 2 below). This process, called "branching diffusion", is an interesting mathematical object, potentially relevant in many fields of science, from the physics of glasses to evolutionary biology (for a review by a mathematician, see Ref. [4]). After some given time, pick the two leftmost particles and trace their most recent common ancestor. Surprisingly enough, we found that the distribution of the age of the latter is identical to the distribution of the gap size in diffraction, for a deep reason related to the very mechanism of how fluctuations build up in these two processes.

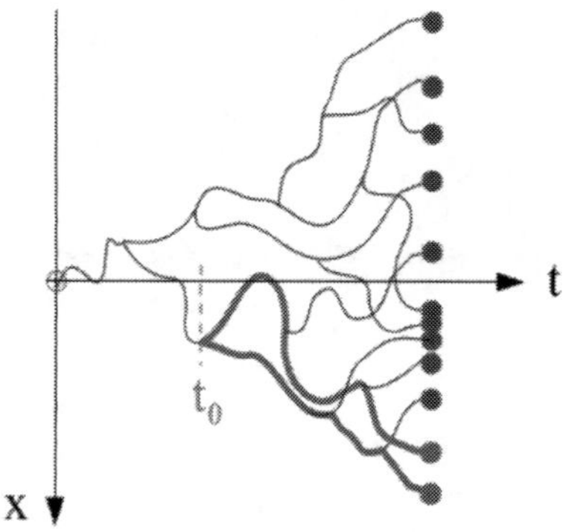

Figure 2. Schematic representation of a realization of a branching Brownian motion process. A particle initially at position $x = 0$ diffuses, and may split into 2 particles at a constant rate. The particles evolve further independently, following the same rules, until the final time t is reached. Then, the two particles at the largest positions are picked, and their most recent common ancestor is traced: Its decay time is denoted by t_0.

Our study contributes to link distinct fields of science, and paves the way for a better understanding of diffraction phenomena in particle physics, that can be tested at a future Electron-Ion Collider (EIC) [5].

This short review summarizes the papers in Refs. [1, 6]. We shall start by recalling the equation for the rapidity gap distribution, first derived by Kovchegov and Levin (KL) [7]. We then elaborate on a picture which enables us to find asymptotic form of the solution to the latter. We discuss the link with genealogies in branching random walks, before presenting the results of numerical calculations which support our analytical results. The final section gathers some prospects.

2 Equation for the rapidity gap distribution

Throughout, we shall not address electron-nucleus diffractive scattering, but rather onium-nucleus scattering, which can however be connected to the former through a well-known method. The link is discussed in more detail in Ref. [6], and is illustrated in Fig. 1.

Let us consider the scattering of an onium of size r off a big nucleus at a total center-of-mass energy corresponding to the rapidity Y. We are going to write down an equation for the evolution of the distribution of the rapidity gap around the nucleus.

We need to start with the forward elastic onium-nucleus S-matrix element. It obeys the Balitsky-Kovchegov (BK) equation [8, 9]

$$\partial_y S(r, y) = \bar{\alpha} \int \frac{d^2 r'}{2\pi} \frac{r^2}{r'^2 (r - r')^2} \left[S(r', y) S(r - r', y) - S(r, y) \right], \tag{1}$$

where the initial condition may be taken from the McLerran-Venugopalan model [10], which we can approximate by $S(r, y = 0) = 1 - e^{-r^2 Q_A^2/4}$, with Q_A the nuclear saturation scale, typically of the order of 1 GeV. Note that the (dimensionless) total, elastic and inelastic cross sections at a given impact parameter *per unit surface* in the transverse plane to the collision axis may be expressed with the help of S:

$$\sigma_{\text{tot}} = 2(1 - S), \quad \sigma_{\text{el}} = (1 - S)^2, \quad \sigma_{\text{in}} = \sigma_{\text{tot}} - \sigma_{\text{el}} = 1 - S^2. \tag{2}$$

We now turn to the diffractive cross section with a given rapidity gap y_0. According to Kovchegov and Levin, the distribution of y_0 can be deduced from an auxiliary function $S_2(r, \tilde{y})$, which also obeys the BK equation:

$$\partial_{\tilde{y}} S_2(r, \tilde{y}) = \bar{\alpha} \int \frac{d^2r'}{2\pi} \frac{r^2}{r'^2(r - r')^2} [S_2(r', \tilde{y}) S_2(r - r', \tilde{y}) - S_2(r, \tilde{y})] \quad \text{for} \quad \tilde{y} > 0, \tag{3}$$

with the peculiar initial condition $S_2(r, \tilde{y} = 0) = [S(r, y_0)]^2$, which couples the two equations for S and S_2. The gap distribution is then evaluated as

$$\frac{d\sigma_{\text{diff}}}{dy_0} = -\frac{\partial}{\partial y_0} S_2(r, Y - y_0). \tag{4}$$

These equations also follow quite straightforwardly [6, 11] from the Good-Walker picture [12].

The solution to the set of equations (1,3) is actually what we are after. But the latter are very difficult to solve: there are no known methods to address such coupled nonlinear evolution equations. In a certain sense, the work presented here may be viewed as an effort to derive a solution to the KL equations. But instead of trying to solve them brute force, with purely mathematical tools, we develop a picture of diffractive scattering, from which what we believe should be the large-rapidity asymptotics of the KL equations straightforwardly follow.

3 Picture of onium-nucleus scattering

3.1 Interpretation of the total cross section in different frames

The total cross section can be deduced from the elastic S-matrix element through the optical theorem, see Eq. (2). S satisfies the BK equation, the solution of which is known for some values of the variables, in a limited range:

$$1 - S \simeq \text{const} \times \ln \frac{1}{r^2 Q_s^2(Y)} \left[r^2 Q_s^2(Y)\right]^{\gamma_0} \quad \text{with} \quad Q_s^2(Y) \simeq Q_A^2 \frac{e^{\chi'(\gamma_0)\bar{\alpha}Y}}{(\bar{\alpha}Y)^{3/2\gamma_0}}, \tag{5}$$

where $\bar{\alpha} \equiv \alpha_s N_c/\pi$, $\gamma_0 \simeq 0.63$ is a number, and the values of the complex function $\chi(\gamma) = 2\psi(1) - \psi(\gamma) - \psi(1-\gamma)$ are the eigenvalues of the kernel of the linearized BK equation. This behavior holds when the variables (r, Y) are chosen such that $1 < |\ln[r^2 Q_s^2(Y)]| < \sqrt{\chi''(\gamma_0)\bar{\alpha}Y}$, which we shall call the *scaling region*.

$S(r, Y)$ is of course a boost-invariant quantity. But to interpret it, it is useful to choose particular Lorentz frames.

In the restframe of the *onium*, the full QCD evolution is in the nucleus. This evolution is *deterministic* since it starts with a highly-occupied state (the nucleus in its groundstate). The larger Y, the denser the nucleus Fock state: $1 - S$, solution to the BK equation, is an estimator of the density of the nucleus.[1]

In the restframe of the *nucleus* instead, the QCD evolution is in the onium. Since the latter is a quark-antiquark pair in its initial (fundamental) state, its evolution is highly stochastic, and there are large event-by-event fluctuations in its content. In this context, $1 - S$ may be interpreted as the fraction of realizations of the QCD evolution which lead to a scattering event. We note that there is an interaction between the onium state and the nucleus whenever there is a gluon in the former which has a transverse momentum of the order of (or smaller than) the saturation scale of the nucleus, Q_A. Thus $1 - S$ can be interpreted as *the probability of having a gluon of transverse momentum smaller than* Q_A in a particular realization of the state of the onium.

3.2 Total and diffractive cross sections in the y_0-frame

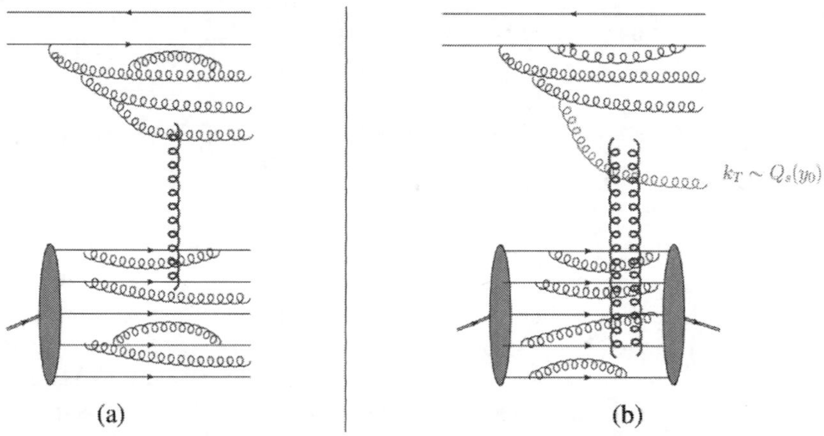

Figure 3. [Adapted from Ref. [1]] Example of graphs contributing to the total (a) and diffractive dissociation (b) amplitude. For the latter, the gap has a size y_0, and the condition for its existence is the presence of a rare fluctuation, a gluon of transverse momentum $k_\perp \sim Q_s(y_0)$, in the Fock state of the onium evolved to rapidity $\tilde{y}_0 = Y - y_0$.

We can also choose an intermediate frame in which the evolution is shared between the onium and the nucleus. Let us boost the nucleus exactly to the rapidity y_0, while the onium carries the remaining rapidity $\tilde{y}_0 \equiv Y - y_0$:[2] Examples of graphs are displayed in Fig. 3 for the total and diffractive cross section. In this particular frame, in order to create a rapidity gap, the scattering of the onium Fock state off the nucleus should be *elastic*. For such an elastic scattering to be probable, there should be at least a gluon in the state of the onium which has a transverse momentum $k_\perp$ comparable to (or smaller than) the characteristic momentum scale of the nucleus in that frame, namely to the saturation scale $Q_s(y_0)$. We choose the onium size r in the scaling region, in which the total cross section is small. Therefore, such a fluctuation is relatively rare: The gluon which interacts is most probably softer than the typical soft gluons present in these states of the onium.

[1]Actually, $1 - S^2$ is the probability that the onium has an interaction. It may be extracted from the data for diffractive elastic vector-meson production [13].

[2]The rapidity y_0 is also related to the invariant mass M_X of the diffracted system: $y_0 \simeq \ln W^2/M_X^2$, where W is the total center-of-mass energy of the onium-nucleus system, see Fig. 1.

So the diffractive cross section conditioned to a gap of size y_0, $d\sigma_{\text{diff}}/dy_0$, is proportional to the probability of having an exceptionally soft gluon, of transverse momentum of the order of $Q_s(y_0)$, in the Fock state of the onium evolved to rapidity $\tilde{y}_0$. According to the discussion given in the previous section, this quantity is the solution of the BK equation. Therefore, it reads

$$\begin{aligned}\frac{d\sigma_{\text{diff}}}{dy_0} &\simeq \text{const} \times \ln \frac{1}{r^2 \tilde{Q}_s^2(\tilde{y}_0)} \left[r^2 \tilde{Q}_s^2(\tilde{y}_0)\right]^{\gamma_0} \quad \text{with} \quad \tilde{Q}_s^2(\tilde{y}_0) = Q_s^2(y_0) \frac{e^{\bar{\alpha}\tilde{y}_0\chi'(\gamma_0)}}{(\bar{\alpha}\tilde{y}_0)^{3/2\gamma_0}} \\ &\simeq \text{const} \times \sigma_{\text{tot}} \left[\frac{\bar{\alpha}Y}{\bar{\alpha}y_0\, \bar{\alpha}(Y-y_0)}\right]^{3/2}. \end{aligned} \tag{6}$$

This result is valid in the scaling window[3] defined above, and under the additional conditions $\bar{\alpha}y_0, \bar{\alpha}(Y-y_0) \gg 1$. In order to arrive at the final expression in Eq. (6), we have just used the expression of the saturation scale as a function of the rapidity given in Eq. (5).

4 Ancestry in branching random walks

It turns out that the structure of diffractive events bears similarities with an apparently completely different problem: The genealogy of particles near the boundary of a branching random walk (BRW) in Ref. [14]. That dipole evolution (at fixed impact parameter) is a peculiar branching random walk has been known for some time (see e.g. [15]), but that the very structure of the evolution may play a role is new and quite surprising.

We may define a branching random walk in the following way (see Fig. 2 for an illustration). Let us consider a set of particles evolving in time t by diffusing in the real variable x, and assume that each of them has a probability to split to two particles, independently of the other particles. (Diffusion may actually be replaced by discrete jumps, which may occur at the same time as the splitting: the fundamental nature of the problem keeps unchanged).

Assume that the mean density of particles $n(x,t)$ obeys the equation $\partial_t n = \chi n$; χ is an appropriate operator acting on n viewed as a function of x. For example, $\chi = \partial_x^2 + 1$ in the case of the branching Brownian motion. χ admits the eigenfunctions $e^{-\gamma x}$ and we denote by $\chi(\gamma)$ the corresponding eigenvalues. After the (large) time t, pick exactly the two leftmost particles, and look for the first common ancestor splitting time t_0 (see Fig. 2).

Then, according to Ref. [14], t_0 is distributed as

$$p(t_0) = c_p \left[\frac{t}{t_0(t-t_0)}\right]^{3/2}, \quad \text{with} \quad c_p = \frac{1}{\gamma_0} \frac{1}{\sqrt{2\pi\chi''(\gamma_0)}}, \tag{7}$$

where γ_0 solves $\chi(\gamma_0) = \gamma_0 \chi'(\gamma_0)$. Hence, up to the overall normalization, which is determined in the case of the genealogies, but not in the case of diffraction, $(1/\sigma_{\text{tot}})(d\sigma_{\text{diff}}/dy_0)$ (see Eq. (6)) corresponds to $p(t_0)$, with the identifications $\bar{\alpha}Y \leftrightarrow t$, $\bar{\alpha}\tilde{y}_0 \leftrightarrow t_0$.

We think that this formal analogy has actually a deep origin, related to a property of the common ancestor. In the same way as in our diffraction calculation, the common ancestor of the boundary particles also stems from a fluctuation, in the form of a particle sent ahead of the *expected* position of the most extreme particle, occurring in the course of the evolution, at time $t - t_0$.

---

[3]The region at the border of the scaling window, around $|\ln[r^2 Q_s^2(Y)]| \sim \sqrt{\chi''(\gamma_0)\bar{\alpha}Y}$, may be described by Eq. (6) supplemented with the factor $\exp\left[-\frac{\ln^2 r^2 Q_s^2(Y)}{2\chi''(\gamma_0)\bar{\alpha}(Y-y_0)}\right]$, see Ref. [6].

5 Numerical checks

We stressed in Sec. 2 that it is difficult to address the Kovchegov-Levin equations analytically. On the other hand, it is quite straightforward to solve them numerically, since it amounts to integrating successively two BK equations (in their fixed-impact parameter version).

We first go to very large rapidities in order to see how well the numerical solution of the KL equation matches our analytical expectations (6). The comparison is quite spectacular, see Fig. 4, and gives us confidence in the validity of our analytical expression. Note that the KL equation had been solved numerically before [16], but too low rapidities were considered to see the asymptotical regime emerging. For more realistic rapidities, maybe attainable at

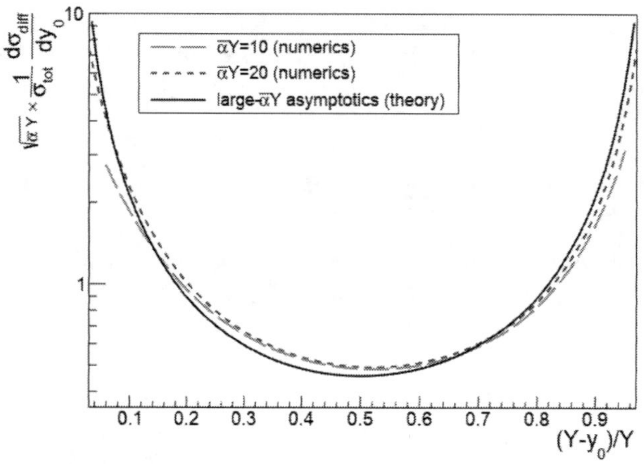

Figure 4. [From Ref. [1]] Numerical solution of the KL equation in the asymptotic regime, namely for $\bar{\alpha}Y = 20$ and $\bar{\alpha}Y = 40$ (see the legend), in order to appreciate the convergence to the asymptotics. The results are presented for the diffractive cross section normalized to the total cross section, and in scaled variables, in such a way that the asymptotic prediction deduced from Eq. (6) (black curve) be rapidity-independent. Its normalization is arbitrary.

a future Electron-Ion Collider (EIC), the rapidity gap distribution still has a similar shape, peaking for minimal and maximal gap sizes; compare Fig. 5 to Fig. 4.

6 Prospects

On the theory side, we intend to study systematically the asymptotics of the distribution of the rapidity gap y_0, in parallel to the distribution of the decay time of the common ancestor of the largest dipoles in an event [17]. Preliminary calculations seem to suggest that the overall constant in the gap distribution may be determined; The first subleading term in the limit $\bar{\alpha}y_0, \bar{\alpha}(Y - y_0) \gg 1$ might also be calculated analytically.

On the phenomenology side, we deem that the distribution of rapidity gaps is an interesting observable to consider at EIC. Arriving at robust predictions will require to examine more in depth all regimes, also outside of the scaling window considered here, and to solve numerically a modified version of the KL equation which would incorporate next-to-leading order effects (running coupling [18], kinematical constraint [19], resummation of collinear logarithms [20]), known to be quantitatively important.

This work was supported in part by the Agence Nationale de la Recherche under the project # ANR-16-CE31-0019.

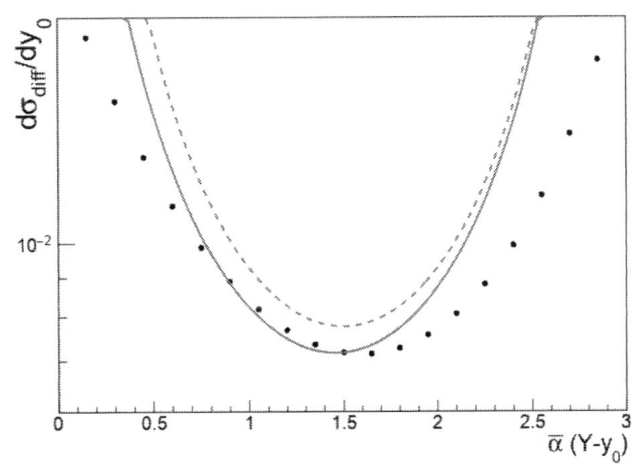

Figure 5. [From Ref. [6]] Numerical solution of the KL equation (dots) for $\bar{\alpha}Y = 3$, compared to the asymptotic prediction (6) (dashed line), and to the asymptotics supplemented by a subleading term (see Ref. [6] and footnote 3; full line). The normalization of the latter is *ad hoc*.

References

[1] A. Mueller, S. Munier, Phys. Rev. Lett. **121**, 082001 (2018), 1805.09417

[2] L. Schoeffel, Prog. Part. Nucl. Phys. **65**, 9 (2010), 0908.3287

[3] K.J. Golec-Biernat, M. Wusthoff, Phys. Rev. **D59**, 014017 (1998), hep-ph/9807513

[4] A. Bovier, *Gaussian Processes on Trees: From Spin Glasses to Branching Brownian Motion*, Cambridge Studies in Advanced Mathematics (Cambridge University Press, 2016), ISBN 9781316871409

[5] A. Accardi et al., Eur. Phys. J. **A52**, 268 (2016), 1212.1701

[6] A. Mueller, S. Munier, Phys. Rev. **D98**, 034021 (2018), 1805.02847

[7] Y.V. Kovchegov, E. Levin, Nucl. Phys. **B577**, 221 (2000), hep-ph/9911523

[8] I. Balitsky, Nucl. Phys. **B463**, 99 (1996), hep-ph/9509348

[9] Y.V. Kovchegov, Phys. Rev. **D61**, 074018 (2000), hep-ph/9905214

[10] L.D. McLerran, R. Venugopalan, Phys. Rev. **D49**, 2233 (1994), hep-ph/9309289

[11] Y. Hatta, E. Iancu, C. Marquet, G. Soyez, D.N. Triantafyllopoulos, Nucl. Phys. **A773**, 95 (2006), hep-ph/0601150

[12] M.L. Good, W.D. Walker, Phys. Rev. **120**, 1857 (1960)

[13] S. Munier, A.M. Stasto, A.H. Mueller, Nucl. Phys. **B603**, 427 (2001), hep-ph/0102291

[14] B. Derrida, P. Mottishaw, EPL (Europhysics Letters) **115**, 40005 (2016)

[15] S. Munier, Sci. China Phys. Mech. Astron. **58**, 81001 (2015), 1410.6478

[16] E. Levin, M. Lublinsky, Phys. Lett. **B521**, 233 (2001), hep-ph/0108265

[17] D. Le Anh, Internship report, École polytechnique (unpublished) (2018)

[18] Y.V. Kovchegov, H. Weigert, Nucl. Phys. **A784**, 188 (2007), hep-ph/0609090

[19] L. Motyka, A.M. Stasto, Phys. Rev. **D79**, 085016 (2009), 0901.4949

[20] E. Iancu, J.D. Madrigal, A.H. Mueller, G. Soyez, D.N. Triantafyllopoulos, Phys. Lett. **B744**, 293 (2015), 1502.05642

EPJ Web of Conferences **192**, 00009 (2018) https://doi.org/10.1051/epjconf/201819200009
QCD@Work 2018

A tale of two Regge limits

Vittorio Del Duca[1,2,*]

[1]Institute for Theoretical Physics, ETH Zürich, 8093 Zürich, Switzerland
[2]INFN, Laboratori Nazionali di Frascati, Italy

Abstract. In light of the strong advances in understanding the mathematical structure of scattering amplitudes, we discuss the Regge limit of QCD and of the $\mathcal{N} = 4$ Super Yang-Mills theory.

1 Regge limit in QCD

In the Regge limit, in which the squared centre-of-mass energy s is much larger than the momentum transfer $|t|$, $s \gg |t|$, any scattering process is dominated by the exchange in the t channel of the highest-spin particle. In the case of QCD or $\mathcal{N} = 4$ Super Yang-Mills (SYM), that entails the exchange of a gluon. The Regge limit of QCD is described by the Balitsky-Fadin-Kuraev-Lipatov (BFKL) theory, which models strong-interaction processes with two large and disparate scales, by resumming the radiative corrections to parton-parton scattering. This is achieved at leading logarithmic (LL) accuracy, in $\ln(s/|t|)$, through the BFKL equation [1–4], *i.e.* an integral equation for the evolution of the t-channel gluon propagator in transverse-momentum space and Mellin moment space. The building blocks of the kernel of the BFKL equation are a real correction – the emission of a gluon along the ladder, *a.k.a.* the central emission vertex – and a virtual correction – the one-loop Regge trajectory [5]. The BFKL equation is then obtained by iterating the one-loop corrections to all orders in α_S, at LL accuracy in $\ln(s/|t|)$, *i.e.* by resumming the terms of $O(\alpha_S^n \ln^n(s/|t|))$. Underpinning the BFKL theory is the ladder structure in the t channel of the scattering amplitudes in the Regge limit. For example, in the case of gluon-gluon scattering in QCD, the gluon ladder is described by the $\mathbf{8}\otimes\mathbf{8}$ colour representation, which is decomposed as $\mathbf{8}\otimes\mathbf{8} = \{\mathbf{1}\oplus\mathbf{8}_s\oplus\mathbf{27}\}\oplus[\mathbf{8}_a\oplus\mathbf{10}\oplus\overline{\mathbf{10}}]$, where the term in curly brackets in the direct sum is even in colour space under $s \leftrightarrow u$ exchange, while the term in square brackets is odd. However, to LL accuracy in $\ln(s/|t|)$, the parton-parton scattering amplitudes are real and only the antisymmetric octet $\mathbf{8}_a$ contributes. That implies the gluon Reggeisation at LL accuracy [5]. For phenomenological purposes, the BFKL equation needs then be endowed with process-dependent impact factors, the simplest case of which are the quark or gluon impact factors [2, 6], by which one can obtain, for example, the dijet cross section at large rapidities [7].

The computation of the next-to-leading-logarithmic (NLL) corrections [8–11] to the BFKL equation, *i.e.* of the terms of $O(\alpha_S^n \ln^{n-1}(s/|t|))$, requires computing the radiative corrections to the kernel of the BFKL equation, *i.e.* the real corrections to the central emission vertex – the emission of two gluons or of a quark-antiquark pair along the ladder [12–15] – and the one-loop corrections to the central emission vertex [16–20], as well as the two-loop

*e-mail: delducav@itp.phys.ethz.ch

© The Authors, published by EDP Sciences. This is an open access article distributed under the terms of the Creative Commons Attribution License 4.0 (http://creativecommons.org/licenses/by/4.0/).

Regge trajectory [21–25]. Underpinning the BFKL theory at NLL accuracy is the fact that although the parton-parton scattering amplitudes develop an imaginary part, the BFKL equation is computed through the real part of the amplitudes, to which only the antisymmetric octet $\mathbf{8}_a$ contributes, which implies the gluon Reggeisation at NLL accuracy [26, 27].

At LL accuracy, the kernel of the BFKL equation is conformally invariant, and thus the leading-order eigenfunctions of the BFKL equation are fixed by conformal symmetry [28]. The NLL corrections [8–11] to the BFKL equation were computed by acting with the next-to-leading-order (NLO) kernel of the equation on the leading-order eigenfunctions. This procedure is not consistent, and it was already clear to Fadin and Lipatov [8] that the terms which make the procedure inconsistent are related to the running of the coupling. The consistent NLO eigenfunctions were constructed by Chirilli and Kovchegov [29, 30], who found indeed that the additional pieces which occur at NLO are proportional to the beta function. However, if the scale of the strong coupling is chosen to be the geometric mean of the transverse momenta at the ends of the ladder, then one can use the leading-order eigenfunctions instead of the NLO ones [31].

1.1 Regge limit and infrared factorisation

The backbone of the BFKL equation are the quark and gluon scattering amplitudes, which are dominated by the exchange of a gluon in the t channel. Because loop-level scattering amplitudes of massless partons are infrared divergent, and so are the impact factors, the Regge trajectory, and the central emission vertex, in which they factorise in the Regge limit, the study of the scattering amplitudes in the Regge limit has benefited from a cross breeding with infrared factorisation [32–42], where the infrared structure of scattering amplitudes for massless partons is known up to three loops [43–48].

For example, the one-loop impact factors [16, 20, 49–51] and the two-loop Regge trajectory have poles in the dimensional regulator ε in $d = 4 - 2\varepsilon$ dimensions, which can be understood, for the one-loop impact factors, in terms of the one-loop cusp anomalous dimension and the one-loop quark and gluon collinear anomalous dimensions [39, 40], and for the two-loop Regge trajectory in terms of the two-loop cusp anomalous dimension [35].

Further, we know that the picture of the Regge-pole factorisation of the amplitudes, based on the gluon Reggeisation, breaks down at two loops by N_c-subleading terms [25]. The violation can be explained through infrared factorisation by showing that the real part of the amplitudes becomes non-diagonal in the t-channel-exchange basis [36, 37]. Accordingly, it can be predicted how the violation propagates to higher loops, and the three-loop prediction for the violation [39, 40], which has NNLL accuracy, has been confirmed by the explicit computation of the three-loop four-point function of full $\mathcal{N} = 4$ SYM [52]. In the Regge factorisation picture, that violation is due to a Regge cut and to the exchange of three Reggeised gluons [41, 53]. Finally, at three loops and at NNLL accuracy, also the $\mathbf{10} \oplus \overline{\mathbf{10}}$ colour representation contributes to the exchange of three Reggeised gluons [41].

The imaginary part of the amplitude is even under $s \leftrightarrow u$ exchange. For gluon-gluon scattering, the $\mathbf{8}_s$ colour representation contributes only at one loop to it through a Regge pole *i.e.* through the exchange of a Reggeised gluon, while more in general at NLL accuracy the **1** and the **27** representations contribute through a Regge cut and the exchange of two Reggeised gluons [42]. The imaginary part of the amplitude contributes then to the squared amplitude, and thus to the BFKL ladder, at NNLL accuracy.

Thus, although a study of the BFKL ladder and firstly of its building blocks at NNLL accuracy is yet to be undertaken – the three-loop Regge trajectory in a particular scheme [41] and the emission of three partons along the gluon ladder [54] being the only ones computed so far – we already have a clear picture of which contributions occur at NNLL accuracy.

While we have a precise knowledge of how the infrared poles occur in the loop corrections to the impact factors and to the Regge trajectory, the finite parts of those corrections are treated as free parameters, which neither infrared nor Regge factorisation can constrain. However, there is a relation [55] between the $O(\varepsilon)$ terms of the one-loop gluon impact factor [20] and the $O(\varepsilon^0)$ terms of the two-loop Regge trajectory [21–25], which hints at an exponentiation of some finite parts of the amplitude in the Regge limit, and thus at more structure in infrared factorisation in the Regge limit than we currently know.

2 Regge limit in $\mathcal{N} = 4$ SYM

In the last few years, it has been realised that the Regge limit of QCD and of $\mathcal{N} = 4$ SYM, and thus the BFKL equation, are endowed with a rich mathematical structure.

In the planar limit of $\mathcal{N} = 4$ SYM, scattering amplitudes exhibit a dual conformal symmetry [56–59]. The dual conformal invariance is broken by infrared divergences [60, 61], but it is possible to construct finite, dual conformally invariant ratios in which all infrared divergences cancel. As a consequence, the analytic structure of scattering amplitudes in planar $\mathcal{N} = 4$ SYM is highly constrained. In particular, the four and five-point amplitudes are fixed to all loop orders by symmetries in terms of the one-loop amplitudes and the cusp anomalous dimension [61–63]. Non-trivial corrections occur for amplitudes with at least six legs [61, 64–67]. The ordinary and dual conformal symmetries are also at the heart of a duality between scattering amplitudes and Wilson loops computed along a light-like polygonal contour [57–61, 65].

It has been observed that the four-point amplitude is Regge exact [58, 68], *i.e.* it is not modified by the Regge limit. The Regge exactness extends to the five-point amplitude [69, 70], if computed in the multi-Regge kinematics (MRK) [5], which features the central-emission vertex; and to the six-point amplitude [66], if computed in the quasi-multi-Regge kinematics [12], which features the emission of two gluons along the ladder [12, 13]. In fact, Regge exactness extends in general to the n-point MHV amplitudes, if computed in the quasi-multi-Regge kinematics of the emission of $(n-4)$ gluons along the ladder [67].

A lot of progress has been done in understanding the mathematical structure of amplitudes in planar $\mathcal{N} = 4$ SYM. It is expected that n-point amplitudes are expressed as iterated integrals of differential one-forms [71] defined on the space of configurations of points in the three-dimensional projective space $\text{Conf}_{\mathcal{N}}(\mathbb{CP}^3)$ [72, 73]. The simplest instance of iterated integrals that one encounters in amplitudes of $\mathcal{N} = 4$ SYM are the multiple polylogarithms (MPL) [74, 75] which correspond to iterated integrals over rational functions. It is believed that all maximally helicity violating (MHV) and next-to-MHV amplitudes in planar $\mathcal{N} = 4$ SYM can be expressed in terms of MPLs of uniform transcendental weight [76].

In the Euclidean region, where all Mandelstam invariants are negative, scattering amplitudes in planar $\mathcal{N} = 4$ SYM in MRK factorise to all orders in perturbation theory into building blocks, which are the impact factor, the Regge trajectory and the central-emission vertex. The last two describe respectively the Reggeised gluons in the t-channel and the emission of gluons along the t-channel ladder formed by the Reggeised gluons. Those building blocks are determined to all orders by the four and five-point amplitudes, and hence scattering amplitudes in MRK are trivial in the Euclidean region [69, 70, 77–80]. In particular, the Regge trajectory is fixed, to all orders in perturbation theory, by the cusp anomalous dimension and the gluon collinear anomalous dimension [77].

Starting from six legs, scattering amplitudes exhibit Regge cuts, if the multi-Regge limit is taken after analytic continuation to a specific Minkowski region [70, 78]. The discontinuity across the cut is described to all orders by a dispersion relation closely related to the BFKL equation, which can be expressed in terms of single-valued functions [81, 82]. In particular, the discontinuity of the six-point amplitude [83] in MRK can be expressed in terms of single-valued harmonic polylogarithms (SVHPL) [84]. More generally, n-point amplitudes in MRK are described by the moduli space $\mathfrak{M}_{0,n-2}$ of Riemann spheres with $(n-2)$ marked points [85]. The algebra of iterated integrals on $\mathfrak{M}_{0,n}$ factors in such a way that the iterated integrals on $\mathfrak{M}_{0,n}$ can always be expressed in terms of MPLs [75] and rational functions with singularities when two marked points coincide. The discontinuity of the n-point amplitude in MRK is parametrised by $(n-5)$ conformally invariant ratios. In the soft limit of any of the transverse momenta of the gluons emitted along the ladder, one of the $(n-5)$ ratios vanishes, and thus two marked points on the Riemann sphere coincide [85]. However, the transverse momenta of the ladder gluons never vanish. This requirement constrains the iterated integrals that can appear in MRK to single-valued functions, *i.e.* linear combinations of products of iterated integrals on $\mathfrak{M}_{0,n}$ (and their complex conjugates) such that all branch cuts cancel, which are single-valued [84, 86, 87] multiple polylogarithms (SVMPL). The statements of Ref. [85], which were made at LL accuracy, can be generalised at NLL accuracy and beyond [88].

3 The Regge limit in $\mathcal{N}=4$ SYM meets the Regge limit in QCD

Underpinning the analysis of the n-point amplitudes in planar $\mathcal{N}=4$ SYM in MRK, is Lipatov's picture [89] of a BFKL-like dispersion relation, which describes the discontinuity in terms of the Hamiltonian of an integrable open Heisenberg spin chain [90, 91]. Previously, Lipatov had characterised the BFKL equation, and more in general the exchange of a ladder of n Reggeised gluons in a singlet configuration in QCD at large N_c, in terms of the Hamiltonian [90, 92, 93] of a closed spin chain. It should come then as no surprise that, just like the six-point amplitude of $\mathcal{N}=4$ SYM in MRK [83], also the analytic structure of the BFKL ladder at LL accuracy in QCD can be described in terms of single-valued iterated integrals on the moduli space $\mathcal{M}_{0,4}$ of Riemann spheres with four marked points, which are SVHPLs [94]. In this case, the single-valuedness can be traced back to the fact that neither of the momenta at the ends of the gluon ladder can vanish. That analysis can be extended at NLL accuracy to the BFKL ladder in QCD, as well as in full $\mathcal{N}=4$ SYM [31], however it requires a generalisation of the SVHPLs, recently introduced by Schnetz [95]. The control of the analytic structure of the BFKL ladder in QCD at NLL accuracy, and the freedom in defining its matter content, let us prove that there is no gauge theory of uniform and maximal transcendental weight such that in momentum space it matches the maximal weight part of QCD, and to identify a set of conditions which allowed us to constrain the field content of theories for which the BFKL ladder has maximal weight [31].

References

[1] V. S. Fadin, E. A. Kuraev, and L. N. Lipatov, Phys. Lett. **B60** (1975) 50–52.
[2] E. A. Kuraev, L. N. Lipatov, and V. S. Fadin, Sov. Phys. JETP **44** (1976) 443–450. [Zh. Eksp. Teor. Fiz.71,840(1976)].
[3] E. A. Kuraev, L. N. Lipatov, and V. S. Fadin, Sov. Phys. JETP **45** (1977) 199–204. [Zh. Eksp. Teor. Fiz.72,377(1977)].
[4] I. I. Balitsky and L. N. Lipatov, Sov. J. Nucl. Phys. **28** (1978) 822–829. [Yad. Fiz.28,1597(1978)].
[5] L. N. Lipatov, Sov. J. Nucl. Phys. **23** (1976) 338–345. [Yad. Fiz.23,642(1976)].
[6] V. Del Duca, Phys. Rev. D **52** (1995) 1527 [hep-ph/9503340].
[7] A. H. Mueller and H. Navelet, Nucl. Phys. B **282** (1987) 727.
[8] V. S. Fadin and L. N. Lipatov, Phys. Lett. B **429** (1998) 127–134 [hep-ph/9802290].

[9] M. Ciafaloni and G. Camici, Phys. Lett. B **430** (1998) 349–354 [hep-ph/9803389].
[10] A. V. Kotikov and L. N. Lipatov, Nucl. Phys. B **582** (2000) 19–43 [hep-ph/0004008].
[11] A. V. Kotikov and L. N. Lipatov, Nucl. Phys. B **661** (2003) 19–61 [hep-ph/0208220]. [Erratum: Nucl. Phys.B685, 405 (2004)].
[12] V. S. Fadin and L. N. Lipatov, JETP Lett. **49** (1989) 352 [Yad. Fiz. **50** (1989) 1141] [Sov. J. Nucl. Phys. **50** (1989) 712].
[13] V. Del Duca, Phys. Rev. D **54** (1996) 989 [hep-ph/9601211].
[14] V. S. Fadin and L. N. Lipatov, Nucl. Phys. B **477** (1996) 767 [hep-ph/9602287].
[15] V. Del Duca, Phys. Rev. D **54** (1996) 4474 [hep-ph/9604250].
[16] V. S. Fadin and L. N. Lipatov, Nucl. Phys. B **406** (1993) 259.
[17] V. S. Fadin, R. Fiore and A. Quartarolo, Phys. Rev. D **50** (1994) 5893 [hep-th/9405127].
[18] V. S. Fadin, R. Fiore and M. I. Kotsky, Phys. Lett. B **389** (1996) 737 [hep-ph/9608229].
[19] V. Del Duca and C. R. Schmidt, Phys. Rev. D **59** (1999) 074004 [hep-ph/9810215].
[20] Z. Bern, V. Del Duca and C. R. Schmidt, Phys. Lett. B **445** (1998) 168 [hep-ph/9810409].
[21] V. S. Fadin, M. I. Kotsky and R. Fiore, Phys. Lett. B **359** (1995) 181.
[22] V. S. Fadin, R. Fiore and M. I. Kotsky, Phys. Lett. B **387** (1996) 593 [hep-ph/9605357].
[23] V. S. Fadin, R. Fiore and A. Quartarolo, Phys. Rev. D **53** (1996) 2729 [hep-ph/9506432].
[24] J. Blumlein, V. Ravindran and W. L. van Neerven, Phys. Rev. D **58** (1998) 091502 [hep-ph/9806357].
[25] V. Del Duca and E. W. N. Glover, JHEP **10** (2001) 035 [hep-ph/0109028].
[26] V. S. Fadin, R. Fiore, M. G. Kozlov and A. V. Reznichenko, Phys. Lett. B **639** (2006) 74 [hep-ph/0602006].
[27] V. S. Fadin, M. G. Kozlov and A. V. Reznichenko, Phys. Rev. D **92** (2015) no.8, 085044 [arXiv:1507.00823 [hep-th]].
[28] L. N. Lipatov, Sov. Phys. JETP **63** (1986) 904 [Zh. Eksp. Teor. Fiz. **90** (1986) 1536].
[29] G. A. Chirilli and Y. V. Kovchegov, JHEP **1306** (2013) 055 [arXiv:1305.1924 [hep-ph]].
[30] G. A. Chirilli and Y. V. Kovchegov, JHEP **1405** (2014) 099 Erratum: [JHEP **1508** (2015) 075] [arXiv:1403.3384 [hep-ph]].
[31] V. Del Duca, C. Duhr, R. Marzucca and B. Verbeek, JHEP **1710** (2017) 001 [arXiv:1705.10163].
[32] M. G. Sotiropoulos and G. F. Sterman, Nucl. Phys. B **419** (1994) 59 [hep-ph/9310279].
[33] G. P. Korchemsky, Phys. Lett. B **325** (1994) 459 [hep-ph/9311294].
[34] I. A. Korchemskaya, G. P. Korchemsky, Nucl. Phys. B **437** (1995) 127 [hep-ph/9409446].
[35] I. A. Korchemskaya, G. P. Korchemsky, Phys. Lett. B **387** (1996) 346 [hep-ph/9607229].
[36] V. Del Duca, C. Duhr, E. Gardi, L. Magnea, and C. D. White, Phys. Rev. D **85** (2012) 071104 [arXiv:1108.5947].
[37] V. Del Duca, C. Duhr, E. Gardi, L. Magnea, and C. D. White, JHEP **12** (2011) 021 [arXiv:1109.3581].
[38] S. Caron-Huot, JHEP **1505** (2015) 093 [arXiv:1309.6521].
[39] V. Del Duca, G. Falcioni, L. Magnea, and L. Vernazza, Phys. Lett. B **732** (2014) 233–240 [arXiv:1311.0304 [hep-ph]].
[40] V. Del Duca, G. Falcioni, L. Magnea, and L. Vernazza, JHEP **02** (2015) 029 [arXiv:1409.8330 [hep-ph]].

[41] S. Caron-Huot, E. Gardi, and L. Vernazza, JHEP **06** (2017) 016 [arXiv:1701.05241 [hep-ph]].
[42] S. Caron-Huot, E. Gardi, J. Reichel and L. Vernazza, JHEP **1803** (2018) 098 [arXiv:1711.04850 [hep-ph]].
[43] S. M. Aybat, L. J. Dixon and G. F. Sterman, Phys. Rev. Lett. **97** (2006) 072001 [hep-ph/0606254].
[44] S. M. Aybat, L. J. Dixon and G. F. Sterman, Phys. Rev. D **74** (2006) 074004 [hep-ph/0607309].
[45] T. Becher and M. Neubert, Phys. Rev. Lett. **102** (2009) 162001 [arXiv:0901.0722].
[46] E. Gardi and L. Magnea, JHEP **0903** (2009) 079 [arXiv:0901.1091].
[47] Ø. Almelid, C. Duhr and E. Gardi, Phys. Rev. Lett. **117** (2016) 172002 [arXiv:1507.00047].
[48] Ø. Almelid, C. Duhr, E. Gardi, A. McLeod and C. D. White, JHEP **1709** (2017) 073 [arXiv:1706.10162].
[49] V. S. Fadin and R. Fiore, Phys. Lett. B **294** (1992) 286.
[50] V. S. Fadin, R. Fiore and A. Quartarolo, Phys. Rev. D **50** (1994) 2265 [hep-ph/9310252].
[51] V. Del Duca and C. R. Schmidt, Phys. Rev. D **57** (1998) 4069 [hep-ph/9711309].
[52] J. M. Henn and B. Mistlberger, Phys. Rev. Lett. **117** (2016) 171601, [arXiv:1608.00850].
[53] V. S. Fadin, AIP Conf. Proc. **1819** (2017), no. 1 060003 [arXiv:1612.04481].
[54] V. Del Duca, A. Frizzo and F. Maltoni, Nucl. Phys. B **568** (2000) 211 [hep-ph/9909464].
[55] V. Del Duca, JHEP **1802** (2018) 112 [arXiv:1712.07030 [hep-ph]].
[56] J. M. Drummond, J. Henn, V. A. Smirnov and E. Sokatchev, JHEP **0701** (2007) 064 [hep-th/0607160].
[57] L. F. Alday and J. M. Maldacena, JHEP **0706** (2007) 064 [arXiv:0705.0303 [hep-th]].
[58] J. M. Drummond, G. P. Korchemsky and E. Sokatchev, Nucl. Phys. B **795** (2008) 385 [arXiv:0707.0243 [hep-th]].
[59] A. Brandhuber, P. Heslop and G. Travaglini, Nucl. Phys. B **794** (2008) 231 [arXiv:0707.1153 [hep-th]].
[60] J. M. Drummond, J. Henn, G. P. Korchemsky and E. Sokatchev, Nucl. Phys. B **795** (2008) 52 [arXiv:0709.2368 [hep-th]].
[61] J. M. Drummond, J. Henn, G. P. Korchemsky and E. Sokatchev, Nucl. Phys. B **826** (2010) 337 [arXiv:0712.1223 [hep-th]].
[62] C. Anastasiou, Z. Bern, L. J. Dixon and D. A. Kosower, Phys. Rev. Lett. **91** (2003) 251602 [hep-th/0309040].
[63] Z. Bern, L. J. Dixon and V. A. Smirnov, Phys. Rev. D **72** (2005) 085001 [hep-th/0505205].
[64] Z. Bern, L. J. Dixon, D. A. Kosower, R. Roiban, M. Spradlin, C. Vergu and A. Volovich, Phys. Rev. D **78** (2008) 045007 [arXiv:0803.1465 [hep-th]].
[65] J. M. Drummond, J. Henn, G. P. Korchemsky and E. Sokatchev, Nucl. Phys. B **815** (2009) 142 [arXiv:0803.1466 [hep-th]].
[66] V. Del Duca, C. Duhr and V. A. Smirnov, JHEP **1003** (2010) 099 [arXiv:0911.5332 [hep-ph]].
[67] V. Del Duca, C. Duhr and V. A. Smirnov, JHEP **1005** (2010) 084 [arXiv:1003.1702 [hep-th]].
[68] S. G. Naculich and H. J. Schnitzer, Nucl. Phys. B **794** (2008) 189 [arXiv:0708.3069 [hep-th]].

[69] R. C. Brower, H. Nastase, H. J. Schnitzer and C. I. Tan, Nucl. Phys. B **814** (2009) 293 [arXiv:0801.3891 [hep-th]].

[70] J. Bartels, L. N. Lipatov and A. Sabio Vera, Phys. Rev. D **80** (2009) 045002 [arXiv:0802.2065 [hep-th]].

[71] K. T. Chen, Bull. Am. Math. Soc. **83** (1977) 831.

[72] J. Golden, A. B. Goncharov, M. Spradlin, C. Vergu and A. Volovich, JHEP **1401** (2014) 091 [arXiv:1305.1617 [hep-th]].

[73] J. Golden, M. F. Paulos, M. Spradlin and A. Volovich, J. Phys. A **47** (2014) no.47, 474005 [arXiv:1401.6446 [hep-th]].

[74] A. B. Goncharov, [math/0103059].

[75] F. C. S. Brown, Annales Sci. Ecole Norm. Sup. **42** (2009) 371 [math/0606419].

[76] N. Arkani-Hamed, J. L. Bourjaily, F. Cachazo, A. B. Goncharov, A. Postnikov and J. Trnka, Cambridge University Press, Cambridge U.K. (2012).

[77] V. Del Duca and E. W. N. Glover, JHEP **0805** (2008) 056 [arXiv:0802.4445 [hep-th]].

[78] J. Bartels, L. N. Lipatov and A. Sabio Vera, Eur. Phys. J. C **65** (2010) 587 [arXiv:0807.0894 [hep-th]].

[79] R. C. Brower, H. Nastase, H. J. Schnitzer and C. I. Tan, Nucl. Phys. B **822** (2009) 301 [arXiv:0809.1632 [hep-th]].

[80] V. Del Duca, C. Duhr and E. W. N. Glover, JHEP **0812** (2008) 097 [arXiv:0809.1822 [hep-th]].

[81] L. N. Lipatov and A. Prygarin, Phys. Rev. D **83** (2011) 125001 [arXiv:1011.2673 [hep-th]].

[82] V. S. Fadin and L. N. Lipatov, Phys. Lett. B **706** (2012) 470 [arXiv:1111.0782 [hep-th]].

[83] L. J. Dixon, C. Duhr and J. Pennington, JHEP **1210** (2012) 074 [arXiv:1207.0186].

[84] F. C. S. Brown, C. R. Acad. Sci. Paris, Ser. I **338** (2004) 527.

[85] V. Del Duca, S. Druc, J. Drummond, C. Duhr, F. Dulat, R. Marzucca, G. Papathanasiou and B. Verbeek, JHEP **1608** (2016) 152 [arXiv:1606.08807].

[86] F. C. S. Brown, [www.ihes.fr/ brown/RHpaper5.pdf].

[87] F. C. S. Brown, SIGMA **2** (2014) e25 [arXiv:1309.5309].

[88] V. Del Duca, S. Druc, J. Drummond, C. Duhr, F. Dulat, R. Marzucca, G. Papathanasiou and B. Verbeek, JHEP **1806** (2018) 116 [arXiv:1801.10605 [hep-th]].

[89] L. N. Lipatov, J. Phys. A **42** (2009) 304020 [arXiv:0902.1444 [hep-th]].

[90] L. N. Lipatov, JETP Lett. **59** (1994) 596 [Pisma Zh. Eksp. Teor. Fiz. **59** (1994) 571] [hep-th/9311037].

[91] L. D. Faddeev and G. P. Korchemsky, Phys. Lett. B **342** (1995) 311 [hep-th/9404173].

[92] L. N. Lipatov, Phys. Lett. B **309** (1993) 394.

[93] L. N. Lipatov, Nucl. Phys. B **548** (1999) 328 [hep-ph/9812336].

[94] V. Del Duca, L. J. Dixon, C. Duhr, and J. Pennington, JHEP **1402** (2014) 086 [arXiv:1309.6647].

[95] O. Schnetz, [arXiv:1606.08598].

Holographic description of total hadronic cross sections at high energies

Akira Watanabe[1,2,*] and *Mei* Huang[1,2,**]

[1]Institute of High Energy Physics and Theoretical Physics Center for Science Facilities, Chinese Academy of Sciences, Beijing 100049, People's Republic of China
[2]University of Chinese Academy of Sciences, Beijing 100049, People's Republic of China

Abstract. We investigate the hadron-hadron total cross sections at high energies in the framework of holographic QCD. In our model setup, the involved strong interaction is described by the Brower-Polchinski-Strassler-Tan Pomeron exchange kernel, and the two form factors are obtained from the bottom-up AdS/QCD models. We show that the resulting nucleon-nucleon total cross section is consistent with the experimental data recently measured by the TOTEM collaboration at the LHC. As examples to see the general versatility of the model, we also present our analysis on the pion-nucleon and pion-pion cases, which can be predicted because all the adjustable parameters are fixed by the nucleon-nucleon data.

1 Introduction

Understanding the partonic structures of the hadrons is one of the most important subjects in hadron physics. High energy scattering phenomena, which can be measured experimentally, have played crucial roles in revealing them for many years. The strong interaction in the scattering processes gives a dominant contribution to the measured cross sections, and the fundamental theory to describe it is knows as quantum chromodynamics (QCD). The expression for the cross section can be factorized into two components, the "hard" (perturbative) and the "soft" (nonperturbative) parts. The perturbative technique in QCD is applicable to perform analysis on the former one, but cannot used for the latter one. The soft part is described by the parton distribution functions (PDFs) which are expressed by the Bjorken variable x and the probe energy scale Q^2 and parameterized with experimental data from various scattering processes. Since the PDFs are not calculable quantities, they are in principle treated as inputs in any analysis of high energy processes.

However, particularly in the small x region, our knowledge is still quite poor and the parameterized PDFs have huge uncertainties, which sometimes make the predictions unreliable, because it is not easy to correctly evaluate such uncertainties in the final results. In this kinematic region, the strong interaction among so many gluons with tiny momenta gives a dominant contribution to the cross section. Since precisely considering such an extremely complicated dynamics based on QCD is impossible in practice, it is much more suitable to

*e-mail: akira@ihep.ac.cn
**e-mail: huangm@ihep.ac.cn

© The Authors, published by EDP Sciences. This is an open access article distributed under the terms of the Creative Commons Attribution License 4.0 (http://creativecommons.org/licenses/by/4.0/).

perform the analysis by effective approaches. The Pomeron exchange is well-known as an effective description of the multi-gluon exchange which may be occurred in various scattering processes in the small x region [1, 2]. This is practically the only way to treat the complicated dynamics, and has been successfully applied to the analysis of various processes such as the hadron-hadron scattering [3] and the deep inelastic scattering (DIS).

The effective description by the Pomeron exchange is realized also in holographic QCD which is an effective approach constructed based on the anti-de Sitter/conformal field theory (AdS/CFT) correspondence (or gauge/string correspondence in general) [4]. This effective method in QCD has gathered a lot of theoretical interests [5–10], and focusing mainly on DIS, many related studies have been done so far [11–23]. In holographic QCD, the Pomeron in the usual Minkowski space is identified as the Reggeized graviton in the higher dimensional curved space, and one of the most successful descriptions of the holographic Pomeron was proposed by Brower-Polchinski-Strassler-Tan (BPST) [13]. The most remarkable result with the BPST kernel may be that the nontrivial scale dependence of the Pomeron intercept, which were measured at HERA, can be well reproduced by applying this kernel to calculate the structure function at small x [21, 22].

In this report, based on Ref. [24] we present our recent analysis on total cross sections of the high energy hadron-hadron scattering in the framework of holographic QCD. So far, the DIS has intensively been studied, in which the density distribution of the probe photon is localized around the ultraviolet (UV) boundary in the AdS space, while the peak position of the target hadron density distribution is located in the infrared (IR) region. Hence, the DIS is a UV-IR scattering, and the extracted Pomeron intercept depends on the four-momentum squared of the probe photon Q^2. On the other hand, the hadron-hadron scattering is interpreted as an IR-IR scattering. Since there is no such a variable scale in this scattering process, the Pomeron intercept becomes constant and its value may be close to that of the well-known soft Pomeron intercept 1.0808 [3]. As to the proton-proton collision, the new experimental data at $\sqrt{s} = 13$ TeV, where s is the Mandelstam variable, were taken recently by the TOTEM collaboration at the LHC [25]. Since the only scale that the involved hadrons have in this process is the proton mass $m_p \sim 1$ GeV, the Bjorken x value at $\sqrt{s} \sim 10$ TeV can be estimated as $x \sim 10^{-8}$ which is much smaller than that was realized at HERA. Therefore, the new data provide us with valuable opportunities to investigate the very small x region which cannot be studied enough before.

In this work, we firstly focus on the nucleon-nucleon case. Employing the BPST Pomeron exchange kernel to describe the involved strong interaction between the two nucleons and the bottom-up AdS/QCD model of the nucleon [26] to calculate its density distribution, which also is identified as the gravitational form factor [27], in the five-dimensional AdS space, and then we obtain the total cross section at high energies. We concentrate on the kinematic range $10^2 < \sqrt{s} < 10^5$ GeV, and explicitly demonstrate the comparison between our calculation and the currently available experimental data. It is shown that the recent TOTEM data are well reproduced within our model, which strongly supports further applications of this framework.

The most important advantage of our model setup is its general versatility. Since all the adjustable parameters contained in the model are fixed by the proton-proton data, other hadron-hadron processes at high energies can be analyzed without any additional parameter. Hence, besides the nucleon-nucleon case, we also consider the pion-nucleon and pion-pion cases as examples in this work. Similar to the nucleon case, the required density distribution of the pion in the AdS space can be obtained from the bottom-up AdS/QCD model of mesons [9, 28]. Although there is no available data for the pion involved processes in the considered kinematic region, we find that our prediction for the total cross section ratio $\sigma_{tot}^{\pi N}/\sigma_{tot}^{NN}$ agrees with that extracted from the well-known work with the soft Pomeron exchange done by Donnachie and Landshoff [3].

2 Model setup

In this work, we investigate total cross sections of the high energy hadron-hadron scattering, and the formalism is in principle applicable for any hadron pair. Considering the two-to-two process, $1 + 2 \to 3 + 4$, and following the preceding studies [13, 15, 17, 21], the scattering amplitude is written with the BPST Pomeron exchange kernel χ in the five-dimensional AdS space as

$$\mathcal{A}(s,t) = 2is \int d^2 b e^{i\boldsymbol{k}_\perp \cdot \boldsymbol{b}} \int dz dz' P_{13}(z) P_{24}(z') \left[1 - e^{i\chi(s,\boldsymbol{b},z,z')}\right], \tag{1}$$

where s and t denote the Mandelstam variables, $\boldsymbol{b}$ represents the two-dimensional impact parameter, and z and z' are fifth coordinates for the involved two hadrons. $P_{13}(z)$ and $P_{24}(z')$ represent density distributions of the hadrons in the AdS space and are normalized, because the considered hadrons are normalizable modes.

Applying the optical theorem to Eq. (1) and picking up the leading contribution from the eikonal representation, the expression for the total cross section can be obtained as

$$\sigma_{tot}(s) = 2 \int d^2 b \int dz dz' P_{13}(z) P_{24}(z') \mathrm{Im}\chi(s, \boldsymbol{b}, z, z'). \tag{2}$$

In the conformal limit, it is known that the analytical form of the imaginary part of χ can be written down, and the impact parameter integration can be performed analytically. As a result, Eq. (2) is rewritten as

$$\sigma_{tot}(s) = \frac{g_0^2 \rho^{3/2}}{8\sqrt{\pi}} \int dz dz' P_{13}(z) P_{24}(z') (zz') \mathrm{Im}[\chi_c(s, z, z')], \tag{3}$$

$$\mathrm{Im}[\chi_c(s, z, z')] \equiv e^{(1-\rho)\tau} e^{-[(\log^2 z/z')/\rho\tau]} / \tau^{1/2}, \tag{4}$$

$$\tau = \log(\rho z z' s/2), \tag{5}$$

where ρ and g_0^2 are adjustable parameters which control the energy dependence and the magnitude of the cross section, respectively.

Although the expressions shown above were rigorously derived, it is also known through the preceding applications of the BPST kernel to the DIS at small x that the inclusion of the confinement effect is required to reproduce the structure function data measured at HERA. Hence, we utilize the modified kernel, in which the added term mimics the confinement effect, for the numerical evaluations in this work, given by

$$\mathrm{Im}[\chi_{mod}(s, z, z')] \equiv \mathrm{Im}[\chi_c(s, z, z')] + \mathcal{F}(s, z, z') \mathrm{Im}[\chi_c(s, z, z_0 z_0'/z')], \tag{6}$$

$$\mathcal{F}(s, z, z') = 1 - 2\sqrt{\rho\pi\tau} e^{\eta^2} \mathrm{erfc}(\eta), \tag{7}$$

$$\eta = \left(-\log \frac{zz'}{z_0 z_0'} + \rho\tau\right) / \sqrt{\rho\tau}, \tag{8}$$

where the functional form of $\mathrm{Im}[\chi_c]$ is the same as that of Eq. (4). Also, z_0 and z_0' appearing here are the cutoffs in the fifth coordinates that characterize the QCD scale, but these are not adjustable because they are uniquely fixed by the hadron masses.

To numerically evaluate the total cross sections, it is also needed to specify the density distributions, in which the involved hadron properties reflect. The density distributions are extracted from the hadron-Pomeron(graviton)-hadron three-point functions which can be computed by utilizing the bottom-up AdS/QCD models. In this work, we evaluate the nucleon-nucleon, pion-nucleon, and pion-pion total cross sections, so we need the density distributions of the nucleon and the pion. Since the authors of Refs. [27, 28] calculated the

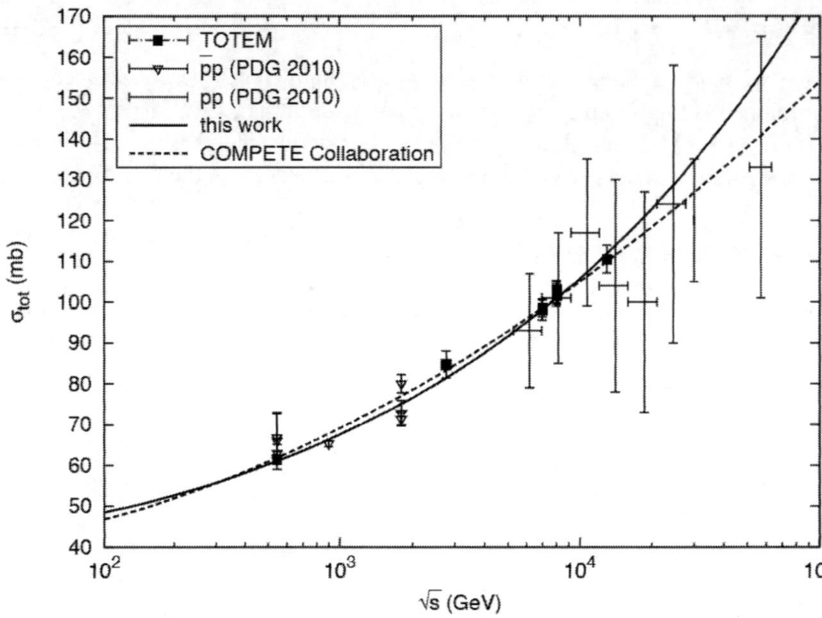

Figure 1. The nucleon-nucleon total cross sections as a function of $\sqrt{s}$. The solid and dashed lines depict our calculation and the empirical fit by the COMPETE collaboration [33], respectively. The recent TOTEM data [25, 29] and other pp [30] and p̄p [31] data are plotted with error bars.

gravitational form factors of those hadrons, using the classical actions of the AdS/QCD models of the nucleon [26] and mesons [9], we utilize their results in this study. Those results were also applied to the analysis of the structure functions of the nucleon and the pion in DIS at small x in the previous studies [22].

3 Numerical results

There are in fact some parameters included in the bottom-up AdS/QCD models which are used to calculate the required density distributions of the nucleon and the pion. However, they can be fixed with the hadronic observables such as the masses and the decay constants. Therefore, the adjustable parameters to be determined with the experimental data here are the two, ρ and g_0^2, which are contained in the BPST Pomeron exchange kernel as seen in the previous section. Besides the proton-proton collision data recently measured by the TOTEM collaboration at LHC [25, 29], other pp [30] and p̄p [31] data, which were summarized in 2010 by the Particle Data Group [32], are considered to determine those parameters in this study.

We display the resulting nucleon-nucleon total cross section in Fig. 1, in which, besides the experimental data, we also plot the empirical fit obtained by the COMPETE collaboration [33] for comparison. It is seen in Fig. 1 that our calculation is in agreement with the data in the whole considered region, and also is consistent with the empirical fit at $\sqrt{s} < 10$ TeV. In the higher energy region, our result becomes larger than the empirical one, and the size of the deviation between the two curves increases with $\sqrt{s}$.

Next, we show in Fig. 2 our results for the pion-nucleon and pion-pion total cross sections. These cross sections can be calculated without any adjustable parameter, because all

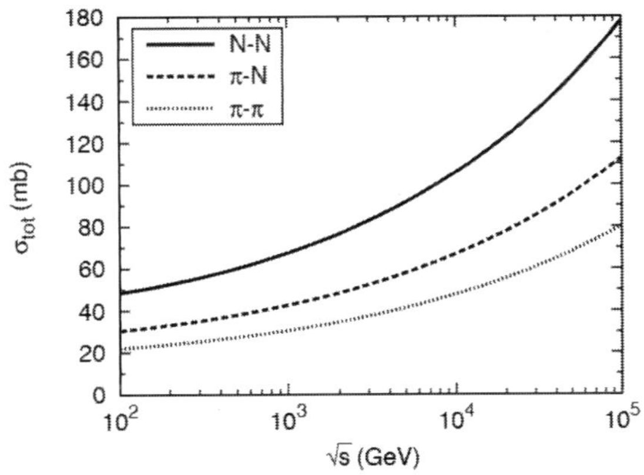

Figure 2. The total cross sections obtained in this work as a function of $\sqrt{s}$. The solid, dashed, and dotted lines depict the nucleon-nucleon, pion-nucleon, and pion-pion results, respectively.

the parameters are fixed with the pp($\bar{p}$p) data. One can see from the figure that the pion-nucleon result is substantially smaller than the nucleon-nucleon one, and the pion-pion result is smallest among the three curves. Since in our model the energy dependence is only from the BPST kernel, the s dependence is common to the three results, which implies the universality of the BPST Pomeron. Therefore, we can predict the total cross section ratios as follows:

$$\frac{\sigma_{tot}^{\pi N}}{\sigma_{tot}^{NN}} = 0.63, \quad \frac{\sigma_{tot}^{\pi\pi}}{\sigma_{tot}^{NN}} = 0.45. \tag{9}$$

Although there is no experimental data for the pion involved processes, an empirical value can be extracted from the work with the soft Pomeron exchange done by Donnachie and Landshoff [3]. The extracted ratio is $\sigma_{tot}^{\pi N}/\sigma_{tot}^{NN} = 0.63$ which agrees with our prediction.

4 Summary

In this work, we have investigated the hadron-hadron total cross sections at high energies in the framework of holographic QCD whose formalism is constructed in the five-dimensional AdS space. Applying the BPST Pomeron exchange kernel to describe the involved strong interaction and utilizing the bottom-up AdS/QCD models to obtain the density distributions of the hadrons, we have calculated the total cross sections.

Our nucleon-nucleon result is in agreement with the experimental data, including the recently measured TOTEM ones. Our calculation is also consistent with the empirical fit obtained by the COMPETE collaboration at $\sqrt{s} < 10$ TeV, although a substantial deviation is observed at higher s. Since the currently available data at $\sqrt{s} > 10$ TeV, except for the recent TOTEM's 13 TeV one, were extracted from the cosmic-ray experiments, they have huge uncertainties. Hence, more precise data in such a very high s regime are needed to pin down the total cross section.

In our model setup, we can consider various hadron pairs as the participants in the hadron-hadron scattering, giving appropriate density distributions for them. Besides the nucleon-nucleon case, we have also analyzed the pion-nucleon and pion-pion cases, and predicted the total cross section ratios. Our predictions could be tested at future experimental facilities with a high intensity pion beam.

Our framework is in principle applicable for any high energy scattering process if the strong interaction involved in the process can be approximated by the Pomeron exchange. Further investigations are certainly needed.

Acknowledgments

This work was supported by the NSFC under Grant Nos. 11725523, 11735007, and 11261130311 (CRC 110 by DFG and NSFC).

References

[1] J.R. Forshaw, D.A. Ross, *Quantum Chromodynamics and the Pomeron* (Cambridge University Press, Cambridge, 1997).
[2] S. Donnachie, G. Dosch, P. Landshoff, O. Nachtmann, *Pomeron Physics and QCD* (Cambridge University Press, Cambridge, 2002).
[3] A. Donnachie, P. Landshoff, Phys.Lett. **B296**, 227 (1992).
[4] J.M. Maldacena, Adv.Theor.Math.Phys. **2**, 231 (1998); S. Gubser, I.R. Klebanov, A.M. Polyakov, Phys.Lett. **B428**, 105 (1998); E. Witten, Adv.Theor.Math.Phys. **2**, 253 (1998).
[5] M. Kruczenski, D. Mateos, R.C. Myers, D.J. Winters, JHEP **0307**, 049 (2003).
[6] D. T. Son, M. A. Stephanov, Phys.Rev. **D69**, 065020 (2004).
[7] M. Kruczenski, D. Mateos, R.C. Myers, D.J. Winters, JHEP **0405**, 041 (2004).
[8] T. Sakai, S. Sugimoto, Prog.Theor.Phys. **113**, 843 (2005); T. Sakai, S. Sugimoto, Prog.Theor.Phys. **114**, 1083 (2005).
[9] J. Erlich, E. Katz, D.T. Son, M.A. Stephanov, Phys.Rev.Lett. **95**, 261602 (2005).
[10] L. Da Rold, A. Pomarol, Nucl.Phys. **B721**, 79 (2005).
[11] J. Polchinski, M.J. Strassler, Phys.Rev.Lett. **88**, 031601 (2002); J. Polchinski, M.J. Strassler, JHEP **0305**, 012 (2003).
[12] H. Boschi-Filho, N. R. F. Braga, H.L. Carrion, Phys.Rev. **D73**, 047901 (2006).
[13] R.C. Brower, J. Polchinski, M.J. Strassler, C.I. Tan, JHEP **0712**, 005 (2007).
[14] Y. Hatta, E. Iancu, A. Mueller, JHEP **0801**, 026 (2008).
[15] R.C. Brower, M.J. Strassler, C.I. Tan, JHEP **0903**, 050 (2009).
[16] C. Ballon Bayona, H. Boschi-Filho, N.R. Braga, JHEP **0810**, 088 (2008).
[17] R.C. Brower, M.J. Strassler, C.I. Tan, JHEP **0903**, 092 (2009).
[18] L. Cornalba, M.S. Costa, Phys.Rev. **D78**, 096010 (2008).
[19] B. Pire, C. Roiesnel, L. Szymanowski, S. Wallon, Phys.Lett. **B670**, 84 (2008).
[20] L. Cornalba, M.S. Costa, J. Penedones, Phys.Rev.Lett. **105**, 072003 (2010).
[21] R.C. Brower, M. Djuric, I. Sarcevic, C.I. Tan, JHEP **1011**, 051 (2010).
[22] A. Watanabe, K. Suzuki, Phys.Rev. **D86**, 035011 (2012); A. Watanabe, K. Suzuki, Phys. Rev. **D89**, 115015 (2014).
[23] A. Watanabe, H.n. Li, Phys. Lett. **B751**, 321 (2015).
[24] A. Watanabe, M. Huang, arXiv:1809.02515 [hep-ph].
[25] G. Antchev et al. (TOTEM Collaboration), arXiv:1712.06153 [hep-ex].
[26] D.K. Hong, T. Inami, H.U. Yee, Phys.Lett. **B646**, 165 (2007).
[27] Z. Abidin, C.E. Carlson, Phys.Rev. **D79**, 115003 (2009).
[28] Z. Abidin, C.E. Carlson, Phys.Rev. **D77**, 115021 (2008).
[29] G. Antchev et al. (TOTEM Collaboration), EPL **101**, 21002 (2013); G. Antchev et al. (TOTEM Collaboration), EPL **101**, 21004 (2013); G. Antchev et al. (TOTEM Collaboration), Phys. Rev. Lett. **111**, 012001 (2013); G. Antchev et al. (TOTEM Collaboration),

Nucl. Phys. **B899**, 527 (2015); G. Antchev et al. (TOTEM Collaboration), Eur. Phys. J. **C76**, 661 (2016); F.J. Nemes, PoS **DIS2017**, 059 (2018).

[30] R.M. Baltrusaitis, G.L. Cassiday, J.W. Elbert, P.R. Gerhardy, S. Ko, E.C. Loh, Y. Mizumoto, P. Sokolsky, D. Steck, Phys. Rev. Lett. **52**, 1380 (1984); M. Honda, M. Nagano, S. Tonwar, K. Kasahara, T. Hara, N. Hayashida, Y. Matsubara, M. Teshima, S. Yoshida, Phys. Rev. Lett. **70**, 525 (1993); P. Abreu et al. (Pierre Auger Collaboration), Phys. Rev. Lett. **109**, 062002 (2012).

[31] R. Battiston et al. (UA4 Collaboration), Phys. Lett. **117B**, 126 (1982); G. Arnison et al. (UA1 Collaboration), Phys. Lett. **128B**, 336 (1983); M. Bozzo et al. (UA4 Collaboration), Phys. Lett. **147B**, 392 (1984); G.J. Alner et al. (UA5 Collaboration), Z. Phys. **C32**, 153 (1986); N.A. Amos et al. (E710 Collaboration), Phys. Rev. Lett. **68**, 2433 (1992); F. Abe et al. (CDF Collaboration), Phys. Rev. **D50**, 5550 (1994); C. Augier et al. (UA4/2 Collaboration), Phys. Lett. **B344**, 451 (1995); C. Avila et al. (E-811 Collaboration), Phys. Lett. **B537**, 41 (2002).

[32] K. Nakamura et al. (Particle Data Group), J. Phys. **G37**, 075021 (2010).

[33] J.R. Cudell, V.V. Ezhela, P. Gauron, K. Kang, Yu.V. Kuyanov, S.B. Lugovsky, E. Martynov, B. Nicolescu, E.A. Razuvaev, N.P. Tkachenko (COMPETE Collaboration), Phys. Rev. Lett. **89**, 201801 (2002).

EPJ Web of Conferences **192**, 00011 (2018) https://doi.org/10.1051/epjconf/201819200011
QCD@Work 2018

Two loop four dimensional formulation of dimensionally regularized amplitudes in quantum chromodynamics

Angelo Raffaele Fazio[1,*]

[1]Departamento de Física, Universidad Nacional de Colombia, Ciudad Universitaria, Bogotá D.C., Colombia

Abstract. We propose a pure four-dimensional formulation (FDF) of the d-dimensional regularization of one and two loop scattering amplitudes in quantum chromodynamics. In our formulation particles propagating inside the loops are represented by massive internal states regulating the divergences. The equivalence between the FDF and the Four Dimensional Helicity scheme is discussed. We present explicit representations of the polarization and helicity states of the four-dimensional particles propagating in the loops. They allow for a complete, four-dimensional, unitarity-based construction of d-dimensional amplitudes. Generalized unitarity within the FDF does not require any higher-dimensional extension of the Clifford and the spinor algebra.

1 Introduction

The ultraviolet (UV) and infrared (IR) infinities arising in the higher order perturbative calculations in quantum field theories can be treated by dimensional regularization [1]. Those divergences are usually nested like in the following integral

$$\int \frac{d^4k}{(2\pi)^4} \frac{1}{(k^2)^2} = \pi^2 \int_0^\infty dk^2 \frac{1}{k^2} = \pi^2 \int_0^\infty \frac{dx}{x} \tag{1}$$

which is UV divergent for $k^2 \to \infty$ and IR divergent for $k^2 \to 0$. Dimensional regularization is based on the continuum dimension method by first generalizing the dimensionality of the space from *four* to d integer, such that the integral converges, and then replacing d by the complex number $4 - 2\epsilon$, which contains the regularizing parameter ϵ. Symbolically

$$\int \frac{d^4k}{(2\pi)^4} \to \mu_{\rm DS}^{4-d} \int \frac{d^dk}{(2\pi)^d} \tag{2}$$

with μ_{DS} the unit mass needed to take into account the change of the engineering dimensions of the couplings of the theory. The d- dimensional integral in (1) can be treated as

$$\int d^dk \frac{1}{(k^2)^2} = \frac{\pi^{\frac{d}{2}}}{\Gamma(\frac{d}{2})} \int_0^\infty dk^2 (k^2)^{\frac{d}{2}-3}. \tag{3}$$

The continuation from 4 to $d = 4-2\epsilon$ dimensions makes all momentum integrals well defined and UV and IR singularities appear in the Laurent expansion of meromorphic functions as $\frac{1}{\epsilon^n}$

*e-mail: arfazio@unal.edu.co

© The Authors, published by EDP Sciences. This is an open access article distributed under the terms of the Creative Commons Attribution License 4.0 (http://creativecommons.org/licenses/by/4.0/).

poles. Actually the integral (1), combined with other UV and IR divergent terms, is usually defined [2] so that to cancel those divergences and not to bring contributions to the finite part.

Consider the physical cross section at the Next-to-Leading Order which receives by the Kinoshita- Lee- Nauenberg theorem the real and virtual contributions

$$\sigma = \sigma_V + \sigma_R = \int d\Phi_V |M_V|^2 + \int d\Phi_R |M_R|^2, \tag{4}$$

identified by the corresponding amplitudes M_R and M_V. A complete dimensional scheme (DS), meaning a complete procedure to make the analytical dimensional continuation in the dynamics of the theory is reflected at the level of the Feynman integrals by the way of treating the gauge boson metric, generically indicated in the following by g:

$$\sigma^{DS} = \underbrace{\int d\Phi_V |M_V(.....,[g],.....)|^2}_{\frac{a}{\epsilon^2}+\frac{b}{\epsilon}+c+d\epsilon+e\epsilon^2+...} + \underbrace{\int d\Phi_R |M_R(....,[g],....)|^2}_{-\frac{a}{\epsilon^2}-\frac{b}{\epsilon}+l+m\epsilon+n\epsilon^2+...}$$

$$= \sigma_{\text{finite}} + \epsilon\sigma_1 + \epsilon^2\sigma_2 + \dots \tag{5}$$

where the scheme independent terms cancel and the analytic part of the expansion is scheme independent except for the first term. Regardless on the scheme the physical cross section should be obtained as

$$\sigma = \lim_{\epsilon\to 0} \sigma^{DS} = \sigma_{\text{finite}} \tag{6}$$

if the adopted regularization scheme is compatible with the unitarity of the theory, meaning with the statement of the the optical theorem. The purely d-dimensional treatment of all objects, the so called conventional (or naive) dimensional regularization is conceptually simpler, but it breaks supersymmetry and it gives ambiguities like $\mathrm{Tr}(\gamma_5\gamma^\mu\gamma^\nu\gamma^\rho\gamma^\tau)$ with chiral symmetries, but it is unitarity compatible [3]. However for the modern techniques of generalized unitarity [4] of the integrand reconstruction by gluing tree level helicity amplitudes, a four dimensional approach in general considerably simplifies the computations. In those cases the 4-dimensional treatment of the gluons is better compatible with supersymmetry and it is more amenable to helicity methods, where we glue amplitudes like the one coming from the Parke-Taylor [5] formula for Maximally Helicity Violating amplitudes

$$A_{\text{MHV}}(1^+,.........,i^-,.......,j^-,.......n^+) = i(-g)^{n-2}\frac{<ij>^4}{<12><23>.........<n1>} \tag{7}$$

written here in terms of the helicity spinors, pedagogically introduced in [6]. The previous considerations imply the need for a dimensional reduction of the d dimensional vector field to a four dimensional one together with a dimensional regularization of the loop integrals.

2 Variants of the dimensional regularization

There is an elegant way to unify essentially all common variants of dimensional regularization in a single framework, where all definitions can easily be formulated and where the differences and relations between the schemes become transparent [7]. Such a framework is based on the introduction of three vector spaces: the original four dimensional space ($4S$) and the infinite dimensional vector spaces, intrinsic in every dimensional scheme to provide the regularization of theories formulated in an arbitrary number of dimensions [8]. The space ($4S$) has metric tensor $g^{\mu\nu}_{[4]}$, then we have the "quasi-d-dimensional" space ($QS_{[d]}$) with metric

	CDR	HV	FDH	DRED
singular VF	$g^{\mu\nu}_{[d]}$	$g^{\mu\nu}_{[d]}$	$g^{\mu\nu}_{[d_s]}$	$g^{\mu\nu}_{[d_s]}$
regular VF	$g^{\mu\nu}_{[d]}$	$g^{\mu\nu}_{[4]}$	$g^{\mu\nu}_{[4]}$	$g^{\mu\nu}_{[d_s]}$

Table 1. Treatment of vector fields in the four different regularization schemes, i.e. prescription which metric tensor has to be used in propagator numerators and polarization sums. The quantity d_s is usually taken to be 4.

tensor $g^{\mu\nu}_{[d]}$, the "quasi-d_s-dimensional" space ($QS_{[d_s]}$) with metric tensor $g^{\mu\nu}_{[d_s]}$. The "quasi-dimensionalities" of those infinite dimensional spaces are related by the following relations between the complex numbers:

$$d_s = d + n_\epsilon = 4 - 2\epsilon + n_\epsilon, \tag{8}$$

implying the following formal direct sum and inclusions

$$QS_{[d_s]} = QS_{[d]} \oplus QS_{[n_\epsilon]} \qquad S_4 \subset QS_{[d]} \subset QS_{[d_s]}. \tag{9}$$

Indeed since only gluons that appear inside a divergent loop or phase space integral need to be regularized, those are called "singular" gluons and all the others are called "regular". Based on that distinction the following variants of dimensional regularization and dimensional reduction are introduced. CDR ("Conventional dimensional regularization") in which singular and regular gluons (and other vector fields) are all treated as quasi-d–dimensional. HV ("'t Hooft Veltman") in which singular gluons are treated as quasi-d–dimensional but the regular ones are treated as 4–dimensional. For the dimensional reduction we have DRED ("original/old dimensional reduction"), where regular and singular gluons are all treated as quasi-d_s-dimensional, and FDH ("four-dimensional helicity scheme") so that singular gluons are treated as quasi-d_s-dimensional but external ones are treated as 4-dimensional.

The treatment of vector fields in the four different regularization schemes and the corresponding prescriptions on which metric tensor has to be used in propagator numerators and vectors polarization sums are summarized in the table 1.

In order to use FDH and DRED, the variants based on the dimensional reduction, the crucial step is to split quasi-d_s-dimensional gluons into d-component gauge fields and $n_\epsilon \to 2\epsilon$ scalar fields, so called ϵ-scalars, as from the decomposition of the metric tensor

$$(g_{[d_s]})^{\mu\nu} = (g_{[d]})^{\mu\nu} + (g_{[n_\epsilon]})^{\mu\nu} \tag{10}$$

with the algebra

$$g^{\mu\nu}_{[d_s]}(g_{[d]})_{\nu\rho} = (g_{[d]})^{\mu}{}_{\rho}, \quad g^{\mu\nu}_{[d_s]}(g_{[n_\epsilon]})_{\nu\rho} = (g_{[n_\epsilon]})^{\mu}{}_{\rho}, \quad g^{\mu\nu}_{[d]}(g_{[n_\epsilon]})_{\nu\rho} = 0, \quad (g_{[n_\epsilon]})_{\nu\rho} g^{\rho\mu}_{[4]} = 0. \tag{11}$$

Since new degrees of freedom enter in the game, the independence of the gauge boson manifest in the couplings of the ϵ-scalars resulting in different renormalization constants and β-functions. In QED the gauge coupling e and the evanescent coupling e_e renormalize differently being not protected by the Lorentz and the gauge invariance on $QS_{[d]}$

$$\beta_e = \mu^2_{\rm DS} \frac{d}{d\mu^2_{\rm DS}} \left(\frac{e}{4\pi}\right)^2 = \frac{4}{3}\left(\frac{e}{4\pi}\right)^4 + \ldots \tag{12}$$

$$\beta_{e_e} = \mu_{\rm DS}^2 \frac{d}{d\mu_{\rm DS}^2} \left(\frac{e_e}{4\pi}\right)^2 = 6\left(\frac{e}{4\pi}\right)^4 - 6\left(\frac{e}{4\pi}\right)^2\left(\frac{e_e}{4\pi}\right)^2 + \ldots \tag{13}$$

Even by imposing the renormalization condition $e = e_e$ the flows of the two couplings are different [7].

3 FDF:Four dimensional formulation of the FDH scheme

FDF is a novel regularization approach that was introduced to reproduce FDH results at the one-loop level [9]. Starting from unregularized expressions in one loop diagram the loop momenta are shifted as $\not{\ell}_{[d]} \to \not{\ell}_{[4]} + i\mu\gamma_5$ before any other algebraic manipulation is performed. The matrix γ_5 provides the chirality in four dimensions. The scale μ corresponds to the other $d-4$ dimensional components of the loop momentum and serves as regulator for the in general divergent quasi-d dimensional loop momentum in the integral. By definition, odd powers of μ are set to zero, resulting in the useful relation

$$\not{\ell}_{[d]}\not{\ell}_{[d]} = \ell_{[d]}^2 = \ell_{[4]}^2 - \mu^2. \tag{14}$$

One main advantage with FDF approach is that the Lorentz algebra is realized in strictly four dimensions. By construction the treatment of γ_5 in FDF unifies those of Breitenlohner-Maison and the anti-commuting chiral prescriptions like in the 't Hooft-Veltman [10].

Due to the previously declared shift the fermionic cutted lines in the diagrams obey to a generalized Dirac equation

$$(\not{\ell}_{[4]} + i\mu\gamma_5 + m)\, u(\ell) = 0. \tag{15}$$

By expressing the four momenta in terms of the massless four-momenta $\ell^{\flat}_{[4]}$ and $q_{\ell[4]}$ the spinors generalized solutions are

$$u_+(\ell) = \left|\ell^{\flat}_{[4]}\right\rangle + \frac{m - i\mu}{[\ell^{\flat}_{[4]}, q_{\ell[4]}]} \left|q_{\ell[4]}\right] \quad u_-(\ell) = \left|\ell^{\flat}_{[4]}\right] + \frac{m + i\mu}{< \ell^{\flat}_{[4]}, q_{\ell[4]} >} \left|q_{\ell[4]}\right\rangle \tag{16}$$

with the usual notation of the helicity spinors. In [9] it is possible to find the wave functions for the cut gluonic and scalar lines for the consequent dimensional reduction involved in this procedure.

At one loop it is known the decomposition of any scattering amplitude in terms of a basis of master scalar integrals á la Passarino-Veltman [11]. By FDF we have a tool to add additional basis integrals at one loop consisting, in the case of having massless internal lines, into two sets of additional boxes $I_4[(\mu^2)^2]$, $I_4[\mu^2]$, additional triangles $I_3[\mu^2]$ and additional bubbles $I_2[\mu^2]$, with the notations of having insertions of μ^2 in the numerators of those additional master integrals. The decomposition of that amplitude in this basis involve coefficients of the additional integrals that are independent of the dimensional regulator, therefore those coefficients can be computed by integration contours using the methods of algebraic geometry [12]. The amplitude in now expressed in terms of larger set of integrals that in principle they could be evaluated as they are, however it is usually easier to evaluate them in terms of standard integrals.When we do that the dependence on the ϵ regulator reappears, in fact at one loop the conversion to the standard integrals amounts to a shift of the dimensionality, one has dimension shift identities like [15]

$$I_4[(\mu^2)^2] = -\epsilon(1-\epsilon)16\pi^2 I_4^{8-2\epsilon}[1]. \tag{17}$$

For a box integral we have now five degrees of freedom instead of four. The so called quadruple cut [13], which is the modern procedure of extracting the leading singularity of one-loop amplitudes (contrast it to [14]) and which allows the calculation of the box coefficients in the scattering amplitudes, is in terms of this basis a one-dimensional surface μ-dependent more than a single point in the space of the momenta. However after gluing the corresponding amplitudes the box coefficients of the two integrals in the extended basis are still found by collecting in the numerators respectively the coefficients of the monomial $(\mu^2)^2$ and μ^2. In fact by removing factors with ϵ exponents from the measure $d^{4-2\epsilon}k = d^4k\, d^{-2\epsilon}k$ and multiplying by a weight function of μ^2 to isolate different integrals we find in terms of the block amplitudes the coefficients of the extra basis integrals as

$$c_i = \oint \frac{d\mu^2}{\mu^2} A^1_{\text{tree}}(\mu^2) A^2_{\text{tree}}(\mu^2) A^3_{\text{tree}}(\mu^2) A^4_{\text{tree}}(\mu^2), \tag{18}$$

with a final result compatible with [16].

4 Foundational work to perform two loops reductions in FDF

First of all we need a basis, this has been provided in pure 4 dimensions for planar diagrams in [17]. The notations for the double box $P^*_{n_1,n_2}$, refer to n_1 lines on the right and n_2 lines on the left of the double box. In this basis we have two integrals for massless, one-mass, diagonal-two mass double box $P^*_{2,2}$. Three integrals for short side two mass, three-mass double box. Four integrals for four masses double box. One integral for massless pentabox $P^*_{3,2}$. A pure four dimensional basis is reduced to this with master integrals containing up to height propagators. More complicated integrals can be reduced to them up to ϵ-terms. The new feature compared with the one-loop case is the presence of irreducible numerators even when it is simply involved the scalar product of a loop momentum with an external leg, meaning that they cannot be written as linear combination of the denominators propagators and external invariants.

To perform the reduction to the previous basis at two loops the tensor reduction is still necessary and useful but it is not sufficient, because some additional technology is needed in order to take care of the irreducible invariants and their powers. That technology is based on the so-called integration by part identities (IBP) [18]. This technology is based on an axiom of dimensional regularization that the integral of a total divergence is zero, because there are no boundary terms

$$0 = \int d^d\ell_1 d^d\ell_2 \frac{\partial}{\partial \ell_1^\mu} \frac{v^\mu}{\text{Denominator}} \tag{19}$$

this gives linear relations between integrals that can be solved in terms of the independent master integrals. In the notation of the previous equation a so called IBP generating vector is included which takes care of the fact that no doubled propagators are generated in the IBP, this simplifies the solution of the system of the IBP. By identifying the IBP generating vectors would provide a quicker approach to solve the system for generating master integrals [17]. By the definition of the vectors v^μ, they are algebraic geometric objects arising from appropriate polynomial divisions, their algebraically geometric properties as well as their connection to the different topologies of the basis need still to be completely studied. In the aim of finding a two-loop basis with coefficients still independent of the regularization parameter, the IBP arise a kind of line-one problem:

$$\frac{\partial}{\partial \ell_1^\mu} \frac{\ell_1^\mu}{\text{Denominator}} = \frac{4-2\epsilon}{\text{Denominator}} + \ldots \tag{20}$$

consequently ϵ seems to be intrinsic to these reductions bringing to a decomposition of the type

$$\text{Ampl} = \sum_{j\in\text{Basis}} c_j(\epsilon)\text{Int}_j \tag{21}$$

as for instance in the one-loop bubble by Passarino-Veltman reduction

$$B^{\mu\nu}(k^2) = \frac{2-\epsilon}{4(3-2\epsilon)} B_0(k^2) k^{\mu}_{[4]} k^{\nu}_{[4]} - \frac{1}{4(3-2\epsilon)} B_0(k^2) g^{\mu\nu}_{[4]}. \tag{22}$$

This is a problem in view of implementing the $4-2\epsilon$ dimensional generalized unitarity program, for which it is desired a basis with coefficients which are ϵ independent. The regularizer ϵ is indeed something deeply related to the integration on the real slice. By the methods of generalized unitarity we want indeed to perform integration on contours in the complex plane to isolate the appropriate residues, clearly we cannot have a basis which is ϵ dependent, because the ϵ dependence is obtained from integration on the real slice. The conjecture considered here [19] is that again at two-loop there is an augmented basis, which now will depend on μ_1^2, μ_2^2, $\mu_1 \cdot \mu_2$. The tree amplitudes to be glued in the integrand reconstruction indeed come from the six dimensional extension of the theory [20]. By FDF we can reconstruct the total dependence of the two-loop integrands in terms of the extra dimensions using an appropriate modifications of FDF in terms of six dimensional ingredients [21]. However the recursive construction of the Dirac gamma matrices in higher dimensions is based on an operation of tensor product having the structure with four dimensional ingredients as seed of generation of the Dirac representation of the Lorentz group in higher dimensions [22]. In this sense FDF seems to be a basic block so that modulo multiplication with appropriate sigma Pauli matrices, the FDF results for the tree amplitudes computations [9] in the generalized unitarity procedure can be exported to two loops.

The way proposed here to reach such an augmented basis is thorough four-dimensional IBP

$$0 = \int d^d\ell_1 d^d\ell_2 \frac{\partial}{\partial \ell^{\mu}_{[4]i}} \frac{v^{\mu}_{[4]}}{\text{Denominator}} \tag{23}$$

The involved IBP generating vectors $v^{\mu}_{[4]}$ will be dependent on the μ_1^2, μ_2^2, $\mu_1 \cdot \mu_2$ and even for simple integrals like the double box they have not been computed yet, however it is expected that their algebraic geometry is very different form the ones coming from the IBP in d dimensions.

The last ingredient that we need is how to convert back to standard integrals. We have the additional components μ_i inside integrand we want to go back to recover ϵ. There are basically two techniques to be combined:

- take a Gram determinants $[G(\{p_i\},\{q_i\})] = \det_{i,j}(p_i \cdot q_j)$, do Feynman parametrization to get relations like this

$$P^{*,*}_{2,2}[(\mu_1^2)^2] = -\frac{\epsilon(1-\epsilon)}{(3-2\epsilon)(1-2\epsilon)G^2_{124}} P^{*,*}_{2,2}[G^2(\ell_1,1,2,4)] \tag{24}$$

  where together with a rational function of ϵ there is the dependence on G^2_{124}, an invariant of the external momenta;

- perform standard (d-dimensional) IBPs with μ_i factors and irreducible products inserted in the numerators

$$0 = \int d^d\ell_1 d^d\ell_2 \frac{\partial}{\partial \ell^{\mu}_{i[d]}} \frac{\{1,\mu_1^2,.....\}(\text{irreducible})^j v^{\mu}_{[4]}}{\text{Denominator}}. \tag{25}$$

5 Conclusions

Alternative dimensional regularization schemes are available for higher order computations in perturbative gauge theories. However there is not a wide use of them. It is needed to provide more practical examples to show the efficiency of those schemes.The four dimensional formulation (FDF) of the four dimensional helicity scheme (FDH) is a proposal to get the cut constructible and the rational part of one-loop amplitudes by just four dimensional cuts. FDF is efficient to find contributions of evanescent operators in perturbative computations. The foundations for d-dimensional unitarity within FDF at two loops have been discussed.

References

[1] G. Leibbrandt, Rev.Mod.Phys. **47**, 849 (1975) .

[2] D. Bardin, G.Passarino, *The Standard Model in the making* (Claredon Press, New York, 1999) pp. 166-169.

[3] G. 't Hooft, M.J.G. Veltman, Nucl.Phys. B**50**, 318 (1972).

[4] Z. Bern, L.J. Dixon, D.A. Kosower, Annals Phys. **322**, 1587 (2007).

[5] I suggest for a review: T.R. Taylor, Phys.Rept. **691**, 1 (2017).

[6] M.E. Peskin, arXiv:1101.2414 [hep-ph].

[7] C. Gnendinger et al., Eur. Phys. J. C**77**, 471 (2017).

[8] J.C. Collins, *Renormalization* (Cambridge University Press, Cambridge, 1984) pp. 62-87.

[9] A.R. Fazio, P. Mastrolia, E. Mirabella, W.J. Torres Bobadilla, Eur. Phys. J. C**74**, 3197 (2014).

[10] C. Gnendiger, A. Signer, Phys.Rev. D**97**, 096006 (2018).

[11] G. Passarino, M. Veltman, Nucl. Phys. B**160** 151 (1979).

[12] D. A. Kosower, K. J. Larsen, Phys.Rev. D**85**, 045017 (2012).

[13] R. Britto, F. Cachazo, B. Feng, Nucl.Phys. B**725** 275 (2005).

[14] Eden, Landshoff, Olive, Polkinghorne *The analytic S-matrix* (Cambridge at University press, Cambrdge, 1966) chap. 2.

[15] Z. Bern, A.G. Morgan, Nucl.Phys. B**467**, 479 (1996).

[16] S.D. Badger, JHEP **0901**, 049 (2009).

[17] J.Gluza, K. Kajida, D. A. Kosower , Phys.Rev. D**83**, 045012 (2011).

[18] For a review: V.A. Smirnov, *Analytic tools for Feynman integrals* (Springer, London, 2012) pp. 127-154.

[19] D.A. Kosower, *Maximal Unitarity and Algebraic Geometry*, talk given at IAS focused program in Scattering Amplitudes in Hong Kong, November 2014.

[20] S. Badger, C. Bronnum-Hansen, H. Bayu Hartanto, T. Peraro, Phys.Rev.Lett. **120** no.9, 092001 (2018).

[21] A.R. Fazio, P. Mastrolia, W.J. Torres, *to appear*.

[22] P. van Nieuwenhuizen, *Six Lectures at the Trieste 1981 Summer School on Supergravity*, in *Supergravity 81* (Cambridge University Press, Cambridge, 1982) p. 151.

Recent results from COMPASS on the GPD program

Andrzej Sandacz[1,*,**], on behalf of the COMPASS Collaboration

[1]National Institute for Nuclear Research, Division of High Energy Physics, PL 00-681 Warsaw, Poland

Abstract. The dedicated COMPASS GPD program started in 2012 with a short pilot 'DVCS run' and the first results from this run on DVCS, exclusive π^0 and exclusive ω channels have been obtained. From the sum of cross sections measured with positive and negative beam polarities the DVCS cross-section and its dependence on the squared four-momentum transfer t have been extracted. It allowed a first estimate of the transverse size of the partonic distribution of the nucleon $\sqrt{\langle r_\perp^2 \rangle}$ in a range of intermediate x_{Bj} with a mean x_{Bj} value of 0.056. We report also on the absolute value of the cross section for exclusive π^0 production and its dependence on t and azimuthal angle ϕ between the scattering plane and π^0 production plane. These measurements are aiming to constrain the GPDs, in particular chiral-odd (transversity) GPDs. Further, we present results on the spin density matrix elements for exclusive ω meson production and discuss the contributions from unnaturaly parity exchanges and from transitions that do not conserve helicity.

1 Introduction

The theoretical framework of General Parton Distributions (GPDs) [1–3] provides a novel description of the nucleon's partonic structure and contains a wealth of new information. The GPDs provide a description of the nucleon as an extended object, sometimes referred to as a 3-dimensional 'nucleon tomography' [4], which correlates (transverse) spatial and (longitudinal) momentum degrees of freedom. The GPDs also allow access to such a fundamental property of the nucleon as the orbital angular momentum of quarks [2]. The mapping of the nucleon GPDs requires a comprehensive program of measuring various hard processes in a broad kinematic range, in particular Deeply Virtual Compton Scattering (DVCS). In addition, Hard Exclusive Meson Production (HEMP) is expected to add an independent and complementary information. A comprehensive experimental overview of DVCS is presented in Ref. [5] and results of a recent phenomenological studies and predictions for DVCS can be found in Ref. [6].

The GPD part of COMPASS-II proposal [7] is devoted to measurements of both DVCS and HEMP with polarised μ^+ and μ^- beams and a liquid hydrogen target. The new detectors, the recoil proton detector CAMERA and the large-angle electromagnetic calorimeter ECAL0, which are essential for measurements of exclusive processes, were constructed and incorporated into the COMPASS set-up in 2012. The commissioning of these new detectors was done in 2012 and was followed by a 4-week pilot run. The results on DVCS, exclusive π^0

*Work supported by the Polish NCN Grant 2017/26/M/ST2/00498.
**e-mail: andrzej.sandacz@ncbj.gov.pl

© The Authors, published by EDP Sciences. This is an open access article distributed under the terms of the Creative Commons Attribution License 4.0 (http://creativecommons.org/licenses/by/4.0/).

production and exclusive ω production, which are presented in this contribution, have been obtained from this pilot run.

2 Experimental set-up

COMPASS is a fixed-target experiment situated at the high-intensity M2 beam-line of the CERN SPS. It can deliver either hadron or naturally polarised muon beams of a given charge in the energy range between 50 and 280 GeV. For the GPD program the data are collected with the muon beam nominal energy of 160 GeV and with muons of both polarities: μ^+ (μ^-) with polarisation of about -0.8 (+0.8).

The COMPASS apparatus [8] consists of a two-stage forward spectrometer comprising various tracking detectors, electromagnetic and hadronic calorimeters, muon identification detectors, and a ring imagining Cherenkov counter grouped around two dipole magnets SM1 and SM2. For the GPD program the set-up was complemented by installing the recoil proton detector CAMERA around the new 2.5 m long liquid hydrogen target and the large-angle electromagnetic calorimeter ECAL0. The 4-meter long CAMERA consists of two concentric barrels, each of 24 scintillator slats with a read out at both ends. The recoil proton detection is based on the ToF measurement between the two barrels. The calorimeter ECAL0 that is situated just downstream of CAMERA allows to extend the accessible kinematic domain for DVCS and exclusive π^0 production towards higher x_{Bj}, improve hermeticity for detection of exclusive events and reduce the background to single-photon events production, which originates from π^0 and other decays. By adding these detectors COMPASS has been converted into a facility measuring exclusive reactions within kinematic domain from $x \sim 0.01$ to ~ 0.15, which cannot be explored at any other existing or planned facility in the near future.

3 Results on DVCS

The event selection exploits the overconstrained kinematics for the measured exclusive process $\mu p \to \mu' p' \gamma$. Events with a single outgoing muon and a single neutral cluster above the DVCS thresholds for the electromagnetic calorimeters are selected first. The muon and photon are then combined with all possible proton candidates reconstructed in CAMERA. The spectrometer information on incoming and scattered muons as well as on the positions and energy for the neutral cluster is used as a prediction to compare with kinematics of all track candidates that are found using the ToF system of the target recoil detector. Exemplary results of this comparison are displayed in Fig. 1 using two variables that characterize the kinematics of the recoiling target particle. Figure 1(left) shows the difference between the measured and the predicted azimuthal angle, $\Delta\phi$, and Fig. 1(right) the difference between the measured and the predicted transverse momentum, Δp_T. Here, ϕ and p_T are given in the laboratory system. The combinatorial ambiguity is resolved by applying cuts on variables that are sensitive to the exclusivity.

The background is further reduced by applying a kinematic constrained fit. The input to the fit is supplied with all measured quantities, their uncertainties and correlations. The fit performs a chi-squared minimisation using constraints on the energy and momentum conservation. In the following the kinematic variables determined by the kinematic fit are used since the fit offers their most precise determination.

In this analysis we determine the differential cross section $\mathrm{d}^4\sigma^{\mu p \to \mu' p' \gamma}/\mathrm{d}Q^2\mathrm{d}\nu\mathrm{d}t\mathrm{d}\phi$ for exclusive muoproduction of single photon. Here, Q^2 and ν denote the virtual photon virtuality and its energy in the lab. system, t is the square of the four-momentum transfer between the target and recoil protons and ϕ is the azimuthal angle between the lepton scattering plane and

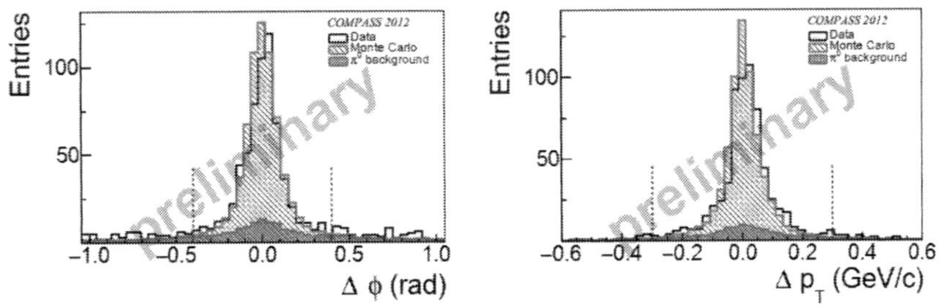

Figure 1: Distributions of the difference between predicted and reconstructed values of the recoiling proton candidates. The dashes blue vertical lines indicate the applied cuts. Here, Monte Carlo also includes π^0 background for the single photon sample.

the photon production plane. In order to extract the differential cross section as a function of Q^2, ν, t and ϕ the data has been corrected for acceptance effects using bin-wise acceptance correction factors.

This cross section was measured separately using either a μ^+ or μ^- beam of 160 GeV/c average momentum. The natural polarisation of the muon beam originates from the parity-violating deacy-in-flight of the parent meson, which implies opposite signs of the polarisation for the used μ^+ and μ^- beams. Denoting charge and helicity of the incident muon by + (−) and ← (→), the sum of the cross sections for μ^+ and μ^- beams reads:

$$2\mathrm{d}\sigma \equiv \mathrm{d}\sigma^{\overset{+}{\leftarrow}} + \mathrm{d}\sigma^{\overset{-}{\rightarrow}} = 2(\mathrm{d}\sigma^{BH} + \mathrm{d}\sigma^{DVCS}_{unpol} - |P_\mu|\mathrm{Im}I), \tag{1}$$

where P_μ denotes polarisation of the muon beam. The single-photon final state in lepton-nucleon scattering can also originate from the BH process (BH). Hence the DVCS and BH interfere, so that the above sum comprises also the contribution from the interference term denoted by ImI.

After subtracting the exactly calculable BH cross section from Eq. (1) and integrating the remainder over ϕ there remains only the dominant contribution from transversely polarised virtual photons to the DVCS cross section $\mathrm{d}^3\sigma^{\mu p\to\mu' p'\gamma}_{DVCS}/\mathrm{d}Q^2\mathrm{d}\nu\mathrm{d}t$.

A major background source for single photon production is the production of π^0 mesons, where one of the decay photons is undetected. The contamination arising from π^0 production was estimated and subtracted from the data inside each bin using LEPTO 6.5.1 together with an exclusive π^0 Monte Carlo based on a parameterisation from Goloskokov-Kroll model [9].

The goal of this measurement is to determine the t-dependence of the unpolarised (ϕ-integrated) DVCS cross section in virtual photon-proton scattering. The $\gamma^* p$ cross section is derived from the measured muon-proton cross section by taking into account the transverse virtual photon flux Γ

$$\frac{\mathrm{d}\sigma^{\gamma^* p}}{\mathrm{d}t} = \frac{1}{\Gamma(Q^2,\nu,E_\mu)}\frac{\mathrm{d}^3\sigma^{\mu p}_{DVCS}}{\mathrm{d}Q^2\mathrm{d}\nu\mathrm{d}t}\ , \tag{2}$$

where according to the Hand's convention

$$\Gamma(Q^2,\nu,E_\nu) = \frac{\alpha_{em}(1-x_{Bj})}{2\pi Q^2 y E_\mu}\left[y^2(1-\frac{2m_\mu^2}{Q^2}) + \frac{2}{1+Q^2/\nu^2}(1-y-\frac{Q^2}{4E_\mu^2})\right]. \tag{3}$$

The values of the differential cross section $\mathrm{d}\sigma^{\gamma^* p\to\gamma p}_{DVCS}/\mathrm{d}|t|$ for four bins of $|t|$ are shown in

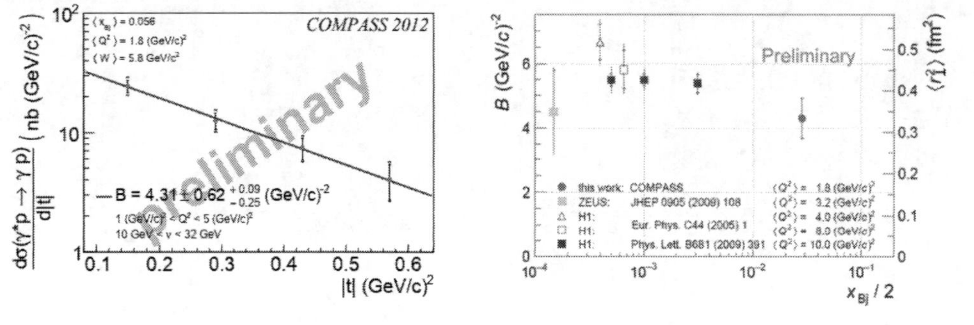

Figure 2: **Left:** The differential DVCS cross section as a function of $|t|$. **Right:** COMPASS results for the t-slope parameter compared to previous results from HERA.

Fig. 2 (left). They have been fitted with an exponential function in order to determine the t-slope parameter B. The value of B is equal to $(4.3 \pm 0.6\ ^{+0.1}_{-0.3})(\mathrm{GeV}/c)^{-2}$ inside the kinematic range that is indicated in the legend of Fig. 2 (left), and which corresponds to $\langle x_{Bj} \rangle = 0.056$, $\langle Q^2 \rangle = 1.8\ (\mathrm{GeV}/c)^2$ and $\langle W \rangle = 5.8$ GeV. The extracted value of B allows us to estimate the transverse size of partonic distibution in the nucleon $r_\perp$ using the relation

$$\langle r_\perp^2(x_{Bj}) \rangle = 2B(x_{Bj}), \tag{4}$$

which for our measurement results in $\sqrt{\langle r_\perp^2 \rangle} = (0.58 \pm 0.04 \pm^{+0.01}_{-0.02})$ fm at a mean x_{Bj} value of 0.056.

The main systematic error of the cross section is due to uncertainties of absolute normalisation and determination of the amount of π^0 background. In case of parameter B the error from the absolute normalisation plays a minor role, while the main systematic uncertainty originates from the normalisation of the π^0 background.

In Fig. 2 (right) the values of parameter B from COMPASS and HERA [10, 11] are plotted as a function of x_{Bj}. As the slope B is related to the transverse size of the partonic distribution (see Eq. (4)), one may conclude that these model-independent results indicate a decrease of the transverse size of the proton with increasing x_{Bj}.

4 Results on exclusive π^0 production

For the selection of the sample of exclusive π^0 production, $\mu p \to \mu' p' \pi^0$, an analogous procedure is used as described for the DVCS sample. In addition to the cuts on exclusivity variables, a cut on the invariant mass of two-photon system is applied. In order to determine background originating from non-exclusive events, the data are modeled with Monte Carlo in both kinematic regions where either the signal or the background dominates. Here, LEPTO 6.5.1 is used for a non-exclusive part. For the signal contribution a dedicated Monte Carlo generator [12] was used with the cross section parameterisation from the Goloskokov-Kroll model. The background is subtracted bin-wise from the data. Other sources of background, like misidentified exclusive $\omega \to \pi^0\gamma \to \gamma\gamma\gamma$ events, where one photon is lost, are found to be negligible.

First, we extract the four-fold differential μp cross sections separately for μ^+ and μ^- data. Then the unpolarised cross section is obtained as the weighted mean of the two. In this way the dependence on the beam polarisation cancels. In order to convert μp cross section into

https://doi.org/10.1051/epjconf/201819200012

$\gamma^* p$ cross section we divide the former one by the tranasverse virtual photon flux (cf. Eq. (3)).

The systematic uncertainty in this measurement is mostly due to scaling, which can be decomposed into effects depending on the absolute normalisation and background estimate, as well as to the threshold effects and kinematic uncertainties.

The unpolarised $\gamma^* p$ cross section for exclusive π^0 production reads

$$\frac{\mathrm{d}^2\sigma^{\gamma^* p}}{\mathrm{d}t\mathrm{d}\phi} = \frac{1}{2\pi}\left[\frac{\mathrm{d}\sigma_T}{\mathrm{d}t} + \epsilon\,\frac{\mathrm{d}\sigma_L}{\mathrm{d}t} + \epsilon\,\cos(2\phi)\frac{\mathrm{d}\sigma_{TT}}{\mathrm{d}t} + \sqrt{2\epsilon(1+\epsilon)}\,\cos(\phi)\frac{\mathrm{d}\sigma_{LT}}{\mathrm{d}t}\right], \qquad (5)$$

where $\sigma_T, \sigma_L, \sigma_{TT}, \sigma_{LT}$ are the structure functions, ϵ is the virtual photon polarisation parameter and ϕ represents the azimuthal angle between lepton scattering plane and π^0 production plane. Here, the subscript $T(L)$ denotes the contribution from transversely (longitudinally) polarised virtual photons, while the subscripts TT and LT denote the contributions from the interference between transversely-transversely and longitudinally-transversely polarised photons.

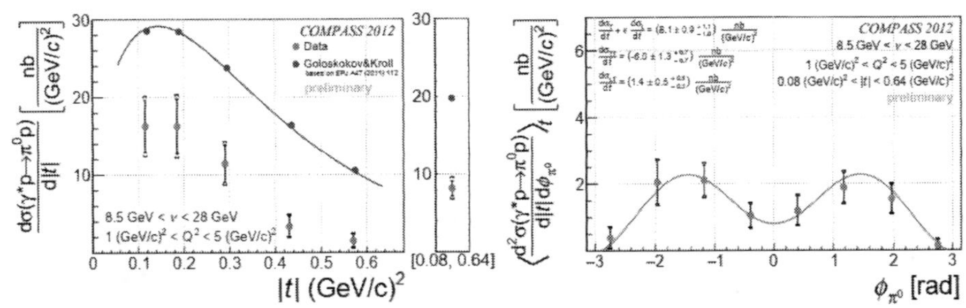

Figure 3: **Left:** The exclusive π^0 cross section as a function of $|t|$. COMPASS results are compared to predictions of the GPD model of Ref. [13]. **Right:** The exclusive π^0 cross section as a function of ϕ. Solid line indicates the result of the fit with the values of the fitted parameters shown in the legend.

After integration over ϕ, we extract the cross section in bins of $|t|$, which is displayed in Fig. 3 (left). It decreases exponentially with increasing $|t|$ at values of $|t|$ larger than about 0.25 $(\mathrm{GeV}/c)^2$, while at smaller $|t|$ the t-dependence becomes weaker. Being the first measurement at energies significantly higher than previous results, it is worth to compare it to the model prediction. The COMPASS result is compared to the prediction of the Goloskokov-Kroll model [13]. The blue curve represents the cross section from the model as a function of t, while the blue points correspond to the cross section averaged over $|t|$ bins of the data. The mean cross sections for the full t-range are compared at the rightmost part of the plot. We observe that the magnitude of the predicted cross section overshoots our measurement by approximately a factor of two.

Furthermore, we extract the cross section as a function of ϕ in the range of measured t and eight bins of ϕ of equal width. The red dots in Fig. 3 (right) show the measured cross section for each bin and a fit of the function given by Eq. (5), which is represented by the red curve. The values of the fitted parameters are indicated in the legend of the plot. We observe a large contribution from σ_{TT} and a smaller positive one from σ_{LT}. This indicates a significant role of transversely polarised photons in exclusive π^0 production.

According to Refs. [13, 14] the structure functions for exclusive pseudo-scalar meson production, which appear in Eq. (5), depend on GPDs $\tilde{H}^q$, $\tilde{E}^q$, H_T^q and $\bar{E}_T^q = 2\tilde{H}_T^q + E_T^q$. For π^0 production a large contribution from transversely polarised virtual photons is expected, which is mainly generated by the chiral-odd ('transversity') GPDs $\bar{E}_T^q$. It manifests itself in a large contribution from σ_{TT} and a dip in the differential cross section $d\sigma/dt$ as $|t|$ decreases to zero. These features are in a qualitative agreement with our results, although the predicted values of cross section overshoot the measured ones by a factor of about 2.

5 Results on exclusive ω production

The considered process is $\mu p \to \mu' p' \omega$ with subsequent decays $\omega \to \pi^+\pi^-\pi^0$ and $\pi^0 \to \gamma\gamma$. Therefore, the selected event should contain a track of scattered muon, two tracks of hadrons with opposite charges and two neutral clusters in an electromagnetic calorimeter. Because in this analysus the information from CAMERA is not used, in order to suppress non-exclusive background, the cut $|E_{\text{miss}}| < 3.0$ GeV was applied, with $E_{\text{miss}} = (M_X^2 - M_p^2)/(2M_p)$. Here M_p is the proton mass and $M_X^2 = (p + q - p_{\pi^+} - p_{\pi^-} - p_{\pi^0})^2$ is the missing mass squared, and p, q, p_{π^+}, p_{π^-} and p_{π^0} are the four-momenta of target nucleon, virtual photon and each of the three pions.

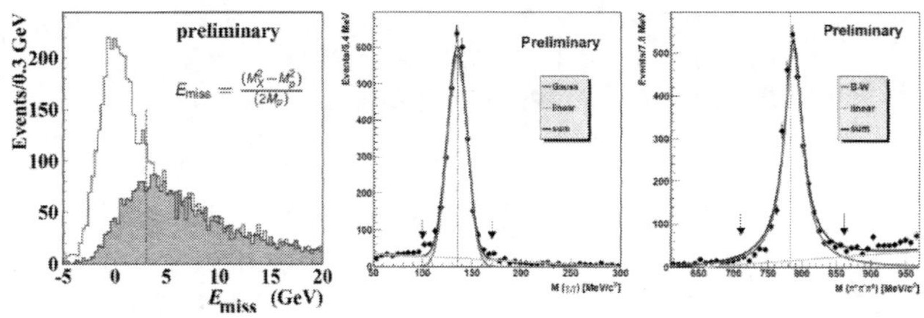

Figure 4: **Left:** E_{miss} distribution. Shaded histogram corresponds to non-exclusive background estimated using LEPTO Monte Carlo. **Middle:** The $\gamma\gamma$ invariant mass distribution fitted by a sum of Gaussian function and a linear dependence for background. **Right:** The $\pi^+\pi^-\pi^0$ invariant mass fitted by a sum of Breit-Wigner function and a linear dependence for background. Dashed vertical lines denote the PDG values.

Selected kinematic distributions for the sample are shown in Fig. 4. These are E_{miss} distribution, the invariant mass of two photons and the invariant mass of $\pi^+\pi^-\pi^0$ system. Clear π^0 and ω signals are observed. In the E_{miss} distribution the data are shown as open histogram, while the shaded one corresponds to the background evaluated using LEPTO Monte Carlo. The amount of background in the signal window $|E_{\text{miss}}| < 3.0$ GeV depends on Q^2, W and p_T^2, and when averaged over the whole sample it amounts to 28%.

The SDMEs describe the spin structure of the reaction. They are related to the bilinear combinations of helicity amplitudes that describe transitions between specified spin states of the virtual photon, target proton, produced vector meson and recoil proton. For an unpolarised nucleon target, after summing up over initial and final spin states of the proton, the SDMEs depend just on helicities of the virtual photon and produced meson. The values of SDMEs allow us to test the *s*-channel helicity conservation (SCHC) hypothesis, as well as to investigate the role of transversity (parton helicity-flip) GPDs and of unnatural parity exchange (UPE) mechanisms in exclusive ω production.

The SDMEs appear as parameters in the three dimensional angular distributions of the vector meson production and decay. Definitions of angles and reference frames are given *e.g.* in Ref. [15]. The SDMEs were determined by using the Unbinned Maximum Likelihood method to fit the experimental three-dimensional angular distribution of ω production and decay. The fitted distribution was a weighted superposition of the angular distributions for exclusive events and non-exclusive background. The fraction of background in the signal window was determined using E_{miss} distribution as illustrated in Fig. 4.

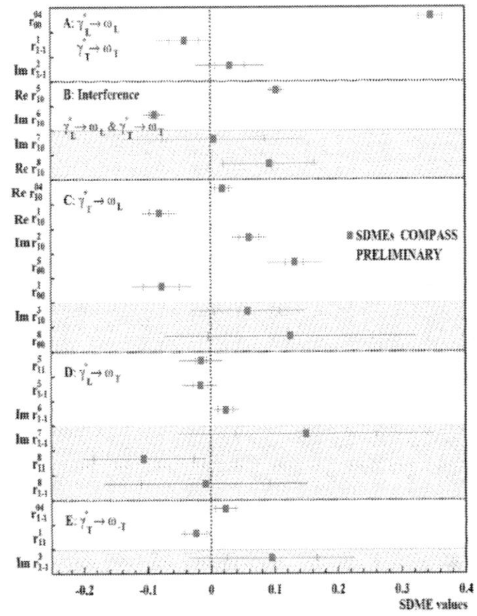

Figure 5: The 23 SDMEs for exclusive ω leptoproduction extracted in the entire COMPASS kinematic region. The inner error bars represent the statistical uncertainties, while the outer ones indicate the statistical and systematic uncertainties added in quadrature.

SDMEs for exclusive ω production extracted for the entire kinematic region are presented in Fig. 5. The average values of kinematic variables are $\langle Q^2 \rangle = 2.13\ (\mathrm{GeV}/c)^2$, $\langle W \rangle =$ 7.6 GeV and $\langle p_T^2 \rangle = 0.16\ (\mathrm{GeV}/c)^2$, where p_T denotes the transverse momentum of ω meson relative to the direction of the virtual photon. The SDMEs are grouped in classes, each class corresponding to dominant transitions amplitudes with given helicities of the virtual photon and ω meson, as indicated in the figure.

In the case of SCHC, all SDMEs of classes C, D and E should be equal to zero. However, for the transitions $\gamma_T^* \to V_L$ (class C) one observes a clear violation of SCHC. In the GK model [16] the SDMEs from this class are related to the contributions from transversity GPDs $\bar{E}_T$ and H_T coupled with the higher-twist wave function of the meson. For instance the results for SDME r_{00}^5 are expected to be mostly affected by GPDs $\bar{E}_T$. Kinematic dependences of these SDMEs on Q^2, W, and p_T^2, which were also extracted in this analysis, may help to further constrain the model.

The existence of processes with unnatural parity exchange (UPE) in ω production on the proton can be tested by examining linear combination of SDMEs such as

$$u_1 = 1 - r_{00}^{04} + 2r_{1-1}^{04} - 2r_{11}^{1} - 2r_{1-1}^{1}. \tag{6}$$

The quantity u_1 can be expressed in terms of helicity amplitudes and is proportional to the relative contribution of UPE processes to the cross section. Therefore, a different from zero value of u_1 would indicate a conribution of UPE processes. The kinematic dependence of u_1 on Q^2, W and p_T^2 is shown in Fig.6. In the GK model the dominant UPE process is the pion pole exchange [17], which implies a decrease of u_1 with W. The data support the expected decrease.

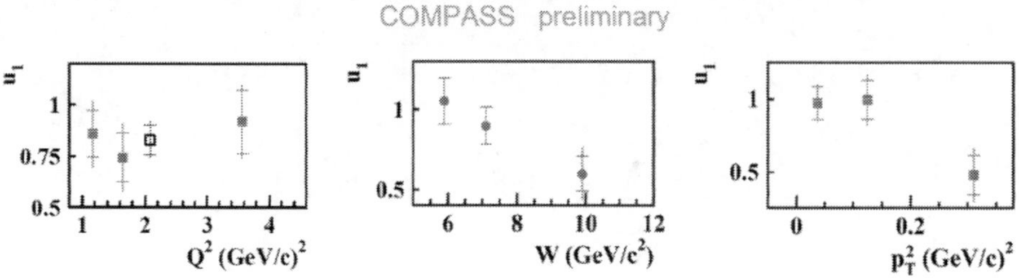

Figure 6: Dependence of quantity u_1 on kinematic variables Q^2, W and p_T^2. The empty square respresents the $u1$ value for the entire kinematic region.

6 Outlook

In 2016 and 2017 the COMPASS Collaboration has collected a large data sample on the GPD program. An expected increase of the statistic is by a factor of about 10 compared to the 2012 data. An investigation of GPDs with DVCS and HEMP on unpolarised protons using the data from 2016-2017 will allow us to determine the x_{Bj}-dependence of t-slope of the differential cross sections. That is related to the spatial transverse distribution of partons and the 'nucleon tomography'. Measurements of the beam charge and spin sum and difference of single-γ cross sections will give access to the real and imaginary parts of the DVCS amplitude, and will allow to further constrain GPDs H. Studies of exclusive production of vector mesons (ρ^0, ω, ϕ) will lead to the quark flavour and gluon separation for GPDs H, while that of exclusive π^0 production will provide constraints on chiral-odd GPDs. The projected accuracies for measurements of various observables within the GPD program can be found in the COMPASS-II proposal [7].

References

[1] D. Mueller *et al*, Fortsch. Phys. **42** (1994) 101.

[2] X. Ji, Phys. Rev. Lett. **78** (1997) 610; Phys. Rev. D **55** (1997) 7114.

[3] A.V. Radyushkin, Phys. Lett. B **385** (1996) 333; Phys. Rev. D **56** (1997) 5524.

[4] M. Burkardt, Phys. Rev. D **62** (2000) 071503; erratum-ibid. D **66** (2002) 119903; Int. J. Mod. Phys. A **18** (2003) 173; Phys. Lett. B **595** (2004) 245.

[5] N. d'Hose, S. Niccolai and A. Rostomyan, Eur. Phys. J. A **52** (2016) 157.

[6] K, Kumerički, S. Liuti and H. Moutarde, Eur. Phys. J. A **52** (2016) 151.

[7] F. Gautheron *et al.* [COMPASS Collaboration], CERN-SPSC-2010-014, SPSC-P-340, May 2010.

[8] COMPASS Collaboration, P. Abbon *et al.*, Nucl. Instr. Meth. A **577** (2007) 455.

[9] S.V. Goloskokov and P. Kroll, Eur. Phys. J. C **65** (2010) 137.

[10] ZEUS Collaboration, S. Chekanov *et al.*, JHEP **0905** (2009) 108.

[11] H1 Collaboration, F.D. Aaron *et al.* Phys. Lett. B **681** (2009) 391; A. Aktas *et al.* Eur. Phys. J. C **44** (2005) 1.

[12] A. Sandacz and P. Sznajder, arXiv:1207.0333 [hep-ph].

[13] S.V. Goloskokov and P. Kroll, Eur. Phys. J. A **47** (2011) 112.

[14] CLAS Collaboration, I. Bedlinsky *et al.*, Phys. Rev. C **90** (2014) 025205.

[15] K. Schilling and G. Wolf, Nucl. Phys. B **61**, 381 (1973).

[16] S.V. Goloskokov, P. Kroll, Eur. Phys. J. C **74** 2725 (2014).

[17] S.V. Goloskokov, P. Kroll, Eur. Phys. J. A **50** 146 (2014).

OPE in Wilson lines with sub-eikonal spin corrections for TMDs and g_1 structure function

Giovanni Antonio Chirilli[1,*]

[1] *Institut für Theoretische Physik, University of Regensburg, D-93040 Regensburg, Germany*

Abstract. Low-x evolution of spin-dependent TMDs and spin g_1 structure function are relevant for the future Electron Ion Collider. To study spin-dynamics at high energies it is necessary to extend the eikonal approximation to include sub-eikonal corrections. I will discuss the Operator Product Expansion (OPE) in terms of Wilson lines with sub-eikonal corrections and derive the sub-eikonal quark propagator.

1 Introduction

Most of the progress in high-energy QCD has been obtained within the eikonal approximation. The high precision reached by the experiments and the possibility to study spin-dynamics in Deep Inelastic (DIS) scattering at the proposed Electron Ion Collider [1], makes the study of sub-eikonal corrections especially timely.

In order to access the sub-eikonal corrections in high-energy QCD one has to study the sub-eikonal corrections of the propagators of the theory extending, in this way, the Operator Product Expansion to include the sub-eikonal corrections. The result will be relevant for the study of TMDs and g_1 structure function.

In the eikonal approximation, the operator which describes the DIS scattering amplitude is a trace of two Wilson lines. Its evolution equation with respect to the rapidity of the fields generates the Balitsky-hierarchy [3] which is equivalent to the Jalilian-Marian–Iancu–McLerran–Weigert–Leonidov–Kovner (JIMWLK) [6–8] evolution equation thus, they are referred to as B–JIMWLK equation. The Balitsky-Kovchegov (BK) equation [4, 5] is the first of the Balitsky hierarchy and it is obtained in the mean field approximation. Its linearization coincides with the BFKL [9, 10] equation. As we will see, the sub-eikonal corrections will introduce new operators which we will result in new evolution equations thus providing further information on the QCD dynamics at high energy.

We will start with the derivation of the scalar propagator in the background of a highly boosted gluonic field generated by the target in order to lay down the formalism and the notation used. We then present the result for the quark propagator and make, as an example, a simple application of the OPE with the new derived operators.

*e-mail: giovanni.chirilli@ur.de

© The Authors, published by EDP Sciences. This is an open access article distributed under the terms of the Creative Commons Attribution License 4.0 (http://creativecommons.org/licenses/by/4.0/).

2 Scalar propagator

2.1 Eikonal approximation

To reach the high-energy kinematics, we will perform a longitudinal boost of the fields generated by the target-particle in which the projectile particle propagates.

We assume that the projectile-particle is propagating along p_1 direction, while the target particle is moving along p_2 direction, where p_1^μ and p_2^μ are two light-cone vectors such that $p_{1\mu}p_2^\mu = p_1 \cdot p_2 = \frac{s}{2}$. We can perform a Sudakov decomposition of the momentum any momentum p^μ as $p^\mu = \alpha p_1^\mu + \beta p_2^\mu + p_\perp^\mu$. The light-cone components are defined as $x_\bullet = x_\mu p_1^\mu = \sqrt{\frac{s}{2}} x^-$ and $x_* = x_\mu p_2^\mu = \sqrt{\frac{s}{2}} x^+$ with $x^\pm = \frac{x^0 \pm x^3}{\sqrt{2}}$.

The gauge field A^μ generated by the target, under a boost, gets rescaled by a large parameter λ as follows

$$\begin{aligned} A_\bullet(x_\bullet, x_*, x_\perp) &\rightarrow \lambda A_\bullet(\lambda^{-1}x_\bullet, \lambda x_*, x_\perp), \\ A_*(x_\bullet, x_*, x_\perp) &\rightarrow \lambda^{-1}A_*(\lambda^{-1}x_\bullet, \lambda x_*, x_\perp), \\ A_\perp(x_\bullet, x_*, x_\perp) &\rightarrow A_\perp(\lambda^{-1}x_\bullet, \lambda x_*, x_\perp). \end{aligned} \tag{1}$$

In Schwinger representation the free scalar propagator is

$$\langle x|\frac{i}{p^2 + i\epsilon}|y\rangle = i\int d\!\!{}^-{}^4k\, \frac{e^{-ik\cdot(x-y)}}{p^2 + i\epsilon}, \tag{2}$$

with $\langle k|x\rangle = e^{ix\cdot k}$. In (2) we used the $\hbar$-inspired notation $d\!\!{}^-{}^4k \equiv \frac{d^4k}{(2\pi)^4}$ and $\delta^{(4)}(x-y) = (2\pi)^4\delta^{(4)}(x-y)$ so that, $\int d\!\!{}^-{}^4k\, \delta^{(4)}(x-y) = 1$.

In the eikonal approximation, the only component surviving the boost is $A_\bullet$ so, $\hat{P}^2 \simeq \hat{p}^2 + 2\alpha g\hat{A}_\bullet$ and we can represent the propagator as a series

$$\langle x|\frac{i}{\hat{P}^2 + i\epsilon}|y\rangle \simeq \langle x|\frac{i}{\hat{p}^2 + i\epsilon}|y\rangle + g\int d^4z\langle x|\frac{i}{\hat{p}^2 + i\epsilon}|z\rangle 2\,i\,\alpha A_\bullet(z_*, z_\perp)\langle z|\frac{i}{\hat{p}^2 + i\epsilon}|y\rangle + \dots . \tag{3}$$

In the shock-wave picture (see Fig. 1) relevant for high-energy scattering, we assume that the particle starts and ends its propagation outside the interval in which the field strength tensor is different then zero (see Fig. 1). With this assumption we can rewrite expansion (3) as

$$\begin{aligned} \langle x|\frac{i}{\hat{P}^2 + i\epsilon}|y\rangle =& \left[\int_0^{+\infty}\frac{d\!\!{}^-\alpha}{2\alpha}\theta(x_* - y_*) - \int_{-\infty}^0\frac{d\!\!{}^-\alpha}{2\alpha}\theta(y_* - x_*)\right]e^{-i\alpha(x_\bullet - y_\bullet)}\int d^2z d^2z'\langle x_\perp|e^{-i\frac{\hat{p}_\perp^2}{\alpha s}x_*}|z_\perp\rangle \\ &\times\langle z_\perp|\mathrm{Pexp}\left\{ig\int_{y_*}^{x_*}d\frac{2}{s}\omega_*\, e^{i\frac{\hat{p}_\perp^2}{\alpha s}\omega_*}A_\bullet(\omega_*)e^{-i\frac{\hat{p}_\perp^2}{\alpha s}\omega_*}\right\}|z'_\perp\rangle\langle z'_\perp|e^{i\frac{\hat{p}_\perp^2}{\alpha s}y_*}|y_\perp\rangle . \end{aligned} \tag{4}$$

Eq. (4) has been obtained assuming the commutation relation $[\hat{\alpha}, \hat{A}] = 0$ and the fact that the most dominant component of the gauge external field is $A_\bullet$.

Under the longitudinal Lorentz boost, the longitudinal distance traveled by the particle in the external filed is rescaled as $\omega_* \rightarrow \frac{1}{\lambda}\omega_*$ while the gauge field is rescaled as $A_\bullet \rightarrow \lambda A_\bullet$ with $\lambda \gg 1$ the boost parameter. So, we can write

$$e^{i\frac{\hat{p}_\perp^2}{\alpha s}\omega_*}A_\bullet(\omega_*)e^{-i\frac{\hat{p}_\perp^2}{\alpha s}\omega_*} = A_\bullet(\omega_*) + O(\lambda^0). \tag{5}$$

Making use of (5) in propagator (4), we obtain

$$\langle x|\frac{i}{\hat{P}^2+i\epsilon}|y\rangle = \left[\int_0^{+\infty}\frac{đ\alpha}{2\alpha}\theta(x_*-y_*) - \int_{-\infty}^{0}\frac{đ\alpha}{2\alpha}\theta(y_*-x_*)\right]e^{-i\alpha(x_\bullet-y_\bullet)}$$
$$\times\int d^2z\langle x_\perp|\,e^{-i\frac{\hat{p}_\perp^2}{\alpha s}x_*}|z_\perp\rangle[x_*,y_*]_z\langle z_\perp|e^{i\frac{\hat{p}_\perp^2}{\alpha s}y_*}|y_\perp\rangle\,, \tag{6}$$

where we have defined the gauge link at fixed transverse position $z_\perp$ as

$$[x_*,y_*]_z = \mathrm{Pexp}\left\{ig\frac{2}{s}\int_{y_*}^{x_*}d\omega_*\,A_\bullet(\omega_*,z_\perp)\right\}\,. \tag{7}$$

Propagator (6) describes the propagation of the particle in the external field along a straight line. To consider deviation from the straight-line propagation we need to take into account higher order terms in Eq. (5).

Now we can trade the finite gauge link with the infinite Wilson line because, under the infinite boost, the dominant component of the field strength tensor, $F_{\bullet i}$, has an infinitesimal thin support in x_* coordinate and we assumed that $F_{\bullet i}$ is peaked at the origin while outside the infinitesimal interval $[-\epsilon_*,\epsilon_*]$, $F_{\bullet i}=0$ (see Fig. 1). Therefore, in the gauge rotated field A^Ω the gauge field outside the external field is zero and we can trade the gauge link $[\epsilon_*,-\epsilon_*]$ with the infinite Wilson line $[\infty p_1,-\infty p_1]$ and obtain

$$\langle x|\frac{i}{P^2+i\epsilon}|y\rangle = \left[\int_0^{+\infty}\frac{đ\alpha}{2\alpha}\theta(x_*-y_*) - \int_{-\infty}^{0}\frac{đ\alpha}{2\alpha}\theta(y_*-x_*)\right]e^{-i\alpha(x_\bullet-y_\bullet)}$$
$$\times\int d^2z\langle x_\perp|\,e^{-i\frac{\hat{p}_\perp^2}{\alpha s}x_*}|z_\perp\rangle U_z\langle z_\perp|e^{i\frac{\hat{p}_\perp^2}{\alpha s}y_*}|y_\perp\rangle\,, \tag{8}$$

where we have defined the infinite Wilson line U_z at fixed transverse position $z_\perp$ as

$$U_z \equiv [p_1\infty,-p_1\infty]_z = \mathrm{Pexp}\left\{ig\frac{2}{s}\int_{-\infty}^{+\infty}dz_*A_\bullet(\frac{2}{s}p_1z_*+z_\perp)\right\}\,. \tag{9}$$

Notice that, we can restore $A_\perp^\mu$ component as a transverse gauge link (see Fig. 1) since it is a pure gauge.

2.2 sub-eikonal approximation

In this section we are interested in deriving the sub-eikonal corrections to the scalar propagator Eq. (6). To this end, we consider a background gauge field with all components different than zero, $A_\mu^{cl}(x_*,x_\perp)=(A_*^{cl}(x_*,x_\perp),A_\bullet^{cl}(x_*,x_\perp),A_\perp^{cl}(x_*,x_\perp))$, and we need to identify the most dominant component and sub-dominant one from the operator $\hat{P}^2=\{\hat{p}_\perp^\mu,\hat{A}_\mu^\perp\}+\{\frac{2}{s}\hat{P}_\bullet,\hat{A}_*\}-g\hat{A}_\perp^2$.

It can be shown that the term $\{\frac{2}{s}\hat{P}_\bullet,\hat{A}_*\}$ is related to the possibility for the fields to depend mildly on $x_\bullet$ but it contributes as sub-sub-correction, so we will neglect it and keep considering fields $x_\bullet$ independent.

It is convenient to define $\hat{O}=\{\hat{p}_\perp^\mu,\hat{A}_\mu^\perp\}-g\hat{A}_\perp^2$ so that $\hat{P}^2=\hat{p}^2+2g\alpha A_\bullet+g\hat{O}$ and the scalar propagator can be written as

$$\langle x|\frac{i}{\hat{P}^2+i\epsilon}|y\rangle = \langle x|\frac{i}{\hat{p}^2+2\alpha g\hat{A}_\bullet+g\hat{O}+i\epsilon}|y\rangle$$
$$= \left[\int_0^{+\infty}\frac{đ\alpha}{2\alpha}\theta(x_*-y_*) - \int_{-\infty}^{0}\frac{đ\alpha}{2\alpha}\theta(y_*-x_*)\right]e^{-i\alpha(x_\bullet-y_\bullet)}$$
$$\times\langle x_\perp|\,e^{-i\frac{\hat{p}_\perp^2}{\alpha s}x_*}\mathrm{Pexp}\left\{ig\int_{y_*}^{x_*}d\frac{2}{s}\omega_*\,e^{i\frac{\hat{p}_\perp^2}{\alpha s}\omega_*}\left(\hat{A}_\bullet(\omega_*)+\frac{\hat{O}(\omega_*)}{2\alpha}\right)e^{-i\frac{\hat{p}_\perp^2}{\alpha s}\omega_*}\right\}e^{i\frac{\hat{p}_\perp^2}{\alpha s}y_*}|y_\perp\rangle\,. \tag{10}$$

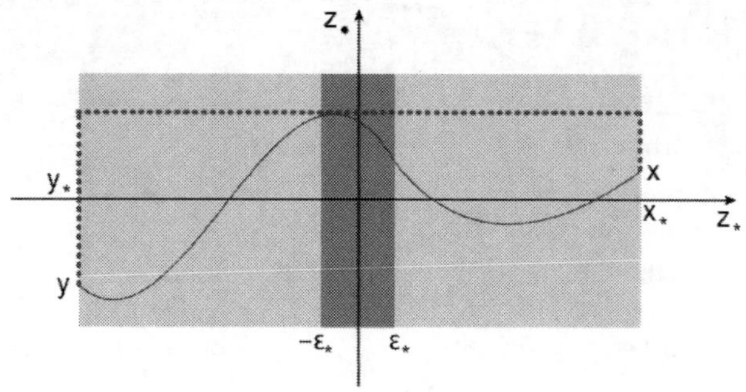

Figure 1. Particle propagating in an external gluonic field. The red strip represents the shock-wave defined in the infinitesimal interval $[-\epsilon_*, \epsilon_*]$ in which $F_{\mu\nu} \neq 0$. The light-red areas to the left and to the right of the shock-wave represent the background filed made of pure gage field. The curvy line is the particle's path from point x to point y. The dotted lines represent the Wilson lines.

We are now ready to consider the sub-dominant contributions we neglected in the previous section. We observe that

$$e^{i\frac{\hat{p}_\perp^2}{\alpha s}\omega_*}\left(\hat{A}_\bullet + \frac{\hat{O}}{2\alpha}\right)e^{-i\frac{\hat{p}_\perp^2}{\alpha s}\omega_*} = \hat{A}_\bullet + \frac{\hat{O}}{2\alpha} + i\frac{\omega_*}{\alpha s}[\hat{p}_\perp^2, \hat{A}_\bullet] + O(\lambda^{-1})\,. \tag{11}$$

Using Eq. (11), after some algebra we arrive at

$$\begin{aligned}\langle x|\frac{i}{\hat{P}^2 + i\epsilon}|y\rangle =& \left[\int_0^{+\infty}\frac{d\!\!{}^{-}\alpha}{2\alpha}\theta(x_* - y_*) - \int_{-\infty}^0\frac{d\!\!{}^{-}\alpha}{2\alpha}\theta(y_* - x_*)\right]e^{-i\alpha(x_\bullet - y_\bullet)}\\ &\times\langle x_\perp|\,e^{-i\frac{\hat{p}_\perp^2}{\alpha s}x_*}\Big\{[x_*, y_*] + \frac{ig}{2\alpha}\Big[\frac{2}{s}x_*\Big(\{P_i, A^i(x_*)\} - gA_i(x_*)A^i(x_*)\Big)[x_*, y_*]\\ &-[x_*, y_*]\frac{2}{s}y_*\Big(\{P_i, A^i(y_*)\} - gA_i(y_*)A^i(y_*)\Big)\\ &+\int_{y_*}^{x_*}d\frac{2}{s}\omega_*\Big(\{P^i, [x_*, \omega_*]\frac{2}{s}\,\omega_*\,F_{i\bullet}(\omega_*)\,[\omega_*, y_*]\}\\ &+g\int_{\omega_*}^{x_*}d\frac{2}{s}\,\omega'_*\,\frac{2}{s}(\omega_* - \omega'_*)[x_*, \omega'_*]F^i{}_\bullet[\omega'_*, \omega_*]\;F_{i\bullet}\,[\omega_*, y_*]\Big)\Big]\Big\}e^{i\frac{\hat{p}_\perp^2}{\alpha s}y_*}|y_\perp\rangle + O(\lambda^{-2})\,.\end{aligned} \tag{12}$$

As we will see in the next section, propagator (12) represent an intermediate useful step in order to obtain the quark propagator with sub-eikonal corrections.

3 Quark propagator with sub-eikonal corrections

The quark propagator in an external field in Schwinger notation is

$$\langle x|\frac{i}{\hat{\not{P}} + i\epsilon}|y\rangle = \langle x|\hat{\not{P}}\,\frac{i}{p^2 + 2\alpha gA_\bullet + gB + i\frac{2}{s}gF_{\bullet i}\not{p}_2\gamma^i + i\epsilon}|y\rangle \tag{13}$$

where we have defined $B \equiv O + \frac{4}{s^2}gF_{\bullet *}\sigma_{*\bullet} + \frac{g}{2}F_{ij}\sigma^{ij}$.

To get the sub eikonal corrections to the quark propagator we have to expand (13) in terms of the sub-dominant contribution and repeat similar steps that we performed to get the sub-eikonal terms for the scalar propagator. Expanding we have

$$\langle x|\hat{\not P}\frac{i}{P^2+\frac{g}{2}F_{\mu\nu}\sigma^{\mu\nu}}|y\rangle = \langle x|\hat{\not P}\Big[\frac{i}{P^2} - \frac{i}{P^2}(ig\frac{2}{s}F_{i\bullet}\gamma^i\not p_2 + g\mathcal{B}_1)\frac{1}{P^2}$$
$$+\frac{i}{P^2}(ig\frac{2}{s}F_{i\bullet}\gamma^i\not p_2 + g\mathcal{B}_1)\frac{1}{P^2}(ig\frac{2}{s}F_{i\bullet}\gamma^i\not p_2 + g\mathcal{B}_1)\frac{1}{P^2}\Big]|y\rangle \tag{14}$$

where we defined $\mathcal{B}_1 = \frac{4}{s^2}F_{\bullet *}\sigma_{*\bullet} + \frac{1}{2}\sigma_{ij}F^{ij}$.

We would like to obtain a gauge invariant expression of Eq. (14) so, after some algebra (see ref. [11] for the details of the derivation) we arrive at

$$\langle x|\frac{i}{\hat{\not P}+i\epsilon}|y\rangle = \Big[\int_0^{+\infty}\frac{d\!\!{}^-\alpha}{2\alpha}\theta(x_*-y_*) - \int_{-\infty}^{0}\frac{d\!\!{}^-\alpha}{2\alpha}\theta(y_*-x_*)\Big]e^{-i\alpha(x_\bullet-y_\bullet)}\frac{1}{\alpha s}$$
$$\times\langle x_\perp|e^{-i\frac{\hat p_\perp^2}{\alpha s}x_*}\Big\{\hat{\not p}\not p_2[x_*,y_*]\hat{\not p} + \hat{\not p}\not p_2\hat{\mathcal{O}}_1(x_*,y_*;p_\perp)\hat{\not p}$$
$$+\hat{\not p}\not p_2\frac{1}{2}\hat{\mathcal{O}}_2(x_*,y_*;p_\perp) - \frac{1}{2}\hat{\mathcal{O}}_2(x_*,y_*;p_\perp)\not p_2\hat{\not p}\Big\}e^{i\frac{\hat p_\perp^2}{\alpha s}y_*}|y_\perp\rangle + O(\lambda^{-2}). \tag{15}$$

where we have defined the operators

$$\hat{\mathcal{O}}_1(x_*,y_*;p_\perp) = \frac{ig}{2\alpha}\int_{y_*}^{x_*}d\frac{2}{s}\omega_*\Big([x_*,\omega_*]\frac{1}{2}\sigma^{ij}F_{ij}[\omega_*,y_*] + \{\hat p^i,[x_*,\omega_*]\frac{2}{s}\omega_*F_{i\bullet}(\omega_*)[\omega_*,y_*]\}$$
$$+g\int_{\omega_*}^{x_*}d\frac{2}{s}\omega'_*\frac{2}{s}(\omega_*-\omega'_*)[x_*,\omega'_*]F^i{}_\bullet[\omega'_*,\omega_*]F_{i\bullet}[\omega_*,y_*]\Big), \tag{16}$$

and

$$\hat{\mathcal{O}}_2(x_*,y_*;p_\perp) = \frac{ig}{2\alpha}\int_{y_*}^{x_*}d\frac{2}{s}\omega_*\Big[[x_*,\omega_*]\frac{1}{4}(i\mathfrak{D}_kF_{ij})\{\sigma^{ij},\gamma^k\}[\omega_*,y_*] + \{\hat p_k,[x_*,\omega_*]iF_{kj}\gamma^j[\omega_*,y_*]\}$$
$$+[x_*,\omega_*]iF_{kj}\gamma^j(i\mathfrak{D}^k[\omega_*,y_*]) - (i\mathfrak{D}^k[x_*,\omega_*])iF_{kj}\gamma^j[\omega_*,y_*]$$
$$+(\hat\alpha\not p_1 - \hat{\not p}_\perp)[x_*,\omega_*]i\frac{2}{s}F_{\bullet *}[\omega_*,y_*] + (i\not{\mathfrak{D}}_\perp[x_*,\omega_*])i\frac{2}{s}F_{\bullet *}[\omega_*,y_*]$$
$$-[x_*,\omega_*]i\frac{2}{s}F_{\bullet *}(i\not{\mathfrak{D}}_\perp[\omega_*,y_*])\Big], \tag{17}$$

From propagator (15), given an observable in terms of operators, one may apply the high-energy OPE and obtain sub-eikonal corrections to the scattering amplitude as a convolution of coefficient functions and matrix elements of Wilson lines with the insertion of the new operators appearing in Eqs. (16) and (17).

4 OPE in Wilson lines with sub-eikonal corrections

As an example, let us apply the high-energy OPE with sub-eikonal corrections to DIS process described by the two electromagnetic currents evaluated in the nucleon (or nucleus) state (See ref. [2]

for a review on Wilson lines formalism at high energy). For simplicity, let us consider only the sub-eikonal term given by the operator $\frac{1}{2}F_{ij}\sigma^{ij}$. We have

$$\langle P,S|\bar\psi(x)\gamma^\mu\psi(x)\bar\psi(y)\gamma^\nu\psi(y)|P,S\rangle \overset{x_*>0>y_*}{=} \int d^2z_1 d^2z_2 \frac{\mathrm{tr}\{\mathcal{X}_1 \not p_2 \mathcal{Y}_1 \gamma^\nu \mathcal{Y}_2 \not p_2 \mathcal{X}_2 \gamma^\mu\}}{4\pi^6\, x_*^4 y_*^4\, \mathcal{Z}_1^3 \mathcal{Z}_2^3} \langle P,S|\mathrm{tr}\{U_{z_1} U^\dagger_{z_2}\}|P,S\rangle$$

$$-g\int d^2z_1 d^2z_2 \frac{\mathrm{tr}\{\mathcal{X}_1 \not p_2 \mathcal{Y}_1 \gamma^\nu \mathcal{Y}_2 \not p_2 \sigma^{ij} \mathcal{X}_2 \gamma^\mu\}}{64\pi^6\, x_*^4 y_*^4\, \mathcal{Z}_1^3 \mathcal{Z}_2^2} \int_{y_*}^{x_*} d\omega_* \Big[\langle P,S|\mathrm{tr}\{U_{z_1}[-\infty p_1,\omega_*]_{z_2} F_{ij}[\omega_*,+\infty p_1]_{z_2}\}|P,S\rangle$$

$$+\langle P,S|\mathrm{tr}\{[+\infty p_1,\omega_*]_{z_2} F_{ij}[\omega_*,-\infty p_1]_{z_2} U^\dagger_{z_1}\}|P,S\rangle\Big] \tag{18}$$

where we have defined

$$\mathcal{Z}_i \equiv \frac{(x-z_i)_\perp^2}{x_*} - \frac{(y-z_i)_\perp^2}{y_*} - \frac{4}{s}(x_\bullet - y_\bullet) \tag{19}$$

and $\mathcal{X}_i \equiv x - z_i$, and $\quad \mathcal{Y}_i \equiv y - z_i$.

The first term after equal sign in Eq. (18) is given by the Leading Order (LO) Impact Factor (the coefficient function of an OPE), which has been known for more than 40 years, and matrix element of two Wilson lines whose evolution equation with respect to the rapidity parameter gives the BK equation [3–5] or the B-JIMWLK equation [6–8] according to whether we apply the mean-field approximation or not.

The second term after equal sign in Eq. (18) is the new term in the OPE which comes from the sub-eikonal corrections in the quark propagator. We would have more terms in OPE (18), had we considered all terms given in operators $\hat{O}_1$ and $\hat{O}_2$.

Let us consider the evolution equation of the Wilson line with the insertion of the operator $\frac{1}{2}F_{ij}\sigma^{ij}$. The one loop diagram are given in Fig. (2) and the evolution equation is

$$\int_{-\infty}^{+\infty} d\omega_* \langle \mathrm{tr}\{[\infty,-\infty]_z[-\infty,\omega_*]_{z'} gF_{ij}(\omega_*)[\omega_*,\infty]_{z'}\}\rangle_{\text{BK-type}}$$

$$= \frac{\alpha_s}{2\pi^2}\int_{-\infty}^{+\infty} d\omega_* \int_0^{+\infty}\frac{d\alpha}{\alpha}\int d^2\omega \frac{(z-z')_\perp^2}{(z-\omega)_\perp^2 (z'-\omega)_\perp^2} \tag{20}$$

$$\times\Big[\mathrm{tr}\{U_z U^\dagger_\omega\}\mathrm{tr}\{U_\omega[-\infty,\omega_*]_{z'}\, gF_{ij}(\omega_*,z'_\perp)[\omega_*,\infty]_{z'}\} - N_c \mathrm{tr}\{U_z[-\infty,\omega_*]_{z'}\, gF_{ij}(\omega_*,z'_\perp)[\omega_*,\infty]_{z'}\}\Big]$$

We observe that, besides the rapidity divergence given by $\int_0^\infty \frac{d\alpha}{\alpha}$ we have an extra divergence for $\omega \to z'_\perp$. The evolution equation (20) indeed, resum $\left(\alpha_s \ln^2 \frac{1}{x_B}\right)$, which is the type of resummation one finds in polarized DIS [12, 13].

5 Conclusions

We have presented the quark propagator with sub-eikonal corrections in high-energy QCD. Such result is relevant for the description, for example, of spin-dynamics at high-energy.

Most of the results obtained in the high-energy QCD have been obtained within the eikonal approximation. Because of the high-precision reached by the experiments and the possibility to access spin-dynamics at the proposed Electron Ion Collider (EIC), the sub-eikonal corrections become relevant.

To obtain information coming from sub-eikonal corrections, we have derived the quark propagator, which is the main result presented here. It is given in Eq. (15) and the new obtained operators are given in Eq. (16) and (17).

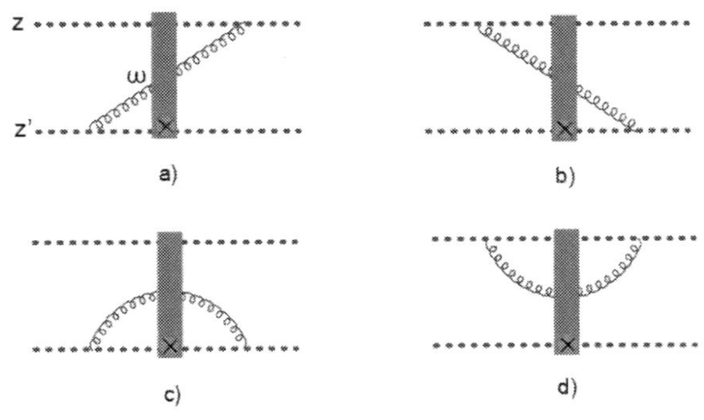

Figure 2. BK-type diagrams for the operator give in Eq. (20)

As an example of the OPE with sub-eikonal corrections, we have considered the sub-eikonal correction described by the operator $\frac{1}{2}F_{ij}\sigma^{ij}$, and have obtained its evolution equation with respect to the rapidity parameter. We noticed that, besides the usual rapidity divergence, which is related to the resummation of $\alpha_s \ln \frac{1}{x_B}$, one has an extra divergence due to the presence of the operator $\frac{1}{2}F_{ij}\sigma^{ij}$. This new divergence, together with the rapidity divergence, makes up the double Log divergence which is characteristic of the spin-dynamics at high energy.

References

[1] A. Accardi *et al.*, Eur. Phys. J. A **52** (2016) no.9, 268 doi:10.1140/epja/i2016-16268-9 [arXiv:1212.1701 [nucl-ex]].

[2] I. Balitsky, *"High-Energy QCD and Wilson Lines"*, In *Shifman, M. (ed.): At the frontier of particle physics, vol. 2*, p. 1237-1342 (World Scientific, Singapore, 2001).

[3] I. Balitsky, Nucl. Phys. B **463** (1996) 99 doi:10.1016/0550-3213(95)00638-9.

[4] Y. V. Kovchegov, *Small-x F_2 structure function of a nucleus including multiple pomeron exchanges, Phys. Rev.* **D60** (1999) 034008,

[5] Y. V. Kovchegov, *Unitarization of the BFKL pomeron on a nucleus, Phys. Rev.* **D61** (2000) 074018,

[6] J. Jalilian-Marian, A. Kovner and H. Weigert, Phys. Rev. D **59**, 014015 (1998) doi:10.1103/PhysRevD.59.014015 [hep-ph/9709432].

[7] J. Jalilian-Marian, A. Kovner, A. Leonidov and H. Weigert, Phys. Rev. D **59**, 014014 (1998) doi:10.1103/PhysRevD.59.014014 [hep-ph/9706377].

[8] E. Iancu, A. Leonidov and L. D. McLerran, Nucl. Phys. A **692** (2001) 583 doi:10.1016/S0375-9474(01)00642-X [hep-ph/0011241].

[9] E. A. Kuraev, L. N. Lipatov and V. S. Fadin, *The Pomeranchuk singlularity in non-Abelian gauge theories, Sov. Phys. JETP* **45** (1977) 199–204.

[10] I. Balitsky and L. Lipatov, *The Pomeranchuk Singularity in Quantum Chromodynamics, Sov.J.Nucl.Phys.* **28** (1978) 822–829.

[11] G. A. Chirilli, arXiv:1807.11435 [hep-ph].

[12] J. Bartels, B. I. Ermolaev and M. G. Ryskin, Z. Phys. C **70** (1996) 273 [hep-ph/9507271].

[13] J. Bartels, B. I. Ermolaev and M. G. Ryskin, Z. Phys. C **72** (1996) 627 doi:10.1007/s002880050285, 10.1007/BF02909194 [hep-ph/9603204].

Forward particle production in proton-nucleus collisions at next-to-leading order

D.N. Triantafyllopoulos[1,*]

[1]European Centre for Theoretical Studies in Nuclear Physics and Related Areas (ECT*) and Fondazione Bruno Kessler, Strada delle Tabarelle 286, I-38123 Villazzano (TN), Italy

Abstract. We consider the next-to-leading order (NLO) calculation of single inclusive particle production at forward rapidities in proton-nucleus collisions and in the framework of the Color Glass Condensate (CGC). We focus on the quark channel and the corrections associated with the impact factor. In the first step of the evolution the kinematics of the emitted gluon is kept exactly (and not in the eikonal approximation), but such a treatment which includes NLO corrections is not explicitly separated from the high energy evolution. Thus, in this newly established "factorization scheme", there is no "rapidity subtraction". The latter suffers from fine tuning issues and eventually leads to an unphysical (negative) cross section. On the contrary, our reorganization of the perturbation theory leads by definition to a well-defined cross section and the numerical evaluation of the NLO correction is shown to have the correct size.

1 Introduction

When doing perturbation theory in Quantum Mechanics, e.g. when calculating the energy change in the states of a system due to a small perturbation, normally it suffices to sum a few terms. That is, $E_n = \sum_{k=0}^{k_0} g^{2k} E_n^{(k)}$, with g^2 the size of the perturbation, $E_n^{(k)}$ fixed (as they depend only on a few parameters of the unperturbed Hamiltonian) and k_0 an integer chosen according to the desired accuracy. In an interacting Quantum Field Theory, like QCD, very often one is interested in a cross section, which perturbatively reads $\sigma = \sigma_0 + g^2\sigma_1 + g^4\sigma_2 \cdots$. Now things can be different since the coefficients σ_k can also depend on the momenta of the outgoing particles: although the QCD coupling $\alpha_s = g^2/4\pi$ is small at short distances, for some momenta one may have $\alpha_s\sigma_{k+1} \sim \sigma_k$. Then the fixed order series does not provide accurate information and one must resum to all orders in α_s in such kinematical domains.

Even one of the most basic emissions in QCD, that of a gluon from a parent parton (quark or gluon), exhibits such features. Starting from the amplitude in Fig. 1.a, we find that the differential probability under consideration in the double logarithmic limit[1] reads

$$\mathrm{d}P = \frac{\alpha_s C_\mathrm{R}}{\pi} \frac{\mathrm{d}k_\perp^2}{k_\perp^2} \frac{\mathrm{d}x}{x}, \tag{1}$$

with C_R the Casimir for the representation of the parent parton. Say we want to measure the number of gluons with a very small longitudinal momentum fraction x. Then, at order $n+1$ in

*e-mail: trianta@ectstar.eu

[1]The logarithmic limit in transverse momentum is taken just for simplicity and in fact should be relaxed.

© The Authors, published by EDP Sciences. This is an open access article distributed under the terms of the Creative Commons Attribution License 4.0 (http://creativecommons.org/licenses/by/4.0/).

Figure 1. (a) Single gluon emission from a parton and (b) strongly ordered gluon emissions.

perturbation theory all intermediate gluons with strongly ordered longitudinal momenta as in Fig. 1.b are integrated over, leading to a large "coefficient" proportional to $\ln^n(1/x)$. When x is sufficiently small, this coefficient becomes comparable to (or larger than) α_s^n, and twe must sum over all n, i.e. over all the diagrams with an arbitrary number of strongly ordered gluons. We find that the result exponentiates and, for example, the integrated gluon distribution (the number of gluons per unit rapidity and with transverse momentum less than Q^2) grows like

$$xG(x, Q^2) \sim \frac{1}{x^\lambda} \quad \text{with} \quad \lambda = 0.2 \div 0.3. \tag{2}$$

Such a growth cannot continue to arbitrary small-x, since gluons start to overlap and the interaction among them becomes strong, even though the coupling is still weak at the transverse scale of interest. "Gluon saturation" will occur at a scale for which the number of gluons per unit phase space becomes of order $1/\alpha_s$, that is when $xG(x, Q_s^2)/Q_s^2 \pi R^2 \sim 1/\alpha_s$, where R is the size of the hadron under study. Then the saturation momentum for a large nucleus of atomic number A reads

$$Q_s^2(x) \sim Q_0^2 A^{1/3} x^{-\lambda_s}, \tag{3}$$

with Q_0^2 being non-perturbative, while λ_s can be determined perturbatively. For small-x and/or large A, the saturation scale is much larger than Λ_{QCD} and indeed weak coupling techniques are applicable. Still, apart from the resummation of large logarithms, one eventually needs to deal with non-linear dynamics due to saturation. A QCD effective theory for describing such phenomena is the CGC [1].

The soft part of the hadronic wavefunction can be probed, for example, in proton-nucleus collisions in the forward region of the proton. A quark collinear with the proton and with a moderate to large momentum fraction $x_p = p^+/Q^+$, multiply scatters with the soft gluons of the nucleus that carry a fraction $X_g = p^-/P^-$, and picks up a transverse momentum $k_\perp$ so that it moves at an angle θ with the collision axis. Here, Q^+ is the longitudinal light-cone momentum of the proton which moves in the positive direction, while P^- is that of the nucleus moving towards the negative one. A simple kinematics exercise shows that the quark rapidity in the COM reads $\eta = -\ln\tan(\theta/2)$, while $x_p = k_\perp e^\eta/\sqrt{s}$ and $X_g = k_\perp e^{-\eta}/\sqrt{s}$, with $s = 2Q^+P^-$ the COM energy squared. Then it is obvious that the forward region $\theta \ll 1$ maps to small-X_g. We also note that in the infinite momentum frame of the nucleus, which shall be used later on, one has $X_g = k_\perp^2/x_p s$.

Resumming both $[\bar{\alpha}_s \ln(1/X_g)]^n$ and $\bar{\alpha}_s[\bar{\alpha}_s \ln(1/X_g)]^n$ terms, with $\bar{\alpha}_s = \alpha_s N_c/\pi$ and N_c the number of colors, in the presence of saturation, one has been able to give the cross section for single inclusive particle production at NLO in the CGC framework [2, 3]. The problem is that such a cross section when evaluated numerically turns out to be negative (although it is positive at LO) for values $k_\perp \gtrsim Q_s$ [4]. Here we shall review the process under consideration, and provide for the solution in the aforementioned problem [5].

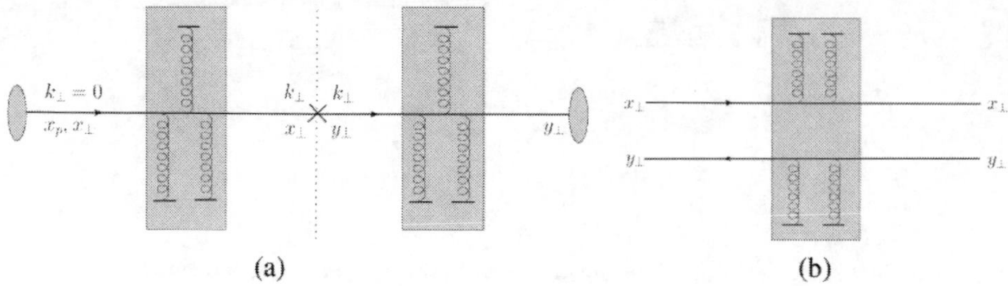

Figure 2. (a) Quark scattering off the nucleus in the CA and DDA and (b) dipole-nucleus scattering.

2 Particle production at LO and the BK equation

In our analysis we shall omit the fragmentation of the outgoing forward quark into hadrons, since it does not play any role in the problem under discussion and as it is anyway straightforward to insert it at any time in the calculation. At LO, the collinear with the proton quark scatters multiply off the target field A^μ, which can be strong for transverse momenta around (or smaller than) Q_s. The quark moves having its transverse coordinate fixed at $\boldsymbol{x}$, while its color undergoes a precession and such an eikonal interaction is given by the Wilson line

$$V(\boldsymbol{x}) = \mathrm{P}\exp\left[ig \int \mathrm{d}x^+ A_a^-(x^+, \boldsymbol{x}) t^a\right], \tag{4}$$

where P stands for path ordering along the light-cone time x^+ of the quark and t^a is in the fundamental representation. The direct amplitude (DA) $\mathcal{M}_{ij}(\boldsymbol{k})$ is just the two-dimensional Fourier transform of V_{ij} and then the single inclusive production for a particle (here the quark) with transverse momentum $\boldsymbol{k}$ and rapidity η is obtained by multiplying with the complex conjugate amplitude (CCA) (see Fig. 2.a) and by averaging/summing over colors. It reads

$$\frac{\mathrm{d}N^{\mathrm{LO}}}{\mathrm{d}\eta \mathrm{d}^2\boldsymbol{k}} = \frac{x_p q(x_p)}{(2\pi)^2} \int \mathrm{d}^2\boldsymbol{r}\, \mathrm{e}^{-\mathrm{i}\boldsymbol{k}\cdot\boldsymbol{r}} S(\boldsymbol{r}, X_g), \tag{5}$$

where $x_p q(x_p)$ is the quark pdf in the proton, $\boldsymbol{r} = \boldsymbol{x} - \boldsymbol{y}$ and $S(\boldsymbol{r}; X_g)$ is the elastic S-matrix for the $q\bar{q}$ dipole $(\boldsymbol{x}, \boldsymbol{y})$ to scatter off the nucleus[2] (see Fig. 2.b). This S-matrix is bilinear in the Wilson lines (as we have taken the product of the DA with the CCA), and is given by

$$S(\boldsymbol{x}, \boldsymbol{y}, X_g) = \frac{1}{N_c}\left\langle \mathrm{tr}[V(\boldsymbol{x}) V^\dagger(\boldsymbol{y})]\right\rangle_{X_g}. \tag{6}$$

It is evaluated at the target scale X_g, which is determined from the kinematics, and resums the aforementioned powers of $\bar{\alpha}_s \ln(1/X_g)$. In the problem at hand, we have already assumed in Eq. (5) that the nucleus is homogeneous so that S does not depend on the impact parameter $\boldsymbol{b} = (\boldsymbol{x} + \boldsymbol{y})/2$. When expanding the gauge field A^μ to second order, i.e. when neglecting the multiple scattering, the cross section becomes proportional to the nuclear gluon pdf as it should. We do not elaborate on the details of the averaging in Eq. (6), but it suffices to say that it amounts to specifying the initial condition of an equation that we now discuss.

The easiest way to implement the resummation of the large logarithms is to write the BK equation [7, 8] which describes the evolution of the elastic dipole S-matrix. For our purposes,

[2] In fact such an expression has been proved to be valid even at NLO [6].

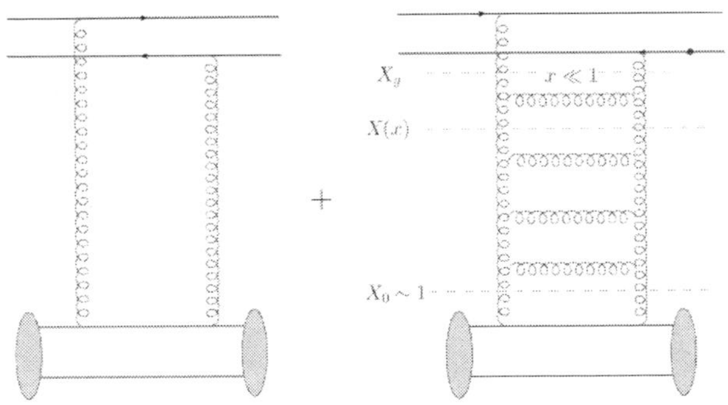

Figure 3. Particle production at LO.

it is more convenient to present such an equation in its integral form which reads

$$S(\boldsymbol{r}, X_g) = S(\boldsymbol{r}, X_0) + \frac{\bar{\alpha}_s}{2\pi} \int_{X_g}^{1} \frac{\mathrm{d}x}{x} \int \mathrm{d}^2 z \, \frac{\boldsymbol{r}^2}{\boldsymbol{z}^2(\boldsymbol{r} - \boldsymbol{z})^2} [S(\boldsymbol{z}) S(\boldsymbol{r} - \boldsymbol{z}) - S(\boldsymbol{r})]_{X(x)}, \tag{7}$$

where $X(x) = X_g/x$ is the longitudinal momentum fraction in the target in the arbitrary intermediate evolution step. The kernel in Eq. (7) is called the dipole kernel [9] as it stands for the transverse dependence of the probability for the dipole $\boldsymbol{r}$ to split into $\boldsymbol{z}$ and $\boldsymbol{r} - \boldsymbol{z}$. $S(\boldsymbol{r}, X_0)$ is the tree level contribution, the one in the absence of any QCD evolution, and typically it is given by the McLerran-Venugopalan (MV) model [10] or some variant of it. Such an equation is better understood in a frame where almost all of the evolution is associated with the nucleus, apart from the emission of the primary gluon (the one closest to the dipole $\boldsymbol{r}$), cf. Fig. 3. It is clear that the solution to Eq. (7) unitarizes, i.e. $S \to 0$ when $X_g \to 0$ for a fixed dipole size $r_\perp = |\boldsymbol{r}|$ (as we assume that S depends only on the magnitude of $\boldsymbol{r}$). In the current setup, it is natural to define the saturation momentum as the line in the $r_\perp$-X_g plane along which the S-matrix is fixed to a value of order (but of course smaller than) 1, e.g.

$$S(r_\perp = 2/Q_s(X_g), X_g) = 1/2. \tag{8}$$

Indeed one finds that $Q_s^2(X_g)$ grows as an inverse power of X_g, although the growth is too strong. Using NLO BK dynamics and resumming collinear logarithms which can get large, the growth is tamed [11–13] and comes to a good agreement with the phenomenology. Moreover, eventually one is able to reasonably describe the relevant data in d-Au collisions using Eq. (7) with running coupling and inserting K-factors to account for the remaining higher order corrections [14, 15].

3 Particle production at NLO

As depicted in Fig. 4.a, NLO corrections to the single inclusive particle production, can be separated in two parts. First, one may have two adjacent gluons in the "ladder" which have fractions such that $x_n \sim x_{n+1} \ll 1$, or one may have a loop. These are corrections associated with the evolution of the color dipole and can be taken into account by using the NLO BK equation. We shall not deal with them, since in principle they can be implemented after collinear resummations have been done without creating any additional issues. Second, one has a NLO correction when the primary gluon is not emitted eikonally, i.e. when its fraction x is of order (but smaller than) 1. Thus, by treating exactly the kinematics one can include this

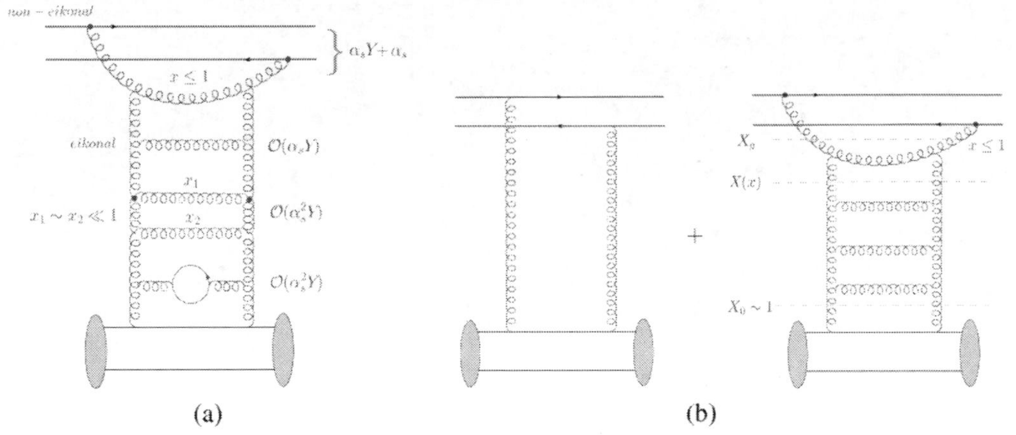

Figure 4. (a) NLO corrections and (b) NLO factorization.

impact factor correction (still, one must notice that the scattering with the nucleus remains eikonal and is given by Wilson lines) into the calculation of the cross section. Then, by extending the LO result in Eqs. (5) and (7), it is not hard to see that at NLO we have the sum of the two terms in Fig. 4.b. In fact there are two NLO contributions, one proportional to N_c and one to C_F, and we focus on the former which reads

$$\frac{\mathrm{d}N}{\mathrm{d}\eta \mathrm{d}^2\boldsymbol{k}} = \frac{x_p q(x_p)}{(2\pi)^2} \mathcal{S}(\boldsymbol{k}, X_0) + \frac{1}{4\pi} \int_0^{1-X_g/X_0} \mathrm{d}\xi \, \frac{1+\xi^2}{1-\xi} \times \left[\Theta(\xi - x_p) \frac{x_p}{\xi} \mathcal{J}(\boldsymbol{k}, \xi, X(\xi)) - x_p q(x_p) \mathcal{J}_v(\boldsymbol{k}, \xi, X(\xi)) \right]. \tag{9}$$

The real $\mathcal{J}$ and virtual $\mathcal{J}_v$ pieces (implicitly proportional to $\bar{\alpha}_s$) are known [2, 3, 16], $\mathcal{S}$ stands for the Fourier transform of S, and $1 - \xi = x$ is the fraction of the primary gluon. In a compact notation, where we omit the quark pdf, suppress the transverse coordinates and trivial prefactors, and portray collectively the scattering of the $q\bar{q}g$ system, we can write

$$\frac{\mathrm{d}N}{\mathrm{d}\eta \mathrm{d}^2\boldsymbol{k}} = \mathcal{S}_0(\boldsymbol{k}) + \bar{\alpha}_s \int_{X_g}^1 \frac{\mathrm{d}x}{x} \mathcal{K}(x) \, \mathcal{S}_{q\bar{q}g}(\boldsymbol{k}, X(x)). \tag{10}$$

Eq. (10) is our main result and being the sum of the tree level contribution and a positive term should not pose any problems if no further approximations are done [5]. Indeed, this will be manifest in the numerical results to be shown in the next section.

4 $k_\perp$-factorization: fine-tuning and negativity issues

Eq. (10) is not in the traditional form of $k_\perp$-factorization. This is so, because the kernel $\mathcal{K}$ depends not only on the transverse coordinates, but also on x, and moreover, the S-matrices of the $q\bar{q}g$ system are evaluated at the "floating" scale $X(x)$. Now the task is to see how we can arrive at a $k_\perp$-factorized expression. First notice that the LO result is recovered, as it should, by letting in Eq. (10) $\mathcal{K}(x) \to \mathcal{K}(0)$: indeed, the two terms in the r.h.s. combine to give $\mathcal{S}(\boldsymbol{k}, X_g)$. Then, we are free to add and subtract this LO piece to the r.h.s. of Eq. (10) and by doing so we get the "subtracted" expression

$$\frac{\mathrm{d}N}{\mathrm{d}\eta \mathrm{d}^2\boldsymbol{k}} = \mathcal{S}(\boldsymbol{k}, X_g) + \bar{\alpha}_s \int_{X_g}^1 \frac{\mathrm{d}x}{x} \left[\mathcal{K}(x) - \mathcal{K}(0)\right] \mathcal{S}_{q\bar{q}g}(\boldsymbol{k}, X(x)). \tag{11}$$

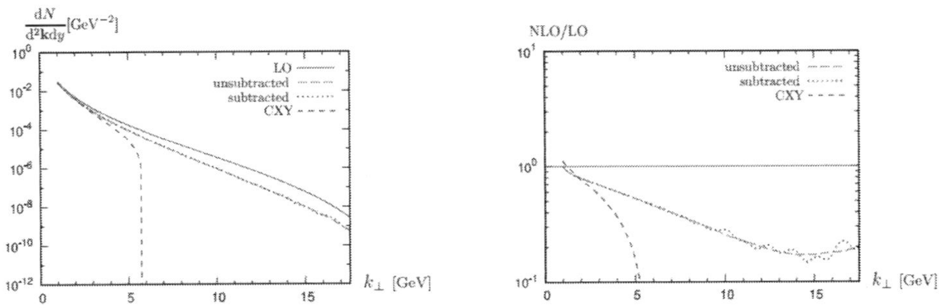

(a) (b)

Figure 5. LO, NLO and $k_\perp$-factorization (CXY) results for the single inclusive "quark production". Adapted from [17].

This is a mathematically correct step, however it is a dangerous one, since there is reshuffling of a very large contribution between the two terms. In order to obtain the correct result from Eq. (11), one must be extremely precise on the numerical evaluation of the two terms [5].

Two more steps, which now involve approximations, are required to bring Eq. (11) in a $k_\perp$-factorized form. Since $\mathcal{K}(x) - \mathcal{K}(0)$ clearly vanishes as $x \to 0$, the integrand on the r.h.s. is not anymore dominated by small values of x, and therefore at NLO accuracy we can let $\mathcal{S}_{q\bar{q}g}(\boldsymbol{k}, X(x)) \to \mathcal{S}_{q\bar{q}g}(\boldsymbol{k}, X_g)$ and furthermore let $X_g \to 0$ in the lower limit of the integration. Then we arrive at [5]

$$\frac{\mathrm{d}N^{k_\perp\text{-fact}}}{\mathrm{d}\eta \mathrm{d}^2\boldsymbol{k}} = \mathcal{S}(\boldsymbol{k}, X_g) + \bar{\alpha}_s \int_0^1 \frac{\mathrm{d}x}{x} \left[\mathcal{K}(x) - \mathcal{K}(0)\right] \mathcal{S}_{q\bar{q}g}(\boldsymbol{k}, X_g), \tag{12}$$

which is what "standard" $k_\perp$-factorization would give: the S-matrices are evaluated locally at X_g, while the kernel depends only in the transverse coordinates (since, in principle, we can now integrate over x).

We repeat that, while the "unsubtracted" and "subtracted" expressions (10) and (11) are mathematically identical, Eq. (12) is not equivalent with them. Due to the approximations performed, it mistreats the kinematics at small-x and leads to a negative cross section. Indeed, all this is confirmed in the numerical evaluation of all the aforementioned formulas for the cross section as shown in Fig. 5. We also note that, even though Eq. (10) and Eq. (11) agree very well with each other, the subtracted expression exhibits some fluctuations for high $k_\perp$ as a result of the large reshuffling. One sees that the NLO correction is negative and with a large but under control magnitude: around 50% at $k_\perp \sim 5$ GeV [17].

The discussion so far has been restricted to using a QCD coupling fixed to a moderately small value. It is not trivial at all to extend Eq. (10) with running coupling included. The dependence on $\bar{\alpha}_s$ there is in two places: the implicit BK-dependence in $\mathcal{S}_{q\bar{q}g}(\boldsymbol{k}, X(x))$ and the explicit impact factor one in front of the integration. Regarding the former, that is the coupling along the small-x ladder, one first solves the BK-equation in coordinate space with the smallest dipole prescription [18], i.e. using $\bar{\alpha}_s(r_{\min})$ with $r_{\min}$ the size of the smallest dipole among the three involved in the splitting process in Eq. (7) and then one transforms the results to momentum space. Regarding the latter, that is the coupling associated with the primary gluon, it is natural to use $\bar{\alpha}_s(k_\perp)$ [5], since $k_\perp$ is the momentum of the "measured" outgoing quark, and all this is leading again to well-defined results. In order to have a more homogeneous description, one may try to move the coupling inside the integration over the transverse coordinates [17]. Without going into much detail here, we find that when the primary gluon has a moderate x and is soft in the transverse space, the scale in the coupling must be set by the "daughter dipole" [19], which is the largest one in the kinematics under consideration. A smallest dipole choice here would lead to heavily unstable results [17], since it generates a fake potential leading to unphysical contributions to the scattering [19]. Moreover, for the NLO terms in the C_F-sector a coordinate space prescription for the coupling either cannot be implemented, or leaves uncanceled some spurious longitudinal logarithms [19], so eventually $\bar{\alpha}_s(k_\perp)$ is the only viable choice.

5 Conclusion

Single inclusive particle production at forward rapidities in proton-nucleus collisions is a process suited to study gluon saturation. Early results at NLO in the CGC effective theory led to a negative cross section and here we have shown that this was the result of approximations which were necessary to arrive at an expression consistent with $k_\perp$-factorization. Thus, we have developed a generalized factorization procedure (which involves a longitudinally dependent kernel and is non local in rapidity) that gives by definition a well-defined cross section [5]. As expected, the NLO correction is negative and reduces the LO result by $\sim 50\%$ when the outgoing quark (that fragments to a hadron) has a transverse momentum of a few GeV.

References

[1] F. Gelis, E. Iancu, J. Jalilian-Marian, R. Venugopalan, Ann.Rev.Nucl.Part.Sci. **60**, 463 (2010), 1002.0333
[2] G.A. Chirilli, B.W. Xiao, F. Yuan, Phys. Rev. Lett. **108**, 122301 (2012), 1112.1061
[3] G.A. Chirilli, B.W. Xiao, F. Yuan, Phys. Rev. **D86**, 054005 (2012), 1203.6139
[4] A.M. Stasto, B.W. Xiao, D. Zaslavsky, Phys. Rev. Lett. **112**, 012302 (2014), 1307.4057
[5] E. Iancu, A.H. Mueller, D.N. Triantafyllopoulos, JHEP **12**, 041 (2016), 1608.05293
[6] A.H. Mueller, S. Munier, Nucl. Phys. **A893**, 43 (2012), 1206.1333
[7] I. Balitsky, Nucl. Phys. **B463**, 99 (1996), hep-ph/9509348
[8] Y.V. Kovchegov, Phys. Rev. **D60**, 034008 (1999), hep-ph/9901281
[9] A.H. Mueller, Nucl.Phys. **B415**, 373 (1994)
[10] L.D. McLerran, R. Venugopalan, Phys. Rev. **D49**, 3352 (1994), hep-ph/9311205
[11] D. Triantafyllopoulos, Nucl.Phys. **B648**, 293 (2003), hep-ph/0209121
[12] G. Beuf, Phys.Rev. **D89**, 074039 (2014), 1401.0313
[13] E. Iancu, J. Madrigal, A. Mueller, G. Soyez, D. Triantafyllopoulos, Phys.Lett. **B744**, 293 (2015), 1502.05642
[14] A. Dumitru, A. Hayashigaki, J. Jalilian-Marian, Nucl. Phys. **A765**, 464 (2006), hep-ph/0506308
[15] J.L. Albacete, A. Dumitru, H. Fujii, Y. Nara, Nucl.Phys. **A897**, 1 (2013), 1209.2001
[16] B. Ducloué, T. Lappi, Y. Zhu, Phys. Rev. **D93**, 114016 (2016), 1604.00225
[17] B. Ducloué, T. Lappi, Y. Zhu, Phys. Rev. **D95**, 114007 (2017), 1703.04962
[18] E. Iancu, J.D. Madrigal, A.H. Mueller, G. Soyez, D.N. Triantafyllopoulos, Phys. Lett. **B750**, 643 (2015), 1507.03651
[19] B. Ducloué, E. Iancu, T. Lappi, A.H. Mueller, G. Soyez, D.N. Triantafyllopoulos, Y. Zhu, Phys. Rev. **D97**, 054020 (2018), 1712.07480

Measurements of open charm production and flow in $\sqrt{s_{\mathrm{NN}}}$ = 200 GeV Au+Au collisions with the STAR experiment at RHIC

Spyridon Margetis[1,*] for the STAR Collaboration

[1]Department of Physics, Kent State University, Ohio 44242, USA

Abstract. We present an improved, higher statistics, measurement with respect to the previously published results of the D^0 meson elliptic flow (v_2) as a function of transverse momentum (p_T) in Au+Au collisions at $\sqrt{s_{\mathrm{NN}}}$ = 200 GeV with the STAR Heavy Flavor Tracker (HFT). The D^0 v_2 results are compared to those of light-flavor hadrons to test the number-of-constituent-quark (NCQ) scaling. They are also compared to recent hydrodynamic and transport model calculations. We also report on the updated measurements of D^0 nuclear modification factors R_{AA} and R_{CP} using the 2014 data. The measured D^0 R_{AA} in central collisions is less than 1 across the entire p_T region. The D^0 yields show strong suppression at high p_T (> 6 GeV/c) in central collisions, consistent with those of light flavor hadrons.

We also report the measurements of collision centrality and p_T dependences of the $\Lambda_c^{\pm}$ production in Au+Au collisions at $\sqrt{s_{\mathrm{NN}}}$ = 200 GeV, using the HFT. The $\Lambda_c^{\pm}$ signal significance is greatly improved with the addition of the high-statistics data set collected in 2016 and the use of a supervised machine learning method for a more efficient topological reconstruction of the decay vertices. The measured $\Lambda_c^{\pm}/D^0$ ratio in 10-80% Au+Au collisions shows a significant enhancement compared to the PYTHIA prediction for $p + p$ collisions, across the measured p_{T} range.

1 Introduction

Relativistic heavy-ion collisions at the RHIC and LHC create a hot dense medium consisting of deconfined quarks and gluons, usually referred to as the Quark Gluon Plasma (QGP) [1, 2]. The production of heavy quarks occurs mainly during the primordial stage of heavy-ion collisions before the QGP is formed. As a consequence, the heavy quarks experience the entire evolution of the system and can be used to access information concerning the early time dynamics. Recent measurements at RHIC, based on 2014 data from the STAR experiment, have shown that D^0 mesons in minimum-bias and mid-central heavy-ion collisions exhibit significant elliptic flow [3]. The flow magnitude follows the same number-of-constituent-quark (NCQ) scaling pattern as observed for light-flavor hadrons in mid-central collisions. It is of particular interest to measure the centrality dependence of these observables and to test the NCQ scaling for charmed hadrons in different centrality classes. During 2016, STAR [4] collected an additional sample of Au+Au collisions at $\sqrt{s_{\mathrm{NN}}}$ = 200 GeV using the Heavy

*e-mail: smargeti@kent.edu

© The Authors, published by EDP Sciences. This is an open access article distributed under the terms of the Creative Commons Attribution License 4.0 (http://creativecommons.org/licenses/by/4.0/).

Flavor Tracker (HFT) [5, 6]. An improved precision for the elliptic flow measurements of heavy-flavor hadrons has been achieved by combining the data samples collected in 2014 and 2016, allowing more quantitative studies of the QGP properties.

The measurements of D^0 nuclear modification factor R_{AA} (R_{CP}), ratio of the spectra measured in heavy-ion collisions to that in $p + p$ (peripheral heavy-ion) collisions scaled by the number of binary collisions (N_{coll}), provide insights into the energy loss mechanism of charm quarks in the QGP medium and help constrain model parameters [7, 8]. The hadronization of charm quarks in the presence of QGP can be studied through the measurement of yield ratios of charm hadrons, particularly the $\Lambda_c^{\pm}/D^0$ (D^0 here denotes both D^0 and $\bar{D}^0$) yield ratio. Enhancement of the ratio, relative to that in $p + p$ collisions, is expected in the intermediate p_{T} region (2 - 6 GeV/c) if charm quarks hadronize via the coalescence mechanism in the QGP [9, 10]. The magnitude of this enhancement is sensitive to the charm quark dynamics and the presence of diquarks in the medium [9].

In these proceedings, we present an improved measurement of the D^0 meson elliptic flow (v_2) as a function of transverse momentum (p_T) in Au+Au collisions at $\sqrt{s_{\text{NN}}}$ = 200 GeV. The D^0 v_2 results are compared to those of light-flavor hadrons and to recent hydrodynamic and transport model calculations. We also present measurements of the $\Lambda_c^{\pm}/D^0$ yield ratio and the D^0 R_{AA} and R_{CP} in $\sqrt{s_{\text{NN}}}$ = 200 GeV Au+Au collisions. The $\Lambda_c^{\pm}/D^0$ yield ratio is measured as a function of p_{T} and centrality and is compared to model calculations. The D^0 R_{AA} and R_{CP} are presented for various centrality classes as a function of p_{T} and compared to similar measurements for light flavor hadrons as well as model calculations.

2 Experiment and Analysis

The STAR detector at RHIC has a full azimuthal acceptance and a pseudorapidity (η) coverage of $|\eta| < 1$ [4]. The HFT [5, 6] is a high-resolution silicon detector installed close to the beam pipe and provides an excellent track pointing resolution, e.g. less than 40 μm for kaons with p_{T} = 1 GeV/c. This allows to place cuts on the quantities describing the decay topology of heavy flavor hadrons, which greatly enhances the signal significance. Particle identification (PID) at STAR is provided by the ionization energy loss (dE/dx) measured in the Time Projection Chamber (TPC) [11] and the velocity measured using the Time of Flight (TOF) detector [12]. About 900 M minimum-bias Au+Au events from the 2014 run and ~1 B minimum-bias Au+Au events from the 2016 run at RHIC are used. Minimum-bias events are defined by a coincidence of signals in the east and west Vertex Position Detectors (VPD) [13] located at pseudorapidity $4.4 < |\eta| < 4.9$. The collision centrality is determined from the number of charged particles within $|\eta| < 0.5$ and corrected for triggering efficiency using a Monte Carlo Glauber simulation [14].

The D^0 and $\overline{D^0}$ mesons are reconstructed via their hadronic decay channels: $D^0\,(\overline{D^0}) \to K^-\pi^+\,(K^+\pi^-)$ (branching ratio: 3.89%, $c\tau \sim 123$ μm) [15] by utilizing the TOF (PID) and TPC along with the HFT for tracking. Good-quality tracks with $p_T > 0.6$ GeV/c and $|\eta| < 1$ are ensured by requiring a minimum of 20 TPC hits (out of possible 45), and with at least one hit in each layer of the Intermediate Silicon Tracker (IST) and PiXeL (PXL) components of the HFT. To enhance the signal-to-background ratio, topological variable cuts are optimized using the Toolkit for Multivariate Data Analysis (TMVA) package [16], in orthogonal cut mode.

The second-order event plane azimuthal angle (Ψ_2) is reconstructed from tracks measured in the TPC. To suppress the non-flow effects in the v_2 measurements, only tracks that are in the opposite rapidity hemisphere with at least $\Delta\eta > 0.05$ with respect to the reconstructed D^0, are used for the Ψ_2 reconstruction. The v_2 coefficients are calculated using the event-plane method [17] measuring the D^0 yields in different azimuthal intervals defined with respect to the event plane angle ($\phi - \Psi_2$). The D^0 yields are weighted by the inverse of the reconstruction efficiency $\times$ acceptance for each interval of collision centrality. The observed v_2 is then calculated by fitting the azimuthal dependence of the D^0 yield using the function $p_0(1 + 2\, v_2^{obs} \cos[2\,(\phi - \Psi_2)])$. The resolution-corrected v_2 is then obtained by dividing v_2^{obs} with the event-plane resolution Ψ_2 [18].

The $\Lambda_c^{\pm}$ candidates are reconstructed via the $\Lambda_c^{\pm} \rightarrow p^{\pm} K^{\mp} \pi^{\pm}$ channel using both the 2014 and 2016 datasets. The supervised learning algorithm, Boosted Decision Trees (BDT), from the TMVA is used for signal and background separation. The BDT is trained using a signal sample of $\Lambda_c^{\pm}$ decayed (using PYTHIA [19]) to $pK\pi$ triplets, with detector effects taken into account, and a background sample from wrong-sign triplets from data. The use of BDT along with the combined 2014+2016 data gives a factor of 2 improvement in the $\Lambda_c^{\pm}$ signal significance compared to the previous measurement [20].

3 Results

The averaged $v_2(p_T)$ of D^0 and $\overline{D^0}$ mesons is measured in 0-10%, 10-40% and 0-80% central Au+Au collisions at $\sqrt{s_{NN}}$ = 200 GeV based on combined datasets recorded during 2014 and 2016. This provides an improvement of about a 30% in the statistical precision com-

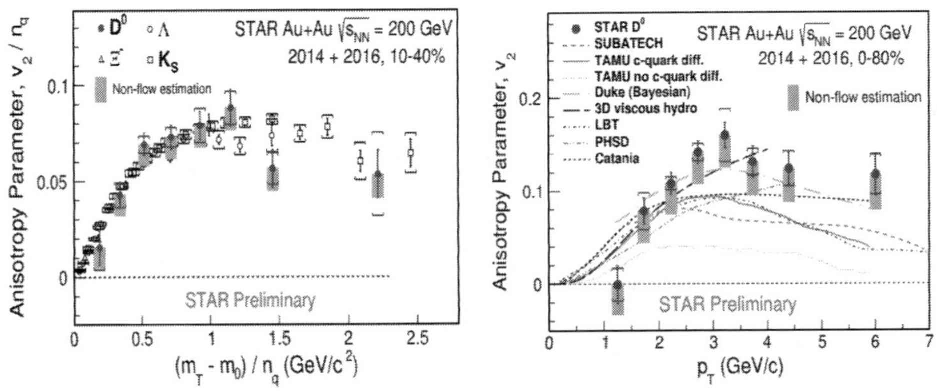

Figure 1. Left panel: v_2/n_q as a function of $(m_T - m_0)/n_q$ for D^0 and $\overline{D^0}$ mesons combined in 10-40% central Au+Au collisions at $\sqrt{s_{NN}}$ = 200 GeV along with K_S^0, Λ, and Ξ [21]. Right panel: v_2 as a function of p_T for D^0 and $\overline{D^0}$ mesons combined in 0-80% Au+Au collisions compared with model calculations [22–28].

pared to previously published results using 2014 data alone [3]. The blue solid markers in the left panel of Fig. 1 present the NCQ-scaled v_2 as a function of NCQ-scaled transverse kinetic energy $(m_T - m_0)$ for D^0 mesons in 10-40% central Au+Au collisions at $\sqrt{s_{NN}}$ = 200 GeV. The results are compared to light hadron species, namely the K_S^0 meson and the Λ and Ξ baryons [21]. The NCQ-scaled D^0 v_2 is compatible within uncertainties with those of light hadrons for $(m_T - m_0)/n_q < 2.5$ GeV/c^2. This observation suggests that charm quarks exhibit the same strong collective behavior as light-flavor quarks, and may be close to thermal equilibrium with the medium in Au+Au collisions at $\sqrt{s_{NN}}$ = 200 GeV. The right panel in Fig. 1 presents the D^0 v_2 results in 0-80% central Au+Au collisions, compared to SUBATECH [22], TAMU [23], Duke [24], 3D viscous hydro [25], LBT [26], PHSD [27], and Catania [28] model calculations. These models include different treatments of the charm quark interactions with the medium. They also differ in initial state conditions, QGP evolution, hadronization, etc. We have performed a statistical significance test of the consistency between the data and each model, quantified by χ^2/NDF and the p value. We have found that the TAMU model without charm quark diffusion cannot describe the data, while the same model with charm quark diffusion turned-on shows better agreement. All the other models can describe the data in the measured p_T region.

The left panel of Fig. 2 shows the $\Lambda_c^{\pm}/D^0$ ratio as a function of p_T for the 10-80% centrality bin. The data shows a significant enhancement of the ratio compared to the PYTHIA prediction for $p + p$ collisions. The SHM model [29] also underpredicts the data. The Ko model (0-5%) [9] and the Greco model (0-20%) [10] calculations include coalescence of thermalized charm quarks and are closer to the measured values, but still underpredict data at higher p_T. The right panel of Fig. 2 shows the $\Lambda_c^{\pm}/D^0$ ratio as a function of centrality. The $\Lambda_c^{\pm}/D^0$ ratio shows an increasing trend towards more central collisions. The measured value in the peripheral collisions is consistent with the value from $p + p$ collisions at 7 TeV measured by ALICE [30].

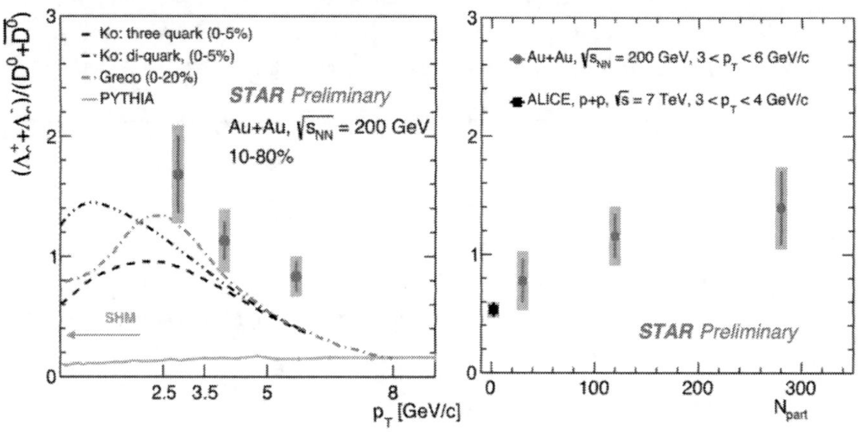

Figure 2. The $\Lambda_c^{\pm}/D^0$ ratio as a function of p_T for 10-80% centrality class (left) and as a function of N_{part} for $3 < p_T < 6$ GeV/c (right), in Au+Au collisions at $\sqrt{s_{NN}} = 200$ GeV. The error bars and gray bands represent statistical and systematic uncertainties, respectively.

Results of the D^0 R_{AA}, extended down to zero p_T using the HFT data from 2014, are shown in the left panels of Fig. 3. The figures also show re-analyzed 2010/2011 results [31], which are consistent with the results from the HFT analysis of 2014 data. The D^0 R_{AA} values are below 1 at all p_T in central collisions. The R_{AA} values for $p_T > 5$ GeV/c show significant suppression in central collisions, and the suppression decreases towards more peripheral collisions, as the system size decreases. The right panels in Fig. 3 show the D^0 R_{CP}, relative to the 40-60% centrality bin. The R_{CP} values are consistent with unity for $p_T < 2$ GeV/c, but decrease towards high p_T (> 6 GeV). The figure also shows the R_{CP} values for different light flavor hadrons in 200 GeV Au+Au collisions [32]. The R_{CP} values from light flavor hadrons are lower than those of D^0 for $p_T < 4$ GeV/c, while they show similar levels of suppression towards higher p_T. Comparisons to two model calculations which incorporate both collisional and medium induced radiative energy losses are also shown in the figure. The Duke model [33] uses a modified Langevin equation to describe the charm quark transport in the medium and radiative energy loss, while the LBT [7] uses a linearized Boltzman transport model with the higher-twist formalism for medium-induced radiative energy loss. Both models can well describe the data. The model parameters had been tuned to describe the previous STAR measurements from [8], and the new high precision results reported here can help better constrain the model parameters.

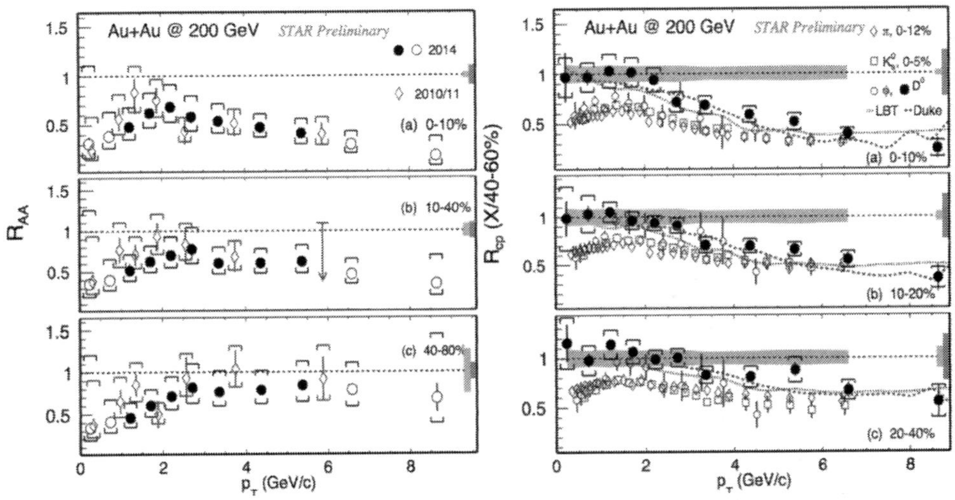

Figure 3. The D^0 R_{AA} (left) and R_{CP} values relative to 40-60% centrality bin (right) as a function of p_T for different centrality intervals. The solid circles in the left panels indicate that measured p+p yields are used for calculating R_{AA} and the open circles indicate that the $p+p$ yields from an extrapolation using a Levy fit to the measured yields are used. The open diamonds are the values from re-analyzed 2010/2011 data. The error bars and brackets on the data points indicate statistical and systematic uncertainties respectively. The dark and light green boxes on the right of each panel show the global uncertainty from the reference ($p+p$ cross section uncertainty for R_{AA} and N_{coll} uncertainty for 40-60% centrality for R_{CP}) and the N_{coll} uncertainty for the given centrality bin, respectively. The shaded gray bands around 1 on the R_{CP} plots indicate the uncertainty from vertex resolution correction for the 40-60% centrality bin.

4 Conclusion

In summary, we report STAR results on the D^0 elliptic flow (v_2) as a function of p_T combining 2014 and 2016 data samples. The D^0 v_2 suggests that charm quarks may be close to thermal equilibrium in the medium. Furthermore, studies are now in progress in determining the D^0 v_2 in the peripheral collisions (40-80%), with an enlarged pseudorapidity gap to reduce non-flow effects. The measurements of the D^0 R_{AA} and R_{CP} are also presented for 200 GeV Au+Au collisions. The R_{AA} and R_{CP} values show strong suppression at high p_T in central collisions and the suppression decreases towards peripheral collisions. The R_{CP} values at high p_T (> 6 GeV/c) are consistent with those of light flavor hadrons, while they are larger for p_T < 4 GeV/c. Transport models with collisional and radiative energy losses are able to describe the measured R_{CP} values.

We also present the measurements of the $\Lambda_c^{\pm}/D^0$ ratio as functions of collision centrality and p_T in Au+Au collisions at $\sqrt{s_{NN}}$ = 200 GeV. The ratio shows a significant enhancement compared to the values from PYTHIA for $p+p$ collisions and also to SHM calculations. Model calculations with coalescence hadronization are closer to the experimental measurements, suggesting that the coalescence mechanism plays an important role in charm quark hadronization in the QGP.

References

[1] R. Rapp and H. van Hees arXiv:0803.0901.

[2] B. Muller, J. Schukraft, and B. Wyslouch, Ann. Rev. Nucl. Part. Sci. **62**, 361 (2012).

[3] L. Adamczyk *et al.* (STAR Collaboration), Phys. Rev. Lett. **118**, 212301 (2017).

[4] STAR Collaboration, Nucl. Inst. Meth. A**499**, 624 (2003).

[5] D. Beavis *et al.*, The HFT Technical Design report as prepared for the July 2011 CD2-3 review, STAR Note SN0600.

[6] G. Contin *et al.*, Nucl. Instrum. Meth. A**907**, 60 (2018)

[7] S. Cao *et al.*, Phys. Rev. C **94**, 014909 (2016).

[8] STAR Collaboraion, Phys. Rev. Lett. **113**, 142301, (2014).

[9] S. H. Lee, K. Ohnishi, S. Yasui, I.-K. Yoo and C. M. Ko Phys. Rev. Lett.**100**, 222301 (2008).

[10] S. Ghosh, S. K. Das, V. Greco, S. Sarkar and J. E. Alam, Phys. Rev. D**90** 054018 (2014).

[11] M. Anderson *et al.*, Nucl. Instr. Meth. A **499**, 659 (2003).

[12] B. Bonner *et al.*, Nucl. Instr. Meth. A **508**, 181 (2003).

[13] W. J. Llope *et al.*, Nucl. Instrum. Meth. A**522**, 252 (2004).

[14] B. Abelev *et al.*, (STAR Collaboration), Phys. Rev. C **79**, 034909 (2009).

[15] Review of Particle Physics, M. Tanabashi *et al.*, Phys. Rev. D **98**, 030001 (2018).

[16] A. Hocker *et al.*, PoS ACAT, **040** (2007).

[17] A. M. Poskanzer and S. A. Voloshin, Phys. Rev. C **58**, 1671 (1998).

[18] H. Masui *et al.*, Nucl. Instrum. Meth. A **833**, 181 (2016).

[19] T. Sjöstrand, S. Mrenna and P. Skands, Journal of High Energy Physics**2006**, 026 (2006).

[20] G. Xie (STAR Collaboration), Nucl. Phys. A **967**, 928-931, 2017.

[21] B. I. Abelev *et al.*, (STAR collaboration), Phys. Rev. C **77**, 054901 (2008).

[22] T. Song *et al.*, Phys Rev C **92**, 014910 (2015).

[23] M. He. *et al.*, Phys Rev C **86**, 014903 (2012); Phys Rev Lett **110**, 112301 (2013).

[24] S. Cao *et al.*, Phys. Rev. C **97**, 014907 (2018).

[25] L. Pang *et al.*, Phys Rev C **86**, 024911 (2012).

[26] S. Cao *et al.*, Phys Rev C **94**, 014909 (2016).

[27] T. Song *et al.*, Phys Rev C **92**, 014910 (2015).

[28] F. Scardina *et al.*, Phys Rev **96**, 044905 (2017).

[29] I. Kuznetsova and J. Rafelski, The European Physical Journal C**51** 113–133 (2007).

[30] ALICE Collaboration, arXiv:1712.09581

[31] L. Adamczyk et al. (STAR Collaboration), Phys. Rev. Lett. **113**, 142301 (2014).

[32] STAR Collaboration, Phys. Rev. Lett. **97**, 152301 (2006); Phys. Rev. C **79**, 064903 (2009); Phys. Rev. Lett. **108**, 072301

[33] Y. Xu *et al.* Phys. Rev. C. **97**, 014907 (2018).

Open heavy-flavour measurements with ALICE at the LHC

Fabio Colamaria[1,*] on behalf of the ALICE Collaboration

[1]Istituto Nazionale di Fisica Nucleare, Sezione di Bari - Via E. Orabona 4, 70125 Bari, Italy

Abstract. Heavy quarks are produced in the early stages of ultra-relativistic heavy-ion collisions, and their number is preserved throughout the subsequent evolution of the system. Therefore, they constitute ideal probes for characterising the Quark–Gluon Plasma (QGP) medium and for the study of its transport properties. In particular, heavy quarks interact with the partonic constituents of the plasma, losing energy, and are expected to be sensitive to the medium collective motion induced by its hydrodynamical evolution. In pp collisions, the measurement of heavy-flavour hadron production provides a reference for heavy-ion studies, and allows also testing perturbative QCD calculations in a wide range of collision energies. Similar studies in p–Pb collisions help in disentangling cold nuclear matter effects from modifications induced by the presence of a QGP medium, and are also useful to investigate the possible existence of collective phenomena also in this system. The ALICE detector provides excellent performances in terms of particle identification and vertexing capabilities. Hence, it is fully suited for the reconstruction of charmed mesons and baryons and of electrons from heavy-flavour hadron decays at central rapidity. Furthermore, the ALICE muon spectrometer allows reconstructing heavy-flavour decay muons at forward rapidity. A review of the main ALICE results on open heavy flavour production in pp, p–Pb and Pb–Pb collisions is presented. Recent, more differential measurements are also shown, including azimuthal correlations of heavy-flavour particles with charged hadrons in p–Pb collisions, and D-meson tagged-jet production in p–Pb and Pb–Pb collisions.

1 Introduction

Heavy quarks (charm and beauty) are excellent probes for the study of the Quark–Gluon Plasma (QGP), a colour-deconfined medium of strongly-interacting matter formed in ultra-relativistic heavy-ion collisions [1]. Due to their large mass, heavy quarks are produced in hard parton scatterings during the initial stages of the collision, and their thermal production and in-medium annihilation rates are negligible at the LHC energies. Hence, they experience the whole evolution of the QGP medium and are subject to its effects.

In particular, charm and beauty quarks are expected to interact with the medium constituents via elastic and inelastic processes. This leads to a loss of their initial energy while crossing the medium [2–4], though to a lesser extent with respect to lighter quarks and gluons, due to the colour-charge and 'dead-cone' effects [5]. A sensitive observable to partonic energy loss is the nuclear modification factor, defined as $R_{\mathrm{AA}}(p_{\mathrm{T}}) = \frac{\mathrm{d}N_{\mathrm{AA}}/\mathrm{d}p_{\mathrm{T}}}{\langle N_{\mathrm{coll}}\rangle \mathrm{d}N_{\mathrm{pp}}/\mathrm{d}p_{\mathrm{T}}}$, i.e. the

*e-mail: fabio.colamaria@ba.infn.it

© The Authors, published by EDP Sciences. This is an open access article distributed under the terms of the Creative Commons Attribution License 4.0 (http://creativecommons.org/licenses/by/4.0/).

ratio of the production yields of a given particle normalised by the production yield in pp collisions and scaled by the average number of binary nucleon-nucleon collisions $\langle N_{\mathrm{coll}} \rangle$.

In heavy-ion collisions with non-zero impact parameter, a collective motion of the medium is also expected to emerge, in case of thermalisation of its constituents. This motion translates the initial spatial anisotropy of the system into a momentum anisotropy, reflected in the final-state particle azimuthal distribution [6]. The degree of participation of heavy quarks to the medium collective motion can be quantified via the 'elliptic-flow' coefficient v_2 at low and intermediate transverse momentum (p_{T}). The elliptic-flow coefficient is defined as the second-order coefficient of the Fourier decomposition of the azimuthal distribution of a given particle.

Among the heavy-flavour species, the study of charm and beauty baryons is of particular interest. It was suggested that, in a QGP environment, heavy-flavour quarks could hadronise also via recombination with other quarks present in the medium, in addition to string fragmentation. This would induce an enhancement of the baryon-to-meson ratio at low and intermediate p_{T} with respect to pp collisions [7, 8]. In addition, for pp collisions, the sensitivity of heavy-flavour baryons to fragmentation allows us to test the universality of the parton fragmentation functions, commonly evaluated from electron-positron data.

The availability of large data samples collected during the latest years has opened the possibility of performing more differential analysis also for rare probes as heavy flavours. In this contest, the study of azimuthal correlation distributions between electrons from heavy-flavour hadron decays (HFe) and charged hadrons in p–Pb collisions can provide a measurement of the HFe v_2 coefficient also in this collision system. A positive v_2 was already measured by ALICE for light-flavour hadrons at midrapidity [9], and its interpretation, either as a final-state or initial-state effect, is currently debated, evidencing the need of additional measurements also in the heavy-flavour sector. Another recent differential study is the measurement of the production cross section of jets with a D-meson constituent (D-jets), in all collision systems. This measurement provides a closer access to the parton kinematics with respect to the study of the D-meson cross section. In addition, comparing the momentum fraction of the jet carried by the D meson in pp and Pb–Pb collisions provides information on the modification of the heavy-quark fragmentation in a QGP medium.

2 Experimental apparatus and results

The ALICE detector [10, 11] provides excellent performances in terms of particle identification, low-p_{T} track reconstruction, and vertexing. Thanks to these capabilities, open heavy-flavour particles at low and intermediate p_{T} can be reconstructed with great precision. In particular, $\mathrm{D}^0, \mathrm{D}^{*+}, \mathrm{D}^+$ and D_s^+ mesons, as well as Λ_c^+ and Ξ_c^0 baryons can be reconstructed from their hadronic (for D-mesons and Λ_c^+ baryons) and semileptonic (for Λ_c^+ and Ξ_c^0) decay channels at midrapidity ($|y| < 0.5$), exploiting their displaced decay topologies, thanks to the possibility of separating primary and secondary vertices with the Inner Tracking System (ITS). To improve the selection, daughter particles can be identified by means of specific energy loss in the Time Projection Chamber (TPC) and time-of-flight in the Time Of Flight (TOF) detector. Heavy-flavour hadron decay electrons can also be reconstructed at central rapidity. The identification of electron at high p_{T} is performed with the Electromagnetic Calorimeter (EMCal), while TPC and TOF are used in the low p_{T} region. Furthermore, the ALICE muon spectrometer, composed of a set of tracking and trigger stations, allows reconstructing heavy-flavour hadron decay muons at forward rapidity ($-4 < y < -2.5$).

2.1 Results for charmed mesons

The left panel of Fig. 1 shows the nuclear modification factor for the average of non-strange D mesons as a function of p_{T}, in 0–10% central Pb–Pb collisions at $\sqrt{s_{\mathrm{NN}}} = 5.02$ TeV [12]. A strong suppression of a factor ≈ 5 is observed for $2 < p_{\mathrm{T}} < 6$ GeV/c, while the R_{AA} increases at lower p_{T}, being compatible with unity for $1 < p_{\mathrm{T}} < 2$ GeV/c. This suppression can be ascribed to final-state effects, since in p–Pb collisions the equivalent observable R_{pPb} was found to be compatible with unity within uncertainties over the full p_{T} range [13]. On the right panel of the same figure, the elliptic-flow coefficient of the average of non-strange D mesons is shown for Pb–Pb collisions at $\sqrt{s_{\mathrm{NN}}} = 5.02$ TeV, in the 30–50% centrality class. A positive v_2 is measured in the range $2 < p_{\mathrm{T}} < 10$ GeV/c, with a strength comparable to that of charged pions. This suggests that low-momentum charm quarks effectively participate to the medium collective motion. Both measurements refer to the prompt component of D-meson yields only. The feed-down contribution was subtracted via a model-driven approach, based on FONLL [14] pQCD calculations for feed-down D-meson production.

In both panels, ALICE measurements are compared with model predictions based on heavy-quark transport in a viscous hydrodynamic medium, which provide a description for both observables. Models in which charm quarks pick up collective flow via recombination or subsequent elastic collisions and include radiative energy loss in an expanding medium better reproduce the two observables. Anyway, a simultaneous description of both observables over the full p_{T} range remains a challenging task, and these measurements are able to set strong constraints to models. The best v_2 description is provided by models predicting a charm thermalisation time of $\tau_c = 3 - 14$ fm/c. This is of the same order of the QGP decoupling time, implying a certain degree of thermalisation of charm quarks in the medium.

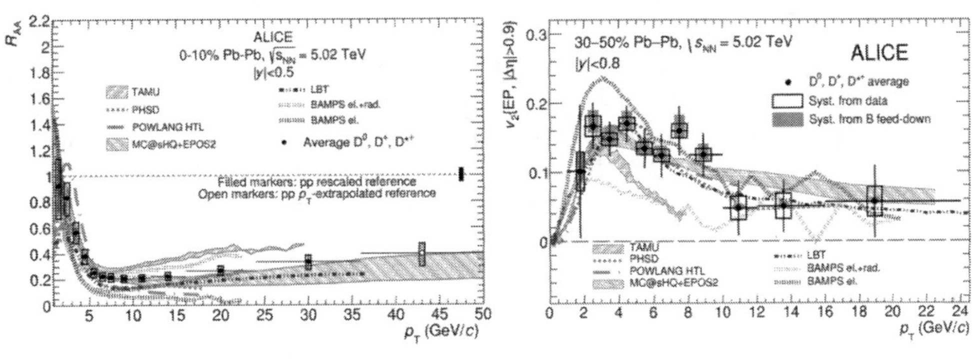

Figure 1. Left: average nuclear modification factor of D^0, D$^+$ and D^{*+} mesons as a function of p_{T} in 0–10% central Pb–Pb collisions. Right: average elliptic-flow coefficient of D^0, D$^+$ and D^{*+} mesons in Pb–Pb collisions, for 30–50% centrality. For both panels, data are compared with model predictions which provide a simultaneous description of both observables.

2.2 Results for charmed baryons

In Fig. 2 ALICE results for the ratio of Λ_c^+ and D^0 prompt production cross sections are presented as a function of p_{T}, for pp collisions at $\sqrt{s} = 7$ TeV [15] and p–Pb collisions at $\sqrt{s_{\mathrm{NN}}} = 5.02$ TeV (preliminary results from LHC Run 2). The two measurements are compatible within uncertainties over the full p_{T} range. They also show the same transverse momentum evolution observed in the light-flavour sector for the Λ/K_s^0 ratio [16, 17], with a

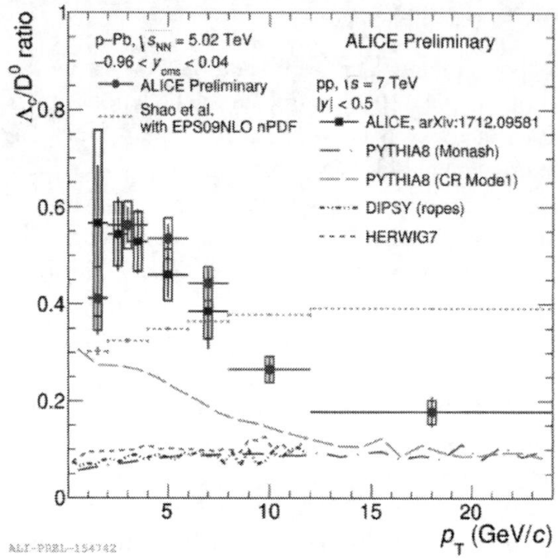

Figure 2. Ratio of Λ_c^+/D^0 production yields in pp and p–Pb collisions as a function of p_T, compared with Monte Carlo predictions.

decreasing trend for $p_T > 4$ GeV/c. ALICE results show larger ratios with respect to measurements performed in e^+e^- collisions at lower centre-of-mass energies, where mechanisms expected to increase the baryon-to-meson ratio as colour reconnection and multi-parton interactions are expected to play a smaller role. The Λ_c^+/D^0 ratios reported by ALICE are also larger, though still compatible, than analogue measurements by LHCb [18] at forward rapidity ($2 < y < 4.5$) in a comparable transverse momentum range and at the same energy.

Measurements in pp collisions are compared to predictions from several Monte Carlo generators employing different hadronisation mechanisms, with fragmentation fractions of charm into hadrons based on electron-positron collision data. The event generator which provides the closest description to data is PYTHIA8 [19, 20] with an enhanced colour reconnection mechanism (orange dashed curve), which is supposed to increase the baryon production. Nevertheless, also this generator underestimates ALICE measurements by a factor ≈ 2 over the full p_T range.

Figure 3 shows ALICE measurements of the nuclear modification factor of non-strange D mesons (average of D^0, D^+ and D^{*+}), D_s^+ meson and Λ_c^+ as a function of p_T, at central rapidity, in p–Pb collisions (left panel) and Pb–Pb collisions (right panel) at $\sqrt{s_{NN}} = 5.02$ TeV. 0–10% (0–80%) central Pb–Pb collisions are considered for D-meson (Λ_c^+) measurement. The values of R_{pPb} of all the charmed hadron species are consistent within uncertainties and compatible with unity, suggesting a limited impact of cold-nuclear-matter effects. The measurements in Pb–Pb show a possible hierarchy for the nuclear modification factor values, with $R_{AA}(\Lambda_c^+) > R_{AA}(D_s^+) > R_{AA}$(non-strange D). This feature could be explained by the onset of coalescence mechanism in Pb–Pb central collisions: D_s^+ production should be increased by recombination of charm quark with strange quarks in a strangeness-rich environment as the QGP, and coalescence shall enhance the production of baryons with respect to the production rate in pp collisions.

Charmed-hadron R_{AA} is also compared to nuclear modification factor measurements for charged tracks and charged pions. A hint of a larger R_{AA} is observed for D mesons with

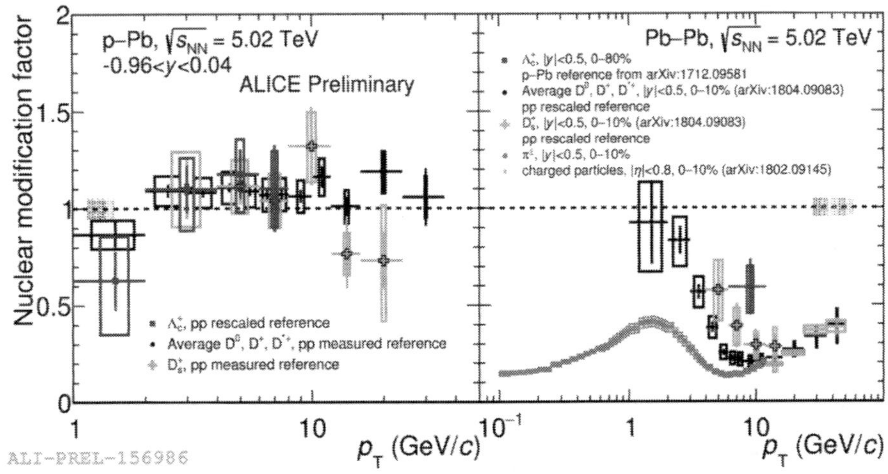

Figure 3. Comparison of nuclear modification factor of average D mesons, D_s^+, Λ_c^+, charged pions and charged particles in p–Pb (left panel) and Pb–Pb (right panel) collisions.

respect to charged pions over the common p_T range. It is not straightforward, anyway, to interpret this feature as a smaller in-medium energy loss for charm quarks rather than for light-flavour quarks and gluons, due to the dead-cone effect. The light hadron production does not scale with the average number of binary collision $\langle N_{coll}\rangle$ for $p_T \leq 3\text{GeV}/c$; a different energy scale from heavy- and light-flavour hadrons to corresponding partons could enter in play due to the different fragmentation of the latter, and the initial p_T spectrum is not the same for light and heavy partons; finally, the radial flow contribution could affect differently light- and heavy-flavour mesons.

2.3 Results for heavy-flavour correlations and jets

The left panel of Figure 4 shows the azimuthal correlation distribution between heavy-flavour decay electrons with $2 < p_T < 4$ GeV/c and charged particles with $0.3 < p_T < 2$ GeV/c, in p–Pb collisions at $\sqrt{s_{NN}} = 5.02$ TeV, normalised by the number of heavy-flavour decay electrons [21]. Two different multiplicity ranges, 0–20% (red points) and 60–100% (black points), classified using the signal amplitude in the V0-A detector, were considered. The correlation distributions are shown after the subtraction of a 'baseline' term, evaluated via a fit with a constant to the three lowest points of the distribution. An enhancement of the peaks at $\Delta\varphi \approx 0$ and $\Delta\varphi \approx \pi$ can be observed for high-multiplicity collisions. The jet contribution is assumed comparable for the two multiplicity classes and removed by subtracting the low-multiplicity from the high-multiplicity distribution. This allows extracting the v_2 coefficient of heavy-flavour decay electrons via a fit to the subtracted correlation distribution with a Fourier decomposition in $\Delta\varphi$.

The values of the elliptic-flow coefficient are shown in the right panel of Fig. 4 as a function of the electron transverse momentum. A positive v_2 is measured for the range $1.5 < p_T < 4$ GeV/c with a significance of 5.1σ. The values of heavy-flavour decay electron v_2 are generally lower, though still comparable, than those measured for charged particles, dominated by light-flavour hadrons. An interpretation of this comparison is anyway not immediate, since the transverse-momentum distribution of heavy-flavour hadrons from which the electrons originate is considerably broader than that of light-flavour hadrons, as well as having a larger average value. The v_2 of heavy-flavour decay electrons has a similar trend of that of inclusive muons, measured by ALICE at forward and backward rapidity [22] and dominated by heavy-flavour decay muons for $p_T > 2$ GeV/c. Drawing conclusions from this comparison is not straightforward because of the different cold-nuclear-matter effects affecting heavy flavours in different rapidity ranges.

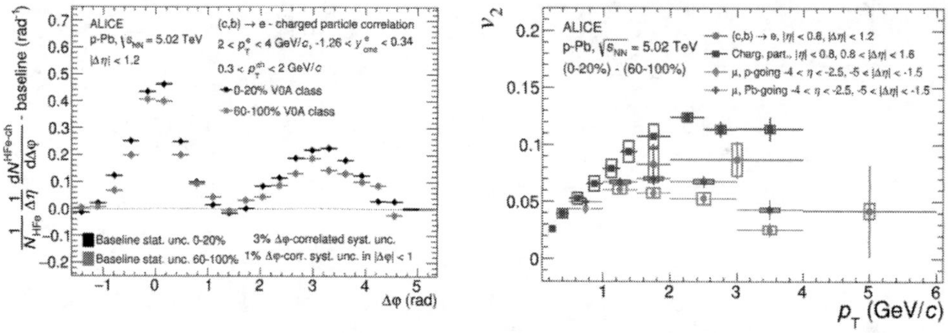

Figure 4. Left: baseline-subtracted azimuthal correlations between HFe and charged particles, for high-multiplicity and low-multiplicity p–Pb collisions. Right: v_2 of HFe as a function of p_T compared to the v_2 of unidentified charged particles.

ALICE has measured the p_T-differential production cross section of prompt D^0-jets in p–Pb and Pb–Pb collisions at $\sqrt{s_{NN}} = 5.02$ TeV. Jets were reconstructed in the pseudorapidity range $|\eta| < 0.6$, using the FASTJET [23] package with the anti-k_T algorithm, after substituting the daughters tracks of the D^0 with the D^0 particle. The feed-down contribution to the D^0-jet spectrum was subtracted exploiting POWHEG+PYTHIA [24, 25] simulations. Detector effects and background fluctuations were corrected for with a Bayes unfolding procedure [26].

In the left panel of Fig. **??**, the D^0-jet production cross section in p–Pb collisions is showed for $5 < p_T(\text{jet}) < 50$ GeV/c. Data are compared to POWHEG+PYTHIA predictions, which are found to be compatible with ALICE measurements within uncertainties, dominated by the theory uncertainty band. In the right panel of the same figure, the nuclear modification factor of D^0-jets in 0–20% central Pb–Pb collisions is presented for $5 < p_T(\text{jet}) < 20$ GeV/c, showing a strong suppression by a factor $\approx$ 5-6. D^0-jet R_{AA} is compared with the inclusive-jet R_{AA}, dominated by gluon and light-flavour jets. A stronger suppression seems to occur for D^0-jets, though the non-overlapping p_T ranges of the two measurements prevent drawing firm conclusions. D^0-jet R_{AA} is also compared with the average of non-strange D-meson R_{AA} [12], showing a similar suppression in the common p_T range. This could signal that the D^0-jets energy loss could be dominated by the energy loss of their leading particle, although the different energy scale for the charm quark transverse momentum in the two cases has to be accounted for.

References

[1] P. Braun-Munzinger, J. Stachel, Nature **448**, 302 (2007)

[2] M. Gyulassy, M. Plumer, Phys. Lett. **B243**, 432 (1990)

[3] R. Baier, Y.L. Dokshitzer, A.H. Mueller, S. Peigne, D. Schiff, Nucl. Phys. **B484**, 265 (1997), hep-ph/9608322

[4] M.H. Thoma, M. Gyulassy, Nucl. Phys. **B351**, 491 (1991)

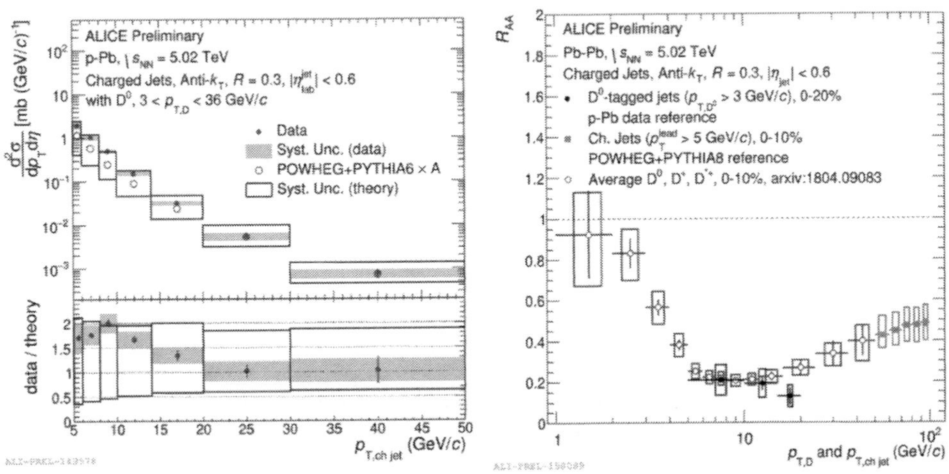

Figure 5. Left: p_{T}-differential production cross section of D^0-jets compared with POWHEG+PYTHIA [24, 25] predictions. Right: Nuclear modification factor of D^0-jets as a function of p_{T}, compared with those of average D-mesons and inclusive jets.

[5] N. Armesto, A. Dainese, C.A. Salgado, U.A. Wiedemann, Phys. Rev. **D71**, 054027 (2005), hep-ph/0501225
[6] J.Y. Ollitrault, Phys. Rev. **D46**, 229 (1992)
[7] V. Greco, C.M. Ko, R. Rapp, Phys. Lett. **B595**, 202 (2004), nucl-th/0312100
[8] Y. Oh, C.M. Ko, S.H. Lee, S. Yasui, Phys. Rev. **C79**, 044905 (2009), 0901.1382
[9] B. Abelev et al. (ALICE), Phys. Lett. **B719**, 29 (2013), 1212.2001
[10] K. Aamodt et al. (ALICE), JINST **3**, S08002 (2008)
[11] B.B. Abelev et al. (ALICE), Int. J. Mod. Phys. **A29**, 1430044 (2014), 1402.4476
[12] S. Acharya et al. (ALICE), Submitted to: JHEP (2018), 1804.09083
[13] J. Adam et al. (ALICE), Phys. Rev. **C94**, 054908 (2016), 1605.07569
[14] M. Cacciari, M. Greco, P. Nason, JHEP **05**, 007 (1998), hep-ph/9803400
[15] S. Acharya et al. (ALICE), JHEP **04**, 108 (2018), 1712.09581
[16] B.B. Abelev et al. (ALICE), Phys. Rev. Lett. **111**, 222301 (2013), 1307.5530
[17] J. Adam et al. (ALICE), Phys. Lett. **B760**, 720 (2016), 1601.03658
[18] R. Aaij et al. (LHCb) (2018), 1809.01404
[19] T. Sjostrand, S. Mrenna, P.Z. Skands, JHEP **05**, 026 (2006), hep-ph/0603175
[20] T. Sjostrand, S. Mrenna, P.Z. Skands, Comput. Phys. Commun. **178**, 852 (2008), 0710.3820
[21] S. Acharya et al. (ALICE) (2018), 1805.04367
[22] J. Adam et al. (ALICE), Phys. Lett. **B753**, 126 (2016), 1506.08032
[23] M. Cacciari, G.P. Salam, G. Soyez, Eur. Phys. J. **C72**, 1896 (2012), 1111.6097
[24] P. Nason, JHEP **11**, 040 (2004), hep-ph/0409146
[25] S. Jadach, W. Płaczek, S. Sapeta, A. Siódmok, M. Skrzypek, JHEP **10**, 052 (2015), 1503.06849
[26] G. D'Agostini, *Improved iterative Bayesian unfolding*, in *Alliance Workshop on Unfolding and Data Correction Hamburg, Germany, May 27-28, 2010* (2010), 1010.0632

Diffusion of charm and beauty in the Glasma

Marco Ruggieri[1,*] and *Santosh Kumar* Das[1,2]

[1]School of Nuclear Science and Technology, Lanzhou University, 222 South Tianshui Road, Lanzhou 730000, China.
[2]School of Physical Science, Indian Institute of Technology Goa, Ponda, Goa 403401, India.

Abstract. Relativistic nuclear collisions offer a unique way to study strong interactions at very high energy. The collision process can be described within the gluon saturation framework as the interaction of two colored glasses, and because of this interaction strong longitudinal gluon fields, namely the Glasma, are produced immediately after the collision. Besides, heavy quarks are also produced in the very early stage and because of their large mass and small concentration, their motion does not affect the evolution of the Glasma, thus behaving as ideal probes of the Glasma itself. We study the evolution of the heavy quarks in the Glasma allegedly produced in high energy p-Pb collisions by solving consistently the equations of motion of the quarks in the evolving Glasma fields. We find that this motion can be understood in terms of diffusion in momentum space, similarly to the random motion of a heavy probe in a hot thermalized medium. We show how the diffusion of heavy probes affects the nuclear modification factor of D and B mesons in p-Pb collisions.

1 Introduction

The study of the initial condition of the system produced by high energy collisions is a difficult but interesting problems related to the physics of relativistic heavy ion collisions (RHICs), as well as to that of high energy proton-proton (pp) and proton-nucleus (pA) collisions. At very high collision energy the two colliding projectiles can be described within the color-glass-condensate (CGC) effective theory [1–3] (see [4] for a review and a more complete reference list), in which fast partons are frozen by time dilation and act like static sources for low momentum gluons: their large occupation number allows for a classical treatment of these fields. The collision of two colored glasses leads to the formation of strong gluon fields in the forward light cone named the Glasma [5]; in weak coupling this consists of longitudinal color-electric and color-magnetic fields and they are described by the Classical Yang-Mills (CYM) theory. Among the high energy collisions mentioned above, pA attract a lot of interest because they allow for both a theoretical and an experimental study of the cold nuclear matter effects (CNME) which are not related to the formation of the quark-gluon plasma (QGP): these include shadowing [6] and gluon saturation [7–9].

Heavy quarks, namely charm and beauty, are excellent probes of the system created in high energy nuclear collisions, see [10–23] and references therein. As a matter of fact, their formation time can be roughly estimated as $\tau_{\rm form} \approx 1/(2m)$ with m the quark mass which gives

*e-mail: ruggieri@lzu.edu.cn

© The Authors, published by EDP Sciences. This is an open access article distributed under the terms of the Creative Commons Attribution License 4.0 (http://creativecommons.org/licenses/by/4.0/).

$\tau_{\rm form} \leq 0.1$ fm/c for charm and $\tau_{\rm form} \leq 0.04$ fm/c for beauty. Moreover, their number is very small compared to that of light quarks, thus charm and beauty carry a very tiny color current and their effect on the background evolving bulk (light quarks and gluons) is negligible. Because of these, charm and beauty can propagate and probe the evolving Glasma produced in high energy collisions bringing no disturbance to it.

In this proceeding we report on our recent study [24] in which we have focused on the diffusion of charm in high energy p-Pb collisions, and we extend the aforementioned study by publishing for the first time a result obtained for the diffusion of beauty. The main purpose of our study is to compute consistenly the propagation of the heavy quarks in the initial gluon fields, then computing the nuclear modification factor, $R_{\rm pPb}$, for $D-$mesons and $B-$mesons that has been measured recently [25, 26]. In fact, in [24] we have found that the propagation charm in the classical gluon field leads to $R_{\rm pPb}$ that reminds qualitatively the one measured for $D-$mesons in p-Pb collisions.

Because in our study we do not have a longitudinal expansion we do not attempt to a direct quantitative comparison with existing experimental data: while this will be the subject of forthcoming publications, we feel it is very important to understand the theoretical problem of the diffusion of heavy colored probes in the Glasma which has been overlooked by the community, and having this purpose in mind a first necessary step is to understand quantitatively the diffusion of these probes in a static gluon medium. It is worth mentioning that propagation of heavy quarks in the Glasma has been studied for the first time in [21] within a Fokker-Planck equation. The main differences with respect to [21] are that we do not rely on the small transferred momentum expansion, and we include the dynamical evolution of the gluon medium, eventually offering also a link to observables. Within our framework we solve consistently the CYM equations and the Wong equations; this approach is equivalent to solve the Boltzmann-Vlasov equations for the heavy quarks in a collisionless plasma. Nevertheless, our main results do not differ drastically by those of [21] and we agree with that reference by describing the evolution of the heavy probes in terms of diffusion in momentum space, similarly to what happens in a thermal medium.

2 The model

In this section we briefly describe the model we implement in our calculations; more details can be found in [24]. For the distribution of the color charges in the two colliding objects we adopt the standard MV model, in which the static color charge densities ρ_a on the nucleus A are assumed to be random variables that are normally distributed with zero mean and variance specified by the equation

$$\langle \rho_A^a(\boldsymbol{x}_T)\rho_A^b(\boldsymbol{y}_T)\rangle = (g^2\mu_A)^2\varphi_A\,(\boldsymbol{X}_T)\,\delta^{ab}\delta^{(2)}(\boldsymbol{x}_T-\boldsymbol{y}_T); \tag{1}$$

here A corresponds to either the proton or the Pb nucleus, a and b denote the adjoint color index; in this work we limit ourselves for simplicity to the case of the $SU(2)$ color group therefore $a, b = 1, 2, 3$. In Eq. (1) $g^2\mu_A$ denotes the color charge density and it is of the order of the saturation momentum Q_s [27].

The function $\varphi_A(\boldsymbol{X}_T)$ in Eq. (1) with $\boldsymbol{X}_T = (\boldsymbol{x}_T+\boldsymbol{y}_T)/2$ allows for a nonuniform probability distribution of the color charge in the transverse plane. For the case of the Pb nucleus in a p-Pb collision we assume a uniform probability and take $\varphi(\boldsymbol{x}_T) = 1$. On the other hand, for the proton we distribute the color charges by virtue of a standard gaussian distribution [28–31]:

$$\varphi_p(\boldsymbol{X}_T) = e^{-(X_T^2)/(2B_{cq})}, \tag{2}$$

with $B_{cq} = 3$ GeV. For the Pb nucleus we limit ourselves to show results obtained for $g^2\mu_{\text{Pb}} = 3.4$ GeV while for the proton we use $g^2\mu_{\text{p}} = 1.41$ GeV: a detailed discussion on how these parameters are fixed can be found in [24].

In order to determine the Glasma fields at $t = 0^+$ we firstly solve the Poisson equations for the gauge potentials, namely

$$-\partial_\perp^2 \Lambda^{(A)}(\boldsymbol{x}_T) = \rho^{(A)}(\boldsymbol{x}_T). \tag{3}$$

Wilson lines are computed as $V^\dagger(\boldsymbol{x}_T) = e^{i\Lambda^{(A)}(\boldsymbol{x}_T)}$, $W^\dagger(\boldsymbol{x}_T) = e^{i\Lambda^{(B)}(\boldsymbol{x}_T)}$, and the pure gauge fields of the two colliding nuclei are given by $\alpha_i^{(A)} = iV\partial_i V^\dagger$, $\alpha_i^{(B)} = iW\partial_i W^\dagger$. In terms of these fields the solution of the CYM in the forward light cone at initial time, namely the Glasma potentials, are given by $A_i = \alpha_i^{(A)} + \alpha_i^{(B)}$ for $i = x, y$ and $A_z = 0$, and the initial Glasma color-electric and color-magnetic fields are

$$E^z = i\sum_{i=x,y}\left[\alpha_i^{(B)}, \alpha_i^{(A)}\right], \tag{4}$$

$$B^z = i\left(\left[\alpha_x^{(B)}, \alpha_y^{(A)}\right] + \left[\alpha_x^{(A)}, \alpha_y^{(B)}\right]\right), \tag{5}$$

while the transverse fields are vanishing.

The dynamical evolution of Glasma is given by the CYM equations, namely

$$\frac{dA_i^a(x)}{dt} = E_i^a(x), \tag{6}$$

$$\frac{dE_i^a(x)}{dt} = \sum_j \partial_j F_{ji}^a(x) + \sum_{b,c,j} f^{abc} A_j^b(x) F_{ji}^c(x). \tag{7}$$

where the magnetic part of the field strength tensor is

$$F_{ij}^a(x) = \partial_i A_j^a(x) - \partial_j A_i^a(x) + \sum_{b,c} f^{abc} A_i^b(x) A_j^c(x); \tag{8}$$

here $f^{abc} = \varepsilon^{abc}$ with $\varepsilon^{123} = +1$.

We initialize the heavy probes in the transverse momentum plane by means of the standard Fixed Order + Next-to-Leading Log (FONLL) QCD calculation which describes the D-mesons spectra in pp collisions after fragmentation [32–34]

$$\left.\frac{dN}{d^2 p_T}\right|_{\text{prompt}} = \frac{x_0}{(x_1 + p_T)^{x_2}}; \tag{9}$$

a similar equation holds for beauty. We also assume that the initial longitudinal momentum vanishes. Initialization in coordinate space is done as follows: the tranverse coordinates distribution is built up by means of the function $\varphi(\boldsymbol{x}_T)$ in Eq. (2), because we expect the heavy quarks to be produced in the overlap region of proton and Pb nucleus that coincides with the transverse area of the proton; on the other hand, we use a uniform distribution for the longitudinal coordinate.

The dynamics of heavy quarks in the Glasma is studied by the Wong equations [35] that for a single quark can be written as

$$\frac{dx_i}{dt} = \frac{p_i}{E}, \tag{10}$$

$$E\frac{dp_i}{dt} = Q_a F_{i\nu}^a p^\nu, \tag{11}$$

$$E\frac{dQ_a}{dt} = -Q_c \varepsilon^{cba} \boldsymbol{A}_b \cdot \boldsymbol{p}, \tag{12}$$

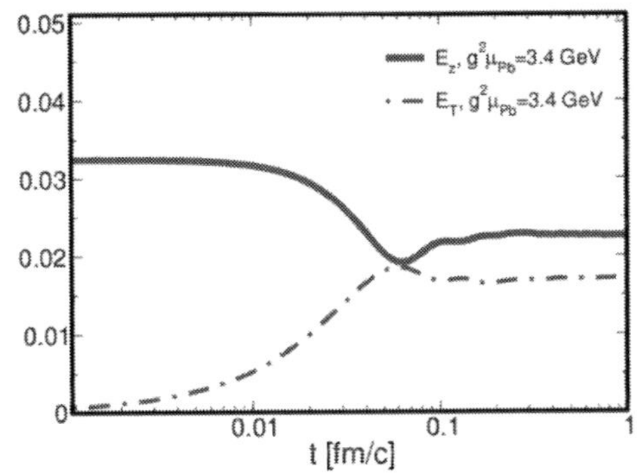

Figure 1. Averaged color-electric fields for p-Pb collision, measured in lattice units. Solid line corresponds to the longitudinal field while dashed line denote the transverse fields; we have put $g^2\mu_{\text{Pb}} = 3.4$ GeV. Lattice spacing is $\delta x = 0.04$ fm. Adapted from [24].

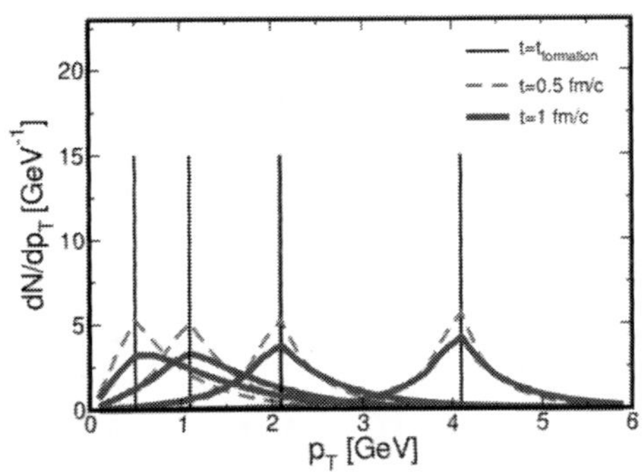

Figure 2. Evolution of δ–distribution functions of c–quarks in the Glasma fields. Black solid lines correspond to the initializations, green dashed lines to $t = 0.5$ fm/c and blue solid lines to $t = 1$ fm/c. We take $g^2\mu_{\text{Pb}} = 3.4$ GeV. Adapted from [24].

where $i = x, y, z$; here, the first two equations are the familiar Hamilton equations of motion for the coordinate and its conjugate momentum, while the third equation corresponds to the gauge invariant color current conservation. Here $E = \sqrt{p^2 + m^2}$ with $m = 1.5$ GeV for charm and $m = 4.5$ GeV for beauty. More details can be found in [24].

3 Results

In Fig. 1 we plot the averaged color-electric fields as a function of time. Solid line corresponds to the longitudinal field while dashed line denote the transverse fields; we have put $g^2\mu_{\text{Pb}} = 3.4$ GeV and lattice spacing is $\delta x = 0.04$ fm [24]. At initial time the system consists of purely longitudinal fields but the gluon dynamics leads to a quick production of transverse fields: in fact, within $t \approx 0.1$ fm/c a bulk made of both longitudinal and transverse fields is formed.

To elucidate the interaction of the c–quarks with the evolving Glasma we prepare initializations in which we put all the c–quarks in a thin p_T bin to obtain a δ–like distribution; the evolution of this distribution is studied by means of the Wong equations. This study is useful to better understand the interaction of the Glasma with different p_T modes. The results of this are shown in Fig. 2 in which we plot the spectrum of charms at initial time (solid black lines),

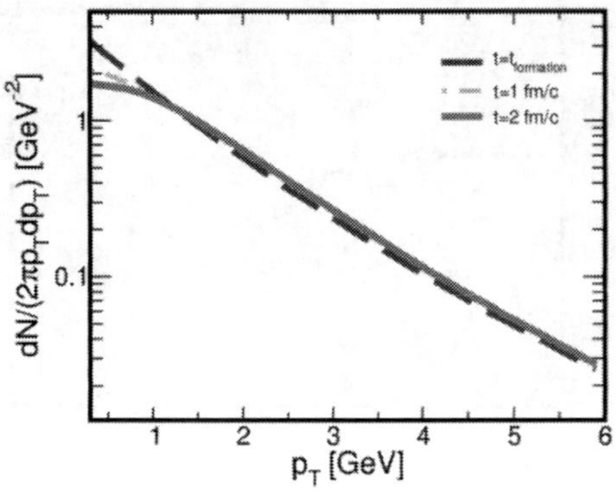

Figure 3. Spectrum of c−quarks at the initial time (dashed maroon line), at t = 1 fm/c (orange dot-dashed line) and at t = 2 fm/c (green solid line). We have put $g^2\mu_{\rm Pb}$ = 3.4 GeV.

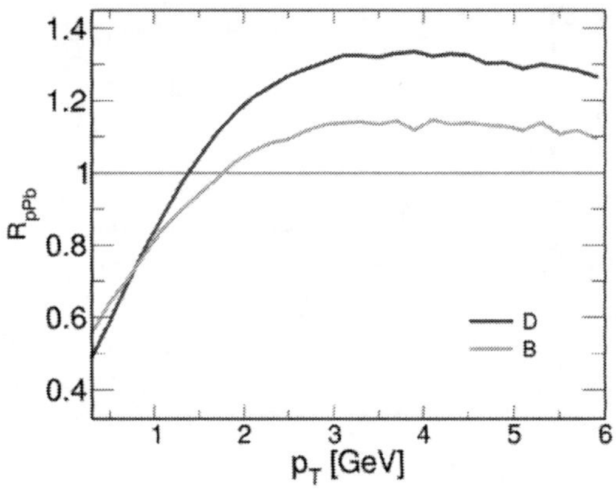

Figure 4. Nuclear modification factor for D−mesons (blue line) and B−mesons (orange line).

at t = 0.5 fm/c (green dashed lines) and at t = 1 fm/c (solid blue lines) for several values of the initial p_T. We notice that in all the cases examined here the interaction with the Glasma leads to the spreading of the spectrum, analogously to the standard diffusion in momentum space encountered in a Brownian motion. In addition to this, for low p_T we find a drift towards higher momenta that can be described as an average acceleration of the c−quarks. We notice that a drift is also present for higher values of p_T and tends to lower the $\langle p_T \rangle$ despite this effect is very small and for large p_T the diffusion dominates over the drift.

In Fig. 3 we plot the evolution of the charm spectrum with time. At the initial time the spectrum is given by the standard FONLL form; the interaction with the evolving Glasma bends downwards the spectrum at low p_T as a result of the diffusion discussed above: particles migrate from low to high p_T similarly to the acceleration that charged particles would experience in a strong electric field.

In Fig. 4 we plot the nuclear modification factor R_{pPb} defined as

$$R_{\text{pPb}} = \frac{\left(dN/d^2p_T\right)_{\text{Glasma}}}{(dN/d^2p_T)_{\text{FONLL}}}, \tag{13}$$

where $\left(dN/d^2p_T\right)_{\text{Glasma}}$ corresponds to the spectrum computed after the evolution of the heavy quarks with the Glasma: blue line denotes R_{pPb} for D−mesons and orange line corresponds to the same quantity for B−mesons. For both the lines we have used the standard fragmentation framework to extract the momentum distribution of mesons from the underlying distributions of quarks [36]. For both charm and beauty we have stopped the calculation at $t = 1$ fm/c. We notice that because of the diffusion in momentum space described above, the R_{pPb} is smaller than one up to $p_T \approx 1$ GeV; on the other hand, for large values of p_T it remains consistend with the perturbative FONLL within a 20%. The deviation of R_{pPb} from one is a direct consequence of the interaction of the heavy probes with the Glasma; this effect is more pronounced for the c−quarks while for beauty the overall effect is much milder because of its larger mass. Since in our calculation the longitudinal expansion is missing, we do not compare our results with experimental data [25, 26], despite we notice that the qualitative shape of these agrees with our theoretical calculations: in the future we will upgrade our calculations in order to take the longitudinal expansion into account in order to produce results that can be directly compared to experimental data.

4 Conclusions

In this talk we have reported on our study about the diffusion of heavy probes, namely charm and beauty, in the early stage of high energy nuclear collisions, focusing on the case of pA collisions. The dynamics of the gluon fields has been studied by means of the classical Yang-Mills equations; that of the heavy probes has been studied by the Wong equations solved on the top of the evolving background gluon field.

We have found that the evolution of the heavy quarks in the Glasma can be understood in terms of diffusion in momentum space [24] in agreement with [21]. This diffusion leads to a nuclear modification factor R_{pPb} different from one, signaling the effect of the interactions with the gluon fields during the evolution. In particular, for low p_T we have found $R_{\text{pPb}} \leq 1$ as a result of the diffusion and drift of small energy quarks towards higher momenta, similarly to what would happen to probe electric charges in a strong electric field. We have computed the nuclear modification factor for both D−mesons and B−mesons, studying the diffusion of both charm and beauty with the background gluon field. The diffusion is qualitatively the same for both charm and beauty, although quantitatively beauty experiences a smaller effect than charm because of its larger mass.

We have not attempted a direct comparison with experimental data because our calculations miss the longitudinal expansion; nevertheless, we feel that our results are an important step towards a more complete understanding of the dynamics of heavy colored probes in the strong gluon fields produced in the early stages of high energy nuclear collisions, and suggest that this dynamics should be further studied in order to have a quantitative estimate of the role of the pre-thermalization epoch on heavy quark observables in pp, pA and AA collisions.

Acknowledgements. M. R. acknowledges the committee of QCD@Work2018 held in Matera (Italy) where this talk has been presented, for their kind ospitality and the beautiful and inspiring environment that they have built up during the workshop, assisted by the marvelous architecture and breathtaking landscapes of the town where the workshop has taken place. The work of the authors is supported by the National Science Foundation of China (grants number 11875153 and 11805087) and by the Fundamental Research Funds for the Central Universities (grant number 862946).

References

[1] L. D. McLerran and R. Venugopalan, Phys. Rev. D **49**, 2233 (1994)
[2] L. D. McLerran and R. Venugopalan, Phys. Rev. D **49**, 3352 (1994)
[3] L. D. McLerran and R. Venugopalan, Phys. Rev. D **50**, 2225 (1994)
[4] F. Gelis *et al.*,Ann. Rev. Nucl. Part. Sci. **60**, 463 (2010).
[5] T. Lappi and L. McLerran, Nucl. Phys. A **772**, 200 (2006)
[6] K. J. Eskola, H. Paukkunen and C. A. Salgado, JHEP **0904**, 065 (2009)
[7] A. H. Rezaeian, Phys. Lett. B **718**, 1058 (2013)
[8] H. Fujii and K. Watanabe, Nucl. Phys. A **920**, 78 (2013)
[9] B. Ducloué, T. Lappi and H. Mäntysaari, Phys. Rev. D **91**, no. 11, 114005 (2015)
[10] R. Rapp *et al.*, arXiv:1803.03824 [nucl-th].
[11] G. Aarts *et al.*, Eur. Phys. J. A **53**, no. 5, 93 (2017)
[12] V. Greco, Nucl. Phys. A **967**, 200 (2017).
[13] S. K. Das *et al.*,Phys. Lett. B **768**, 260 (2017)
[14] S. K. Das, F. Scardina, S. Plumari and V. Greco, Phys. Lett. B **747**, 260 (2015)
[15] S. K. Das *et al.*,J. Phys. G **44**, no. 9, 095102 (2017)
[16] S. K. Das *et al.*,J. Phys. G **42**, no. 9, 095108 (2015)
[17] A. Beraudo *et al.*,JHEP **1603**, 123 (2016)
[18] Y. Xu *et al.*,Nucl. Part. Phys. Proc. **276-278**, 225 (2016)
[19] V. Ozvenchuk *et al.*,J. Phys. Conf. Ser. **779**, no. 1, 012033 (2017).
[20] S. K. Das, F. Scardina, S. Plumari and V. Greco, Phys. Rev. C **90**, 044901 (2014)
[21] S. Mrowczynski, Eur. Phys. J. A **54**, no. 3, 43 (2018)
[22] F. Prino and R. Rapp, J. Phys. G **43**, no. 9, 093002 (2016)
[23] A. Andronic *et al.*, Eur. Phys. J. C **76**, no. 3, 107 (2016)
[24] M. Ruggieri and S. K. Das, arXiv:1805.09617 [nucl-th].
[25] B. B. Abelev *et al.* [ALICE Collaboration], Phys. Rev. Lett. **113**, no. 23, 232301 (2014)
[26] R. Aaij *et al.* [LHCb Collaboration], JHEP **1710**, 090 (2017)
[27] T. Lappi, Eur. Phys. J. C **55**, 285 (2008)
[28] B. Schenke and R. Venugopalan, Phys. Rev. Lett. **113**, 102301 (2014)
[29] B. Schenke, S. Schlichting and R. Venugopalan, Phys. Lett. B **747**, 76 (2015)
[30] H. Mäntysaari, B. Schenke, C. Shen and P. Tribedy, Phys. Lett. B **772**, 681 (2017)
[31] H. Mäntysaari and B. Schenke, Phys. Rev. D **94** (2016) no.3, 034042
[32] M. Cacciari, M. Greco and P. Nason, JHEP **9805** (1998) 007; M. Cacciari, S. Frixione and P. Nason, JHEP **0103** (2001)
[33] M. Cacciari *et al.*,JHEP **1210** (2012) 137
[34] M. Cacciari, M. L. Mangano and P. Nason, arXiv:1507.06197 [hep-ph].
[35] S. K. Wong, Nuovo Cim. A **65**, 689 (1970)
[36] C. Peterson *et al.*, Phys. Rev. D **27**, 105 (1983).

Nuclear modification of prompt and non-prompt J/ψ production in p–Pb collisions at $\sqrt{s_{NN}}$ = 5.02 TeV with ALICE

Giuseppe Trombetta[1,*] on behalf of the ALICE Collaboration

[1]Dipartimento Interateneo di Fisica 'M. Merlin' and Sezione INFN, Bari, Italy

Abstract. Various Cold Nuclear Matter (CNM) effects, such as nuclear shadowing or partonic energy loss, can modify the production of J/ψ in heavy-ion collisions with respect to what is measured in elementary colliding systems. The study of p–Pb collisions at the Large Hadron Collider (LHC) energy scale represents a crucial tool to assess the influence of Cold Nuclear Matter on J/ψ production in order to achieve a more correct interpretation of Pb–Pb collision results. The ALICE detector at the LHC is capable of reconstructing J/ψ mesons at central rapidity through their e^+e^- decay channel down to zero transverse momentum (p_T), and has measured the fraction of J/ψ produced from the decay of beauty-flavoured hadrons (non-prompt J/ψ) in p–Pb collisions down to p_T = 1.3 GeV/c. In this paper, the results obtained by ALICE from the measurement of the prompt and non-prompt J/ψ yields at mid-rapidity in p–Pb collisions at $\sqrt{s_{NN}}$ = 5.02 TeV will be discussed in comparison to different theoretical predictions including CNM effects.

1 Introduction

Quarkonia and open heavy-flavoured hadrons have long been the subject of an intense theoretical and experimental effort. Their production represents a challenging testing ground for models based on quantum chromodynamics (QCD), and several initial- or final-state effects can contribute to modify the yields measured over different colliding systems. While heavy-quarks are considered as excellent tools for probing the transition of hadronic matter to a Quark–Gluon Plasma (QGP) phase in ultra-relativistic heavy-ion collisions, several mechanisms not related to the formation of QGP, and referred as Cold Nuclear Matter (CNM) effects, can contribute to modify the observed yields with respect to elementary nucleon-nucleon (NN) collisions. For heavy quarks produced at the LHC, the most relevant are the parton-density shadowing and gluon saturation effects, which can be described using modified nuclear parton distribution functions (nPDFs) [1] or in the framework of the Color-Glass Condensate (CGC) effective theory [2]. Other effects account for the energy loss or momentum broadening of partons [3, 4], which can occur in the early stages of the collision.

The measurement of heavy-flavoured hadrons produced in proton-nucleus (p–A) collisions and its comparison to pp results provides a crucial tool to constrain these mechanisms and

*e-mail: giuseppe.trombetta@ba.infn.it

© The Authors, published by EDP Sciences. This is an open access article distributed under the terms of the Creative Commons Attribution License 4.0 (http://creativecommons.org/licenses/by/4.0/).

disentangle the hot and cold nuclear effects envisioned in nucleus-nucleus (A–A) collisions. The inclusive production of J/ψ mesons at the LHC in p–Pb collisions at $\sqrt{s_{\rm NN}} = 5.02$ TeV was studied by ALICE at backward, mid- and forward rapidity down to zero transverse momentum ($p_{\rm T}$) [5]. Besides prompt J/ψ directly produced in the p-Pb collision or via the decay of heavier charmonium states (such as $\chi_{\rm c}$ and ψ(2S)), a significant contribution comes from non-prompt J/ψ which are produced after the weak decay of beauty-hadrons. ALICE has recently reported the measurement of such component [6], allowing an assessment of CNM effects on beauty quark production as well as a more direct comparison with models describing prompt charmonium production in heavy-ion collisions.

2 Data sample and analysis

The analysis for the measurement of the prompt and non-prompt J/ψ yields was carried out on the Minimum Bias (MB) data sample collected during the LHC p–Pb run in 2013, consisting of about 10^8 events and corresponding to an integrated luminosity $L_{\rm int} = 51.4 \pm 1.9\,\mu{\rm b}^{-1}$. In order to detect J/$\psi$ at central rapidity ($-0.43 < y_{\rm cms} < 1.37$), the resonance was reconstructed through its e^+e^- decay channel exploiting ALICE central barrel detectors, which cover the pseudorapidity range $|\eta| < 0.9$ [7]. The main detectors used are the Time Projection Chamber (TPC), which allows tracking and charged particle identification via specific ionisation energy loss (dE/dx) measurements, and the Inner Tracking System (ITS), which provides track and vertex reconstruction close to the interaction point. Other detectors, such as the V0 and T0 scintillators, are involved in the triggering of minimum bias events and in the rejection of beam-induced background.
The reconstruction of J/ψ candidates in the e^+e^- channel was performed by combining opposite-charge tracks reconstructed in the TPC and ITS and fulfilling dedicated selection criteria. In particular, the dE/dx signal in TPC was required to be compatible with that of an electron, while tracks compatible with the pion and proton assumptions were rejected. Furthermore, all tracks were required to have at least one hit in the innermost layer of the ITS in order to enhance the resolution of secondary decay vertices needed for the non-prompt signal extraction.
The measurement of the fraction $f_{\rm b}$ of the J/ψ yield originated from beauty-hadron decays was performed on a statistical basis, following the same approach used in previous analyses for the pp and Pb–Pb colliding systems [8, 9]. Such approach relies on the discrimination of J/ψ produced far from the primary p-Pb collision vertex by means of the pseudo-proper decay length (x) variable, defined as

$$x = \frac{\vec{L} \cdot \vec{p}_{\rm T}^{\,{\rm J}/\psi}}{p_{\rm T}^{{\rm J}/\psi}} \cdot \frac{c \cdot m^{{\rm J}/\psi}}{p_{\rm T}^{{\rm J}/\psi}} ,$$

with $\vec{L}$ being the vector from the primary vertex to the J/ψ decay vertex and $m^{{\rm J}/\psi}$ being the J/ψ pole mass value. An un-binned maximum likelihood fit to the two-dimensional distribution of invariant mass $m_{e^+e^-}$ and x of the candidate di-electron pairs is then performed to extract $f_{\rm B}$. As shown in Figure 1, prompt and non-prompt J/ψ exhibit well-distinguished x distributions, allowing their statistical separation down to $p_{\rm T}$ as low as ~ 1.3 GeV/c. The main systematic uncertainties affecting the determination of $f_{\rm B}$ arise from the estimation of the x resolution function and from the assumptions on the x and $m_{e^+e^-}$ probability density functions employed in the likelihood fit for the description of the background, prompt and non-prompt components. Figure 2 shows the values of the fraction of non-prompt J/ψ measured by ALICE as a function of transverse momenta compared to the results of ATLAS [10] covering the high $p_{\rm T}$

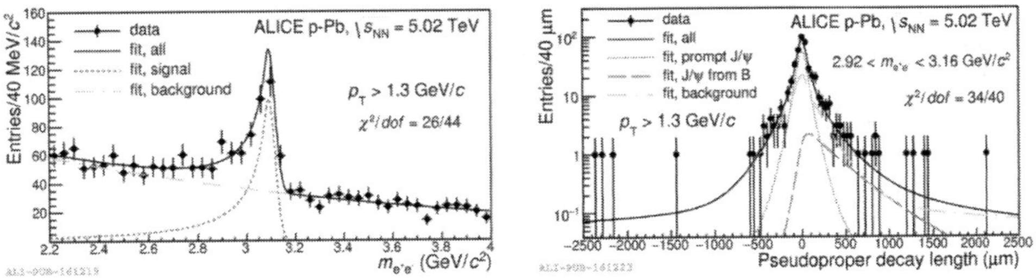

Figure 1. e^+e^- invariant mass (left panel) and pseudo-proper decay length (right panel) distributions of J/ψ candidates, with $p_T > 1.3$ GeV/c. The projections of the prompt, non-prompt and background components resulting from the likelihood fit are shown superimposed.

region ($p_T > 8$ GeV/c) in a similar rapidity range. Similar measurements performed in NN collisions are also shown.

3 Results

The modifications affecting J/ψ production due to the presence of the nuclear medium were quantified by means of the nuclear modification factor R_{pPb}. For a given y and p_T of the J/ψ, it is defined as the ratio of the differential production cross section in p–Pb collisions to that in pp collisions, scaled by the Pb atomic mass number A_{Pb}:

$$R_{\mathrm{pPb}}(y, p_T) = \frac{\mathrm{d}^2\sigma^{\mathrm{J}/\psi}_{\mathrm{pPb}}/\mathrm{d}y\mathrm{d}p_T}{A_{\mathrm{Pb}} \cdot \mathrm{d}^2\sigma^{\mathrm{J}/\psi}_{\mathrm{pp}}/\mathrm{d}y\mathrm{d}p_T}.$$

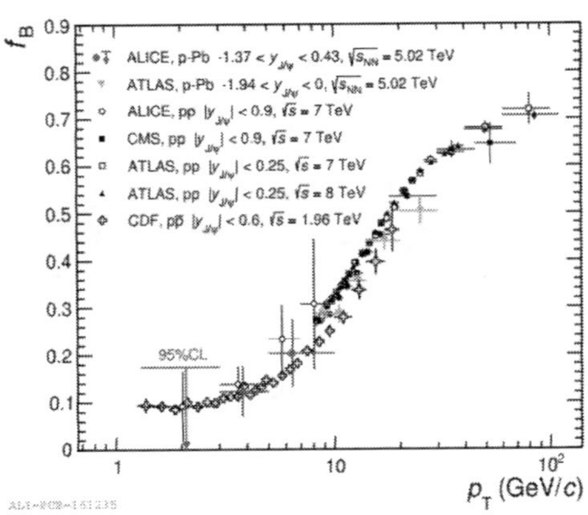

Figure 2. Fraction of J/ψ from the decay of b-hadrons at mid-rapidity as a function of the p_T of the J/ψ in p–Pb collisions at $\sqrt{s_{\mathrm{NN}}} = 5.02$ TeV compared with results from ATLAS [10] and with previous measurements performed in NN collisions at different energies. Vertical error bars represent the quadratic sum of the statistical and systematic errors. For the interval $1.3 < p_T < 3$ GeV/c, the upper limit at the 95% confidence level on the ALICE data point is shown.

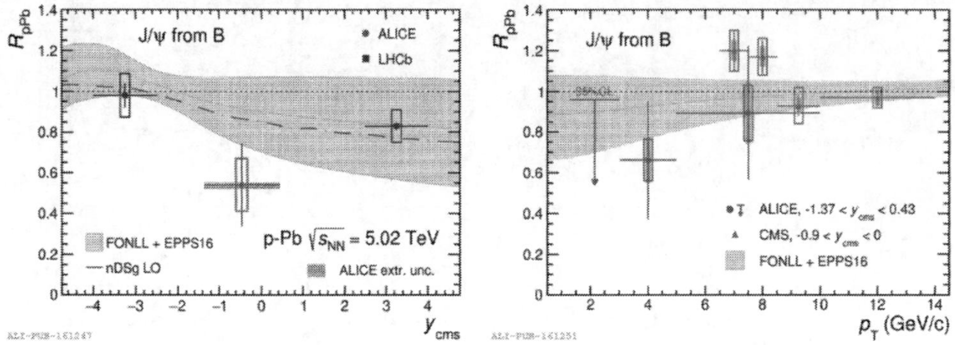

Figure 3. R_{pPb} of non-prompt J/ψ as a function of rapidity for $p_{\mathrm{T}} > 0$ (left panel) and as a function of p_{T} at mid-rapidity (right panel). Error bars and boxes indicate the statistical and systematic uncertainties, respectively. For the ALICE data points, the systematic uncertainty due to the extrapolation down to $p_{\mathrm{T}} = 0$ is depicted as a red box in the left panel, whereas the upper confidence limit at 95% confidence level for the interval $1.3 < p_{\mathrm{T}} < 3$ GeV/c is shown as an arrow in the right panel.

In the absence of nuclear effects, R_{pPb} is expected to equal unity, whereas $R_{\mathrm{Pb}} < 1$ ($R_{\mathrm{Pb}} > 1$) indicates a suppression (enhancement) with respect to what expected in the case of a scaling with the number of binary NN collisions.

By combining the R_{pPb} of inclusive J/ψ measured from the same data sample [5] with the interpolated value of f_{B} in pp collisions at $\sqrt{s} = 5.02$ TeV and with the measurements of f_{B} in p–Pb collisions, the nuclear modification factor for prompt and non-prompt J/ψ mesons at mid-rapidity was determined either p_{T}-integrated or as a function of transverse momentum in three p_{T} intervals.

Figure 3 shows the R_{pPb} of non-prompt J/ψ as measured by ALICE for $p_{\mathrm{T}} > 0$ in comparison to LHCb measurements [11] at backward and forward rapidity, and as a function of p_{T} in comparison to CMS results [12]. The measurements appear to be in reasonable agreement with theoretical calculations including gluon shadowing from the nuclear modification of the parton distribution functions [13, 14]. In Figure 4, the R_{pPb} of prompt J/ψ is similarly reported as a function of rapidity and transverse momentum, compared to predictions from various models implementing CNM effects. In particular, data appear in fair agreement with models combining parton shadowing with energy loss mechanisms [15] or with the effects from the interaction with a nuclear medium [16], while purely energy loss calculations [17] appear disfavoured. Recent models based on different implementations of the CGC effective theory [18, 19] are also in good agreement with the measurements at forward rapidity.

4 Summary

The production of J/ψ mesons in p-Pb collisions at the LHC has been measured by ALICE at mid-rapidity and down to J/ψ p_{T} of 1.3 GeV/c. The nuclear modification factor has been computed for both prompt and non-prompt J/ψ in order to quantify the effects of the cold nuclear medium on charmonium and beauty production. ALICE results, covering the low p_{T} region at mid-rapidity, complement the data of the other LHC experiments and suggest that the reduced production observed at mid-rapidity for both prompt and non-prompt J/ψ with respect to expectations from pp collisions is a low-p_{T} effect. While the suppression is

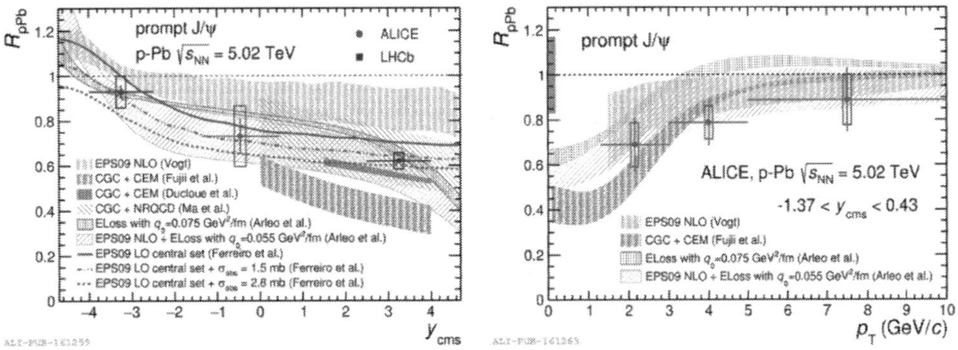

Figure 4. R_{pPb} of prompt J/ψ as a function of rapidity for $p_T > 0$ (left panel) and as a function of p_T at mid-rapidity (right panel). Statistical and systematic uncertainties are shown as vertical error bars and open boxes, respectively. The size of the p_T-correlated relative systematic uncertainties is shown as a grey box around R_{pPb} = 1 in the right panel.

compatible within uncertainties to theoretical predictions employing CNM effects, present uncertainties still do not allow a precise discrimination among the different models.

References

[1] N. Armesto, J. Phys. G **32** (2006) R367.
[2] F. Gelis, Int. J. Mod. Phys. A **28** (2013) 1330001.
[3] I. Vitev, Phys. Rev. C **75** (2007) 064906.
[4] B. Z. Kopeliovich, J. Nemchik, A. Schafer and A. V. Tarasov, Phys. Rev. Lett. **88** (2002) 232303.
[5] J. Adam *et al.* [ALICE Collaboration], JHEP **1506** (2015) 055.
[6] S. Acharya *et al.* [ALICE Collaboration], Eur. Phys. J. C **78** (2018) no.6, 466.
[7] K. Aamodt *et al.* [ALICE Collaboration], JINST **3** (2008) S08002.
[8] B. Abelev *et al.* [ALICE Collaboration], JHEP **1211** (2012) 065.
[9] J. Adam *et al.* [ALICE Collaboration], JHEP **1507** (2015) 051.
[10] G. Aad *et al.* [ATLAS Collaboration], Phys. Rev. C **92** (2015) no.3, 034904.
[11] R. Aaij *et al.* [LHCb Collaboration], JHEP **1402** (2014) 072.
[12] A. M. Sirunyan *et al.* [CMS Collaboration], Eur. Phys. J. C **77** (2017) no.4, 269.
[13] K. J. Eskola, P. Paakkinen, H. Paukkunen and C. A. Salgado, Eur. Phys. J. C **77** (2017) no.3, 163.
[14] D. de Florian, R. Sassot, P. Zurita and M. Stratmann, Phys. Rev. D **85** (2012) 074028.
[15] F. Arleo, R. Kolevatov, S. Peigné and M. Rustamova, JHEP **1305** (2013) 155.
[16] E. G. Ferreiro, F. Fleuret, J. P. Lansberg and A. Rakotozafindrabe, Phys. Rev. C **88** (2013) no.4, 047901.
[17] R. Vogt, Phys. Rev. C **81** (2010) 044903.
[18] B. Ducloué, T. Lappi and H. Mäntysaari, Phys. Rev. D **91** (2015) no.11, 114005.
[19] Y. Q. Ma, R. Venugopalan and H. F. Zhang, Phys. Rev. D **92** (2015) 071901.

EPJ Web of Conferences **192**, 00019 (2018) https://doi.org/10.1051/epjconf/201819200019
QCD@Work 2018

QCD critical end point from a realistic PNJL model

Kun Xu[1,2,*], *Zhibin* Li[1,2,**], and *Mei* Huang[1,2,***]

[1]Institute of High Energy Physics, Chinese Academy of Sciences, Beijing 100049, China
[2]University of Chinese Academy of Sciences, Beijing 100049, China

Abstract. With parameters fixed by critical temperature and equation of state at zero baryon chemical potential, a realistic Polyakov–Nambu–Jona-Lasinio (rPNJL) model predicts a critical end point of chiral phase transition at (μ_B^E = 720MeV, T^E = 93MeV). The extracted freeze-out line from heavy ion collisions is close to the chiral phase transition boundary in the rPNJL model, and the kurtosis $\kappa\sigma^2$ of baryon number fluctuations from the rPNJL model along the experimental freeze-out line agrees well with the BES-I measurement. Our analysis shows that the dip structure of measured $\kappa\sigma^2$ is determined by the relationship between the freeze-out line and chiral phase transition line at low baryon density region, and the peak structure can be regarded as a clean signature for the existence of CEP.

1 Introduction

Exploring Quantum Chromodynamics (QCD) phase structure and understanding properties of QCD matter under extreme conditions are main topics of heavy ion collisions. It is generally believed that there is a QCD critical end point (CEP) for chiral phase transition at finite baryon density, and searching for the CEP and locating the CEP become one of the central goals of beam energy scan (BES) program at Relativistic Heavy Ion Collisions (RHIC), the Facility for Antiproton and Ion Research (FAIR) in Darmstadt and Nuclotron-based Ion Collider Facility (NICA) in Dubna.

The higher order fluctuations of conserved charges carries the divergency feature of the correlation length at the CEP thus are regarded as good observables of CEP [1]. The measurement on the higher cumulants of conserved charges from BES-I at RHIC for Au+Au collisions shows a non-monotonic energy dependent behavior for the kurtosis of the net proton number distributions $\kappa\sigma^2$ in the collision energy from $\sqrt{s_{NN}}$ = 200 to 7.7 GeV, corresponding to the baryon chemical potential μ_B from 0.1 to 0.4 GeV [2–4]. It is observed that $\kappa\sigma^2$ of net proton number distributions starts from about 1 at $\sqrt{s_{NN}}$ = 200 GeV, decreases to around 0.1 at $\sqrt{s_{NN}}$ = 20 GeV and rises quickly up to 3.5 at $\sqrt{s_{NN}}$ = 7.7 GeV. It calls for theoretical understanding whether this non-monotonic structure is related to the existence of the QCD CEP.

The QCD CEP has been investigated from lattice theory [5], and the recent lattice result excluded the existence of CEP in the low baryon density region $\mu_B/T < \pi$ [6]. The QCD CEP

*e-mail: xukun@ihep.ac.cn
**e-mail: lizb@ihep.ac.cn
***e-mail: huangm@ihep.ac.cn

© The Authors, published by EDP Sciences. This is an open access article distributed under the terms of the Creative Commons Attribution License 4.0 (http://creativecommons.org/licenses/by/4.0/).

has also been widely analyzed through symmetry analysis [7], and by using effective chiral models, e.g., the Nambu–Jona-Lasinio (NJL) models including the Polyakov-loop improved NJL (PNJL) model [8], quark-meson (QM) models including the Polyakov-loop improved QM model [9], the Dyson-Schwinger equations (DSE) [10], as well as the 5-dimension holographic QCD model [11, 12]. The model studies lose the power on predicting the exact location of the CEP, because different models even the same model with different parameter sets give various location of CEP. However, still we can extract some useful information from model studies, e.g., the peak structure of $\kappa\sigma^2$ along the freeze-out line can be used as a clean signature for the existence of CEP, and the peak location of the measured $\kappa\sigma^2$ is close to the real QCD CEP mountain [12, 13].

In this talk, we show the results from a realistic Polyakov–Nambu–Jona-Lasinio (rPNJL) model [13]. It is interesting to observe that the kurtosis $\kappa\sigma^2$ produced from the rPNJL model along the experimental freeze-out line agrees with BES-I data well!

2 Model setup

The 3-flavor rPNJL model takes into account 8-quark interaction [14], and its effective potential is given below:

$$\begin{aligned}\Omega &= g_S\sum_f \sigma_f^2 - \frac{g_D}{2}\sigma_u\sigma_d\sigma_s + 3\frac{g_1}{2}(\sum_f \sigma_f^2)^2 + 3g_2\sum_f \sigma_f^4 - 6\int_0^\Lambda \frac{d^3p}{(2\pi)^3}E_f \\ &-2T\int\frac{d^3p}{(2\pi)^3}\ln[1+3(\Phi+\bar{\Phi}e^{-(E_f-\mu_f)/T})e^{-(E_f-\mu_f)/T}+e^{-3(E_f-\mu_f)/T}] \\ &-2T\int\frac{d^3p}{(2\pi)^3}\ln[1+3(\Phi+\bar{\Phi}e^{-(E_f+\mu_f)/T})e^{-(E_f+\mu_f)/T}+e^{-3(E_f+\mu_f)/T}] \\ &+U'(\Phi,\bar{\Phi},T). \end{aligned} \tag{1}$$

Where f takes u, d for two light flavors while s for strange quark, $\sigma_f = \langle\bar{\psi}_f\psi_f\rangle$ is quark-antiquark condensate for different flavors, and $E_f = \sqrt{p^2 + M_f^2}$ with M_f the dynamically generated constituent quark mass taking the form of

$$M_f = m_f - 2g_S\sigma_f + \frac{g_D}{4}\sigma_{f+1}\sigma_{f+2} - 2g_1\sigma_f(\sum_{f'}\sigma_{f'}^2) - 4g_2\sigma_f^3. \tag{2}$$

Here if $\sigma_f = \sigma_u$, then $\sigma_{f+1} = \sigma_d$ and $\sigma_{f+2} = \sigma_s$. U' takes the form of [15]

$$\frac{U'}{T^4} = \frac{U}{T^4} - \kappa\ln[J(\Phi,\bar{\Phi})], \tag{3}$$

which describes the self-interaction of the Polyakov-loop Φ and $\bar{\Phi}$, κ is a dimensionless parameter,

$$\frac{U}{T^4} = -\frac{b_2(T)}{2}\bar{\Phi}\Phi - \frac{b_3}{6}(\Phi^3+\bar{\Phi}^3) + \frac{b_4}{4}(\Phi\bar{\Phi})^2, \tag{4}$$

with $b_2(T) = a_0 + a_1\frac{T_0}{T}\exp(-a_2\frac{T}{T_0})$, and

$$J = (\frac{27}{24\pi^2})(1 - 6\Phi\bar{\Phi} + 4(\Phi^3+\bar{\Phi}^3) - 3(\Phi\bar{\Phi})^2). \tag{5}$$

The parameters of the NJL part shown in Table 1 are fixed by vacuum properties, and the parameters of Polyakov loop part listed in Table 2 are fixed by global fitting of the pressure density at zero chemical potential. With these parameters, the critical temperature and equation state at zero chemical potential in the rPNJL model agree well with lattice data [14].

$m_{u,d}$(MeV)	m_s(MeV)	Λ(MeV)	$g_S\Lambda^2$	$g_D\Lambda^5$	g_1(MeV^{-8})	g_2(MeV^{-8})
5.5	183.468	637.720	2.914	75.968	2.193×10^{-21}	-5.890×10^{-22}

Table 1. Parameters for the NJL part in the rPNJL model.

T_0 (MeV)	a_0	a_1	a_2	b_3	b_4	κ
175	6.75	-9.8	0.26	0.805	7.555	0.1

Table 2. Parameters for the Polyakov loop part in the rPNJL model.

3 Numerical results

From the pressure $P = -\Omega$, the minus thermodynamical potential, one can obtain the net-baryon number fluctuations [4]

$$\chi_n^B = \frac{\partial^n[P/T^4]}{\partial[\mu_B/T]^n}, \tag{6}$$

which gives the cumulants of baryon number distributions $C_n^B = VT^3\chi_n^B$. By introducing the variance $\sigma^2 = C_2^B$ and kurtosis $\kappa = C_4^B/(\sigma^2)^2$, we can write the ratio of fourth and second order cumulants of net-baryon number fluctuations as

$$\kappa\sigma^2 = \frac{C_4^B}{C_2^B} = \frac{\chi_4^B}{\chi_2^B}. \tag{7}$$

The value of $\kappa\sigma^2$ of baryon number fluctuation is 1 in the hadron resonance gas (HRG) phase and it takes the value of $\kappa\sigma^2 \simeq 0.068$ in the ideal free quark gas (FQG) limit at very high temperature [16].

With the parameters fixed in Table 1 and 2, we can obtain $\kappa\sigma^2$ of baryon number fluctuation in the rPNJL model as shown in Fig.1 and compare with lattice data in Ref.[16]. The HRG limit and FQG limit are also shown in Fig.1. It is noticed that above the chiral phase transition T_c, the result of $\kappa\sigma^2$ agrees well with lattice result, and below T_c, $\kappa\sigma^2$ of baryon number fluctuation is only about half of the HRG limit. However, in the regular NJL model, the magnitude of $\kappa\sigma^2$ of baryon number fluctuation is much smaller than the the lattice result [13]. This indicates that gluodynamics contribution plays dominant role in the baryon number fluctuations.

The chiral phase boundary in the rPNJL model is shown in Fig. 2, with the CEP located at ($\mu_B^E = 720$MeV$, T^E = 93$MeV). We also show the freeze-out temperatures and baryon number chemical potentials extracted from experiment in Fig. 2, where the freeze-out temperatures and baryon number chemical potentials extracted from BES-I at RHIC [17] are shown in dots, and the freeze-out temperatures and baryon number chemical potentials extracted from lower energy heavy-ion collisions [18] are shown in triangles. We use two fitted freeze-out lines

$$\begin{aligned} f1: &\quad T(\mu) = 0.158 - 0.14\mu^2 - 0.04\mu^4 - 0.013(0.948-\mu)^2, \\ f2: &\quad T(\mu) = 0.158 - 0.14\mu^2 - 0.04\mu^4. \end{aligned} \tag{8}$$

which are shown in dashed and dashed-dotted lines, respectively. The second freeze-out line is taken from Ref.[4]. Both fitted experimental freeze-out lines are very close to the chiral phase boundary from the rPNJL model. The only difference is that at low baryon density

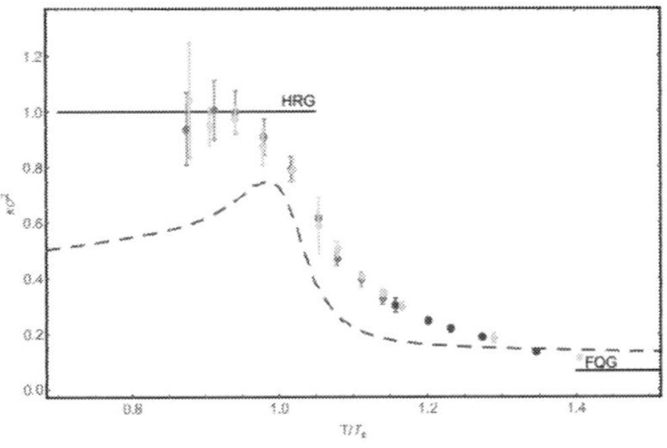

Figure 1. $\kappa\sigma^2$ of baryon number fluctuation in the rPNJL model as a function of the temperature nomalized by the chiral phase transition temperature $T_c = 161\text{MeV}$ at zero baryon chemical potential.

region, the fitted experimental freeze-out line $f1$ starts from above the chiral phase boundary, and $f2$ starts from below the phase boundary. The 3D plot of the kurtosis $\kappa\sigma^2$ as functions of the temperature and chemical potential calculated from the rPNJL model is shown in the right of Fig. 2, from which it can be clearly seen that the first freeze-out line $f1$ starts from the back ridge of the phase boundary and $f2$ starts from the front ridge of the phase boundary.

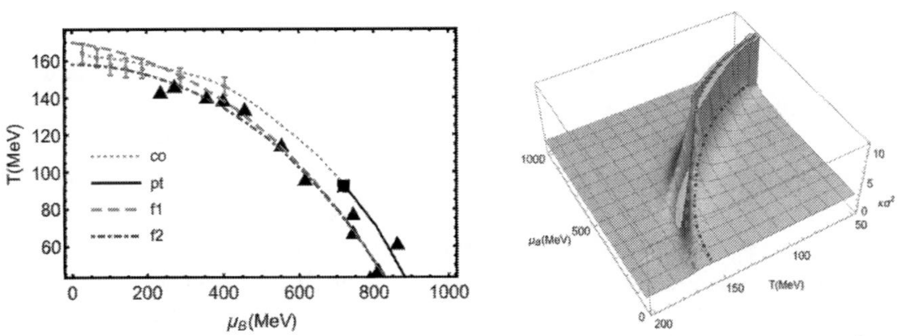

Figure 2. (Left) The chiral phase boundary for the u, d quarks in the rPNJL model with the CEP located at ($\mu_B^E = 720\text{MeV}, T^E = 93\text{MeV}$) (marked by a square), and two freeze-out lines $f1, f2$ defined in Eq.(8) by fitting the freeze-out temperatures and baryon number chemical potentials extracted from BES-I at RHIC [17] (in dots), and for lower energy heavy-ion collisions [18] (in triangles). (Right) The 3D plot of $\kappa\sigma^2$ for the baryon number fluctuations as functions of the temperature and baryon number chemical potential in the rPNJL model, two freeze-out lines $f1, f2$ are shown in dashed and dashed-dotted lines, respectively.

We show the kurtosis $\kappa\sigma^2$ of baryon number fluctuations calculated from the rPNJL model along the two fitted freeze-out lines $f1, f2$ from heavy ion collisions in Fig. 3 and compare with BES-I measurement of $\kappa\sigma^2$ (shown in dots with error bars) [2–4]. The left figure is as a function of the baryon number chemical potential and the right figure is as a function of the collision energy, where the relation between the chemical potential and the collision energy [18] $\mu_B(\sqrt{s}) = \frac{1.477}{1+0.343\sqrt{s}}$ has been used. As a reference, the $\kappa\sigma^2$ calculated from a realistic regular NJL model along the freeze-out line $f2$ [19] is also shown as long dashed line. Because there is no gluon contribution in the regular NJL model, the value of kurtosis in general is much smaller than that in the rPNJL model.

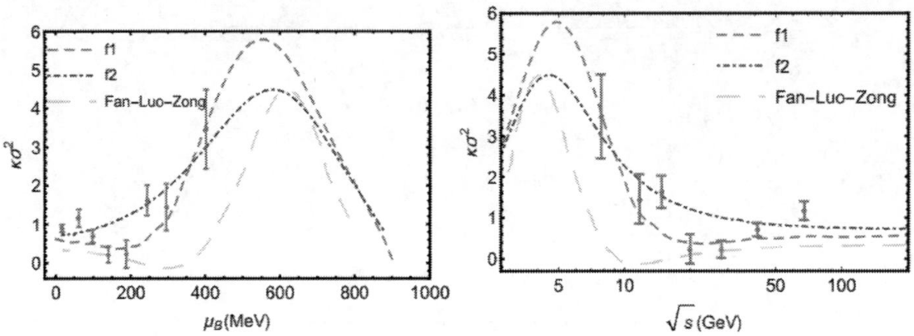

Figure 3. $\kappa\sigma^2$ produced from the rPNJL model as a function of the baryon chemical potential μ_B (Left) and the collision energy $\sqrt{s}$ (Right) along the experimental freeze-out lines $f1$ (dashed line) and $f2$ (dashed-dotted line) comparing with BES-I measurement. As a reference, $\kappa\sigma^2$ from a realistic NJL model [19] along the experimental freeze-out line $f2$ is shown in long dashed line.

It is observed that the kurtosis $\kappa\sigma^2$ produced from the realistic PNJL model along the freeze-out line $f1$ develops a dip structure around μ_B = 200MeV ($\sqrt{s}$ = 20GeV) and a peak structure at around μ_B = 500MeV ($\sqrt{s}$ = 5GeV, while along the freeze-out line $f2$, $\kappa\sigma^2$ only develops a peak structure at around μ_B = 500MeV ($\sqrt{s}$ = 5GeV and no dip structure is developed. As we have discussed that the only difference between the two freeze-out lines is at the low baryon density region: $f1$ crosses the chiral phase boundary from above, i.e. starts from the back ridge of the phase boundary, from the 3D plot in Fig. 2, it is natural to see how the dip structure is formed. For the case along the freeze-out line $f2$, which always lies below the phase boundary, there is no chance to form the dip structure. The peak structure along the freeze-out line is solely determined by the existence of the CEP. From this analysis, we can see that the dip structure of measured $\kappa\sigma^2$ is determined by the relationship between the freeze-out line and chiral phase transition line at low baryon density region, and the peak structure can be regarded as a clean signature for the existence of CEP.

Another unexpected result is that the kurtosis $\kappa\sigma^2$ from the equilibrium result in the rP-NJL model along the experimental freeze-out line $f1$ agrees with BES-I measurement very well. This is a surprising result, because in general, one should take into account the non-equilibrium evolution of the system in heavy ion collisions. However, as shown in Ref.[20] that the kurtosis of baryon number fluctuation even changes sign and becomes negative. It is worthy of mentioning that the measurement of $\kappa\sigma^2$ is always positive, therefore it deserves further understanding on how the "little bang" after collision evolves with time. As shown in [21], one possible scenario is that the system after collision reaches thermalization very quickly in quite high temperature and then evolves in equilibrium state. From our analysis, to form the dip structure, the freeze-out temperature at low baryon density region should be higher than the phase transition temperature! If this scenario is correct, the equilibrium result from a realistic PNJL model might work.

4 Summary

In summary, we have investigated the kurtosis of the baryon number fluctuations $\kappa\sigma^2$ in a reparameterized realistic PNJL model, in which the critical temperature, equation of state and baryon number fluctuations are in good agreement with lattice data at zero chemical potential. This rPNJL model predicts a CEP located at (μ_B^E = 720MeV, T^E = 93MeV). The extracted experimental freeze-out line from heavy ion collisions is very close to the chiral phase transition boundary in the rPNJL model. It is surprised to see that the kurtosis $\kappa\sigma^2$ of baryon number fluctuations from the rPNJL model along the experimental freeze-out line agrees well with the BES-I measurement. We also analyzed the formation of the dip structure and peak structure of measured $\kappa\sigma^2$ along the freeze-out line. Our analysis shows that the dip structure of $\kappa\sigma^2$ is determined by the relationship between the freeze-out line and chiral phase transition line at low baryon density region, and the peak structure is solely determined by the divergency at CEP, and the peak structure can be regarded as a clean signature for the existence of CEP. It is worthy of mentioning that at low baryon density region, the extracted freeze-out temperatures from BES-I measurement at RHIC [17] are indeed higher than the critical temperatures, this supports our analysis on the formation of dip structure of $\kappa\sigma^2$ along the freeze-out line.

Our result shows that the equilibrium result can explain the BES-I data on baryon number fluctuations, this may indicate that the system after collision reaches thermalization quickly and evolve in equilibrium before the freeze-out and phase transition, which is in agreement with the analysis in [21]. For example, in the collision energy of $\sqrt{s} = 200$GeV, the system reaches thermalization at around the temperature of $T \simeq 210 - 230$MeV , which is much higher than the freeze-out temperature and the phase transition temperature around 160MeV.

Acknowledgments

This work was supported in part by the NSFC under Grant Nos. 11725523, 11735007, and 11261130311 (CRC 110 by DFG and NSFC).

References

[1] M. A. Stephanov, "Non-Gaussian fluctuations near the QCD critical point," *Phys. Rev. Lett.* **102** (2009) 032301, arXiv:0809.3450 [hep-ph]; M. A. Stephanov, "On the sign of kurtosis near the QCD critical point," *Phys. Rev. Lett.* **107** (2011) 052301, arXiv:1104.1627 [hep-ph]; M. Asakawa, S. Ejiri, and M. Kitazawa, "Third moments of conserved charges as probes of QCD phase structure," *Phys. Rev. Lett.* **103** (2009) 262301, arXiv:0904.2089 [nucl-th]; C. Athanasiou, K. Rajagopal, and M. Stephanov, "Using Higher Moments of Fluctuations and their Ratios in the Search for the QCD Critical Point," *Phys. Rev.* **D82** (2010) 074008, arXiv:1006.4636 [hep-ph].

[2] **STAR** Collaboration, L. Adamczyk *et al.*, "Energy Dependence of Moments of Net-proton Multiplicity Distributions at RHIC," *Phys. Rev. Lett.* **112** (2014) 032302, arXiv:1309.5681 [nucl-ex].

[3] **STAR** Collaboration, M. M. Aggarwal *et al.*, "Higher Moments of Net-proton Multiplicity Distributions at RHIC," *Phys. Rev. Lett.* **105** (2010) 022302, arXiv:1004.4959 [nucl-ex].

[4] X. Luo and N. Xu, "Search for the QCD Critical Point with Fluctuations of Conserved Quantities in Relativistic Heavy-Ion Collisions at RHIC : An Overview," *Nucl. Sci. Tech.* **28** no. 8, (2017) 112, arXiv:1701.02105 [nucl-ex].

[5] Z. Fodor and S. D. Katz, "A New method to study lattice QCD at finite temperature and chemical potential," *Phys. Lett.* **B534** (2002) 87–92, arXiv:hep-lat/0104001 [hep-lat]; H.-T. Ding, F. Karsch, and S. Mukherjee, "Thermodynamics of strong-interaction matter from Lattice QCD," *Int. J. Mod. Phys.* **E24** no. 10, (2015) 1530007, arXiv:1504.05274 [hep-lat]; C. Schmidt and S. Sharma, "The phase structure of QCD," *J. Phys.* **G44** no. 10, (2017) 104002, arXiv:1701.04707 [hep-lat].

[6] V. Vovchenko, J. Steinheimer, O. Philipsen, and H. Stoecker, "Cluster Expansion Model for QCD Baryon Number Fluctuations: No Phase Transition at $\mu_B/T < \pi$," arXiv:1711.01261 [hep-ph].

[7] R. D. Pisarski and F. Wilczek, "Remarks on the Chiral Phase Transition in Chromodynamics," *Phys. Rev.* **D29** (1984) 338–341; Y. Hatta and T. Ikeda, "Universality, the QCD critical / tricritical point and the quark number susceptibility," *Phys. Rev.* **D67** (2003) 014028, arXiv:hep-ph/0210284 [hep-ph];

[8] T. M. Schwarz, S. P. Klevansky, and G. Papp, "The Phase diagram and bulk thermodynamical quantities in the NJL model at finite temperature and density," *Phys. Rev.* **C60** (1999) 055205, arXiv:nucl-th/9903048 [nucl-th];
P. Zhuang, M. Huang, and Z. Yang, "Density effect on hadronization of a quark plasma," *Phys. Rev.* **C62** (2000) 054901, arXiv:nucl-th/0008043 [nucl-th];
J.-W. Chen, J. Deng, and L. Labun, "Baryon susceptibilities, non-Gaussian moments, and the QCD critical point," *Phys. Rev.* **D92** no. 5, (2015) 054019, arXiv:1410.5454 [hep-ph];
J.-W. Chen, J. Deng, H. Kohyama, and L. Labun, "Robust characteristics of non-gaussian fluctuations from the NJL model," *Phys. Rev.* **D93** no. 3, (2016) 034037, arXiv:1509.04968 [hep-ph];
W.-j. Fu and Y.-l. Wu, "Fluctuations and Correlations of Conserved Charges near the QCD Critical Point," *Phys. Rev.* **D82** (2010) 074013, arXiv:1008.3684 [hep-ph];
K. Fukushima, "Phase diagrams in the three-flavor Nambu-Jona-Lasinio model with the Polyakov loop," *Phys. Rev.* **D77** (2008) 114028, arXiv:0803.3318 [hep-ph]. [Erratum: Phys. Rev.D78,039902(2008)];
N. Weiss, "The Effective Potential for the Order Parameter of Gauge Theories at Finite Temperature," *Phys. Rev.* **D24** (1981) 475;
C. Ratti, S. Roessner, M. A. Thaler, and W. Weise, "Thermodynamics of the PNJL model," *Eur. Phys. J.* **C49** (2007) 213–217, arXiv:hep-ph/0609218 [hep-ph];
G.-y. Shao, Z.-d. Tang, X.-y. Gao, and W.-b. He, "Baryon number fluctuations and QCD phase structure," arXiv:1708.04888 [hep-ph];
X.-y. Xin, S.-x. Qin, and Y.-x. Liu, "Improvement on the Polyakov–Nambu–Jona-Lasinio model and the QCD phase transitions," *Phys. Rev.* **D89** no. 9, (2014) 094012;
G.-y. Shao, Z.-d. Tang, M. Di Toro, M. Colonna, X.-y. Gao, and N. Gao, "Phase transition of strongly interacting matter with a chemical potential dependent Polyakov loop potential," *Phys. Rev.* **D94** no. 1, (2016) 014008, arXiv:1603.09033 [nucl-th];
M. Dutra, O. Louren?o, A. Delfino, T. Frederico, and M. Malheiro, "Polyakov–Nambu–Jona-Lasinio phase diagrams and quarkyonic phase from order parameters," *Phys. Rev.* **D88** no. 11, (2013) 114013, arXiv:1312.1130 [hep-ph].

[9] E. S. Bowman and J. I. Kapusta, "Critical Points in the Linear Sigma Model with Quarks," *Phys. Rev.* **C79** (2009) 015202, arXiv:0810.0042 [nucl-th];
H. Mao, J. Jin, and M. Huang, "Phase diagram and thermodynamics of the Polyakov linear sigma model with three quark flavors," *J. Phys.* **G37** (2010) 035001, arXiv:0906.1324 [hep-ph];
B. J. Schaefer and M. Wagner, "QCD critical region and higher moments for three flavor models," *Phys. Rev.* **D85** (2012) 034027, arXiv:1111.6871 [hep-ph];
B.-J. Schaefer and M. Wagner, "Higher-order ratios of baryon number cumulants," *Central Eur. J. Phys.* **10** (2012) 1326–1329, arXiv:1203.1883 [hep-ph];
B.-J. Schaefer, J. M. Pawlowski, and J. Wambach, "The Phase Structure of the Polyakov–Quark-Meson Model," *Phys. Rev.* **D76** (2007) 074023, arXiv:0704.3234 [hep-ph].

[10] S.-x. Qin, L. Chang, H. Chen, Y.-x. Liu, and C. D. Roberts, "Phase diagram and critical endpoint for strongly-interacting quarks," *Phys. Rev. Lett.* **106** (2011) 172301, arXiv:1011.2876 [nucl-th]; J. Luecker, C. S. Fischer, L. Fister, and J. M. Pawlowski, "Critical Point and Deconfinement from Dyson-Schwinger Equations," *PoS* **CPOD2013** (2013) 057, arXiv:1308.4509 [hep-ph]; W.-j. Fu, J. M. Pawlowski, F. Rennecke, and B.-J. Schaefer, "Baryon number fluctuations at finite temperature and density," *Phys. Rev.* **D94** no. 11, (2016) 116020, arXiv:1608.04302 [hep-ph].

[11] R. Critelli, J. Noronha, J. Noronha-Hostler, I. Portillo, C. Ratti, and R. Rougemont, "Critical point in the phase diagram of primordial quark-gluon matter from black hole physics," *Phys. Rev.* **D96** no. 9, (2017) 096026, arXiv:1706.00455 [nucl-th];

[12] Z. Li, Y. Chen, D. Li, and M. Huang, "Locating the QCD critical end point through the peaked baryon number susceptibilities along the freeze-out line," *Chin. Phys.* **C42** no. 1, (2018) 013103, arXiv:1706.02238 [hep-ph].

[13] Z. Li, K. Xu, X. Wang and M. Huang, arXiv:1801.09215 [hep-ph].

[14] A. Bhattacharyya, S. K. Ghosh, S. Maity, S. Raha, R. Ray, K. Saha and S. Upadhaya, Phys. Rev. D **95**, no. 5, 054005 (2017) doi:10.1103/PhysRevD.95.054005 [arXiv:1609.07882 [hep-ph]].

[15] S. K. Ghosh, T. K. Mukherjee, M. G. Mustafa and R. Ray, Phys. Rev. D **77**, 094024 (2008) doi:10.1103/PhysRevD.77.094024 [arXiv:0710.2790 [hep-ph]].

[16] A. Bazavov *et al.*, "The QCD Equation of State to $O(\mu_B^6)$ from Lattice QCD," *Phys. Rev.* **D95** no. 5, (2017) 054504, arXiv:1701.04325 [hep-lat].

[17] S. Das [STAR Collaboration], EPJ Web Conf. **90**, 08007 (2015) doi:10.1051/epjconf/20159008007 [arXiv:1412.0499 [nucl-ex]].

[18] V. V. Begun, V. Vovchenko and M. I. Gorenstein, J. Phys. Conf. Ser. **779**, no. 1, 012080 (2017) doi:10.1088/1742-6596/779/1/012080 [arXiv:1609.04827 [nucl-th]].

[19] W. Fan, X. Luo, and H.-S. Zong, "Mapping the QCD phase diagram with susceptibilities of conserved charges within Nambu–Jona-Lasinio model," *Int. J. Mod. Phys.* **A32** no. 11, (2017) 1750061, arXiv:1608.07903 [hep-ph];

[20] S. Mukherjee, R. Venugopalan and Y. Yin, Phys. Rev. C **92**, no. 3, 034912 (2015); S. Mukherjee, R. Venugopalan and Y. Yin, Phys. Rev. Lett. **117**, no. 22, 222301 (2016);

[21] P. Braun-Munzinger, K. Redlich and J. Stachel, In *Hwa, R.C. (ed.) et al.: Quark gluon plasma* 491-599, [nucl-th/0304013].

Dual chiral density waves in nuclear matter

Hiroaki Abuki[1,*], *Yusuke* Takeda[2,**], and *Masayasu* Harada[2,***]

[1]Department of Education, Aichi University of Education, 1 Hirosawa, Kariya 448-8542, Japan
[2]Department of Physics, Nagoya University, Nagoya, 464-8602, Japan

Abstract. We study inhomogeneous chiral phases in nuclear matter using a hadronic model with the parity doublet structure. With an extended ansatz for the dual chiral density wave off the chiral limit, we numerically determine the phase structure. A new type of dual chiral density wave where the condensate has nonvanishing space average is confirmed and it comes to occupy a wide range of low density region as the chiral invariant mass parameter is lowered.

1 Introduction

The chiral symmetry breaking serves as a key ingredient to understand the phase structure of QCD, since it is responsible for the mass generation of hadrons as well as the mass splittings of chiral partners. The chiral symmetry is expected to restore at sufficiently high baryon density. An interesting possibility opens up when one allows the chiral condensate to vary in space; Nakano and Tatsumi demonstrated that the symmetry restoration may take place via several steps [1]; going up in density from the vacuum, the system first goes into an intriguing state named as dual chiral density wave (DCDW), that is, a particular type of inhomogeneous chiral phases, making a spiral in the (σ_0, π_0) chiral plane along z direction.

There are number of approaches to chiral inhomogeneous phases [2]. These include; the mean-field approximation [3–6], the Ginzburg-Landau expansion [7–10], a self-consistent mean-field approach [11]. These are based on quark-based models such as an NJL-type model. One of the advantages is that these models are capable of realizing the QCD vacuum properties as well as the color-flavor locked phase of quark matter which is known to be the densest phase of QCD. But there is also a disadvantage, that is, the lack of the ability to reproduce properties of nuclear matter at saturation point, namely, the QCD phase next to the vacuum phase being followed right after the liquid-gas phase transition.

We here report on our recent work on this topic [12] where we adopted a hadronic model with parity doublet picture (mirror assignment) [13, 14], with vector mesons included in a manner guided by the hidden local symmetry [15, 16]. We first extend the ansatz for the DCDW phase so as to take into account the effect of the explicit symmetry breaking. The extended ansatz smoothly interpolates between the DCDW phase and a nearly symmetry-restored phase. With this setup, we construct the effective potential by diagonalizing the Bogoliubov-de Gennes (BdG) Hamiltonian for nucleons, and determine phases via numerically minimizing the potential. Our main finding is the emergence of another type of DCDW phase which occupies the lower density region according to the value of chiral invariant mass.

*e-mail: abuki@auecc.aichi-edu.ac.jp
**e-mail: yusuketakeda4370@gmail.com
***e-mail: harada.masayasu@nagoya-u.jp

© The Authors, published by EDP Sciences. This is an open access article distributed under the terms of the Creative Commons Attribution License 4.0 (http://creativecommons.org/licenses/by/4.0/).

2 Model

In our analysis, we introduce $N^*(1535)$ as the chiral partner to the ordinary nucleon based on the parity doublet structure. Along the line described in [17], we construct a relativistic mean field model to describe nuclear matter, which includes the scalar (σ), pseudoscalar (π) mesons, and also the vector (ω) meson within the unitary gauge of the hidden local symmetry.

The pure mesonic part for the Lagrangian is then given by

$$\begin{aligned}\mathcal{L}_{\text{mes.}} &= \frac{m_\omega^2}{2}\omega_\mu\omega^\mu - \frac{1}{4}\omega_{\mu\nu}\omega^{\mu\nu} + \frac{1}{2}\partial_\mu\sigma\partial^\mu\sigma + \frac{1}{2}\partial_\mu\boldsymbol{\pi}\partial^\mu\boldsymbol{\pi} \\ &+\frac{\bar{\mu}^2}{2}(\sigma^2+\boldsymbol{\pi}^2) - \frac{\lambda_4}{4}(\sigma^2+\boldsymbol{\pi}^2)^2 + \frac{\lambda_6}{6}(\sigma^2+\boldsymbol{\pi}^2)^3 - f_\pi m_\pi^2\sigma,\end{aligned} \tag{1}$$

where $\bar{\mu}^2$, λ_4 and λ_6 are model parameters. ω_μ and $\omega_{\mu\nu} = \partial_\mu\omega_\nu - \partial_\nu\omega_\mu$ are the field and field strength for ω meson, which are singlet under the chiral transformation. The mass and the decay constant of the pion, as well as the mass of omega meson, are to be set to physical values $m_\pi = 139$ MeV, $f_\pi = 92.2$ MeV, and $m_\omega = 783$ MeV, respectively.

In the present model based on the parity doublet structure, the transformation properties of the positive and negative parity nucleon fields are given by

$$\begin{aligned}\psi_{1r} &\to g_R\psi_{1r}, & \psi_{2r} &\to g_L\psi_{2r}, \\ \psi_{1l} &\to g_L\psi_{1l}, & \psi_{2l} &\to g_R\psi_{2l},\end{aligned} \tag{2}$$

where g_R (g_L) is an element of $SU(2)_R$ ($SU(2)_L$) chiral symmetry group, and ψ_{1r} and ψ_{2r} (ψ_{1l} and ψ_{2l}) are the right-handed (left-handed) fields. Based on the transformation properties, the Lagrangian relevant for nucleons is expressed as [17]

$$\begin{aligned}\mathcal{L}_{\text{N-mes}} &= \bar{\psi}_1(i\partial\!\!\!/ + \mu\!\!\!/_B - g_\omega\omega\!\!\!/)\psi_1 + \bar{\psi}_2(i\partial\!\!\!/ + \mu\!\!\!/_B - g_\omega\omega\!\!\!/)\psi_2 - m_0(\bar{\psi}_2\gamma_5\psi_1 - \bar{\psi}_1\gamma_5\psi_2) \\ &-g_1\bar{\psi}_1(\sigma + i\gamma_5\boldsymbol{\pi}\cdot\boldsymbol{\tau})\psi_1 - g_2\bar{\psi}_2(\sigma - i\gamma_5\boldsymbol{\pi}\cdot\boldsymbol{\tau})\psi_2,\end{aligned} \tag{3}$$

where the baryon chemical potential is included via $\mu\!\!\!/_B = \mu_B\gamma_0$. $\boldsymbol{\tau} = (\tau^1, \tau^2, \tau^3)$ is a set of Pauli matrices, g_1, g_2 and g_ω are the coupling constants. m_0 is the chiral invariant mass parameter which survives as a nonvanishing nucleon mass even when the chiral symmetry is restored.

In the present analysis, we adopt the following extended DCDW ansatz, $\sigma = \delta\sigma + \sigma_0\cos(2fz)$, $\pi_3 = \sigma_0\sin(2fz)$, that reads in the complex representation:

$$\langle\sigma + i\boldsymbol{\pi}\cdot\boldsymbol{\tau}\rangle = \delta\sigma + \sigma_0 e^{2ifz\tau^3} \equiv M(z), \tag{4}$$

where $\delta\sigma$, σ_0 and f are variational parameters with dimension one. Space independent part $\delta\sigma$ accommodates the possibility that the space average of DCDW condensate would get nonvanishing shift into σ-direction due to the explicit chiral symmetry breaking. In the mean-field approximation, the nucleon contribution to the effective potential can be written as

$$\Omega_N = \frac{i}{V_4}\text{Tr}\,\text{Log}(i\partial_0 - (\mathcal{H}(z) - \mu_B^*)), \tag{5}$$

where V_4 is the space-time volume and $\mathcal{H}(z)$ is the single particle Bogoliubov-de Gennes (BdG) Hamiltonian defined in the space of fermion bispinor $\psi = (\psi_1, \psi_2)$ as

$$\mathcal{H} = \begin{pmatrix} i\gamma^0\boldsymbol{\gamma}\cdot\nabla + g_1\gamma^0(M(z)^\dagger P_L + M(z)P_R) & m_0\gamma^0(P_R - P_L) \\ m_0\gamma^0(P_L - P_R) & i\gamma^0\boldsymbol{\gamma}\cdot\nabla + g_2\gamma^0(M(z)P_L + M(z)^\dagger P_R)\end{pmatrix},$$

where $P_{R(L)} = \frac{1\pm\gamma_5}{2}$ is the chirality projector. This is nothing but the Dirac Hamiltonian in the presence of a periodic potential field $M(z)$ $(= M(z+\lambda)$ with a wavelength $\lambda = \pi/f)$. Then the functional trace in Eq. (5) can be evaluated by finding eigenvalues of the operator $\mathcal{H}(z)$ [18]. The eigenvalue has a discrete label as well as continuous three-momentum $\mathbf{p}$ in addition to internal quantum numbers; This is because of the Bloch theorem which states that the eigenfunctions in the presence of a periodic potential are the plane waves distorted by periodic functions. To be specific, we decompose the bispinor as

$$\psi(\mathbf{x}) = \sum_{\ell=-\infty}^{\infty} \sum_{\mathbf{p}_\perp} \sum_{|p_z|\le f} e^{i\mathbf{p}\cdot\mathbf{x}+i\boldsymbol{K}_\ell\cdot\mathbf{x}} \psi_{\mathbf{p},\ell}\,,$$

where $\boldsymbol{K}_\ell = (0,0,2f\ell)$ is the reciprocal lattice vector. Moving on to the quasimomentum base $\{\psi_{\mathbf{p},\ell}\}$, the BdG Hamiltonian becomes block-diagonalized, $H_{\ell\ell'}(\mathbf{p})$. Since the isospin remains a good quantum number, we can simply double the proton contribution in the full effective potential. Then omitting the antiprotons which would not contribute at zero temperature, and diagonalizing $H_{\ell\ell'}(\mathbf{p})$ results in an infinite tower of eigenvalues at each $\mathbf{p}$, which repeatedly appears for every Brillouin Zone (BZ), $\mathbf{p} \to \mathbf{p} + \boldsymbol{K}_\ell$ $(\ell = \cdots, -1, 0, 1, \cdots)$:

$$\sum_{\ell'} H_{\ell\ell'}(\mathbf{p})\psi^{(i)}_{n,\mathbf{p},\ell'} = E^{(i)}_{n,\mathbf{p}}\psi^{(i)}_{n,\mathbf{p},\ell} \qquad (n = 0, 1, \cdots, \infty),$$

with $i(= 1,2,3,4)$ labeling the internal quantum number $(p, p^*)\otimes(\uparrow,\downarrow)$, where p^* implies the $I_3 = +1/2$ part of $N^*(1535)$. The total thermodynamic potential is now evaluated as

$$\begin{aligned}\Omega \;=\; & \sum_{n=0}^{\infty}\sum_{i=1}^{4}\int_{-f}^{f}\frac{dp_z}{\pi}\int\frac{d\mathbf{p}_\perp}{(2\pi)^2}(E^{(i)}_{n,\mathbf{p}} - \mu_B^*)\theta(\mu_B^* - E^{(i)}_{n,\mathbf{p}}) \\ & -\tfrac{1}{2}m_\omega^2\omega_0^2 - \tfrac{1}{2}\bar{\mu}^2\left(\delta\sigma^2 + \sigma_0^2\right) + 2\sigma_0^2 f^2 + \tfrac{1}{4}\lambda_4\left[\left(\delta\sigma^2+\sigma_0^2\right)^2 + 2\delta\sigma^2\sigma_0^2\right] \\ & -\tfrac{1}{6}\lambda_6\left[\left(\delta\sigma^2+\sigma_0^2\right)^3 + 6\left(\delta\sigma^2+\sigma_0^2\right)\delta\sigma^2\sigma_0^2\right] - m_\pi^2 f_\pi\delta\sigma.\end{aligned} \tag{6}$$

Parameter setting

We treat m_0 as a free parameter. Parameters g_1 and g_2 can be determined by QCD vacuum property. Once m_0 is given, $m_+^{(0)} = 939$ MeV and $m_-^{(0)} = 1535$ MeV together with $\sigma_0 = f_\pi = 92.2$ MeV fix the values of g_1 and g_2 via

$$m_\pm^{(0)} = \frac{1}{2}\left[\sqrt{(g_1+g_2)^2\sigma_0^2 + 4m_0^2} \mp (g_1 - g_2)\sigma_0\right]. \tag{7}$$

Table 1. Determined parameters for given chiral invariant mass

m_0	500	600	700	800	900
g_1	9.03	8.49	7.82	7.00	5.97
g_2	15.5	15.0	14.3	13.5	12.4
g_ω	11.3	9.13	7.30	5.66	3.52
$\bar{\mu}$ [MeV]	441	437	406	320	114
λ_4	42.2	40.6	35.7	23.2	4.47
$\lambda_6 \cdot f_\pi^2$	17.0	15.8	14.0	8.94	0.644

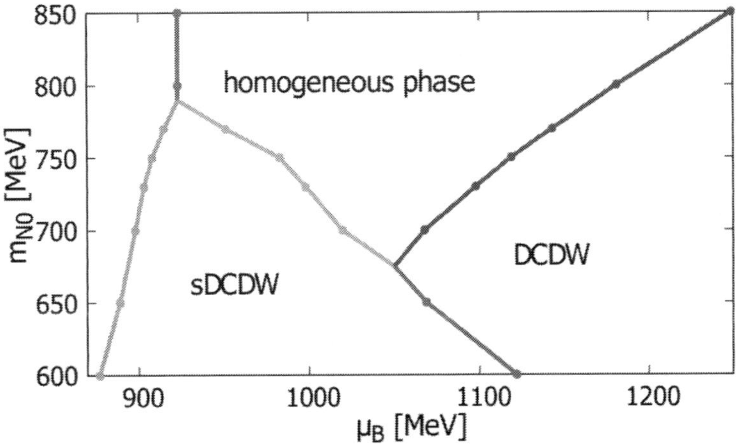

Figure 1. Phase structure in $\mu_B - m_0$ plane. Red line is $\mu_B = 923$ MeV corresponding to the nuclear liquid-gas phase transition point.

From the stationary condition for σ at vacuum, we have $\bar{\mu}^2 = \lambda_4 f_\pi^2 - \lambda_6 f_\pi^4 - m_\pi^2$. Then we are left with three unknown parameters to be fixed, $\{\lambda_4, \lambda_6, g_\omega\}$. In order to fix these parameters, we use nuclear matter property at saturation density as done in [17, 19]:

$$\begin{aligned} \rho_0 &= 0.16\ \mathrm{fm}^{-3}, \\ \left(\frac{\epsilon}{\rho}\right)\Big|_{\rho=\rho_0} - m_+^{(0)} = \mu_B^*|_{\rho=\rho_0} - m_+^{(0)} &= -16\ \mathrm{MeV}, \\ K = 9\rho_0^2 \frac{\partial^2(\epsilon/\rho)}{\partial^2 \rho}\Big|_{\rho=\rho_0} = 9\rho_0 \frac{\partial P}{\partial \rho}\Big|_{\rho=\rho_0} &= 240\ \mathrm{MeV}. \end{aligned} \tag{8}$$

The first and second equations determine the value of g_ω, and ω_0 as a function of density, $\omega_0 = g_\omega \rho / m_\omega^2$. The saturation condition $\frac{\partial}{\partial\rho}(\epsilon/\rho)\big|_{\rho=\rho_0} = P|_{\rho=\rho_0}/\rho_0^2 = 0$, and the condition for the incompressibility, the last equation of Eq. (8), together with the stationary condition $\frac{\partial\Omega}{\partial\sigma}|_{\rho=\rho_0} = 0$ determines parameters $\{\lambda_4, \lambda_6\}$ and the scalar condensate at saturation density, $\sigma_0|_{\rho=\rho_0}$. The parameters are summarized in Table 1. As will be shown later, however, the saturated nuclear matter exists only as a metastable state once chiral invariant mass becomes smaller than some critical value, $m_0 \lesssim 800$ MeV.

3 phase structure

Figure 1 shows the phase structure in $\mu_B - m_0$ plane. We notice that there are two kinds of DCDW phase, "DCDW" and "sDCDW". Depending on the range where m_0 resides, qualitative picture of phase transitions changes.

For $m_0 \gtrsim 780$ MeV, there are three phases; (1) the vacuum phase for $\mu_B \leq 923$ MeV, (2) the homogeneously chiral symmetry broken phase, and (3) the DCDW phase at high density. In the upper panel of Fig. 2, we show order parameters as a function of μ_B for $m_0 = 800$ MeV. Just for comparison we also depict by magenta curve, the solution for the case where the condition $\delta\sigma = 0$ is forced (that is, the case of the standard DCDW ansatz).

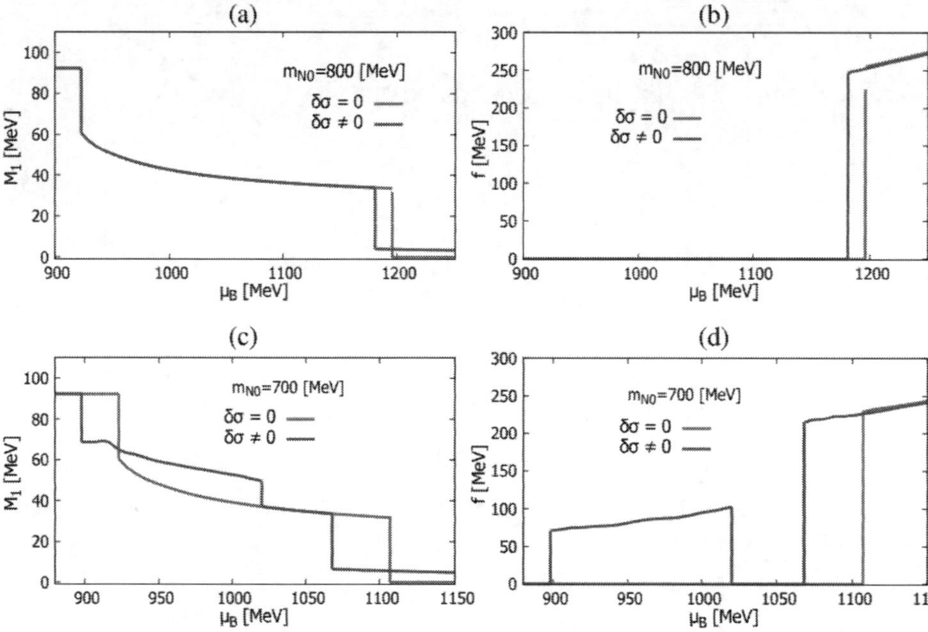

Figure 2. Left: the order parameter M_1 (see text) as a function of μ_B for $m_0 = 800$ MeV (upper panel) and that for $m_0 = 700$ MeV (lower panel). Right: the wavenumber f as a function of μ_B for $m_0 = 800$ MeV (upper panel) and that for $m_0 = 700$ MeV (lower panel). Magenta curves correspond to the solutions obtained under the forced condition $\delta\sigma = 0$.

Plotted in Fig. 2(a) is the space-averaged order parameter M_1, defined by

$$M_1 \equiv \frac{1}{2V}\int_{-\infty}^{\infty} d^3x \mathrm{tr}\,[\langle M\rangle] = \begin{cases} \delta\sigma + \sigma_0 & (f = 0), \\ \delta\sigma & (f \neq 0), \end{cases} \tag{9}$$

which provides a guide for the strength of chiral symmetry breaking. Figure 2(b) shows wavenumber f as a function of μ_B. Nonvanishing wavenumber ($f \neq 0$) means the DCDW phase where the translational symmetry is also broken. We note that, when a shift $\delta\sigma$ is properly taken into account, the onset density the DCDW phase is brought to a lower density since it stabilizes the DCDW solution through the explicit symmetry breaking source, $-f_\pi m_\pi^2 \delta\sigma$.

For 670 MeV $\lesssim m_0 \lesssim$ 780 MeV, we find four phases. These are, going up in density, (1) the vacuum phase, (2) the "sDCDW" phase, (3) the homogeneously chiral-symmetry broken phase, and (4) the DCDW phase. What we call by "sDCDW" phase is the "shifted" DCDW phase, for a reason we will describe shortly. In order to see how phase transitions take place, we display in the lower panel of Fig. 2, the order parameters as a function of μ_B for $m_0 = 700$ MeV. From these we see that, as the baryon density increases, the chiral symmetry restores through several steps. Moreover, we clearly see from the Fig. 2(d), that there are two DCDW phases; 'one at low density, and the other at high density, being separated by the homogeneously chiral-symmetry broken phase. We call the former "sDCDW", and the latter "ordinary" DCDW or simply DCDW.

Let us now have a closer look at the sDCDW which shows up next to the vacuum phase, in the region 900 MeV $\lesssim \mu_B \lesssim$ 1020 MeV for $m_0 = 700$ MeV. In contrast to the ordinary DCDW for $\mu_B \gtrsim 1070$ MeV the phase is not smoothly connected to the DCDW phase found at high density side for $m_0 = 800$ MeV as clearly seen in Fig. 1. In the ordinary DCDW, the magnitude of the wavenumber f is much larger than $\delta\sigma$ (and also than σ_0). The chiral density wavelength is roughly $\lambda = \pi/f \sim 2 - 3$ fm which is still larger than averaged inter-nucleon spacing $\rho^{-1/3} \sim 1.2 - 1.3$ fm. In the sDCDW phase, on the other hand, the wavenumber $f \sim 70 - 100$ MeV is comparable with $\delta\sigma \sim 50 - 70$ MeV. The resulting chiral density wavelength $\lambda = \pi/f \sim 6 - 9$ fm, which is much larger than the averaged inter-nucleon spacing $\rho^{-1/3} \sim 1 - 2$ fm. Most interesting fact is that in the sDCDW phase the amplitude of condensate is smaller than the magnitude of the shift, namely, $\sigma_0 < \delta\sigma$. Accordingly, the center of the chiral spiral in the (σ, π^0) chiral plane, which is located near origin in the ordinary DCDW phase, is significantly shifted to the σ direction. This is why we name the phase the "shifted" DCDW (abbreviated to "sDCDW") phase. We would like to stress that, on the contrary to the ordinary DCDW phase, $\delta\sigma$ in sDCDW phase is not due to the symmetry breaking source term, but rather spontaneously generated. This means, the solution to the stationary condition for $\delta\sigma$ would not vanish even in chiral limit.

As the chiral invariant mass m_0 is decreased, the density window for the homogeneously chiral-symmetry broken phase shrinks as the pressure of two kinds of DCDW phase gets stronger. And once the condition $m_0 \lesssim 670$ MeV is met, it does no longer exist.

4 Conclusions and outlook

We studied the inhomogeneous phase structure in nuclear matter using a nucleon-based model with parity doublet structure where $N^*(1535)$ is introduced as the chiral partner of $N(939)$. Adopting an extended ansatz for DCDW, Eq. (4), we found that, the sDCDW phase shows up in addition to the ordinary DCDW phase when the value of chiral invariant mass m_0 is below some threshold, ~ 780 MeV.

The ordinary DCDW phase appears at high density, where the space average of chiral condensate M_1, Eq. (9), is typically less than 10 MeV. This implies that this phase smoothly transforms into the standard DCDW with $\delta\sigma = 0$ as the chiral limit is approached. The wavenumber f has value of $200 \sim 300$ MeV, corresponding to the density wavelength $\lambda \sim 2 - 3$ fm, which is in fair agreement with the result obtained in [20].

On the other hand, when $m_0 \lesssim 780$ MeV, the sDCDW phase appears at low density. This phase is characterized by a smaller wavenumber f and a large shift of chiral condensate, $\delta\sigma$. It is noteworthy that it is not the effect of explicit chiral symmetry breaking but the dynamical symmetry breaking that produces this large shift of chiral condensate. So we expect that this sDCDW phase survives in the chiral limit.

The parameter range of chiral invariant mass where the sDCDW is stabilized, fails to realize nuclear matter as the pressure of sDCDW is so strong that it washes out the liquid-gas phase transition structure. Then, one might think that the present model for m_0 less than 780 MeV is ruled out. However, the chiral invariant mass m_0 can have density dependence as in Ref. [20] which shows that m_0 decreases against increasing density. In such a case, the sDCDW phase may be realized in high density nuclear matter in the real world.

Exploring the elementary excitations in the sDCDW phase deserves further investigations in future. In the ordinary DCDW phase, a particular combination of the space translation and chiral rotation remains unbroken [21, 22]. As a result, there is no extra Nambu-Goldstone boson other than three pions. In contrast, there is no such unbroken combination of symmetry in the sDCDW phase. We then expect that a phonon mode appears in the sDCDW phase which may signal the phase. An inclusion of external magnetic field may provide another interesting direction of extending current work. For quark matter, several studies were already devoted to the topic of inhomogeneous phases under the magnetic field [23–27]. The analysis within the nucleon-based model with including both $\delta\sigma$ and magnetic field would be an interesting subject worth exploring.

Acknowledgement. This work was partially supported by JPSP KAKENHI Grant Number JP16K05346 (H.A.) and 16K05345 (M.H.).

References

[1] E. Nakano and T. Tatsumi, Phys. Rev. D **71**, 114006 (2005).
[2] See for a review, M. Buballa and S. Carignano, Prog. Part. Nucl. Phys. **81**, 39 (2015).
[3] D. Nickel, Phys. Rev. D **80**, 074025 (2009).
[4] S. Carignano, D. Nickel and M. Buballa, Phys. Rev. D **82**, 054009 (2010).
[5] S. Karasawa and T. Tatsumi, Phys. Rev. D **92**, 116004 (2015).
[6] P. Adhikari, J. O. Andersen and P. Kneschke, Phys. Rev. D **96**, 016013 (2017).
[7] D. Nickel, Phys. Rev. Lett. **103**, 072301 (2009).
[8] H. Abuki, D. Ishibashi and K. Suzuki, Phys. Rev. D **85**, 074002 (2012).
[9] H. Abuki, Phys. Lett. B **728**, 427 (2014).
[10] S. Carignano, M. Mannarelli, F. Anzuini and O. Benhar, Phys. Rev. D **97**, 036009 (2018).
[11] T. G. Lee, K. Nishiyama, N. Yasutake, T. Maruyama and T. Tatsumi, JPS Conf. Proc. **14**, 020808 (2017).
[12] Y. Takeda, H. Abuki and M. Harada, Phys. Rev. D **97**, 094032 (2018).
[13] C. E. Detar and T. Kunihiro, Phys. Rev. D **39**, 2805 (1989).
[14] D. Jido, M. Oka and A. Hosaka, Prog. Theor. Phys. **106**, 873 (2001).
[15] M. Bando, T. Kugo and K. Yamawaki, Phys. Rept. **164**, 217 (1988).
[16] M. Harada and K. Yamawaki, Phys. Rept. **381**, 1 (2003).
[17] Y. Motohiro, Y. Kim and M. Harada, Phys. Rev. C **92**, 025201 (2015); Erratum: [Phys. Rev. C **95**, 059903 (2017)].
[18] See for example, D. Nickel and M. Buballa, Phys. Rev. D **79**, 054009 (2009).
[19] Y. Takeda, Y. Kim and M. Harada, Phys. Rev. C **97**, 065202 (2018).
[20] A. Heinz, F. Giacosa and D. H. Rischke, Nucl. Phys. A **933**, 34 (2015).
[21] T. G. Lee, E. Nakano, Y. Tsue, T. Tatsumi and B. Friman, Phys. Rev. D **92**, 034024 (2015).
[22] Y. Hidaka, K. Kamikado, T. Kanazawa and T. Noumi, Phys. Rev. D **92**, 034003 (2015).
[23] I. E. Frolov, V. C. Zhukovsky and K. G. Klimenko, Phys. Rev. D **82**, 076002 (2010).
[24] K. Nishiyama, S. Karasawa and T. Tatsumi, Phys. Rev. D **92**, 036008 (2015).
[25] R. Yoshiike, K. Nishiyama and T. Tatsumi, Phys. Lett. B **751**, 123 (2015).
[26] G. Cao and A. Huang, Phys. Rev. D **93**, 076007 (2016).
[27] H. Abuki, Phys. Rev. D **98**, 054006 (2018); JPS Conf. Proc. **20**, 011017 (2018); EPJ Web Conf. **129**, 00036 (2016).

The improved Ginzburg-Landau technique

Massimo Mannarelli[1,*]

[1]INFN, Laboratori Nazionali del Gran Sasso, Via G. Acitelli, 22, I-67100 Assergi (AQ), Italy

Abstract. We discuss an innovative method for the description of inhomogeneous phases designed to improve the standard Ginzburg-Landau expansion. The method is characterized by two key ingredients. The first one is a moving average of the order parameter designed to account for the long-wavelength modulations of the condensate. The second one is a sum of the high frequency modes, to improve the description of the phase transition to the restored phase. The method is applied to compare the free energies of 1D and 2D inhomogeneous structures arising in the chirally symmetric broken phase.

1 Introduction

In this contribution we discuss the method presented in [1] for the description of the inhomogeneous phases. The emphasis is on cold quark matter and, more in detail, on the properties of the chiral symmetry broken (χSB) phase featuring an inhomogeneous condensate.

The Lagrangian describing Quantum Chromodynamics (QCD) for three flavor massless quarks has the following symmetries

$$SU(3)_c \times \underbrace{SU(3)_L \times SU(3)_R}_{\supset [U(1)_{e.m.}]} \times U(1)_B \,, \tag{1}$$

where $SU(3)_c$ is the gauge color symmetry and the other symmetries are global, apart for the $U(1)_{e.m.}$ subgroup of the chiral rotations describing the electromagnetic interactions. The ground state may have a lower symmetry because the QCD interaction leads to the formation of various condensates occurring with a specific symmetry breaking pattern. Schematically, the three quark condensates that are the most relevant ones for phenomenology are the followings:

$\langle\bar\psi\psi\rangle$	Chiral condensate: Locks the chiral symmetries
$\langle\bar\psi\sigma_2\gamma_5\psi\rangle$	Pion condensate: Locks chiral rotations and breaks $U(1)_{e.m.}$
$\langle\psi C\gamma_5\psi\rangle$	Diquark condensate: Breaks the $SU(3)_c$ gauge group ,

where ψ is the quark field, with spinorial, color and flavor indices suppressed. The various condensates should be realized in different regions of the so-called QCD phase diagram, see the cartoon in Fig. 1, corresponding to different physical situations. For every phase we have indicated the possible condensates and the corresponding phases. The region at high temperature where no quark condensate is expected corresponds to the so-called quark-gluon

*e-mail: massimo.mannarelli@lngs.infn.it

© The Authors, published by EDP Sciences. This is an open access article distributed under the terms of the Creative Commons Attribution License 4.0 (http://creativecommons.org/licenses/by/4.0/).

plasma and is the one relevant for the Quark epoch of the early Universe, corresponding to a time $t \sim 10^{-12}s$ after the big bang. This is as well the region studied by relativistic heavy-ion colliders. Instead, at low-temperature and high quark chemical potentials the diquark condensate should be realized, see [2, 3] for reviews, and at asymptotic densities the color flavor locked (CFL in the figure) phase [4] should be favored. The region where the chiral condensate and the pion condensate are favored correspond to the "hadron gas" and "pion condensed" phases, respectively.

As indicated in the figure, there are different theoretical methods available for studying the various phases. Perturbative methods (pQCD), relying on the asymptotic freedom of QCD, can only be applied at very large energy scales. Chiral perturbation (χPT) theory is instead valid for low energy scales. Whenever doable, lattice simulations (LQCD) are an extremely useful tool, especially for the description of the static properties of matter. These methods are grounded on QCD and have a well defined range of validity. When these methods fails the alternative is to study hadronic matter using the Nambu-Jona-Lasinio (NJL) model [5] or similar, based on a contact interaction having the global symmetries of QCD, but lacking of the gauge part. In this way one obtains a qualitative and semiquantitative description of the various phases, although the results depend on the parameters characterizing the model. Unfortunately, the region between the chirally broken phase and the deconfined phase at large μ can only be studied by NJL-like models.

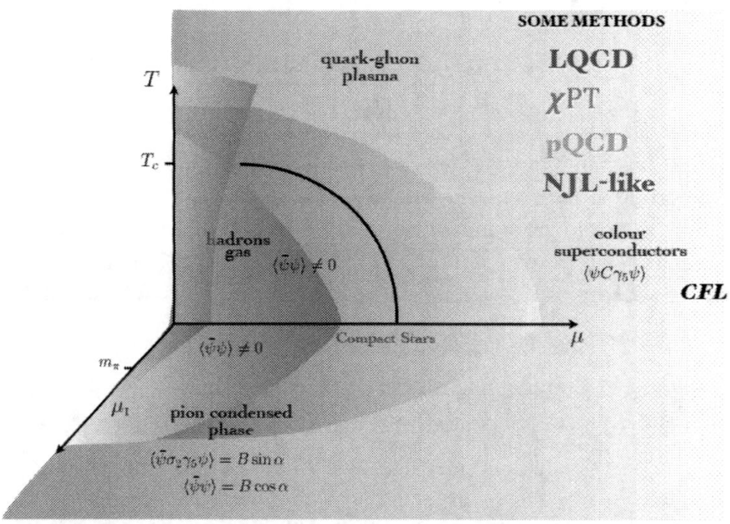

Figure 1. Cartoon of the so-called QCD phase diagram: a grand-canonical description of the properties of hadronic matter as a function of the average quark chemical potential, μ, of the isospin chemical potential, μ_I, and of the temperature, T. The quark condensates for every phases are indicated. The solid line corresponds to a possible first order phase transition, which is currently investigated by relativistic heavy-ion colliders, between the confined and the deconfined phases. Shaded areas correspond to the regions of applicability of the corresponding theoretical method.

2 Inhomogeneous phases

In Fig. 1 we have only indicated the homogeneous phases, but inhomogeneous phases are expected to be favored in various regions. One important example is the Pasta phase [6] in nuclear matter at high densities. Regarding cold quark matter, inhomogeneous phases are expected close to the phase transition between the chirally broken/chirally restored phases and in the deconfined phase. Model calculations indicate that two inhomogeneous phases may arise: the crystalline color superconducting phase see [7] for a review, and the inhomogeneous χSB phase see [8] for a review. The former is realized in the deconfined phase for neutral and beta equilibrated matter, while the latter is expected to arise between the homogeneous χSB phase and the chirally restored phase if the quark-antiquark coupling strength is sufficiently large. Unfortunately, both phases are expected to be in the gray-region of Fig. 1, where no first-principle methods can be applied and one has to rely on NJL-like models. Even within the NJL framework, studying the inhomogeneous phases is an extremely challenging task, because one needs to perform a diagonalization of the full quark Hamiltonian, which is a complicated numerical problem. Alternatively, one can perform a Ginzburg-Landau (GL) expansion in the order parameters and its derivatives. Hereafter we focus on the inhomogeneous χSB phase, with the standard GL expansion [9, 10] given by

$$\begin{aligned}\Omega_{\rm GL} =&\Omega[0] + \int \frac{d\mathbf{x}}{V}\left[\alpha_2 M^2 + \alpha_4\left(M^4 + (\nabla M)^2\right) + \alpha_6\left(M^6 + 3(\nabla M)^2M^2 + \frac{1}{2}(\nabla M^2)^2 + \frac{1}{2}(\nabla^2 M)^2\right)\right.\\ &\left.+\alpha_8\left(14M^4(\nabla M)^2 - \frac{1}{5}(\nabla M)^4 + \frac{18}{5}M(\nabla^2 M)(\nabla M)^2 + \frac{14}{5}M^2(\nabla^2 M)^2 + \frac{1}{5}(\nabla^3 M)^2\right) + \dots\right],\end{aligned} \tag{2}$$

where V is the total volume and the various coefficients are independent of the particular modulation considered. Sometimes these are called "universal" coefficients because if they were derived by the appropriate microscopic high energy theory, in the present case from QCD, they would only depend on the grand-canonical variables. Since we consider the system at vanishing temperature, they should only depend on μ. However, the χSB inhomogeneous phase is realized in the gray region of Fig. 1, meaning that we are forced to use the NJL model. In this case the coefficients not only depend on μ, but also on the regularization scheme. We use the Pauli-Villars regularization scheme [5, 11], which allows to keep into account the high energy contributions. In this way one obtains

$$\begin{aligned}\alpha_2 &= \frac{1}{4G} - \frac{N_f N_c}{8\pi^2}\left(3\Lambda^2\log\left(\frac{4}{3}\right) - 2\mu^2\right), & \alpha_4 &= -\frac{N_f N_c}{16\pi^2}\log\left(\frac{32\mu^2}{3\Lambda^2}\right),\\ \alpha_6 &= \frac{N_f N_c}{96\pi^2}\left(\frac{11}{3\Lambda^2} + \frac{1}{\mu^2}\right), & \alpha_8 &= \frac{N_f N_c}{256\pi^2}\left(\frac{1}{2\mu^4} - \frac{85}{27\Lambda^4}\right),\end{aligned} \tag{3}$$

where Λ is an ultraviolet regulator and N_f and N_c are the numbers of quark flavors and colors, respectively.

The GL expansion is designed to work close to the Lifshitz point where both M and ∇M are small. Indeed, one implicitly assumes that the terms in Eq. (2) proportional to the same α_n are equally important. However, if one wants to extend the study far from the Lifshitz point this assumption it is not in general true. In particular, at vanishing temperature, approaching the second order phase transition M is expected to vanish, but $(\nabla M)/M$ is nonzero. Improving the GL free energy by brute force, including higher order terms is not easy because the calculation of the corresponding coefficients becomes increasingly hard.

In order to extend the GL scheme away from the Lifshitz point we propose the 'improved Ginzburg-Landau" (IGL) expansion [1], which for a real order parameter reads

$$\Omega_{\rm IGL} = \frac{1}{V}\int d\mathbf{x}\left[\Omega_{\rm hom}(\overline{M(z)^2}) + \alpha_6\left(3(\nabla M)^2 M^2 + \frac{1}{2}(\nabla M^2)^2\right) + \alpha_8\left(14M^4(\nabla M)^2 - \frac{1}{5}(\nabla M)^4 + \frac{18}{5}M(\nabla^2 M)(\nabla M)^2 + \frac{14}{5}M^2(\nabla^2 M)^2\right) + \sum_{n\geq 1}\tilde{\alpha}_{2n+2}(\nabla^n M)^2\right], \tag{4}$$

and differs from the standard GL expansion for the first and the last terms in the square bracket. The $\Omega_{\rm hom}(\overline{M(z)^2})$ term is the free energy for an homogeneous order parameter, evaluated considering the moving average of the mass function

$$\overline{M(z)^2} = \frac{1}{\lambda}\int_{z-\lambda/2}^{z+\lambda/2} M^2(\xi)d\xi\,, \tag{5}$$

where $\lambda \lesssim 1/\mu$ is an ultraviolet cutoff. The $\Omega_{\rm hom}(\overline{M(z)^2})$ term resums all the M^n terms and by construction gives the free energy of the homogeneous phases when $\nabla M = 0$. In particular if M is space independent, it gives the free energy of the homogeneous χSB phase. Then, if by increasing the quark chemical potential one enters into the inhomogeneous χSB phase by a second order phase transition, it effectively resums the long-wavelength fluctuations of the order parameter, up to the scale λ.

The second novel term of the IGL expansion, corresponding to the last term in the square bracket of Eq. (4), does instead resum all the gradients of the M^2 terms. This term is the leading one when M is small but its derivatives are large, therefore it is dominant close to the phase transition to the chirally restored phase. The effect of the $\tilde{\alpha}$ terms is shown in Fig. 2 for the chiral density wave (CDW) ansatz

$$M(z) = \Delta e^{2iqz}\,, \tag{6}$$

corresponding to a plane wave along the $z-$axis. As is clear from the figure, the inclusion of more $\tilde{\alpha}$ terms systematically improves the approximation close to the phase transition to the normal phase, where q is large and M is small. In order to compute the $\tilde{\alpha}$ terms one can use any modulation for which the free energy solution can be easily computed. The procedure is explained in detail in [1].

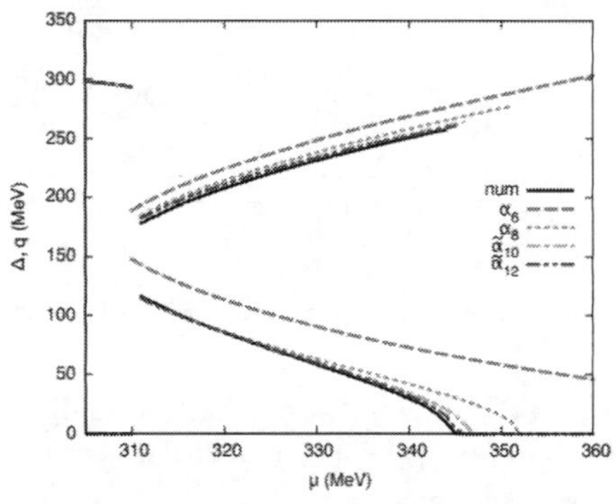

Figure 2. Analysis of the IGL approximation for the plane wave ansatz in Eq. (6). We report the values of Δ (curves with a decreasing values) and of q (curves with an increasing values) in the ground state as a function of the quark chemical potential. The solid lines are obtained by a numerical method, the other lines correspond to the IGL expansion in Eq. (4), including gradient terms of different orders. With increasing number of gradient terms the second order phase transition is increasingly well described.

Once the coefficients of the IGL free energy are determined, one can consider more complicated structures as the real kink crystal (RKC) [8, 11]

$$M(z) = \Delta\sqrt{\nu}\,\text{sn}(\Delta z|\nu)\,, \tag{7}$$

which is believed to be the favored spatial modulation in the χSB inhomogeneous phase. In Fig. 3 we compare the numerical results with those obtained by the standard GL expansion and by the IGL expansion including up to $\tilde{\alpha}_{10}$ terms.

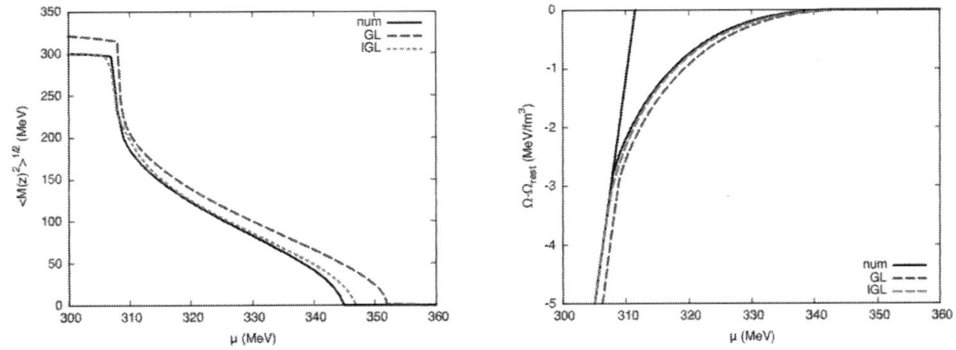

Figure 3. Comparison of the numerical results with the GL expansion and the IGL approximation for the 1D RKC, Eq. (7). Left: average value of the condensate, $\sqrt{\langle M(z)^2\rangle}$. Right: difference between the free energy at the minimum and the free energy of the chirally restored phase.

As in the previous case the IGL improves the description to the normal phase, thanks to the $\tilde{\alpha}$ terms. Moreover, since in this case the onset of the inhomogeneous χSB phase happens by a second order phase transition, the long-wavelength term $\Omega_{\text{hom}}(\overline{M(z)^2})$ allows a better describe of this region.

By using the IGL method we are in a position to easily test different 2D modulations and to compare the corresponding free energies in a reliable way. We consider several structures including a square lattice with two RKC-type modulations along the x and y directions, that is

$$M(x, y) = \Delta \nu \mathrm{sn}(\Delta x, \nu)\mathrm{sn}(\Delta y, \nu)\,, \tag{8}$$

and the two-dimensional square lattice with a sinusoidal ansatz, that is,

$$M(x, y) = \Delta \cos(qx)\cos(qy)\,, \tag{9}$$

both depending on only two parameters.

The numerical evaluation of the free energies of these modulations is extremely complicated [12], as it requires an expansion of the order parameter in a large number of Fourier harmonics. Then, the minimization procedures amounts to a minimization of the free energy with respect to all of their amplitudes. Instead, within the IGL approximation the minimization requires the evaluation of few integrals and can be straightforwardly implemented

When computing the free energy associated with these two modulations we find that they are almost degenerate, apart in the region close to the onset of the inhomogeneous phase, see Fig. 4. In that figure we also see that the 2D modulations are disfavored with respect to the 1D RKC in Eq. (7). We performed a further check by considering the ansatz

$$M(x, y) = \Delta\left[\sqrt{\nu_x}\,\mathrm{sn}(\Delta x, \nu_x) + \sqrt{\nu_y}\,\mathrm{sn}(\Delta y, \nu_y)\right]\,, \tag{10}$$

which can interpolate between the 1D RKC modulation and a more involved two-dimensional structure. Consistent with our other results, we find that the ground state always corresponds to one in which one of the two modulations disappears with ν_x or ν_y being zero.

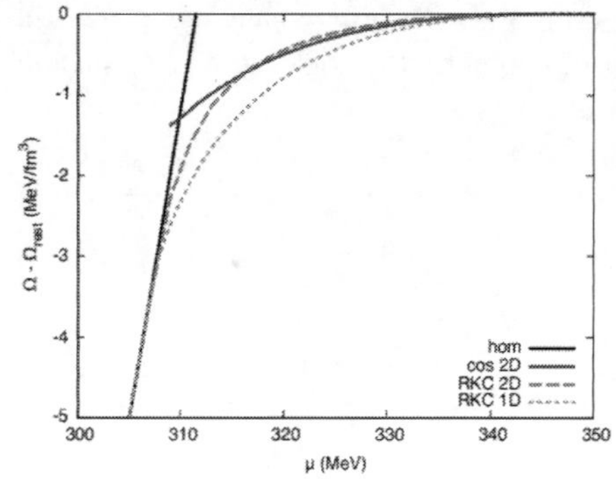

Figure 4. Free energies of three different modulations as a function of the quark chemical potential. The dotted (purple) line corresponds to the 1D RKC, see Eq. (7), the dashed (red) line indicates the 2D RKC, see Eq. (8), and the solid (blue) line corresponds to the 2D cosine, see Eq. (9). The 1D RKC is the energetically favored inhomogeneous phase. All the results have been obtained within the IGL approximation in Eq. (4) including up to $\tilde{\alpha}_{10}$ terms.

3 Conclusions

This IGL approach can be viewed as an effective field theory based on the scale separation between long-wavelength fluctuations, dominating the transition between the homogeneous and the inhomogeneous broken phases, and rapid fluctuations governing the transition to the restored phase. By construction the IGL reproduces the homogeneous limits and allows for a description of the transition to the restored phase with arbitrarily high precision by a controlled gradient expansion.

The IGL approach has been developed in [1] and applied to the inhomogeneous χSB phase at $T = 0$. However, it can be extended to study other phases. For cold quark matter, the IGL method can be used to rapidly evaluate the free energy of the various crystalline color superconducting configurations considered in [13], eventually extending the analysis to different modulations. Clearly, one needs to obtain an expression of the IGL free energy analogous to the one in Eq. (4). This requires some work, but seems doable; in particular it should be possible to obtain the $\tilde{\alpha}$ coefficients by the method described in [1] and using the simple Fulde-Ferrell type plane wave [7, 14, 15]. Numerical results (for few cases) are also available [16] and can be used as a check of the method.

A more challenging task is to modify the IGL method to simultaneously include the chiral and diquark condensates. This is an interesting topic, because both phases could be realized in the region of the QCD phase diagram across the solid line of Fig. 1. It is reasonable to expect that the crystalline color superconducting phase arises in the spatial regions where the chiral condensate is small and the baryonic density is large. As a first step one could consider 1D chiral modulations coexisting with a cosine color superconducting modulation, see for example [7]. But more complicated structures might actually be favored.

A different important issue is the role of fluctuations. One should include them to detect possible instabilities of the space modulation of the condensate [17–19], which seem to be particularly important for 1D modulations [8, 20, 21]. The inclusion of fluctuations in the IGL framework could be implemented following the procedure of [22–24].

References

[1] S. Carignano, M. Mannarelli, F. Anzuini, O. Benhar, Phys. Rev. **D97**, 036009 (2018), 1711.08607

[2] K. Rajagopal, F. Wilczek, *The Condensed Matter Physics of QCD* (2000), arXiv:hep-ph/0011333

[3] M. Alford, A. Schmitt, K. Rajagopal, T. Schäfer, Rev. Mod. Phys. **80**, 1455 (2008), 0709.4635

[4] M.G. Alford, K. Rajagopal, F. Wilczek, Nucl.Phys. **B537**, 443 (1999), hep-ph/9804403
[5] S.P. Klevansky, Rev. Mod. Phys. **64**, 649 (1992)
[6] D.G. Ravenhall, C.J. Pethick, J.R. Wilson, Phys. Rev. Lett. **50**, 2066 (1983)
[7] R. Anglani, R. Casalbuoni, M. Ciminale, N. Ippolito, R. Gatto et al., Rev.Mod.Phys. **86**, 509 (2014), 1302.4264
[8] M. Buballa, S. Carignano, Prog. Part. Nucl. Phys. **81**, 39 (2015), 1406.1367
[9] D. Nickel, Phys. Rev. Lett. **103**, 072301 (2009), 0902.1778
[10] H. Abuki, D. Ishibashi, K. Suzuki, Phys.Rev. **D85**, 074002 (2012), 1109.1615
[11] D. Nickel, Phys.Rev. **D80**, 074025 (2009), 0906.5295
[12] S. Carignano, M. Buballa, Phys. Rev. **D86**, 074018 (2012), 1203.5343
[13] K. Rajagopal, R. Sharma, Phys.Rev. **D74**, 094019 (2006), hep-ph/0605316
[14] P. Fulde, R.A. Ferrell, Phys. Rev. **135**, A550 (1964)
[15] A.I. Larkin, Y.N. Ovchinnikov, Zh. Eksp. Teor. Fiz. **47**, 1136 (1964), sov. Phys. JETP **20** 762 (1965).
[16] D. Nickel, M. Buballa, Phys. Rev. **D79**, 054009 (2009), 0811.2400
[17] T.G. Lee, E. Nakano, Y. Tsue, T. Tatsumi, B. Friman, Phys. Rev. **D92**, 034024 (2015), 1504.03185
[18] R. Yoshiike, T.G. Lee, T. Tatsumi, Phys. Rev. **D95**, 074010 (2017), 1702.01511
[19] Y. Hidaka, K. Kamikado, T. Kanazawa, T. Noumi, Phys. Rev. D **92**, 034003 (2015), 1505.00848
[20] L.D. Landau, E.M. Lifshitz, *Statistical physics*, Number Bd. 1 in Course of theoretical physics (Pergamon Press, 1969), http://books.google.de/books?id=COEIAQAAIAAJ
[21] R.D. Pisarski, V.V. Skokov, A.M. Tsvelik (2018), 1801.08156
[22] R. Casalbuoni, R. Gatto, M. Mannarelli, G. Nardulli, Phys.Lett. **B511**, 218 (2001), hep-ph/0101326
[23] R. Casalbuoni, R. Gatto, M. Mannarelli, G. Nardulli, Phys.Rev. **D66**, 014006 (2002), hep-ph/0201059
[24] M. Mannarelli, K. Rajagopal, R. Sharma, Phys.Rev. **D76**, 074026 (2007), hep-ph/0702021

Non-abelian anomalies and transport *

Juan L. Mañes[1,**], *Eugenio* Megías[2,3,***], *Manuel* Valle[3,****], and *Miguel Á.* Vázquez-Mozo[4,†]

[1]Departamento de Física de la Materia Condensada, Universidad del País Vasco UPV/EHU, Apartado 644, 48080 Bilbao, Spain
[2]Departamento de Física Atómica, Molecular y Nuclear and Instituto Carlos I de Física Teórica y Computacional, Universidad de Granada, Avenida de Fuente Nueva s/n, 18071 Granada, Spain
[3]Departamento de Física Teórica, Universidad del País Vasco UPV/EHU, Apartado 644, 48080 Bilbao, Spain
[4]Departamento de Física Fundamental, Universidad de Salamanca, Plaza de la Merced s/n, 37008 Salamanca, Spain

Abstract. We study the role of quantum anomalies in relativistic fluids of non-abelian theories within the hydrodynamic expansion. To this end, we compute the local functional that solves the anomaly equations, and obtain the Bardeen-Zumino terms that covariantize the currents. We particularize these results to a background with an electromagnetic field and chiral imbalance for two flavors.

1 Introduction

Hydrodynamics is an effective description of out-of equilibrium systems in which local thermodynamical equilibrium is assumed [1]. The equations of motion correspond to the conservation laws of the energy-momentum tensor and charge currents, which are supplemented by expressions of these quantities written in terms of fluids variables, the so-called constitutive relations. In presence of anomalies the currents are no longer conserved, and this has important effects in the hydrodynamics. In the general case, in addition to the ideal hydrodynamical contributions, there are extra terms in the constitutive relations which lead to dissipative and anomalous effects, i.e. for the charge currents $\langle J^\mu \rangle = nu^\mu + \langle J^\mu \rangle_{\text{diss \& anom}}$. Two relevant phenomena appear at first order in the hydrodynamic expansion as a consequence of chiral anomalies: the *chiral magnetic effect*, which is responsible for the generation of an electric current induced by an external magnetic field in the fluid $B^\mu = \varepsilon^{\mu\nu\rho\lambda} u_\nu \partial_\rho \mathbb{V}_\lambda$ [2], and the *chiral vortical effect*, in which the electric current is induced by a vortex $\omega^\mu = \varepsilon^{\mu\nu\rho\lambda} u_\nu \partial_\rho u_\lambda$ [3], i.e. $\langle J^\mu \rangle_{\text{anom}} = \sigma^{\mathscr{B}} B^\mu + \sigma^{\mathscr{V}} \omega^\mu + \cdots$. The corresponding susceptibilities $\sigma^{\mathscr{B}}$ and $\sigma^{\mathscr{V}}$ are even under time reversal symmetry, so that they cannot contribute to entropy production, i.e. $\partial_t s_{\text{anom}} = 0$, and then they lead to non-dissipative effects [4]. Recently it has been proposed a convenient formalism to obtain the non-dissipative part of the anomalous constitutive relations, which is based on the existence of an equilibrium partition function in a stationary

*Talk given by E. Megías at the QCD@Work: International Workshop on QCD, 25-28 June 2018, Matera, Italy.
**e-mail: wmpmapaj@lg.ehu.es
***e-mail: emegias@ugr.es
****e-mail: manuel.valle@ehu.es
†e-mail: Miguel.Vazquez-Mozo@cern.ch

© The Authors, published by EDP Sciences. This is an open access article distributed under the terms of the Creative Commons Attribution License 4.0 (http://creativecommons.org/licenses/by/4.0/).

background [5, 6]. The dissipative effects like, for instance, the shear viscosity or the electric conductivity, demand the use of alternative methods, e.g. kinetic theory, Kubo formulae or fluid/gravity correspondence [7–9]. In this work we will apply the equilibrium partition function formalism to study the anomalous part of the constitutive relations in non-abelian theories.

2 Equilibrium partition function formalism and hydrodymamics

We present in this section a summary of the equilibrium partition function formalism introduced in [5, 6] (see also e.g. [10, 11]). Let us consider a relativistic invariant Quantum Field Theory with a time independent $\mathrm{U}(1)$ gauge connection on the manifold [1]

$$ds^2 = G_{\mu\nu}dx^\mu dx^\nu = -e^{2\sigma(\vec{x})}(dt + a_i(\vec{x})dx^i)^2 + g_{ij}(\vec{x})dx^i dx^j\,, \tag{1}$$

$$\mathscr{A} = \mathscr{A}_0(\vec{x})dx^0 + \mathscr{A}_i(\vec{x})dx^i\,. \tag{2}$$

As usual, the partition function of the system writes

$$Z = \mathrm{Tr}\, e^{-\frac{H-\mu_0 Q}{T_0}}\,, \tag{3}$$

where H is the Hamiltonian of the theory, and Q is the charge associated to the gauge connection. T_0 and μ_0 are the temperature and chemical potential at equilibrium. The most general partition function should be consistent with: i) 3-dim diffeomorphism invariance; ii) Kaluza-Klein (KK) invariance, i.e. $t \to t + \phi(\vec{x})\,,\ \vec{x} \to \vec{x}$; and iii) $\mathrm{U}(1)$ time-independent gauge invariance (up to an anomaly).

From the partition function we can compute the energy-momentum tensor and $\mathrm{U}(1)$ charged current by performing the appropriate t-independent variations, i.e.

$$\delta \log Z = \frac{1}{T_0}\int d^3x\,\sqrt{g_3}\left(-\frac{1}{2}T_{\mu\nu}\delta g^{\mu\nu} + J^\mu \delta\mathscr{A}_\mu\right)\,, \tag{4}$$

where $g_3 = \det(g_{ij})$. In particular, for a general dependence $\log Z = \mathscr{W}(e^\sigma, A_0, a_i, A_i, g^{ij}, T_0, \mu_0)$, where

$$A_0 = \mathscr{A}_0\,, \qquad A_i = \mathscr{A}_i - a_i\mathscr{A}_0\,, \tag{5}$$

are KK invariant quantities, one gets the consistent currents and energy-momentum tensor

$$\begin{aligned} \langle J^i\rangle_{\mathrm{cons}} &= \frac{T_0}{\sqrt{-G}}\frac{\delta\mathscr{W}}{\delta A_i}\,, & \langle J_0\rangle_{\mathrm{cons}} &= -\frac{T_0 e^{2\sigma}}{\sqrt{-G}}\frac{\delta\mathscr{W}}{\delta A_0}\,, \\ \langle T_0{}^i\rangle &= \frac{T_0}{\sqrt{-G}}\left(\frac{\delta\mathscr{W}}{\delta a_i} - A_0\frac{\delta\mathscr{W}}{\delta A_i}\right)\,, & \langle T_{00}\rangle &= -\frac{T_0 e^{2\sigma}}{\sqrt{-G}}\frac{\delta\mathscr{W}}{\delta\sigma}\,. \end{aligned} \tag{6}$$

It is clear from these expressions that $\mathscr{W}$ plays the role of a generating functional for the hydrodynamic constitutive relations.

Let us discuss now the properties of the partition function at first order in the hydrodynamic expansion. The most general form of the partition function in this case, compatible with the symmetries mentioned above, is [5, 11]:

$$\mathscr{W}^{(1)} = \int d^3x\,\sqrt{g_3}\Big[\alpha_1(T,\nu)\varepsilon^{ijk}A_iA_{jk} + \alpha_2(T,\nu)\varepsilon^{ijk}A_if_{jk} + \alpha_3(T,\nu)\varepsilon^{ijk}a_if_{jk}\Big]\,, \tag{7}$$

[1] For simplicity, we will be restricted to abelian theories in this section. However, we will generalize this analysis to the non-abelian case in Sec. 3.

where

$$T = e^{-\sigma}T_0 \,, \qquad \nu = \frac{A_0}{T_0} \,, \tag{8}$$

while $A_{ij} = \partial_i A_j - \partial_j A_i$ and $f_{ij} = \partial_i a_j - \partial_j a_i$. The specific values of the functions $\alpha_i(T,\nu)$ depend on the particular theory considered. In practice, one way to compute α_i is to insert Eq. (7) into (6), and compare the result with the constitutive relations for that theory. In particular, if one considers an ideal gas of Dirac fermions, a suitable computation of the constitutive relations leads to the following expressions for the chiral magnetic and vortical conductivities

$$\sigma^{\mathscr{B}} = C\mu \,, \qquad \sigma^{\mathscr{V}} = \frac{1}{2}C\mu^2 + C_2 T^2 \mu \,, \tag{9}$$

where we have used that $\nu = \mu/T$ with $\mu = e^{-\sigma}A_0$. These results have been obtained in a wide variety of methods, see e.g. [2, 3, 9, 11–19]. In these expressions $C = 1/(4\pi^2)$ and $C_2 = 1/24$ are related to the axial anomaly [3, 20] and mixed gauge-gravitational anomaly [13], respectively. [2] From this, the result for the partition function in this theory turns out to be

$$\alpha_1(T,\nu) = -\frac{C}{6}\nu \,, \quad \alpha_2(T,\nu) = -\frac{1}{2}\left(\frac{C}{6}\nu^2 - C_2\right) \,, \quad \alpha_3(T,\nu) = 0 \,. \tag{10}$$

In the following we will operate in the opposite way, i.e. starting from the partition function, we will compute the constitutive relations of the theory. $\mathscr{W}$ will be obtained by solving the anomaly equations.

3 Non-abelian anomalies

In this section we will study the partition function in non-abelian theories. Our goal is to obtain from it the constitutive relations. We refer to [21, 22] for full details in the computation.

3.1 The chiral anomaly

We begin by studying the theory of a chiral fermion coupled to an external gauge field $\mathscr{A}_\mu^a$ described by the Lagrangian

$$\mathscr{L}_{\rm YM} = i\overline{\psi}\gamma^\mu\left(\partial_\mu - it_a\mathscr{A}_\mu^a\right)\psi \,, \tag{11}$$

where $t_a = t_a^\dagger$ are the Hermitian generators of the Lie algebra satisfying the commutation relations $[t_a, t_b] = if_{abc}t_c$. To study gauge anomalies it is convenient to work with the fermion effective action functional obtained by integrating out the fermion field

$$e^{i\Gamma[\mathscr{A}]} \equiv \int \mathscr{D}\overline{\psi}\mathscr{D}\psi\, e^{iS_{\rm YM}[\mathscr{A},\psi,\overline{\psi}]} \,. \tag{12}$$

Under a general shift $\mathscr{A}_\mu^a \to \mathscr{A}_\mu^a + \delta\mathscr{A}_\mu^a$, one gets that the first order variation of $\Gamma[\mathscr{A}]$ can be expressed as

$$\delta\Gamma[\mathscr{A}] = \int d^D x\, \delta\mathscr{A}_\mu^a(x)\, J_a^\mu(x)_{\rm cons} \,, \tag{13}$$

[2] If we have a set of chiral fermions transforming under a Lie group generated by t_a in the background of arbitrary gauge fields and metric, then the axial and mixed gauge-gravitational anomalies are proportional to $\sim {\rm Tr}(t_a\{t_b,t_c\})$ and $\sim {\rm Tr}(t_a)$, respectively.

where $J_a^\mu(x)_{\rm cons}$ is the consistent current. The axial anomaly is given by the failure of the effective action to be invariant under axial gauge transformations, which are defined as

$$\mathscr{A}_\mu \longrightarrow U\mathscr{A}_\mu U^{-1} - i\partial_\mu U U^{-1}, \qquad U(x) = \exp\left(i\Lambda_a^{\rm A}(x) t_a\right). \tag{14}$$

Under such a transformation,

$$\delta_{\rm gauge}\Gamma[\mathscr{A}] = -\int d^D x \Lambda_a^{\rm A}(x)\, G_a[\mathscr{A}(x)], \tag{15}$$

where $G_a[\mathscr{A}(x)]$ is the consistent anomaly. Particularizing now Eq. (13) to a gauge transformation $\delta\mathscr{A}_\mu^a = (D_\mu\Lambda^{\rm A})^a$, we arrive at the anomalous (non)-conservation law for the consistent current

$$D_\mu J_a^\mu(x)_{\rm cons} = G_a[\mathscr{A}(x)]. \tag{16}$$

We will analyze in the following the consequences of these results for the hydrodynamics of fluids affected by gauge anomalies.

3.2 Non-abelian anomaly and anomalous functional

In the following we will consider a non-abelian theory with symmetry group $\mathrm{U}(N_f)\times\mathrm{U}(N_f)$. The Bardeen form of the anomaly in this theory is [23]

$$\begin{aligned} G_a[\mathscr{V},\mathscr{A}] = \frac{iN_c}{16\pi^2}\varepsilon^{\mu\nu\rho\sigma}\mathrm{Tr}\Big\{t_a\big[&\mathscr{V}_{\mu\nu}\mathscr{V}_{\rho\sigma} + \tfrac{1}{3}\mathscr{A}_{\mu\nu}\mathscr{A}_{\rho\sigma} - \tfrac{32}{3}\mathscr{A}_\mu\mathscr{A}_\nu\mathscr{A}_\rho\mathscr{A}_\sigma \\ &+ \tfrac{8}{3}i(\mathscr{A}_\mu\mathscr{A}_\nu\mathscr{V}_{\rho\sigma} + \mathscr{A}_\mu\mathscr{V}_{\rho\sigma}\mathscr{A}_\nu + \mathscr{V}_{\rho\sigma}\mathscr{A}_\mu\mathscr{A}_\nu)\big]\Big\}, \end{aligned} \tag{17}$$

where t_a are the generators of the Lie algebra of $\mathrm{U}(N_f)$, while

$$\begin{aligned} \mathscr{V}_{\mu\nu} &= \partial_\mu\mathscr{V}_\nu - \partial_\nu\mathscr{V}_\mu - i[\mathscr{V}_\mu,\mathscr{V}_\nu] - i[\mathscr{A}_\mu,\mathscr{A}_\nu], \\ \mathscr{A}_{\mu\nu} &= \partial_\mu\mathscr{A}_\nu - \partial_\nu\mathscr{A}_\mu - i[\mathscr{V}_\mu,\mathscr{A}_\nu] - i[\mathscr{A}_\mu,\mathscr{V}_\nu], \end{aligned} \tag{18}$$

are the field strengths for the vector and axial gauge fields. Notice that the anomaly includes not only triangle, but also square and pentagon one-loop diagram contributions.

Our first goal is to compute the effective action compatible with the anomaly (17). The anomaly arises from the breaking of gauge invariance under axial gauge transformations of the effective action $\Gamma[\mathscr{V},\mathscr{A}]$, so that such an action should satisfy

$$\mathscr{Y}_a(x)\Gamma[\mathscr{V},\mathscr{A}] = 0, \qquad \mathscr{X}_a(x)\Gamma[\mathscr{V},\mathscr{A}] = G_a[\mathscr{V},\mathscr{A}], \tag{19}$$

where $\mathscr{Y}_a(x)$ and $\mathscr{X}_a(x)$ are the local generators of vector and axial gauge transformations, respectively. Then the computation of the local functional $\Gamma[\mathscr{V},\mathscr{A}]$ can be performed, for instance, by solving Eq. (19) by trial and error. The details of this computation will be presented in [22], and we will just show here the final result, which reads

$$\begin{aligned} \Gamma[V,A] = -\frac{N_c}{32\pi^2}\int dt\, d^3x \sqrt{g_3}\,\varepsilon^{ijk}\,\mathrm{Tr}\Big\{&\frac{32}{3} i V_0 A_i A_j A_k \\ &+ \frac{4}{3}(A_0A_i + A_iA_0)A_{jk} + 4(V_0A_i + A_iV_0)V_{jk} \\ &+ \frac{8}{3}(A_0^2 + 3V_0^2)A_i\partial_j a_k\Big\} + C_2 T_0^2\int dt\, d^3x\sqrt{g_3}\,\varepsilon^{ijk}\,\mathrm{Tr}A_i\partial_j a_k, \end{aligned} \tag{20}$$

where V_μ and A_μ are KK invariant vector and axial gauge fields, cf. Eq. (5). Let us mention at this point that Γ can be determined also from differential geometry methods, i.e. starting from the anomaly polynomial, one can compute the Chern-Simons effective action and to obtain from there the t-independent anomalous effective action after a suitable dimensional reduction, see e.g. [21, 24] for details. As it has been mentioned in Sec. 2, the coefficient C_2 is related to the mixed gauge-gravitational anomaly $\sim \mathrm{Tr}(t_a)\,\varepsilon^{\mu\nu\rho\sigma} R_{\mu\nu\lambda\kappa} R_{\rho\sigma}{}^{\lambda\kappa}$. Following similar differential geometry methods, it would be possible to compute this contribution by taking into account the Riemann tensor effect in the anomaly polynomial $\sim \mathrm{Tr}\,\mathscr{F}_A R^\mu{}_\nu R^\nu{}_\mu$. However, in the following we will neglect these contributions as this analysis goes beyond the scope of the present work. [3]

3.3 Covariant currents and constitutive relations

The charged currents obtained from the functional derivatives of the effective action are consistent currents, cf. Eq. (6). However, only covariant currents can enter in the constitutive relations, and these are defined by adding to the consistent currents the Bardeen-Zumino (BZ) polynomials, i.e. [4]

$$J^\mu_{\rm cov} = J^\mu_{\rm cons} + J^\mu_{\rm BZ}\,. \tag{21}$$

The BZ polynomials for this theory have been computed in [21, 22], leading to the following result

$$\begin{aligned} J^\mu_{\rm BZV} &= \frac{N_c}{8\pi^2}\varepsilon^{\mu\nu\rho\sigma}\mathrm{Tr}\Big\{t_a\big(\mathscr{A}_\nu\mathscr{V}_{\rho\sigma} + \mathscr{V}_{\rho\sigma}\mathscr{A}_\nu + \tfrac{8}{3}i\,\mathscr{A}_\nu\mathscr{A}_\rho\mathscr{A}_\sigma\big)\Big\}\,, \\ J^\mu_{\rm BZA} &= \frac{N_c}{24\pi^2}\varepsilon^{\mu\nu\rho\sigma}\mathrm{Tr}\Big\{t_a\left(\mathscr{A}_\nu\mathscr{A}_{\rho\sigma} + \mathscr{A}_{\rho\sigma}\mathscr{A}_\nu\right)\Big\}\,. \end{aligned} \tag{22}$$

Then the covariant currents and stress tensor in equilibrium can be obtained from $\mathscr{W} = i\Gamma$, and using Eqs. (21) and (22). This procedure leads in the vector currents to contributions of the form [22]

$$\langle J^i_{{\rm V}a}\rangle_{\rm cov} \propto e^\sigma\varepsilon^{ijk}\mathrm{Tr}\Big\{t_a A_0 V_{jk}\Big\} \qquad \text{and} \qquad \langle J^i_{{\rm V}a}\rangle_{\rm cov} \propto e^\sigma\varepsilon^{ijk}\mathrm{Tr}\Big\{t_a A_0 V_0 \partial_j a_k\Big\}\,, \tag{23}$$

which remind the chiral magnetic and vortical effects, respectively. [5] Similar contributions appear in the axial current $\langle J^i_{{\rm A}a}\rangle_{\rm cov}$ and energy-momentum tensor. It is interesting to note that the lower temporal indices, $\langle J_{{\rm V}a0}\rangle_{\rm cov}$ and $\langle J_{{\rm A}a0}\rangle_{\rm cov}$, as well as the components of the energy-momentum tensor, $\langle T_{00}\rangle$ and $\langle T^{ij}\rangle$, are vanishing.

While these results are valid for a non-abelian theory with symmetry group $\mathrm{U}(N_f)\times\mathrm{U}(N_f)$, it would be interesting to particularize them for a specific background. It is widely accepted that, in presence of non-abelian charges, the maximal number of chemical potentials to be consistently introduced corresponds to the dimension of the Cartan subalgebra. Then, for $N_f = 2$, we can consider the background given by a combination of $t_0 = \frac{1}{2}\mathbb{1}_{2\times 2}$ and $t_3 = \frac{1}{2}\sigma_3$, where σ_i are the Pauli matrices, i.e.

$$V_\mu(\boldsymbol{x}) = V_{0\mu}(\vec{x})\,t_0 + V_{3\mu}(\vec{x})\,t_3\,, \qquad A_0 = A_{50}\,t_0\,, \qquad A_i = 0\,. \tag{24}$$

For simplicity, we assume that A_{50} is constant. Then, in addition to the equilibrium velocity field $u_\mu = -e^\sigma(1,\vec{a})$, we can define the equilibrium baryonic, isospin and axial chemical potentials, as

$$\mu_0 = \mathscr{V}_{00}\,e^{-\sigma}\,, \quad \mu_3 = \mathscr{V}_{30}\,e^{-\sigma}\,, \quad \mu_5 = \mathscr{A}_{50}\,e^{-\sigma}\,, \tag{25}$$

[3] See [25] for an alternative analysis of these contributions in a group theoretic formulation.

[4] In fact, it is not possible to define an effective action whose derivation with respect to the gauge field gives the covariant current [15].

[5] A reliable identification of each term of (23) with a specific chiral effect can only be done after a computation of the out-of-equilibrium currents in a particular background, as it is done below.

where μ_5 controls the chiral imbalance of the system. In heavy ion collisions, the chiral imbalance is induced by a difference between the numbers of left-handed and right-handed quarks in the fireball [26, 27]. Accordingly, we can distinguish between three currents. For Quantum Chromodynamics (QCD) with u and d quarks, with electric charges $+\frac{2}{3}e$ and $-\frac{1}{3}e$, respectively, these are

$$\begin{aligned} J^{\mu}_{\mathrm{V,electromagnetic}} &= \frac{e}{3}\overline{\psi}\gamma^{\mu}t_0\psi + e\overline{\psi}\gamma^{\mu}t_3\psi\,, \\ J^{\mu}_{\mathrm{V,baryonic}} &= \frac{2}{3}e\overline{\psi}\gamma^{\mu}t_0\psi\,, \\ J^{\mu}_{\mathrm{V,isospin}} &= e\overline{\psi}\gamma^{\mu}t_3\psi\,, \end{aligned} \tag{26}$$

in the vector sector, which obviously fulfill the Gell-Mann–Nishijima formula. Finally, the out-of-equilibrium expressions of the currents follow from a Lorentz covariantization of the equilibrium currents (23). We will present here only partial results, and leave the detailed analysis together with the complete results for [22]. In particular, for the baryonic current one finds

$$\begin{aligned} \langle J^{\mu}_{\mathrm{V,baryonic}}\rangle_{\mathrm{cov}} &= -\frac{N_c e^2}{72\pi^2}\mu_5\varepsilon^{\mu\nu\lambda\rho}u_\nu\mathbb{V}_{\lambda\rho}\,, \\ \langle J^{\mu}_{\mathrm{A,baryonic}}\rangle_{\mathrm{cov}} &= \frac{N_c e}{24\pi^2}(\mu_5^2-\mu_0^2-\mu_3^2)\varepsilon^{\mu\nu\lambda\rho}u_\nu\partial_\lambda u_\rho+\cdots\,, \end{aligned} \tag{27}$$

in the vector and axial sector, respectively. We have used that $\mathcal{V}_{0\mu} = \frac{e}{3}\mathbb{V}_\mu$ and $\mathcal{V}_{3\mu} = e\mathbb{V}_\mu$, where $\mathbb{V}_\mu$ is the electromagnetic field and $\mathbb{V}_{\mu\nu} = \partial_\mu\mathbb{V}_\nu - \partial_\nu\mathbb{V}_\mu$ its field strength. We can easily identify in Eq. (27) the chiral magnetic and vortical contributions in the baryonic current, as well as the explicit results for the corresponding conductivities. Similar results can be obtained for the other currents and for the energy-momentum tensor. [6]

4 Conclusions

In this work we have studied non-dissipative transport effects of non-abelian theories up to first order in the hydrodynamic expansion. In particular, we have analyzed the effects induced by external electromagnetic fields and vortices in a relativistic fluid. The computation has been performed by using the equilibrium partition function formalism which, by construction, is suitable for the study of time reversal properties, i.e. those effects that cannot lead to entropy production. The local functional relevant for the computation of the constitutive relations has been obtained by solving the anomaly equations. Finally, we have found the results of the constitutive relations for the covariant currents after adding the corresponding Bardeen-Zumino polynomials. It would be interesting to extend this study to the case of spontaneous symmetry breaking, as this would lead to relevant information about the hydrodynamics of the corresponding Goldstone bosons interacting with external fields, with an obvious application to QCD at low temperatures. These and other issues will be addressed in a forthcoming publication [22].

Acknowledgement

This work has been supported by Plan Nacional de Altas Energías Spanish MINECO grants FPA2015-64041-C2-1-P, FPA2015-64041-C2-2-P, and by Basque Government grant IT979-16. The research of E.M. is also supported by Spanish MINEICO and European FEDER funds grant FIS2017-85053-C2-1-P, Junta de Andalucía grant FQM-225, as well as by Spanish MINEICO Ramón y Cajal Program and

[6] See e.g. [28] for a related study of chiral anomalies in superfluids.

by Universidad del País Vasco UPV/EHU, Bilbao, Spain, through a Visiting Professor appointment. M.A.V.-M. gratefully acknowledges the hospitality of the KEK Theory Center and the Department of Theoretical Physics of the University of the Basque Country during the early stages of this work.

References

[1] P. Kovtun, J. Phys. **A45**, 473001 (2012).
[2] K. Fukushima, D.E. Kharzeev, H.J. Warringa, Phys. Rev. **D78**, 074033 (2008).
[3] D.T. Son, P. Surowka, Phys. Rev. Lett. **103**, 191601 (2009).
[4] D.E. Kharzeev, H.U. Yee, Phys. Rev. **D84**, 045025 (2011).
[5] N. Banerjee, J. Bhattacharya, S. Bhattacharyya, S. Jain, S. Minwalla, T. Sharma, JHEP **09**, 046 (2012).
[6] K. Jensen, P. Kovtun, A. Ritz, JHEP **10**, 186 (2013).
[7] P.B. Arnold, G.D. Moore, L.G. Yaffe, JHEP **05**, 051 (2003).
[8] G. Policastro, D.T. Son, A.O. Starinets, Phys. Rev. Lett. **87**, 081601 (2001).
[9] J. Erdmenger, M. Haack, M. Kaminski, A. Yarom, JHEP **01**, 055 (2009).
[10] S. Bhattacharyya, JHEP **08**, 165 (2014).
[11] E. Megías, M. Valle, JHEP **11**, 005 (2014).
[12] N. Banerjee, J. Bhattacharya, S. Bhattacharyya, S. Dutta, R. Loganayagam, P. Surowka, JHEP **01**, 094 (2011).
[13] K. Landsteiner, E. Megías, F. Pena-Benitez, Phys. Rev. Lett. **107**, 021601 (2011).
[14] K. Landsteiner, E. Megías, L. Melgar, F. Pena-Benitez, JHEP **09**, 121 (2011).
[15] K. Landsteiner, E. Megías, F. Pena-Benitez, Lect. Notes Phys. **871**, 433 (2013).
[16] K. Jensen, R. Loganayagam, A. Yarom, JHEP **02**, 088 (2013).
[17] E. Megías, K. Landsteiner, F. Pena-Benitez, Acta Phys. Polon. Supp. **6**, 45 (2013).
[18] E. Megías, F. Pena-Benitez, JHEP **05**, 115 (2013).
[19] K. Landsteiner, E. Megías, F. Pena-Benitez, Phys. Rev. **D90**, 065026 (2014).
[20] D.E. Kharzeev, H.J. Warringa, Phys. Rev. **D80**, 034028 (2009).
[21] J.L. Mañes, E. Megías, M. Valle, M.Á. Vázquez-Mozo (2018), 1806.07647.
[22] J.L. Mañes, E. Megías, M. Valle, M.Á. Vázquez-Mozo, work in progress (2018).
[23] W.A. Bardeen, Phys. Rev. **184**, 1848 (1969).
[24] K. Jensen, R. Loganayagam, A. Yarom, JHEP **05**, 110 (2014).
[25] V.P. Nair, R. Ray, S. Roy, Phys. Rev. **D86**, 025012 (2012).
[26] R. Gatto, M. Ruggieri, Phys. Rev. **D85**, 054013 (2012).
[27] X. Planells, A.A. Andrianov, V.A. Andrianov, D. Espriu, PoS **QFTHEP2013**, 049 (2013), 1310.4434.
[28] Y. Neiman, Y. Oz, JHEP **09**, 011 (2011).

Measurement of hadron form factors at BESIII

Christoph Florian Redmer[1,*] for the BESIII Collaboration

[1]Institut für Kernphysik, Johannes Gutenberg-Universität Mainz, Johann-Joachim-Becher Weg 45, 55128 Mainz, Germany

Abstract. The BESIII experiment, operated at the BEPCII e^+e^- collider in Beijing, has acquired large data sets at center-of-mass energies between 2.0 GeV and 4.6 GeV. One of the key aspects of the physics program of the BESIII collaboration is to test the understanding of QCD at intermediate energies. Applying different experimental techniques, form factors of hadrons are measured. Among these are the pion form factor, as an important input to the $(g-2)_\mu$ puzzle, and the electro-magnetic form factors of nucleons and hyperons in the time-like regime. An overview of the recent results and some ongoing studies at BESIII is provided.

1 Introduction

Hadrons are not point-like. Their spatial extension has been studied e.g. in electron nucleon scattering since the 1960s, and it is still a topic of interest as manifested in the proton radius puzzle [1]. In addition to the size of hadrons, also their internal structure is of interest. It is still not fully understood how the masses of hadrons are generated. While the lightest mesons have masses of the order of 100 MeV and Baryon masses start from approximately 1 GeV, their two or three constituents have masses of only a few MeV, respectively. The difference in mass is assumed to be generated due to the internal dynamics of the hadrons. Observables, which allow to parameterize the structure and the internal dynamics are electromagnetic form factors (EMFF). At the BESIII experiment in Beijing hadron EMFFs are studied to test and enhance the understanding of non-perturbative quantum chromodynamics (QCD), where hadrons and not quarks and gluons are the relevant degrees of freedom.

2 BESIII experiment

The BESIII detector is a magnetic spectrometer [2] located at the Beijing Electron Positron Collider (BEPCII) [3]. The cylindrical core of the BESIII detector consists of a helium-based multilayer drift chamber (MDC), a plastic scintillator time-of-flight system (TOF), and a CsI(Tl) electromagnetic calorimeter (EMC), which are all enclosed in a superconducting solenoidal magnet providing a 1.0 T magnetic field. The solenoid is supported by an octagonal flux-return yoke with resistive plate counter muon identifier modules (MUC) interleaved with steel. The acceptance of charged particles and photons is 93% over 4π solid angle. The charged-particle momentum resolution at 1 GeV/c is 0.5%, and the dE/dx resolution is 6%

*e-mail: redmer@uni-mainz.de

© The Authors, published by EDP Sciences. This is an open access article distributed under the terms of the Creative Commons Attribution License 4.0 (http://creativecommons.org/licenses/by/4.0/).

for the electrons from Bhabha scattering. The EMC measures photon energies with a resolution of 2.5% (5%) at 1 GeV in the barrel (end cap) region. The time resolution of the TOF barrel part is 68 ps, while that of the end cap part is 110 ps. The end cap TOF system was upgraded in 2015 with multi-gap resistive plate chamber technology, providing a time resolution of 60 ps [4, 5].

BEPCII provides e^+e^- collisions at center-of-mass energies from 2.0 GeV to 4.6 GeV. It is optimized for the operation at $\sqrt{s} = 3.773$ GeV, the peak of the $\psi(3770)$ resonance, where the design luminosity of 10^{33} cm^{-2}s^{-1} has been achieved. The energy dependence of cross sections can be studied in conventional scan measurements in the energy range covered by BEPCII. The method of initial state radiation (ISR) allows to extend the energy range down to the two pion mass threshold. The emission of a photon from the initial state lowers the effective center-of-mass energy $\sqrt{s'} = \sqrt{s - 2\sqrt{s}E_\gamma}$, where E_γ is the energy of the ISR photon. The cross sections measured in this way are radiative cross sections. They are related to the cross section of the process without ISR by the radiator function $H(s, E_\gamma, \theta_\gamma)$, which describes the probability to emit an ISR photon of the energy E_γ at the polar angle θ_γ at a specific $\sqrt{s}$. The relation of radiative and non-radiative cross sections is given by $\frac{d\sigma_{\text{had}+\gamma}}{dm_\gamma} = \frac{2m_{\text{had}}}{s} H(s, E_\gamma, \theta_\gamma)\sigma_{\text{had}}$.

The radiator function produces a characteristic shape in the angular distribution of the ISR photons. Most of the photons are emitted collinear to the lepton beams. The produced hadronic system is boosted in the direction opposite to the photon. Considering the acceptance of the BESIII detector, two different analysis strategies for ISR events can be defined. On the one hand, the hadronic system as well as the ISR photon can be registered in the detector. This case is referred to as "tagged ISR measurement". It allows to measure cross sections from the two-pion mass threshold up to the actual center-of-mass energy of the accelerator. However, with increasing masses of the produced hadronic system, i.e. decreasing photon energies, it is more likely to mistake random signals in the EMC for an ISR photon, leading to a high background contamination. On the other hand, the hadronic system can be measured in the detector, while the ISR photon escapes detection being emitted along the beam axis. By exploiting energy and momentum conservation the photon four-vector can be determined. Requiring its polar angle to be small rejects most of the background events and selects the peaking part of the differential radiative cross section. This scenario is referred to as "untagged ISR measurement". Due to the boost, the acceptance for the hadronic system is limited. The higher the photon energy, i.e. in case of small masses of the hadronic system, the higher is the probability for the final state particles to be emitted close to the beam pipe, where they cannot be registered by a detector. The acceptance of the BESIII detector imposes a threshold of approximately 1 GeV/c^2 on the mass of the hadronic system in an untagged measurement.

Due to the emission of the ISR photon, the radiative process is suppressed by a factor $\frac{\alpha}{\pi}$. In order to perform precision studies, large data samples are required. In the recent years the BESIII collaboration has acquired large data sets of more than 10 fb^{-1} at center-of-mass energies at and above the $\psi(3770)$ resonance [6, 7]. These are exploited to measure meson and baryon form factors using both, the tagged and untagged ISR methods.

3 Meson form factors

A special interest in meson form factors comes from the anomalous magnetic moment of the muon. There is a long standing discrepancy of more than three standard deviations between the experimental value obtained in a direct measurement [8] and the theory prediction within the Standard Model (SM) [9]. In order to settle this difference new measurements are performed at FNAL with the aim of a fourfold improvement of the current accuracy [10]. A first result with the accuracy of the previous measurement is expected for spring 2019. Another direct measurement is prepared at J-PARC, using a different approach [11]. The accuracy of the SM prediction is currently completely limited by the hadronic contributions to a_μ. The largest of these contributions comes from leading order hadronic vacuum polarization effects. Due to the asymptotic freedom in the strong interaction, the contribution cannot be evaluated by perturbative means. Besides the recent progress in lattice QCD [12], information from experiment has been used to improve the calculations. The optical theorem relates the leading order vacuum polarization contribution $a_\mu^{hVP,LO}$ to hadronic cross sections measured in e^+e^- collisions. The approach allows to systematically improve the accuracy of $a_\mu^{hVP,LO}$ with high precision measurements of hadronic cross sections. The dispersive integral used to evaluate the cross section also contains a kernel function. Both, the cross sections as well as the kernel function show an energy dependence of $1/s$, putting special emphasis on the cross sections at $\sqrt{s} < 1\,\mathrm{GeV}$. As a consequence, the cross section $\sigma(e^+e^- \to \pi^+\pi^-)$, which includes the production and decay of the ρ resonance, makes up for about 70% of the absolute value of $a_\mu^{hVP,LO}$. To evaluate its uncertainty, the cross sections of higher multiplicities $e^+e^- \to \pi^+\pi^-\pi^0, e^+e^- \to \pi^+\pi^-2\pi^0$ and $e^+e^- \to 2\pi^+2\pi^-$ also play an important role.

The cross section $\sigma(e^+e^- \to \pi^+\pi^-)$ has been measured with sub-percent accuracy by the KLOE [13] and BaBar [14] collaborations. However, their results differ by more than 3%, which is reflected in the uncertainty of the evaluation of $a_\mu^{hVP,LO}$. In order to clarify the situation, a new high precision measurement has been performed at BESIII [15]. Based on $2.93\,\mathrm{fb}^{-1}$ of data taken at $\sqrt{s} = 3.773\,\mathrm{GeV}$, the method of tagged ISR has been used to study the cross section of two-pion production at center of mass energies between 600 and $900\,\mathrm{MeV/c^2}$. Due to the similar masses of the final state particles, the radiative production of muon pairs $e^+e^- \to \mu^+\mu^-$ constitutes the dominating background contribution. Combining the information from different sub-detectors, like energy loss in the MDC, energy deposits and shower shapes in the EMC, and the penetration depth in the MUC, an artificial neural network has been trained with Monte Carlo samples to separate effectively muons from pions. Systematic differences between data and simulations have been carefully studied and corrected for. The validity of the corrections has been tested by comparing the muon event yield with the QED prediction as implemented in the PHOKHARA event generator [16, 17], where an accuracy of 0.5% is claimed. Excellent agreement with the selected data was found.

Finally, the $e^+e^- \to \pi^+\pi^-\gamma_{\mathrm{ISR}}$ signal event yield is corrected for detection efficiency, and is normalized to the integrated luminosity of the data set to obtain the radiative cross section. By correcting for vacuum polarization and FSR effects, and by dividing out the radiator function, the bare cross section, relevant to determine the contribution to $a_\mu^{hVP,LO}$, is calculated. A total uncertainty of the cross section of 0.9% is achieved. The dominating systematic uncertainties are due to the luminosity determination and the knowledge of the radiator function, with 0.5% each.

The BESIII result is compared with previous measurements based on a fit of the pion FF with a Gounaris-Sakurai parameterization, illustrated in Fig. 1. Though systematic deviations can be observed for both, the KLOE results as well as the BaBar results, the value of $a_\mu^{hVP,LO}$ obtained from the evaluation of the dispersive integral agrees nicely with the values obtained by the KLOE collaboration, as shown in Fig. 1. The result of the BESIII measurement is $a_\mu^{\pi\pi,LO}(600-900\,\mathrm{MeV}) = 368.2 \pm 2.5_{\mathrm{stat}} \pm 3.3_{\mathrm{syst}} \cdot 10^{-10}$. It confirms the deviation between the direct measurement and the SM prediction of a_μ to be on the level of more than three standard deviations. Recent dispersive evaluations of compilations of hadronic cross sections [9, 18],

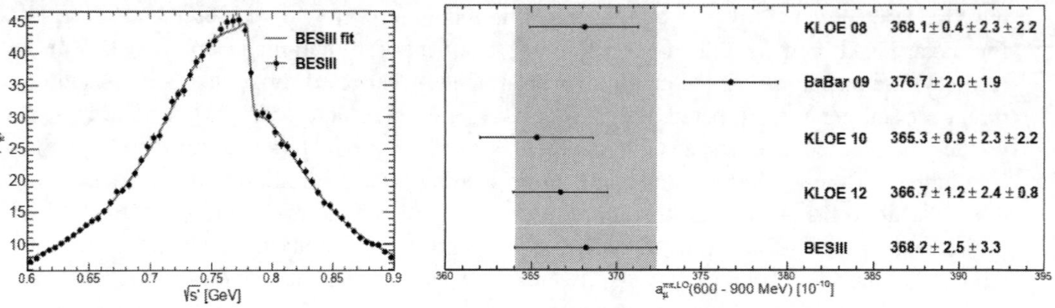

Figure 1. (From Ref. [15]) **left:** Pion FF determined at BESIII with the result of a Gounaris-Sakurai fit. **right:** Comparison of $a_\mu^{\pi\pi,LO}(600-900\,\text{MeV})$ determined at BESIII and by previous experiments.

which include the BESIII result, were able to reduce the uncertainty of the hadronic vacuum contribution to a_μ by more than 20%.

4 Baryon form factors

The number of form factors (FF) of baryons depends on their spin s. Due to their fermionic nature baryons have more than one time-like EMFF, where the number is given by $(2s+1)$. Thus, nucleons and Λ hyperons have two EMFF.

The vertex related to the baryons produced in e^+e^- annihilation can be described as

$$\Gamma^\mu = \gamma^\mu F_1(q^2) + \frac{i\sigma^{\mu\nu}q_\nu}{2M_B}\kappa F_2(q^2), \tag{1}$$

where $F_1(q^2)$ and $F_2(q^2)$ are the Pauli and Dirac FF, respectively. M_B is the mass of the baryon, κ is the anomalous magnetic moment of the baryon, and q^2 is the momentum transfer. Experimentally more convenient is the use of the Sachs FF G_E and G_M, which are linear combinations of the Pauli and Dirac FF:

$$G_\text{E} = F_1(q^2) + \frac{q^2}{4M_B^2}\kappa F_2(q^2) \qquad G_\text{M} = F_1(q^2) + \kappa F_2(q^2) \tag{2}$$

The Sachs FF are EMFFs, describing the charge and magnetic distributions of the baryon. They can be determined by the angular analysis of the differential cross section, which is given as:

$$\frac{d\sigma^\text{Born}}{d\Omega} = \frac{\alpha^2\beta^2 C}{4q^2}\left[(1+\cos^2\theta_B^{CM})|G_\text{M}|^2 + \frac{1}{\tau}|G_\text{E}|^2\sin^2\theta_B^{CM}\right], \tag{3}$$

where C is the Coulomb factor, with $C=1$ for neutral and $C = \frac{\pi\alpha}{\beta(1-e^{\pi\frac{\alpha}{\beta}})}$ for charged particles, respectively, and $\beta = \sqrt{1-\frac{1}{\tau}}$ with $\tau = \frac{q^2}{4M_B^2}$.

Assuming the absolute values of the electric and magnetic FF to be equal, information on the effective FF can be extracted from the total cross section of baryon production in e^+e^- annihilation:

$$\sigma_{B\bar{B}}^\text{Born}(q^2) = \frac{4\pi\alpha^2\beta^2 C}{3q^2}\left[|G_\text{M}|^2 + \frac{1}{2\tau}|G_\text{E}|^2\right] \qquad |G_\text{eff}(q^2)| = \sqrt{\frac{\sigma_{B\bar{B}}^\text{Born}(q^2)}{(1+\frac{1}{2\tau})\frac{4\pi\alpha^2\beta^2 C}{3q^2}}} \tag{4}$$

In the following, recent results and selected ongoing studies of nucleon and hyperon EMFF at BESIII are discussed.

4.1 Nucleon form factors

A first measurement of proton EMFFs has been performed at BESIII in 2012 as a scan experiment [19]. At 12 different center of mass energies from 2232.4 MeV to 3671.0 MeV at total integrated luminosity of 157 pb^{-1} has been collected. (Anti-)proton candidates are selected by requiring two oppositely charged tracks per event. By evaluating the energy loss of the tracks in the MDC as well as their associated time of flight, protons are identified. Additional kinematic constraints on the relative opening angles of the tracks in the center-of-mass frame allow to suppress further background. The Born cross section is determined for each energy point by normalizing the background subtracted event yield to the integrated luminosity and reconstruction efficiency, and by correcting for radiative effects. The effective FF extracted from the Born cross section is in agreement with previous measurements, however, the accuracy is improved by approximately 30% due to the BESIII result.

At the three scan point with largest statistics, the differential cross section is studied in order to determine the FF ratio $R = \frac{|G_E|}{|G_M|}$. The ratio has been determined in two different ways, using a fit to the angular distributions, and by using the method of moments. Both methods yield compatible results, showing the value of the FF ratio to be consistent with $R = 1$, which supports the assumption used in the extraction of the effective FF. The extracted FF ratio is also in agreement with the BaBar result, however, it is clearly limited by statistics. The result motivated a new, high statistics energy scan at BESIII, which was performed in 2015. The data is still under evaluation.

At the same time, the high statistics samples taken at center-of-mass energies at and above the peak of the $\psi(3770)$ resonance have been exploited to study the proton FF. In total a data set of 7.4 fb^{-1} at energies between $\sqrt{s} = 3.773$ GeV and 4.6 GeV [6, 7] is analyzed using the tagged and untagged ISR methods.

For the tagged ISR method, events with two oppositely charged tracks identified as (anti-)protons and one high energetic photon candidate are selected. Performing a $4C$ kinematic fit constrained by energy and momentum conservation most background contributions can be suppressed. The dominating background contribution of $e^+e^- \rightarrow p\bar{p}\pi^0$ can be suppressed by using a successful $5C$ kinematic fit with one additional photon in the event as a veto criterion. The additional constraint comes from enforcing the π^0 mass in the fit. In this way, the tagged ISR method allows to study the Born cross section and the effective FF in 31 bins from the $p\bar{p}$ threshold up to 3.0 GeV. In order to extract the FF ratio in an angular analysis with reasonable accuracy, the number of bins is reduced to six.

The untagged ISR method is exploited by selecting events with two oppositely charged tracks, which have been identified as protons. The missing momentum to the $p\bar{p}$ par is calculated and required to have a small polar angle, assuming it is due to a photon that escaped detection along the beam pipe. Correspondingly, the event candidates must yield a small missing mass. With this analysis scheme, the Born cross section of $e^+e^- \rightarrow p\bar{p}$ and the effective proton FF can be investigated in 30 bins from $\sqrt{s} = 2.0$ GeV to 3.0 GeV. The angular analysis of the FF ratio is performed in three bins of this mass range.

The preliminary results of the angular analysis of both ISR methods are shown in Fig. 2. The accuracy is competitive with previous measurements. The FF ratios are in agreement with the BaBar result as well as the published BESIII scan measurement [19].

Also the preliminary results on the Born cross section and the effective FF of the tagged as well as untagged ISR methods are shown in Fig. 2. They are consistent with previous measurements. The untagged BESIII result is competitive in its statistical accuracy with the BaBar result, which has been obtained with the tagged ISR methods on a significantly larger data sample. The result clearly demonstrates the advantage of BESIII being able to access the small angle ISR events in the untagged method.

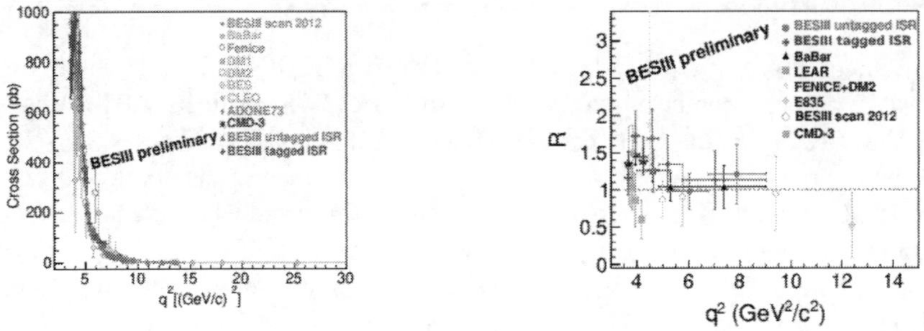

Figure 2. left: $e^+e^- \to p\bar{p}$ Born cross section results from BESIII scan [19] (grey stars), previous measurements, and the preliminary results of the BESIII ISR analyses using the tagged (blue) and untagged (red) method. **right:** Proton EMFF ratio results from BESIII scan [19] (yellow open circles), previous measurements, and the preliminary results of the BESIII ISR analyses using the tagged (blue) and untagged (red) method.

Both the tagged and untagged result of the effective FF support the observation of structures resembling the pattern of a damped oscillation, when the FF is investigated as a function of the three-momentum of the relative motion of the protons. The pattern had been reported in the analysis of the BaBar data [20, 21] and is considered as a sign for a more complex dynamic effect, which still needs clarification.

4.2 Hyperon form factors

Compared to nucleons, the hyperon EMFF are less well investigated experimentally. Thus, the BESIII data of the 2012 energy scan have also been used to study the Λ cross section and EMFF [22]. The data taken at $\sqrt{s} = 2.2324\,\text{GeV}, 2.4\,\text{GeV}, 2.8\,\text{GeV}$, and $3.08\,\text{GeV}$ allow to put emphasis on the threshold region, with the first data point being only 1 MeV above the $\Lambda\bar{\Lambda}$ threshold. The hyperons are generally reconstructed from their decay to $p\pi$. Only at threshold another strategy is necessary, since the low energetic nucleons from the decay are already stopped in the beam pipe. Here, the decay mode $n\pi^0$ is exploited in addition, and the annihilation vertices of $\bar{p}$ and $\bar{n}$ at the beam pipe are used to tag the decay of $\bar{\Lambda}$. Based on this strategy, the cross section of $\Lambda\bar{\Lambda}$ production and the effective FF have been determined. The results, illustrated in Fig. 3 are consistent with previous measurements, however, with improved accuracy. In contrast to the expectation of a vanishing cross section at threshold, the BESIII result reveals a steep increase, which hints at a more complex underlying physics scenario, requiring further detailed studies of hyperon production.

As a part of these studies a new high statistics energy scan has been carried out at BESIII in 2015. A data set of $66.9\,\text{pb}^{-1}$ at $\sqrt{s} = 2.396\,\text{GeV}$ has been acquired, aiming at the determination of the FF ratio as well as the relative phase of the EMFF of Λ. The hyperon pairs are exclusively reconstructed in the final state $e^+e^- \to \Lambda\bar{\Lambda} \to p\bar{p}\pi^+\pi^-$. Protons and pions can be identified by their different momenta. The preliminary results of the extracted cross section and effective FF are consistent with previous measurements, including the BESIII results [22]. The EMFF ratio R and the relative phase $\Delta\Phi$ are determined following the approach suggested by Fäldt and Kupść [23]. The decay distributions of the complete final state are described by a set of six functions $\mathcal{T}_i(\xi)$, where $\xi = (\theta, \theta_1, \phi_1, \theta_2, \phi_2)$ is state vector of the events, containing the scattering angle θ of the hyperon, and the decay angles θ_i and ϕ_i of the respective pair of daughter particles. The free parameters in the approach are $\eta = \frac{\tau - R^2}{\tau + R^2}$, related to the EMFF ratio, and the relative phase $\Delta\Phi$. They are determined by performing an unbinned maximum likelihood fit to the data. The preliminary results are $R = 0.94 \pm 0.16_{\rm stat} \pm 0.03_{\rm syst} \pm 0.02_{\alpha_\Lambda}$ and $|\Delta\Phi| = 42^\circ \pm 16^\circ_{\rm stat} \pm 8^\circ_{\rm syst} \pm 6^\circ_{\alpha_\Lambda}$, where the third uncertainty reflects the deviation of the recent determination of the Λ decay constant [24] from the PDG value [25]. This is the first measurement of the relative phase of hyperon EMFF in e^+e^- collisions.

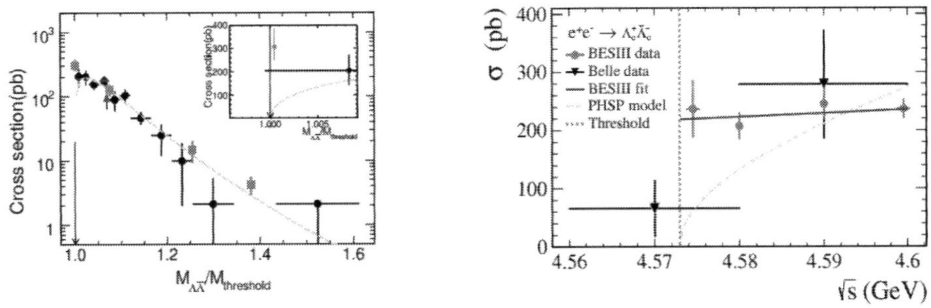

Figure 3. left: (From Ref. [22]) The Born cross section of $e^+e^- \to \Lambda\bar{\Lambda}$ from BESIII (red), BaBar (black) and DM2 (blue). The green line shows a fit with a pQCD prediction. **right:** (From Ref. [26]) The Born cross section of $e^+e^- \to \Lambda_c^+\bar{\Lambda}_c^-$ from BESIII (red) and Belle (black).

The center-of-mass energies accessible at BESIII barely allow to study the threshold region of the charmed hyperon Λ_c. The reaction $e^+e^- \to \Lambda_c^+\bar{\Lambda}_c^-$ has been studied at the four energies $\sqrt{s} = 4.5745\,\text{GeV}, 4.580\,\text{GeV}, 4.590\,\text{GeV}$ and $4.5995\,\text{GeV}$ [26]. The hyperons are reconstructed from ten Cabbibo-favored decay modes. The result cross sections are determined as the weighted average of the individual decay mode analyses. In addition to the Born cross section and the effective FF, the statistics at $\sqrt{s} = 4.5745\,\text{GeV}$ and $4.5995\,\text{GeV}$ is sufficient to determine the FF ratio in an angular analysis of the differential cross section, which is the first determination of the FF ratios for charmed baryons. As can be seen in Fig. 3, there seems to be a tension between the results for the effective FF obtained by the Belle collaboration [27] and the BESIII result. While the Belle measurement is compatible with a vanishing cross section at threshold, the BESIII result suggests a plateau in the cross section. However, the binning of the Belle result, which was obtained using the ISR method, is rather coarse. The lowest mass bin also extends below the $\Lambda_c\bar{\Lambda}_c$ threshold, which might distort the distribution. Further, high precision measurement are necessary to improve the understanding of charmed hyperon production at threshold.

5 Summary and outlook

Form factors are an essential tool to understand the structure and dynamics of hadrons in the non-perturbative regime. The BESIII experiment can exploit the methods of the conventional energy scan as well as the tagged and untagged ISR techniques to provide valuable data, which allow to gain deeper insights.

The measurement of the pion form factor with sub-percent accuracy is an important input to the calculation of the hadronic contributions to the anomalous magnetic moment of the muon a_μ. Additional investigations of higher pion multiplicities, e.g. $e^+e^- \to \pi^+\pi^-\pi^0$ and $e^+e^- \to \pi^+\pi^-\pi^0\pi^0$, which are important to further reduce the theoretical uncertainty of a_μ, are in preparation and results will be published soon.

The investigations of the nucleon EMFF will be complemented by a new high statistics scan measurement, which does not only provide additional data on the proton EMFF G_E and G_M, but also the neutron EMFF are studied. The same data are also under evaluation for hyperon FF, providing more detailed information for Λ, Σ^0, and $\Sigma^{\pm}$ EMFF.

Studies of the charmed hyperon EMFF at BESIII are currently limited by the accessible center-of-mass energies of the BEPCII accelerator. A possible upgrade of the machine is under investigation. The goal is to push the accessible energy range to $\sqrt{s} = 4.9$ Gev, which allows to study charmed hyperons and their decays in more details.

References

[1] C.E. Carlson, Prog. Part. Nucl. Phys. **82**, 59 (2015), 1502.05314

[2] M. Ablikim et al. (BESIII), Nucl. Instrum. Meth. **A614**, 345 (2010), 0911.4960

[3] C. Yu et al., *BEPCII Performance and Beam Dynamics Studies on Luminosity*, in *Proceedings, 7th International Particle Accelerator Conference (IPAC 2016): Busan, Korea, May 8-13, 2016* (2016), p. TUYA01

[4] X. Li et al., Radiation Detection Technology and Methods **1**, 13 (2017)

[5] Y.X. Guo et al., Radiation Detection Technology and Methods **1**, 15 (2017)

[6] M. Ablikim et al. (BESIII), Chin. Phys. **C37**, 123001 (2013), 1307.2022

[7] M. Ablikim et al. (BESIII), Chin. Phys. **C39**, 093001 (2015), 1503.03408

[8] G.W. Bennett et al. (Muon g-2), Phys. Rev. **D73**, 072003 (2006), hep-ex/0602035

[9] A. Keshavarzi, D. Nomura, T. Teubner, Phys. Rev. **D97**, 114025 (2018), 1802.02995

[10] J. Grange et al. (Muon g-2) (2015), 1501.06858

[11] T. Mibe (J-PARC g-2), Nucl. Phys. Proc. Suppl. **218**, 242 (2011)

[12] T. Blum, P.A. Boyle, V. Gulpers, T. Izubuchi, L. Jin, C. Jung, A. Juttner, C. Lehner, A. Portelli, J.T. Tsang (RBC, UKQCD), Phys. Rev. Lett. **121**, 022003 (2018), 1801.07224

[13] A. Anastasi et al. (KLOE-2), JHEP **03**, 173 (2018), 1711.03085

[14] B. Aubert et al. (BaBar), Phys. Rev. Lett. **103**, 231801 (2009), 0908.3589

[15] M. Ablikim et al. (BESIII), Phys. Lett. **B753**, 629 (2016), 1507.08188

[16] G. Rodrigo, H. Czyz, J.H. Kuhn, M. Szopa, Eur. Phys. J. **C24**, 71 (2002), hep-ph/0112184

[17] H. Czyz, J.H. Kuhn, A. Wapienik, Phys. Rev. **D77**, 114005 (2008), 0804.0359

[18] M. Davier, A. Hoecker, B. Malaescu, Z. Zhang, Eur. Phys. J. **C77**, 827 (2017), 1706.09436

[19] M. Ablikim et al. (BESIII), Phys. Rev. **D91**, 112004 (2015), 1504.02680

[20] A. Bianconi, E. Tomasi-Gustafsson, Phys. Rev. Lett. **114**, 232301 (2015), 1503.02140

[21] A. Bianconi, E. Tomasi-Gustafsson, Phys. Rev. **C93**, 035201 (2016), 1510.06338

[22] M. Ablikim et al. (BESIII), Phys. Rev. **D97**, 032013 (2018), 1709.10236

[23] G. Faldt, A. Kupsc, Phys. Lett. **B772**, 16 (2017), 1702.07288

[24] M. Ablikim et al. (BESIII) (2018), 1808.08917

[25] C. Patrignani et al. (Particle Data Group), Chin. Phys. **C40**, 100001 (2016)

[26] M. Ablikim et al. (BESIII), Phys. Rev. Lett. **120**, 132001 (2018), 1710.00150

[27] G. Pakhlova et al. (Belle), Phys. Rev. Lett. **101**, 172001 (2008), 0807.4458

Overview of recent LHCb results

Stefania Vecchi[1,*] on behalf of the LHC*b* collaboration
[1]INFN, sezione di Ferrara

Abstract. A selection of the most recent and relevant measurements by the LHC*b* experiment is presented. It includes the determination of the CKM angle γ, obtained from a combination of a number of measurements of B meson decays, the measurement of the $D^0 - \overline{D}^0$ mixing parameters and the results of several tests of the lepton-flavour universality of the electroweak couplings.

1 Introduction

The LHC*b* detector is a forward spectrometer designed for precision studies of b- and c-hadron decays [1]. Of particular relevance for the measurements discussed in this proceeding is its excellent precision in reconstructing vertices associated to primary pp interactions and to the decays of long-lived particles such as B and D mesons and τ leptons. The momenta of charged particles are reconstructed with high precision by a system of tracking detectors. Different types of charged hadrons, such as protons, kaons and pions, are distinguished using information from two ring-imaging Cherenkov detectors, while electromagnetic particles are identified by means of a calorimeter system and a muon detector.

Unless specified, the measurements presented in the following rely on a data sample of pp collisions at centre-of-mass energies of 7 and 8 TeV collected by the LHC*b* detector during Run1 and corresponding to an integrated luminosity of 3 fb^{-1}.

2 Measurement of the CKM angle γ

The angle $\gamma \equiv \arg(-(V_{ud}V^*_{ub})/(V_{cd}V^*_{cb}))$ of the CKM unitarity triangle can be measured in tree level b-hadron decays exploiting the interference of $b{\to}c$ and $b{\to}u$ transitions. Among the different angles, γ is the least well measured: the current world average value is $\gamma = (65.3^{+1.0}_{-2.5})^\circ$ and $(72.1^{+5.4}_{-5.8})^\circ$ based on indirect and direct measurements, respectively [2].

The measurement of γ is considered as a Standard Model (SM) benchmark: from a theoretical point of view it is very clean as it is dominated by tree level processes, where higher contributions from loops are very suppressed ($O(10^{-4})$). Unfortunately, the experimental determination of γ is quite challenging since the visible branching fractions of the concerned B decays and the typical size of these interferences are quite small ($\mathcal{B}_{\text{vis}} \sim O(10^{-7})$, $I \sim O(10\%)$). The best precision on the angle γ results from a combination of all the possible measurements that exploit different analysis methods (namely GLW [3], ADS [4], GGSZ [5], from the name of the authors that proposed them, and Dalitz and time-dependent, TD) and external inputs, such as amplitude ratios and strong phases of the involved particle decays.

*e-mail: vecchi@fe.infn.it

© The Authors, published by EDP Sciences. This is an open access article distributed under the terms of the Creative Commons Attribution License 4.0 (http://creativecommons.org/licenses/by/4.0/).

Table 1. Measurements contributing to the latest determination of the angle γ by the LHCb collaboration. New and updated results with respect to the previous combination are marked with the (*) and (†) symbols, respectively. References to each measurement can be found in Ref. [6].

B decay	D decay	Method	Dataset
$B^+ \to DK^+$	$D \to h^+h^-$	GLW	Run1&Run2 (†)
$B^+ \to DK^+$	$D \to h^+h^-$	ADS	Run1
$B^+ \to DK^+$	$D \to h^+\pi^-\pi^+\pi^-$	GLW/ADS	Run1
$B^+ \to DK^+$	$D \to h^+h^-\pi^0$	GLW/ADS	Run1
$B^+ \to DK^+$	$D \to K^0_S h^+h^-$	GGSZ	Run1
$B^+ \to DK^+$	$D \to K^0_S h^+h^-$	GGSZ	Run2 (*)
$B^+ \to DK^+$	$D \to K^0_S K^+\pi^-$	GLS	Run1
$B^+ \to D^*K^+$	$D \to h^+h^-$	GLW	Run1&Run2 (†)
$B^+ \to DK^{*+}$	$D \to h^+h^-$	GLW/ADS	Run1&Run2
$B^+ \to DK^{*+}$	$D \to h^+\pi^-\pi^+\pi^-$	GLW/ADS	Run1&Run2 (*)
$B^+ \to DK^+\pi^+\pi^-$	$D \to h^+h^-$	GLW/ADS	Run1
$B^0 \to DK^{*0}$	$D \to K^+\pi^-$	ADS	Run1
$B^0 \to DK^+\pi^-$	$D \to h^+h^-$	GLW-Dalitz	Run1
$B^0 \to DK^{*0}$	$D \to K^0_S\pi^+\pi^-$	GGSZ	Run1
$B^0_s \to D^{\mp}_s K^{\pm}$	$D^+_s \to h^+h^-\pi^+$	TD	Run1 (†)
$B^0 \to D^{\mp}\pi^{\pm}$	$D^+ \to K^+\pi^-\pi^+$	TD	Run1 (*)

The LHCb collaboration performed several measurements of different 2 and 3 or 4-body B or D decays corresponding to different analysis methods listed in Table 2 and a new, improved combination of the results, using a frequentist approach, was recently released [6]. Figure 2 shows the result of the new combination, where the partial contributions from the measurements using B^+, B^0 and B^0_s are also shown individually. The best value corresponds to $\gamma = (74.0^{+5.0}_{-5.8})^\circ$ and represents the most precise determination of γ from a single experiment.

Improving further the precision of γ with additional and more accurate measurements is particularly important to possibly reveal new physics phenomena beyond the SM that could affect its value at a level just below current experimental accuracy. For instance, a precision of 4° is expected for the combination of γ measurements based on the full Run1 and Run2 statistics.

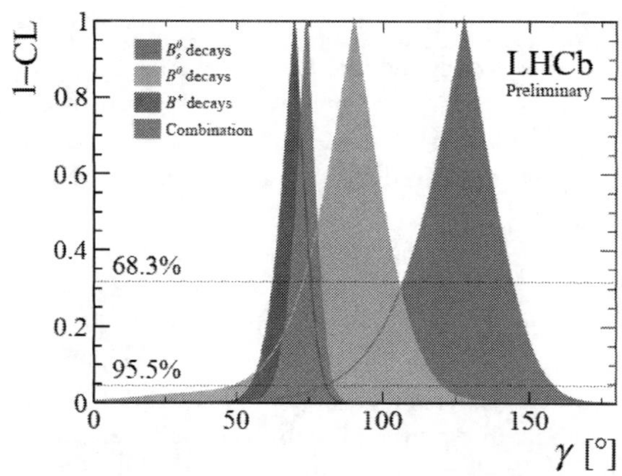

Figure 1. Confidence level plots for combinations of the CKM angle γ measurements split by the initial B meson flavour. Figure is taken from Ref. [6].

3 Measurement of the $D^0 - \overline{D^0}$ mixing parameters

The study of D mesons offers a unique opportunity to access *up*-type quarks in neutral current processes. It offers complementary constraints on possible NP contributions to those arising from the measurements of corresponding processes of *down*-type quarks (B or K mesons). In this context a precise measurement of $D^0 - \overline{D^0}$ mixing and searches for evidences of CP-violating processes play a key role. The $D^0 - \overline{D^0}$ mixing was established by previous measurements [7–10]. The mixing rate is consistent with, but at the upper end of the range of SM expectations [11] and constrains many NP models [12]. At present, no evidence of CP violation, either direct, indirect or in the mixing, has been found.

The parameters describing the $D^0 - \overline{D^0}$ mixing, x' and y',[1] can be determined from a fit of the decay-time-dependent ratio of D^0 decay rates to a wrong sign (WS) sample of $K^+\pi^-$ and to a right sign (RS) sample of $K^-\pi^+$. In particular, the WS decays receive both the contributions of the double Cabibbo-suppressed decay, $D^0 \to K^+\pi^-$, and of the Cabibbo-favored one following the $D^0 - \overline{D^0}$ oscillation, $D^0 \to \overline{D^0} \to K^+\pi^-$. The ratio of yields of WS to RS signal events expressed as a function of the D^0 decay time depends on the parameters x' and y'. Large statistics samples of D^0 mesons from prompt $D^*(2010)^+ \to D^0\pi^+$ decays [2] collected by LHCb during Run1 and part of Run2, corresponding to an integrated luminosity of 5 fb^{-1}, were recently analysed [13]. Depending on the charge of the pion, separate ratios for the D^0 and $\overline{D^0}$ decay rates were determined, their difference being sensitive to possible CP violation effects in the mixing or in the decay. The resulting distributions are consistent with the hypothesis of CP conservation setting the most stringent bounds on the parameters that describe it. Under the assumption of CP conservation, the best determination of the mixing parameters from a single measurement $x'^2 = (3.9 \pm 2.7) \times 10^{-5}$, $y' = (5.28 \pm 0.52) \times 10^{-3}$, was obtained.

4 Tests of lepton flavour universality

In the SM the gauge boson couplings to the different charged lepton species are assumed to be the same. Differences among the semileptonic decays to electron, muon or tau leptons can only arise due to their different masses. Several tests on this fundamental property have been performed by different experiments using charged- and neutral-current processes and have revealed hints of deviations from the expectations of the SM that could potentially be due to NP. In the following, the LHCb results concerning tests of LFU in b-hadron decays are presented.

4.1 Charged current transitions

Semileptonic decays of b-hadrons are an optimal laboratory to test LFU given their relatively large branching fractions and due to the availability of precise SM predictions. Such tests are performed by comparing branching fractions of semileptonic b-hadron (H_b) decays to a common c-hadron (H_c) with different lepton species in the final state:

$$\mathcal{R}(H_c) \equiv \frac{\mathcal{B}(H_b \to H_c \tau \nu_\tau)}{\mathcal{B}(H_b \to H_c \ell \nu_\ell)}, \tag{1}$$

where ℓ represents a light lepton, which is a muon in the case of the LHCb measurements.

[1] $x \equiv \Delta m/\Gamma$, $y \equiv \Delta\Gamma/2\Gamma$, $x' = x\cos\delta + y\sin\delta$, $y' = y\cos\delta - x\sin\delta$, where δ is the strong phase difference between CS and CF amplitudes. For further details see Ref. [13].

[2] Charge conjugated states and decays are also implied.

The LHC*b* collaboration performed two independent measurements of $\mathcal{R}(D^{*+})$, using either the muonic or the three-prong τ decays, and a measurement of $\mathcal{R}(J/\psi)$, exploiting the non negligible production of B_c^+ mesons at the LHC.

In the first measurement of $\mathcal{R}(D^{*+})$ [14], the tau leptons are reconstructed in the muonic decay channel $\tau^- \to \mu^-\overline{\nu}_\mu\nu_\tau$, so that both the decays involved in the ratio are reconstructed in the same visible final state given by a muon and a charged $D^*(2010)^+$ meson decaying to $D^0(\to K^-\pi^+)\pi_s^+$. The yields of the signal and normalisation channels are extracted from a simultaneous maximum-likelihood fit to three kinematic observables that show a discrimination power among the different contributions. These are: the muon energy in the B rest frame, E_μ^*, the missing mass squared, $m_{miss}^2 = (p_B - p_{D^*} - p_\mu)^2$, and the momentum transferred to the lepton pairs, $q^2 = (p_B - p_{D^*})^2$, where p_i stands for the four-momenta of the i-th particle. Since the kinematics of the B decay is not well defined both for the signal and the normalisation channels, such quantities are approximated. In particular, the B momentum is computed using the B direction and the visible momentum $p_{\text{vis}} = p_{D^*} + p_\mu$. In the fit, the distributions of signal, normalisation and background channels are derived from simulation and control data samples. The resulting ratio is $\mathcal{R}(D^{*+}) = 0.336 \pm 0.027(\text{stat.}) \pm 0.030(\text{syst.})$, in agreement with previous results by BaBar [16] and Belle [17] collaborations but 2.1σ above the SM prediction, $\mathcal{R}^{\text{SM}}(D^*) = 0.252 \pm 0.003$ [18].

In the second measurement of $\mathcal{R}(D^{*+})$ [15], the tau lepton is reconstructed in the three-prong decay channel $\tau^- \to \pi^+\pi^-\pi^-\nu_\tau$. The $\overline{B^0} \to D^{*+}\pi^+\pi^-\pi^-$ decay channel is chosen as normalisation, being its topology similar to the one of the signal. As a consequence the determination of $\mathcal{R}(D^{*+})$ relies on external branching fraction measurements of the $\overline{B^0} \to D^{*+}\pi^+\pi^-\pi^-$ and $\overline{B^0} \to D^{*+}\mu^-\overline{\nu_\mu}$ decays. For the signal selection the three-pion vertex from the τ lepton candidate is required to be downstream with respect to the B-candidate decay vertex. In this way the large background from $B \to D^*3\pi$ is highly suppressed. Furthermore a Boosted Decision Tree (BDT) is optimised to discriminate between the signal and the $B \to D^*D_{(s)}$ decays which have similar topologies. Similarly to the previous measurement, the yield of the signal is determined from a multidimensional fit to three observables using templates to model the contributions of the signal and of different backgrounds. The templates are taken from simulations and validated with control samples whenever possible. In this case the observables are the τ lepton decay time, the four-momentum transferred to the lepton pairs, q^2, and the output of the BDT mentioned above. The normalisation channel is reconstructed exclusively and its yield extracted from a fit to the invariant mass distribution. The result is $\mathcal{R}(D^{*+}) = 0.291 \pm 0.019(\text{stat.}) \pm 0.026(\text{syst.}) \pm 0.013(\mathcal{B})$, which is in agreement with previous experimental results and compatible with the SM at 1.1σ level.

An overview of the current experimental situation for $\mathcal{R}(D)$ and $\mathcal{R}(D^*)$ ratios is represented in Fig. 2, where the measurements from different experiments are shown, combined and compared to the SM predictions. Both $\mathcal{R}(D)$ and $\mathcal{R}(D^*)$ combinations exceed the SM expectations. The global compatibility with the SM is of 3.8σ, of which 3.0σ is due to the $\mathcal{R}(D^*)$ measurement [19]. Though still not statistically significative to claim LFU violation, a common pattern of anomalies seems to be present in all these measurements. More precise and additional measurements are needed to clarify the picture.

For this purpose LHC*b* performed for the first time the measurement of the $\mathcal{R}(J/\psi)$ ratio [20], exploiting the non negligible production of B_c^+ mesons at the LHC and the high efficiency of LHC*b* in triggering events with a J/ψ meson decaying to $\mu^+\mu^-$ pairs. The analysis follows the same strategy of the first $\mathcal{R}(D^{*+})$ measurement described above. After the event selection, the yields of signal and normalisation are determined from the same final state $J/\psi(\to \mu^+\mu^-)\mu^+$ by means of a multidimensional fit to the distributions of the B_c^+ decay time, the missing mass and of a variable Z that defines different ranges

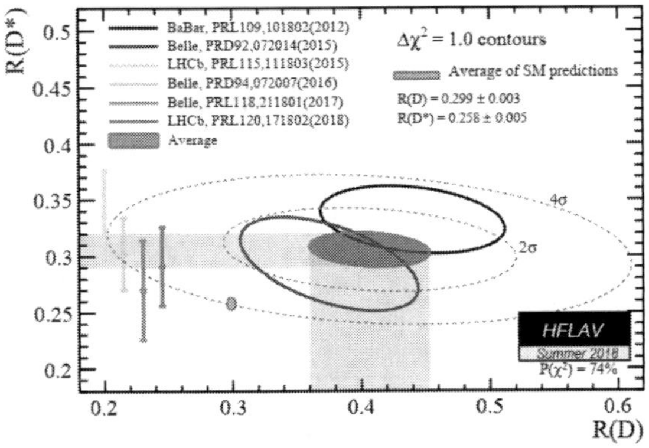

Figure 2. Measurements of $\mathcal{R}(D)$ and $\mathcal{R}(D^*)$ ratios from different experiments, average values and SM predictions. The plot corresponds to the summer 2018 update of Ref. [19] available on the HFLAV webpage.

of dilepton momentum transfer, q^2, and muon energy, E^*_μ. For these semileptonic decays the form factors are not well determined, so they are extracted from data and used to determine both the signal and the normalisation templates. The resulting value of the ratio is $\mathcal{R}(J/\psi) = 0.71 \pm 0.17(\text{stat.}) \pm 0.18(\text{syst.})$ which is above but compatible at 2σ level with the range of predicted SM values, $\mathcal{R}^{\text{SM}}(J/\psi) = 0.25 - 0.28$ [21].

More analyses are ongoing within the LHCb collaboration aiming to extend the tests of LFU to other b-hadron decays, such as Λ_b and B^0_s mesons.

4.2 Neutral current transitions

Tests of LFU can be performed by exploiting $b \to s\ell^+\ell^-$ neutral current transitions, which are mediated by loops in the SM and are highly suppressed. By measuring the ratio of branching fractions of $B \to H_s\mu^+\mu^-$ and $B \to H_s e^+e^-$ decays, where H_s represents a strange hadron,

$$\mathcal{R}(H_s) = \frac{\mathcal{B}(B \to H_s\mu^+\mu^-)}{\mathcal{B}(B \to H_s e^+e^-)}, \tag{2}$$

the universality of the couplings to the second and first generation of leptons can be tested. The ratio is predicted with a very high precision in the SM and expected to be close to unity.

The LHCb collaboration performed the measurement of the $\mathcal{R}(K^{*0})$ [22] and $\mathcal{R}(K^+)$ [23] ratios using exclusively reconstructed decays of B^0 and B^+ mesons. To minimise the experimental uncertainties, both the analyses measured the double ratio

$$\mathcal{R}(H_s) = \frac{\mathcal{B}(B \to H_s\mu^+\mu^-)}{\mathcal{B}(B \to H_s e^+e^-)} \times \frac{\mathcal{B}(B \to H_s J/\psi(\to e^+e^-))}{\mathcal{B}(B \to H_s J/\psi(\to \mu^+\mu^-))}, \tag{3}$$

where the additional normalisation term, corresponding to the two leptonic decays of the J/ψ meson, allows to cancel some systematic uncertainties and to perform stringent checks of the reconstruction efficiency of the signal channels.

The ratio is determined in different regions of dilepton momentum transferred, q^2, using the yields of signal events obtained from a fit to the invariant mass distributions of the selected candidates. The reconstruction of the dilepton final states exhibits different challenges for the

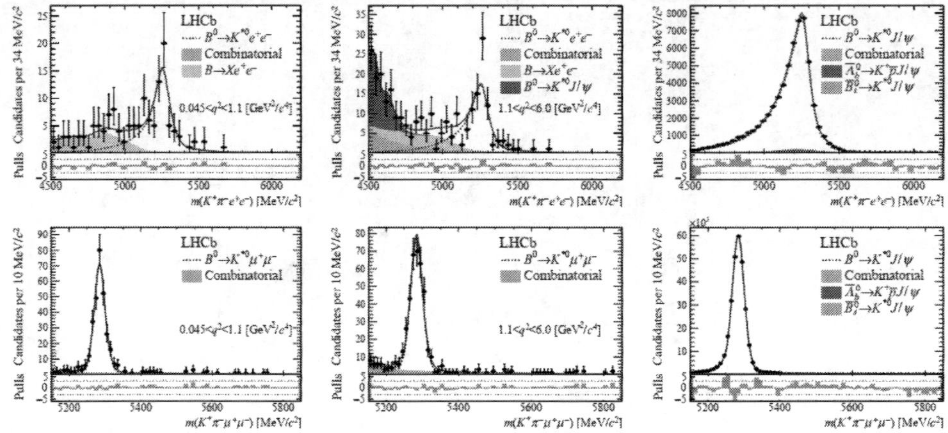

Figure 3. Fit to the invariant mass distributions of $B^0 \to K^{*0}\ell^+\ell^-$ ($\ell = e, \mu$ for top and bottom) candidate decays in different intervals of momentum transferred to the dilepton pairs. Figure is taken from Ref. [22].

electron and muon modes. In particular, for the final states containing e^+e^- pairs, it accounts for the energy loss due to Bremsstrahlung that affects the B mass and momentum resolutions as well as the reconstruction efficiency and the background contamination of these channels. Figure 3 shows the invariant mass distributions of the selected candidates in the measurement of $\mathcal{R}(K^{*0})$ as an example; similar plots are available for the measurement of $\mathcal{R}(K^+)$.

The measured ratios are

$$\mathcal{R}(K^{*0}) = 0.66\,^{+0.11}_{-0.07} \pm 0.03 \quad (0.045 < q^2 < 1.1\ \text{GeV}^2/c^4),$$
$$\mathcal{R}(K^{*0}) = 0.69\,^{+0.11}_{-0.07} \pm 0.05 \quad (1.1 < q^2 < 6\ \text{GeV}^2/c^4),$$
$$\mathcal{R}(K^+) = 0.745^{+0.090}_{-0.074} \pm 0.036 \quad (1.0 < q^2 < 6\ \text{GeV}^2/c^4),$$

which are compatible with the SM predictions [24] within 2.1-2.6σ, depending on the q^2 range and the model prediction, but consistently smaller. The LHC*b* results are considerably more precise than the previous measurements at the "B factories".

The experimental picture that emerges from these measurements become even more intriguing considering that additional anomalies arose in other observables concerning $b \to s\ell^+\ell^-$ transitions. Among them it is worth to recall the smaller differential branching fraction compared to SM predictions observed by LHC*b* in $B^+ \to K^+\mu^+\mu^-$ [25], $B^0 \to K^{*0}\mu^+\mu^-$ [26] and $B^0_s \to \phi(1020)\mu^+\mu^-$ [27] decays. Furthermore, the most relevant deviation was observed in two bins of q^2 for the coefficient P'_5 describing the angular distributions of $B^0 \to K^{*0}\mu^+\mu^-$ decays [28].

If such anomalies are confirmed with more precise measurements, it would be a clear sign of new physics contributions. Several global fits have been performed trying to obtain a consistent interpretation of the different anomalies observed introducing new physics contributions by means of the effective Hamiltonian expansion and modification of the Wilson coefficients. The outcome of these studies are promising, but further efforts are still required, both from the experimental and theoretical side, to clarify the present scenario.

5 Conclusions and outlook

The LHC*b* experiment is approaching the end of Run2 data taking, aiming at collecting a total integrated luminosity of 9 fb^{-1}. In 2019 the detector will be upgraded in many parts in order to improve the data taking capabilities, such as the instantaneous luminosity and the trigger efficiency. In the next running periods LHC*b* is expected to collect a total integrated luminosity of around 23 fb^{-1} (end of Run3, 2023) and 50 fb^{-1} (end of Run4, 2029) [29, 30].

The results presented at this workshop clearly indicate that LHC*b* is contributing to precision tests of the SM. The measurement of the CKM angle γ has now approached the precision of 5° using a multitude of decay channels and analysis methods and represents the best determination from a single experiment. The current precision is expected to improve to 4° once the full Run2 data sample will have been analysed and to 1° after the upgrade.

The mixing of neutral D mesons is established and the result presented at the workshop constitutes the best single measurement. Currently the results are compatible with CP conservation. The improvements foreseen with the detector upgrade are expected to have a strong impact on the precision of the CP measurements in the charm sector.

Intriguing discrepancies from the SM are observed in measurements of semileptonic B decays to leptons in the context of tests of the lepton flavour universality. Several analyses by the LHC*b* collaboration are ongoing exploiting larger data sets currently available and different decay channels. If the experimental picture will be confirmed with more data, there are good prospects to test the consistency of the anomalies and to perform test of NP models.

References

[1] A. Augusto Alves Jr., LHC*b* Collaboration, JINST **3** S08005 (2008).

[2] CKM fitter Group, http://ckmfitter.in2p3.fr/, results shown at ICHEP2016 conference.

[3] Gronau, London and Wyler, Phys. Lett. **B253** 483, (1991), Gronau, London and Wyler, Phys. Lett. **B265** 172, (1991).

[4] Atwood, Dunietz and Soni, Phys. Rev. Lett. **78** 3257, (1997), Atwood, Dunietz and Soni, Phys. Rev. **D63** 036005, (2001).

[5] A. Giri, Y. Grossman, A. Soffer and J. Zupan, Phys. Rev. **D68** 054018, (2003)

[6] R. Aaij et al, LHC*b* Collaboration, LHCb-CONF-2018-002.

[7] B. Aubert et al, BaBar Collaboration, Phys. Rev. Lett. **98**, 211802 (2007).

[8] M. Staric et al, Belle Collaboration, Phys. Rev. Lett. **98**, 211803 (2007).

[9] T. Aaltonen et al., CDF Collaboration, Phys. Rev. Lett. **100**, 121802 (2008).

[10] R. Aaij et al, LHC*b* Collaboration, Phys. Rev. Lett. **111**, 251801 (2013).

[11] A.F. Falk et al., Phys. Rev. **D 69**, 114021 (2004).

[12] E. Golowich, J. Hewett, S. Pakvasa, A.A. Petrov, Phys. Rev. **D 76**, 095009 (2007).

[13] R. Aaij et al, LHC*b* Collaboration, Phys. Rev. **D97** 031101 (2018).

[14] R. Aaij et al., LHC*b* Collaboration, Phys. Rev. Lett. **115** 111803 (2015).

[15] R. Aaij et al., LHC*b* Collaboration, Phys. Rev. **D97** 072013 (2018), R. Aaij et al., LHC*b* Collaboration, Phys. Rev. Lett. **120** 171802 (2018).

[16] J. P. Lees et al., BaBar Collaboration, Phys. Rev. Lett. **109** 101802 (2012).

[17] M. Huschle et al., Belle Collaboration, Phys. Rev. **D92** 072014 (2015), Y. Sato et al., Belle Collaboration, Phys. Rev. **D94** 072007 (2016), S. Hirose et al., Belle Collaboration, Phys. Rev. Lett. **118** 211801 (2017).

[18] J. F. Kamenik and F. Mescia, Phys. Rev. **D78**, 014003 (2008), S. Fajfer, J. F. Kamenik and I. Nisandzic, Phys. Rev. **D85** 094025, (2012).

[19] Y. Amhis et al., Heavy Flavour Averaging Group, Eur. Phys. J. **C77** (2017) 895 and summer 2018 updates available online: https://hflav-eos.web.cern.ch/hflav-eos/semi/summer18/RDRDs.html

[20] R. Aaij et al., LHC*b* Collaboration, Phys. Rev. Lett. **120** 121801 (2018).

[21] A. Yu. Anisimov, I. M. Narodetskii, C. Semay, and B. Silvestre-Brac, Phys. Lett. **B 452**, 129 (1999), V. Kiselev, arXiv:hep-ph/0211021, M. A. Ivanov, J. G. Korner, and P. Santorelli, Phys. Rev. **D73**, 054024 (2006), E. Hernández, J. Nieves, and J. M. Verde-Velasco, Phys. Rev. **D 74**, 074008 (2006).

[22] R. Aaij et al., LHC*b* Collaboration, Journal of High Energy Physics **08** 055 (2017).

[23] R. Aaij et al., LHC*b* Collaboration, Phys. Rev. Lett. **113** 151601 (2014).

[24] G. Hiller, F. Krüger, Phys. Rev. **D69**, 074020 (2004), C. Bouchard et al., HPQCD Collaboration, Phys. Rev. Lett. **111** (2013), and References 26–36 of Ref. [22].

[25] R. Aaij et al., LHC*b* Collaboration, Journal of High Energy Physics **06** 133 (2014).

[26] R. Aaij et al., LHC*b* Collaboration, Journal of High Energy Physics **11** 047 (2016).

[27] R. Aaij et al., LHC*b* Collaboration, Journal of High Energy Physics **09** 179 (2015).

[28] R. Aaij et al., LHC*b* Collaboration, Journal of High Energy Physics **02** 104 (2016).

[29] LHC*b* Collaboration et al., Eur. Phys. J. **C 73** (2013) 2373

[30] I. Bediaga, LHC*b* Collaboration, arXiv:1808.08865, LHCB-PUB-2018-009.

 https://doi.org/10.1051/epjconf/201819200025

Scalar leptoquarks: From GUT to B anomalies

Svjetlana Fajfer[1,*]

[1]Department of Physics, University of Ljubljana, Jadranska 19, 1000 Ljubljana, Slovenia and Jožef Stefan Institute, Jamova 39, P. O. Box 3000, 1001 Ljubljana, Slovenia

Abstract. Leptoquarks are often used to explain B meson anomalies. The minimalistic approach assumes only one light leptoquark explaining both puzzles. However, one scalar leptoquark cannot explain both anomalies. Two relatively light leptoquarks can be accommodated within $SU(5)$ GUT and might nicely explain both anomalies not being in a conflict with any low-energy flavour constraints or LHC results.

1 Introduction

Several experimental results points towards the possible violation of the lepton number universality in the B meson semileptonic decays. The charge current anomaly indicates the deviation from the measured and theoretically predicted ratios $R_{D^{(*)}} = \mathcal{B}(B \to D^{(*)}\tau\bar{\nu}_\tau)/\mathcal{B}(B \to D^{(*)}l\bar{\nu}_l)$, $l = e, \mu$. It was found that the measured ratios are larger than the Standard Model predictions $R^{\rm SM}_{D^{(*)}}$. The experimental results on R_D [1–3] differ by $\sim 2\,\sigma$ with respect to the SM prediction [4] and by $\sim 3\,\sigma$ in the case of R_{D^*} [5–7]. The similar deviations were found in transitions $b \to c\ell\nu$ ($\ell = \mu, \tau$), observed in $R_{J/\psi}$ ratio between $B_c \to J/\psi\ell\nu$ decay widths [8].

In the case of flavour changing neutral current transition (FCNC) $b \to s\mu^+\mu^-$ the LHCb experiment has measured ratios $R_{K^{(*)}} = \mathcal{B}(B \to K^{(*)}\mu\mu)/\mathcal{B}(B \to K^{(*)}ee)$ at low di-lepton invariant mass. Interestingly, these ratios were found to be systematically lower than expected in the SM. In the case of the K meson in the final state, the ratio R_K was measured in the kinematical region $q^2 \in [1.1, 6]$ GeV^2 [9], while R_{K^*} was measured in the region $q^2 \in [0.045, 1.1]$ GeV^2 [10]. The three measured $R_{K^{(*)}}$ deviate from the SM predictions at $\sim 2.5\,\sigma$ level [11, 12].

2 B meson anomalies and leptoquarks

Both B anomalies are usually explained by contributions of New Physics (NP). The effective Lagrangian studies suggest that NP appears at TeV scale [12–15]. Most of these approaches favour the left-handed fermionic interactions. In addition, this new interaction prefers interactions of third fermion generation. As suggested in Ref. [16] the scale of such NP, accommodating the $R_{D^{(*)}}$ anomaly, can be introduced as

$$\mathcal{L}^{D^{(*)}}_{NP} = \frac{2}{\Lambda_{D^{(*)}}}\,\bar{c}_L\gamma_\mu b_L\,\tau_L\gamma^\mu\nu_L\,, \tag{1}$$

*e-mail: svjetlana.fajfer@ijs.si

© The Authors, published by EDP Sciences. This is an open access article distributed under the terms of the Creative Commons Attribution License 4.0 (http://creativecommons.org/licenses/by/4.0/).

The experimental data can be explained for the scale $\Lambda_{D^{(*)}} \sim 3$ TeV. The $R_{K^{(*)}}$ anomaly can be explained by the Lagrangian

$$\mathcal{L}_{NP}^{K^{(*)}} = \frac{1}{\Lambda_{K^{(*)}}} \bar{s}_L \gamma_\alpha b_L \bar{\mu}_L \gamma^\alpha \mu_L \,, \tag{2}$$

from which one finds that the scale of NP is $\Lambda_{K^{(*)}} \sim 30$ TeV. Obviously, by requiring that the same model of NP explains both anomalies, that the NP in $R_{K^{(*)}}$ should be suppressed in comparison with $R_{D^{(*)}}$. The suppression can be realised in a number of different ways. For example $R_{D^{(*)}}$ receives contribution of NP at tree level, while in $R_{K^{(*)}}$ NP enters at loop level in particular model of leptoquarks [17, 18]. The approach of [12, 13] assumes that the second generation of quarks and leptons couples to a NP mediator weakly in comparison with the coupling to third generation. The third possibility is a "Cabibbo mechanism" [19, 20] in which $\cos\theta$ ($\sin\theta$) can determine size of NP contribution. Among many NP proposals, leptoquarks (LQs) explanations are very appealing due to their interactions with quarks and leptons they are natural candidates. Leptoquarks can have spin-0 or spin-1.

Leptoquarks are listed by quantum numbers of $SU(3)_c$ colour, electroweak quantum numbers of $SU(2)_L$ and hypercharge $U(1)_Y$ [21]. The hypercharge is defined as $Y = Q - I_3$. According to their SM quantum numbers, scalar leptoquarks are: one weak triplet $S_3 = (\bar{3}, 3, 1/3)$, two weak doublets $R_2 = (3, 2, 7/6)$, $\tilde{R}_2 = (3, 2, 1/6)$ and three singlets $\tilde{S}_1 = (\bar{3}, 1, 4/3)$, $S_1 = (\bar{3}, 1, 1/3)$ and $\bar{S}_1 = (\bar{3}, 1, -2/3)$. Among all these states only weak doublet R_2 and $\tilde{R}_2$ do not cause proton decay at tree level [21, 22]. The rest of leptoquarks, might mediate proton decay at tree level if there is no additional symmetry forbidding it. For example it was noticed in [22, 23] that within particular $SU(5)$ GUT model S_3 does not cause proton decay. Spin-1 leptoquarks are following: $U_3 = (3, 3, 2/3)$, $V_2 = (\bar{3}, 2, 5/6)$, $\tilde{V}_2 = (\bar{3}, 2, -1/6)$, $\tilde{U}_1 = (3, 1, 5/3)$, $U_1 = (3, 1, 2/3)$ and $\bar{U}_1 = (3, 1, -1/3)$. Among all these states it was found that U_1 leptoquarks coupling to the left handed quarks and leptons [12, 13] can relatively good explain both anomalies. This state gives only V-A form of the interaction with quarks and leptons.

The scalar leptoquark S_3, being a weak triplet, can couple only to the left-handed quarks and leptons. It can alone well explain $R_{K^{(*)}}$ puzzle. In the model presented in [24] we succeeded to accommodate $b \to s\ell^+\ell^-$ sector. Because of its weak triplet nature it couples also to up-type quarks and neutrinos which is a couplings needed to address $R_{D^{(*)}}$. If one explains $R_{D^{(*)}}$ puzzle, S_3 then inevitably contributes to other well constrained flavour observables. For example, after detailed study we have demonstrated [24] that the most constraining are bounds from $R_{\nu\nu}^{(*)} = \mathcal{B}(B \to K^{(*)}\bar{\nu}\nu)/\mathcal{B}(B \to K^{(*)}\bar{\nu}\nu)_{SM}$ and Δm_s, allow only for minor improvement of $R_{D^{(*)}}$ puzzle.

The $\tilde{R}_2 = (3, 2, 1/6)$ leptoquark can accommodate $R_{D^{(*)}}$ by the coupling to a right-handed neutrino [26]. Such mechanism induces contributions which is not interfering with SM. In this year the idea of right-handed neutrino contributing to $R_{D^{(*)}}$ was further explored by [27–29]. The $\tilde{R}_2 = (3, 2, 1/6)$ leptoquark can explain R_K while it cannot accommodate R_{K^*} [26]. Namely R_K, due to the matrix element of two pseudoscalar is blind on the axial part of the current and therefore the measurement of R_{K^*} helps to distinguish between the chiralities of NP operators [30].

The $S_1 = (\bar{3}, 1, 1/3)$ letoquark was extensively explored by [17]. Being a weak singlet this leptoquark has left-handed and right-handed couplings. Unfortunately, this state is severely constrained by the low-energy flavour dynamics [18]. Among scalar letoquarks remaining states are $\tilde{S}_1 = (\bar{3}, 1, 4/3)$ and $\bar{S}_1 = (\bar{3}, 1, -2/3)$. The chirality of the charged leptons of the $\tilde{S}_1$ is not adequate for the $R_{K^{(*)}}$explanation, while $\bar{S}_1$ cannot couple to charge leptons (see e.g. [21]).

As noticed in the studies of [13] and [25] one scalar leptoquark cannot satisfactory described both anomalies without being in a conflict with all low-energy constraints. Two light leptoquark can also generate neutrino masses [23]. Even more, two relatively light leptoquarks can be accommodated within GUT theories [21]. These are additional arguments why scenarios with two leptoquarks are appealing.

3 $SU(5)$ GUT scenario with R_2 and S_3

In order to build an UV complete model the simplest proposal is to construct $SU(5)$ GUT models [19]. In the simplest $SU(5)$ scenario one can accommodate both R_2 and S_3 to be relativrey light, masses being in the TeV regime. In this scenario scalar sector needs to contain 45 and 50 representations, whereas the SM fermions are set in $\bar{5}$ and 10 (for details see [19]).The interactions of R_2 and S_3 with the SM fermions are

$$\mathcal{L} \supset Y_R^{ij} \bar{Q}'_i \ell'_{Rj} R_2 + Y_L^{ij} \bar{u}'_{Ri} \widetilde{R}_2^\dagger L'_j + Y^{ij} \bar{Q}_i'^C i\tau_2 (\tau_k S_3^k) L'_j, \tag{3}$$

where Y_L, Y_R, and Y are Yukawa matrices, τ_k denote the Pauli matrices.For the Yukawa matrices we use $Y_R = Y_R^T$, $Y = -Y_L$ and assume that the matrix $(Y_R E_R^\dagger)_{33} = y_R^{b\tau}$, and $(U_R Y_L)_{22} = y_L^{c\mu}$ and $(U_R Y_L)_{23} = y_L^{c\tau}$, with the rest being negligible small.

Following [19], we allow the two R_2 scalars be members of representations 45 in 50. They can mix (for details see [19]) allowing us to end up with the two light scalars, i.e., R_2 and S_3, and one heavy R_2. The heavy R_2 state is completely decoupled from the low-energy spectrum. The relevant lagrangian, in the mass eigenstate basis of the two light LQs, is we write Lagrangian

$$\begin{aligned}
\mathcal{L} \supset \; & + \; s_\phi (V b E_R^\dagger)^{ij} \bar{u}_{Li} \ell_{Rj} R_2^{\frac{5}{3}} + s_\phi (b E_R^\dagger)^{ij} \bar{d}_{Li} \ell_{Rj} R_2^{\frac{2}{3}} \\
& + \; c_\phi (U_R a U)^{ij} \bar{u}_{Ri} \nu_{Lj} R_2^{\frac{2}{3}} - c_\phi (U_R a)^{ij} \bar{u}_{Ri} \ell_{Lj} R_2^{\frac{5}{3}} \\
& + \; 2^{-\frac{1}{2}} (aU)^{ij} \bar{d}_{Li}^C \nu_{Lj} S_3^{\frac{1}{3}} - (V^* a U)^{ij} \bar{u}_{Li}^C \nu_{Lj} S_3^{-\frac{2}{3}} \\
& + \; a^{ij} \bar{d}_{Li}^C \ell_{Lj} S_3^{\frac{4}{3}} + 2^{-\frac{1}{2}} (V^* a)^{ij} \bar{u}_{Li}^C \ell_{Lj} S_3^{\frac{1}{3}},
\end{aligned} \tag{4}$$

where we define the mixing angle between the two R_2's to be ϕ. It is now easy to derive $Y = -a/\sqrt{2}$, $Y_L = c_\phi a$, and $Y_R = Y_R^T = s_\phi b$. For the maximal mixing $s_\phi = c_\phi = 1/\sqrt{2}$ for both R_2 states we can recover relation $Y_R = Y_R^T$, $Y = -Y_L$. Obviously, the two $SU(5)$ operators proportional to a and b might to generate three Yukawa matrices Y, Y_L, and Y_R. After such set-up, we consider the NP parameters m_{R_2}, m_{S_3}, $y_R^{b\tau}$, $y_L^{c\mu}$, $y_L^{c\tau}$, and θ. With this set-up we find out that the low-energy Yukawa couplings can originate from $SU(5)$. Even more, we checked weather the Yukawa couplings have perturbative nature, meaning that $y_R^{b\tau}$, $y_L^{c\mu}$, and $y_L^{c\tau}$ remain below $\sqrt{4\pi}$. Therefore, we have checked the renormalization group running from the electroweak to the GUT scale 5×10^{15} GeV. For example, as we mentioned in [19] the running of the Yukawa coupling $y_R^{b\tau}$, crucial in explaining the $R_{D^{(*)}}$ puzzle is given by

$$16\pi^2 \frac{d \ln y_R^{b\tau}}{d \ln \mu} = |y_L^{c\mu}|^2 + |y_L^{c\tau}|^2 + \frac{9}{2}|y_R^{b\tau}|^2 + \frac{1}{2} y_t^2 + \ldots, \tag{5}$$

where y_t is the top Yukawa coupling.

4 B anomalies in GUT scenario

4.1 Charged-current decays

The relevant effective Lagrangian for (semi-)leptonic decays is

$$\mathcal{L}_{eff}^{d\to u\ell\bar{\nu}} = -\frac{4G_F}{\sqrt{2}} V_{ud}[(1+g_{V_L})(\bar{u}_L\gamma_\mu d_L)(\bar{\ell}_L\gamma^\mu \nu_L)$$
$$+g_{S_L}(\mu)\,(\bar{u}_R d_L)(\bar{\ell}_R\nu_L) + g_T(\mu)\,(\bar{u}_R\sigma_{\mu\nu}d_L)(\bar{\ell}_R\sigma^{\mu\nu}\nu_L)]\,, \tag{6}$$

where neutrinos are in the flavor basis. The effective Wilson coefficients of $d \to u\ell\bar{\nu}_{\ell'}$ are related to the LQ couplings at the matching scale $\Lambda(\approx 1\,\text{TeV})$ via the expression

$$g_{S_L}(\Lambda) = 4\,g_T(\Lambda) = \frac{y_L^{u\ell'}\,y_R^{d\ell*}}{4\sqrt{2}\,m_{R_2}^2\,G_F V_{ud}}\,, \tag{7}$$

$$g_{V_L} = -\frac{y^{d\ell'}\,(Vy^*)^{u\ell}}{4\sqrt{2}\,m_{S_3}^2 G_F V_{ud}}\,. \tag{8}$$

From the above equations we learn that the only transitions affected by R_2 in our scenario are $b \to c\tau\bar{\nu}_\ell$. S_3, on the other hand, contributes to processes involving u, c, s, b, and $\ell, \ell' = \mu, \tau$, but gives a negligibly small contribution to $R_{D^{(*)}}$.

To compute R_D we employ the $B \to D$ form factors calculated using the lattice QCD [4, 31], resulting in prediction $R_D^{\text{SM}} = 0.29(1)$ which is $\approx 2\,\sigma$ below the experimental average $R_D^{\text{exp}} = 0.41(5)$ [32–34]. On the other hand, the $B \to D^*$ form factors have never been computed on the lattice at nonzero recoil. Thus, for R_{D^*} we consider the leading form factors extracted from the $B \to D^* l\bar{\nu}$ $(l = e, \mu)$ spectra [35], which are combined with the ratios $A_0(q^2)/A_1(q^2)$ and $T_{1-3}(q^2)/A_1(q^2)$ computed in Ref. [6]. We obtain the value $R_{D^*}^{\text{SM}} = 0.26(1)$ which is $\approx 3\,\sigma$ below the experimental average $R_{D^*}^{\text{exp}} = 0.30(2)$ [35]. Moreover, the scalar (tensor) effective coefficients in Eq. 7 have to account for the SM running from the matching scale $\mu = \Lambda$ down to $\mu = m_b$. In the case of the vector coefficient there is no QCD running effects.

Following analysis of Ref. [24] our fit includes constraints from several (semi-)leptonic decays. Interestingly, the ratios for the light leptons, the analog of the one with τ and μ, $R_{D^{(*)}}^{\mu/e} = \mathcal{B}(B \to D^{(*)}\mu\bar{\nu})/\mathcal{B}(B \to D^{(*)}e\bar{\nu})$ gives very important constraint. [18]. We also consider $\mathcal{B}(B \to \tau\bar{\nu})$ as well as the kaon ratio $R_{e/\mu}^K = \Gamma(K^- \to e^-\bar{\nu})/\Gamma(K^- \to \mu^-\bar{\nu})$, which both agrees with their SM predictions (see for details [24]).

4.2 Neutral-current decays

The standard effective Hamiltonian for the $b \to s$ (semi-)leptonic transition can be written as

$$\mathcal{H}_{eff}^{b\to sll} = -\frac{4G_F\lambda_t}{\sqrt{2}} \sum_{i=7,9,10} C_i(\mu)O_i(\mu)\,, \tag{9}$$

where $\lambda_t = V_{tb}V_{ts}^*$. The relevant operators for our discussion are

$$O_{9(10)} = \frac{e^2}{(4\pi)^2}\,(\bar{s}_L\gamma_\mu b_L)(\bar{l}\gamma^\mu(\gamma^5)l)\,. \tag{10}$$

In our scenario, only S_3 contributes at tree-level [24]

$$\delta C_9^{\mu\mu} = -\delta C_{10}^{\mu\mu} = \frac{\pi}{\lambda_t\alpha_{\text{em}}}\frac{v^2}{m_{S_3}^2}\,y^{b\mu}(y^{s\mu})^*\,. \tag{11}$$

Relying on the results of [36, 37] the $1\,\sigma$ interval $C_9^{\mu\mu} = -C_{10}^{\mu\mu} \in (-0.85, -0.50)$ is obtained by performing a fit to the clean $b \to sll$ observables, namely, R_K, R_{K^*}, and $\mathcal{B}(B_s \to \mu\mu)$ [36, 37].

Contributions to the left-handed current operators in $b \to sll$ transition immediately contribute to $B \to K^{(*)}\nu\nu$ decay amplitudes which are well constrained by experiments. These decays are governed by

$$\mathcal{L}_{\text{eff}}^{b\to s\nu\nu} = \frac{\sqrt{2}G_F\alpha_{\text{em}}\lambda_t}{\pi}\, C_L^{ij}\,(\bar{s}_L\gamma_\mu b_L)(\bar{\nu}_{L\,i}\gamma^\mu\nu_{L\,j}), \tag{12}$$

with $C_L^{ij} = \delta_{ij}C_L^{\text{SM}} + \delta C_L^{ij}$ being the Wilson coefficient with the SM contributions $C_L^{\text{SM}} = -6.38(6)$ [38] and the contribution δC_L^{ij} from NP. Similarly as in the $b \to sll$ transition, the only tree-level contribution to $b \to s\nu\bar{\nu}$ comes from the S_3 state and reads [24]

$$\delta C_L^{ij} = \frac{\pi}{2\alpha_{\text{em}}\lambda_t}\frac{v^2}{m_{S_3}^2}\, y^{bj}(y^{si})^*, \quad i,j = \mu,\tau\,. \tag{13}$$

These effective coefficients modify the ratios $R_{\nu\nu}^{(*)} = \mathcal{B}(B \to K^{(*)}\nu\nu)/\mathcal{B}(\mathcal{B} \to \mathcal{K}^{(*)}\nu\nu)^{SM}$ in the following way

$$R_{\nu\nu}^{(*)} = \frac{\sum_{ij}|\delta_{ij}C_L^{SM} + \delta C_L^{ij}|^2}{3|C_L^{\text{SM}}|^2}. \tag{14}$$

In our fit we use experimental bounds $R_{\nu\nu} < 3.9$ and $R_{\nu\nu}^* < 2.7$ [39] and data on the $B_s - \bar{B}_s$ mixing amplitude as in [19], modified by the box-diagrams containing two S_3, leading to the proportionality to $[(yy^\dagger)^{bs}]^2/m_{S_3}^2$. Then we explore the experimental limits on LFV τ decays, $\mathcal{B}(\tau \to \mu\phi)^{\text{exp}} < 8.4 \times 10^{-8}$ and $\mathcal{B}(\tau \to \mu\gamma)^{\text{exp}} < 4.4 \times 10^{-8}$ [40], muon anomalous magnetic moment with the $\approx 3.6\,\sigma$ discrepancy with respect to the SM [41]. In out approach, however, $(g-2)_\mu$ receives very small corrections. Then we explored bounds on the $Z \to l^+l^-$ couplings measured at LEP [42]. Namely, both R_2 and S_3 contribute at on loop level to these decays. The strong constraints come from the measured $D - \bar{D}$ mixing parameters and we have checked that our model is not in a conflict with experimental results obtained for the charm meson mixing.

5 Constraints and predictions

The observables we described above were used in the fit described in [19] by varying the parameters $y_R^{b\tau}$, $y_L^{c\mu}$, $y_L^{c\tau}$ and θ. First we learn that our fit requires S_3 to have larger mass than than R_2. The masses m_{R_2} and m_{S_3} we consider to be the lowest values allowed by projected LHC constraints with $m_{R_2} = 800\,\text{GeV}$ and $m_{S_3} = 2\,\text{TeV}$. Very interesting feature of our model is that $y_R^{b\tau}$ should be a complex number [19]. The complex phase does not induce any observable CP violation. In our study the LHC constraints are taken into account [19]. Our best fit points are

$$\text{Re}[g_{S_L}] = 0, \ |\text{Im}[g_{S_L}]| = 0.59\left(^{+0.13}_{-0.14}\right)_{1\,\sigma}\left(^{+0.20}_{-0.29}\right)_{2\,\sigma}. \tag{15}$$

We find very interesting prediction for the lepton flavour violating process $\mathcal{B}(B \to K\mu\tau)$ which is found to be bounded, as presented in Fig. 2. At $1\,\sigma$ level we find

$$1.1 \times 10^{-7} \le \mathcal{B}(B \to K\mu^\pm\tau^\mp) \le 6.5 \times 10^{-7}\,. \tag{16}$$

This value is bellow the current bound $\mathcal{B}(B \to K\mu\tau)^{\text{exp}} < 4.8 \times 10^{-5}$ [32].

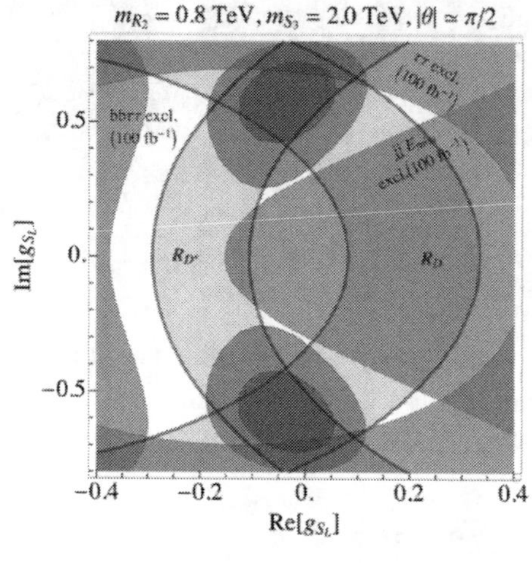

Figure 1. The flavor fit constraints on the g_{S_L} plane for the transition $b \to c\tau\bar{\nu}_\tau$. The allowed $1\,\sigma(2\,\sigma)$ regions are rendered in red (orange). Separate constraints from R_D and R_{D^*} to $2\,\sigma$ accuracy are shown by the blue and purple regions, respectively. The LHC exclusion area are presented by the grey colour.

It is easy to obtain the bound $\mathcal{B}(B \to K^*\mu\tau) \approx 1.9 \times \mathcal{B}(B \to K\mu\tau)$ and $\mathcal{B}(B_s \to \mu\tau) \approx 0.9 \times \mathcal{B}(B \to K\mu\tau)$ [43–45]. We also predict about $\sim 50\%$ enhancement of the ratio $\mathcal{B}(B \to K^{(*)}\nu\nu)$, which can be tested in the near future at Belle-II. Remarkably, these two observables are highly correlated as depicted in Fig. 2. Within our scenario we make prediction for $\mathcal{B}(\tau \to \mu\gamma)$, which lies just below the current experimental limit

$$\mathcal{B}(\tau \to \mu\gamma) \sim 1.5 \times 10^{-8}\,. \tag{17}$$

Finally, our description of the B-physics anomalies, and most particularly $R_{D^{(*)}}$, strongly depends on the assumption that the LQ states are not too far from the TeV scale. Thus, these particles are necessarily accessible at the LHC, yielding also predictions for the direct searches [19].

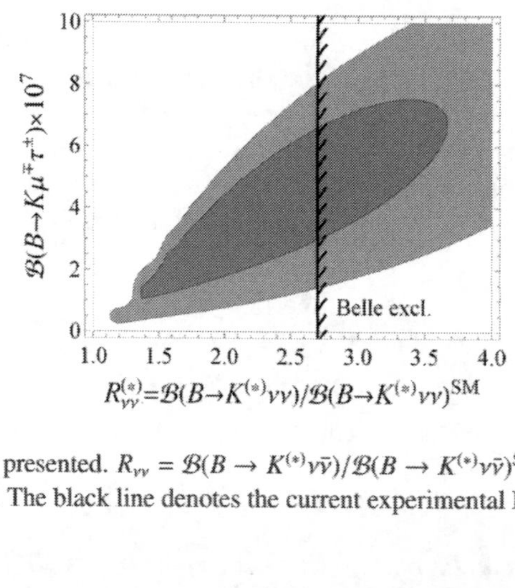

Figure 2. $\mathcal{B}(B \to K\mu\tau)$ is presented. $R_{\nu\nu} = \mathcal{B}(B \to K^{(*)}\nu\bar{\nu})/\mathcal{B}(B \to K^{(*)}\nu\bar{\nu})^{\rm SM}$ for the $1\,\sigma$ (red) and $2\,\sigma$ (orange) regions of Fig. 1. The black line denotes the current experimental limit, $R^*_{\nu\nu} < 2.7$ [39].

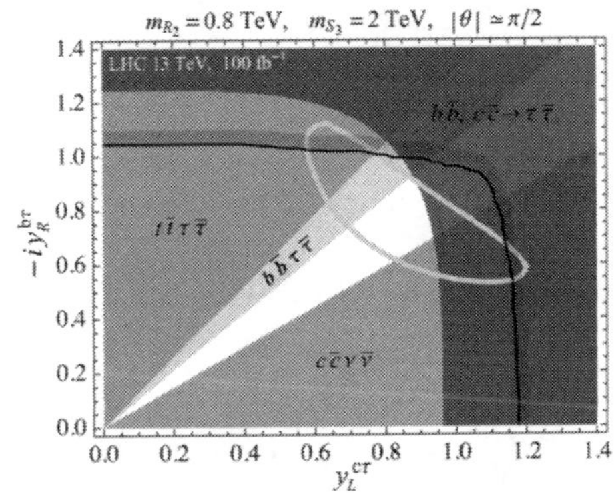

Figure 3. The flavor fit constraints on the g_{S_L} plane for the transition $b \to c\tau\bar{\nu}_\tau$. The allowed $1\,\sigma(2\,\sigma)$ regions are rendered in red (orange). Separate constraints from R_D and R_{D^*} to $2\,\sigma$ accuracy are shown by the blue and purple regions, respectively. The LHC exclusion area are presented by the gray colour.

6 Conclusions

We find out that the simple $SU(5)$ GUT model might accommodate two light scalar leptoquarks that can explain all measured B meson anomalies. Therefore, our scenario gives the explanation of the possible lepton flavour universality violation. The model is not in a conflict with low-energy phenomenology or with direct search constraints at the LHC. It is significant that all Yukawa couplings remain perturbative up to the unification scale. The model predicts the branching ratio for the lepton flavour violating decays $B \to K\mu\tau$, $B \to K^{(*)}\nu\nu$ and $\tau \to \mu\gamma$ as well as indicates their correlation.

Acknowledgements

My thanks goes to Damir Bečirević, Ilja Doršner, Darius Faroughy, Nejc Košnik and Olcyr Sumensari from whom I benefit so much about the topic considered in this contribution. I also acknowledge support of the Slovenian Research Agency under the core funding grant P1-0035.

References

[1] J. P. Lees *et al.* [BaBar Collaboration], Phys. Rev. D **88**, no. 7, 072012 (2013) [arXiv:1303.0571 [hep-ex]].

[2] S. Hirose *et al.* [Belle Collaboration], Phys. Rev. Lett. **118**, no. 21, 211801 (2017) [arXiv:1612.00529 [hep-ex]].

[3] R. Aaij *et al.* [LHCb Collaboration], Phys. Rev. Lett. **115**, no. 11, 111803 (2015) Erratum: [Phys. Rev. Lett. **115**, no. 15, 159901 (2015)] [arXiv:1506.08614 [hep-ex]].

[4] J. A. Bailey *et al.* [MILC Collaboration], Phys. Rev. D **92**, no. 3, 034506 (2015) [arXiv:1503.07237 [hep-lat]].

[5] S. Fajfer, J. F. Kamenik and I. Nisandzic, Phys. Rev. D **85**, 094025 (2012) [arXiv:1203.2654 [hep-ph]].

[6] F. U. Bernlochner, Z. Ligeti, M. Papucci and D. J. Robinson, Phys. Rev. D **95**, no. 11, 115008 (2017) Erratum: [Phys. Rev. D **97**, no. 5, 059902 (2018)] [arXiv:1703.05330 [hep-ph]].

[7] D. Bigi, P. Gambino and S. Schacht, Phys. Lett. B **769**, 441 (2017) [arXiv:1703.06124 [hep-ph]].

[8] R. Aaij *et al.* [LHCb Collaboration], Phys. Rev. Lett. **120**, no. 12, 121801 (2018) [arXiv:1711.05623 [hep-ex]].

[9] R. Aaij *et al.* [LHCb Collaboration], Phys. Rev. Lett. **113**, 151601 (2014) [arXiv:1406.6482 [hep-ex]].

[10] R. Aaij *et al.* [LHCb Collaboration], JHEP **1708**, 055 (2017) [arXiv:1705.05802 [hep-ex]].

[11] G. Hiller and F. Kruger, Phys. Rev. D **69**, 074020 (2004) [hep-ph/0310219].

[12] M. Bordone, G. Isidori and A. Pattori, Eur. Phys. J. C **76**, no. 8, 440 (2016) [arXiv:1605.07633 [hep-ph]].

[13] D. Buttazzo, A. Greljo, G. Isidori and D. Marzocca, JHEP **1711**, 044 (2017) [arXiv:1706.07808 [hep-ph]].

[14] B. Bhattacharya, A. Datta, D. London and S. Shivashankara, Phys. Lett. B **742**, 370 (2015) [arXiv:1412.7164 [hep-ph]].

[15] F. Feruglio, P. Paradisi and A. Pattori, Phys. Rev. Lett. **118**, no. 1, 011801 (2017) [arXiv:1606.00524 [hep-ph]].

[16] L. Di Luzio and M. Nardecchia, Eur. Phys. J. C **77**, no. 8, 536 (2017) doi:10.1140/epjc/s10052-017-5118-9 [arXiv:1706.01868 [hep-ph]].

[17] M. Bauer and M. Neubert, Phys. Rev. Lett. **116**, no. 14, 141802 (2016) doi:10.1103/PhysRevLett.116.141802 [arXiv:1511.01900 [hep-ph]].

[18] D. Bečirević, N. Košnik, O. Sumensari and R. Zukanovich Funchal, JHEP **1611**, 035 (2016) doi:10.1007/JHEP11(2016)035 [arXiv:1608.07583 [hep-ph]].

[19] D. Bečirević, I. Doršner, S. Fajfer, N. Košnik, D. A. Faroughy and O. Sumensari, Phys. Rev. D **98**, 055003 (2018) doi:10.1103/PhysRevD.98.055003 [arXiv:1806.05689 [hep-ph]].

[20] L. Di Luzio, J. Fuentes-Martin, A. Greljo, M. Nardecchia and S. Renner, arXiv:1808.00942 [hep-ph].

[21] I. Doršner, S. Fajfer, A. Greljo, J. F. Kamenik and N. Košnik, Phys. Rept. **641**, 1 (2016) [arXiv:1603.04993 [hep-ph]].

[22] I. Doršner, S. Fajfer and N. Košnik, Phys. Rev. D **86**, 015013 (2012) [arXiv:1204.0674 [hep-ph]].

[23] I. Doršner, S. Fajfer and N. Košnik, Eur. Phys. J. C **77**, no. 6, 417 (2017) [arXiv:1701.08322 [hep-ph]].

[24] I. Doršner, S. Fajfer, D. A. Faroughy and N. Košnik, JHEP **1710**, 188 (2017) [arXiv:1706.07779 [hep-ph]].

[25] A. Crivellin, D. Müller and T. Ota, JHEP **1709**, 040 (2017) [arXiv:1703.09226 [hep-ph]].

[26] D. Bečirević, S. Fajfer, N. Koňik and O. Sumensari, Phys. Rev. D **94**, no. 11, 115021 (2016) [arXiv:1608.08501 [hep-ph]].

[27] A. Greljo, D. J. Robinson, B. Shakya and J. Zupan, arXiv:1804.04642 [hep-ph].

[28] D. Robinson, B. Shakya and J. Zupan, arXiv:1807.04753 [hep-ph].

[29] A. Azatov, D. Bardhan, D. Ghosh, F. Sgarlata and E. Venturini, arXiv:1805.03209 [hep-ph].

[30] G. Hiller and M. Schmaltz, JHEP **1502**, 055 (2015) doi:10.1007/JHEP02(2015)055 [arXiv:1411.4773 [hep-ph]].

[31] H. Na *et al.* [HPQCD Collaboration], **92**, no. 5, 054510 (2015) Erratum: [Phys. Rev. D **93**, no. 11, 119906 (2016)] [arXiv:1505.03925 [hep-lat]].

[32] J. P. Lees *et al.* [BaBar Collaboration], Phys. Rev. D **86**, 012004 (2012) [arXiv:1204.2852 [hep-ex]].

[33] M. Huschle *et al.* [Belle Collaboration], Phys. Rev. D **92**, no. 7, 072014 (2015) [arXiv:1507.03233 [hep-ex]].

[34] R. Aaij *et al.* [LHCb Collaboration], Phys. Rev. D **97**, no. 7, 072013 (2018) [arXiv:1711.02505 [hep-ex]].

[35] Y. Amhis *et al.* [HFLAV Collaboration], Eur. Phys. J. C **77**, no. 12, 895 (2017) [arXiv:1612.07233 [hep-ex]].

[36] G. D'Amico, M. Nardecchia, P. Panci, F. Sannino, A. Strumia, R. Torre and A. Urbano, JHEP **1709**, 010 (2017) [arXiv:1704.05438 [hep-ph]].

[37] B. Capdevila, A. Crivellin, S. Descotes-Genon, J. Matias and J. Virto, JHEP **1801**, 093 (2018) [arXiv:1704.05340 [hep-ph]].

[38] W. Altmannshofer, A. J. Buras, D. M. Straub and M. Wick, JHEP **0904**, 022 (2009) [arXiv:0902.0160 [hep-ph]].

[39] J. Grygier *et al.* [Belle Collaboration], Phys. Rev. D **96**, no. 9, 091101 (2017) Addendum: [Phys. Rev. D **97**, no. 9, 099902 (2018)] [arXiv:1702.03224 [hep-ex]].

[40] C. Patrignani *et al.* [Particle Data Group], Chin. Phys. C **40**, no. 10, 100001 (2016).

[41] J. Beringer *et al.* [Particle Data Group], Phys. Rev. D **86**, 010001 (2012).

[42] S. Schael *et al.* [ALEPH and DELPHI and L3 and OPAL and SLD Collaborations and LEP Electroweak Working Group and SLD Electroweak Group and SLD Heavy Flavour Group], Phys. Rept. **427**, 257 (2006) [hep-ex/0509008].

[43] D. Bečirević, O. Sumensari and R. Zukanovich Funchal, Eur. Phys. J. C **76**, no. 3, 134 (2016) [arXiv:1602.00881 [hep-ph]].

[44] S. L. Glashow, D. Guadagnoli and K. Lane, Phys. Rev. Lett. **114** (2015) 091801 [arXiv:1411.0565 [hep-ph]].

[45] D. Guadagnoli and K. Lane, Phys. Lett. B **751** (2015) 54 [arXiv:1507.01412 [hep-ph]].

EPJ Web of Conferences **192**, 00026 (2018) https://doi.org/10.1051/epjconf/201819200026
QCD@Work 2018

Exploring the pattern of Lepton Flavour Universality violation at hadron colliders

Abhishek M. Iyer[1,*]

[1]INFN-Sezione di Napoli, Via Cintia, 80126 Napoli, Italia

Abstract. We discuss the B anomalies in scenario with warped extra dimensions characterized by additional neutral heavy bosons. Depending on the coupling of the fermions (leptons in particular) different scenarios for the explanation of the anomalies can be obtained. We focus on a specific case which involves contributions due to both the electron and the muon leading to a four dimensional fits in the space of Wilson coefficients. With this as a motivation, in a generic Z' we develop correlations with direct search experiments, wherein the pattern of the Wilson coefficients can be uniquely extracted at the LHC.

1 Introduction

Flavour observables offer a strong probe towards the indirect detection for NP effects with flavour structures different from that of the SM. The recent measurements of the semi-leptonic decay of the B mesons in the $b \to sll$ sector has generated a lot of interest. The observed anomalies were quoted through the measurement of the following hadronicall clean ratio [1]

$$\begin{aligned} R_K &= \left.\frac{\mathcal{B}(B^+ \to K^+\mu^+\mu^-)}{\mathcal{B}(B^+ \to K^+e^+e^-)}\right|_{q^2=1-6\ GeV^2} \\ &= 0.745^{+0.090}_{-0.074}\,(stat) \pm 0.036\,(syst) \end{aligned} \quad (1)$$

The SM expectation for the ratio is $R_K^{SM} = 1.003$ [2] indicating a a $\sim 2.6\ \sigma$ deviation as a possible evidence of lepton flavour non-universality. Another measurement through the ratio $R(K^*)$ exhibited a pattern similar to $R(K)$ as

$$\begin{aligned} R_{K^*} &= \frac{\mathcal{B}(B^0 \to K^{*0}\mu^+\mu^-)}{\mathcal{B}(B^0 \to K^{*0}e^+e^-)} \\ &= 0.660^{+0.110}_{-0.070}(stat) \pm 0.024(syst), \quad \text{low } q^2 \quad (2) \\ &= 0.685^{+0.113}_{-0.069}(stat) \pm 0.047(syst), \quad \text{mid } q^2 \quad (3) \end{aligned}$$

For this, the low q^2 corresponds to $0.045 \le q^2 \le 1.1$ GeV^2 while mid q^2 corresponds to $1.1 \le q^2 \le 6.0$ GeV^2. The corresponding SM values for the q^2 bins are: $R_{K^*}^{SM} \simeq 0.93$ for low q^2 while $R_{K^*}^{SM} = 1$ elsewhere. This translates into a 2.4σ and $\sim 2.5\ \sigma$ deviation for the

*e-mail: iyera@na.infn.it

© The Authors, published by EDP Sciences. This is an open access article distributed under the terms of the Creative Commons Attribution License 4.0 (http://creativecommons.org/licenses/by/4.0/).

low q^2 and for medium q^2 bins respectively. These deviations are typically parametrized by additional contributions to the Wilson coefficients C_i of the following effective operators: [3]:

$$\mathcal{L} \supset \frac{G_F \alpha}{\sqrt{2}\pi} \sum_i C_i O_i \tag{4}$$

where $C_i = C_i^{SM} + C_i^{NP}$.

$$\begin{aligned} O_9 &= (\bar{s}_L \gamma^\mu b_L)(\bar{l}\gamma_\mu l) \qquad O_{9'} = (\bar{s}_R \gamma^\mu b_R)(\bar{l}\gamma_\mu l) \\ O_{10} &= (\bar{s}_L \gamma^\mu b_L)(\bar{l}\gamma_\mu \gamma^5 l) \quad O_{10'} = (\bar{s}_R \gamma^\mu b_R)(\bar{l}\gamma_\mu \gamma^5 l) \end{aligned} \tag{5}$$

Here C_i^{NP} determines the NP contributions to the Wilson coefficients and l denotes a lepton. In this talk we will consider a scenario where NP effects exits in both the muon and the electron sector due to the exchange of the neutral heavy boson. These exotic states can arise to gauge extensions of the SM. However, we consider these fields arising as Kaluza-Klein (KK) excitations of extra-dimensional models. In particular we consider a model with a single warped extra-dimension with the following line element: [5]:

$$ds^2 = e^{-2A(y)} \eta_{\mu\nu} dx^\mu dx^\nu - dy^2 \tag{6}$$

where $A(y) = k|y|$, $k \sim \frac{M_{Pl}}{4\pi}$ and $0 \le y \le \pi R$. The coordinates $y = 0, \pi R$ correspond to the location of the (UV, IR) brane respectively. The minimal model, with brane localized fields and SM gauge symmetry is severely constrainted by EWPT. A generalization of this setup with bulk fields and a gauged custodial symmetry [6] was then proposed which not only relaxed the bounds from large contributions to the T parameter but embedding the fermions in non singlet representations of the gauge group served to evade constraints from $Z \to f\bar{f}$. As mentioned earlier, the presence of bulk fields leads to KK excitations in the effective low energy theory. Along with the KK excitations of the SM W, Z, the model is also characterized by additional 'custodial' heavy gauge bosons thereby leading to a distinct phenomenology, in the flavour sector in particular. A detailed analysis of different flavour transitions in this setup was considered in [7–10]. We explore the parameter space admitted by the current anomalies [11]. We demonstrate fits with the following two scenarios which are characterized by different extent of contributions to the Wilson coefficients of the electron and the muon. We then setup a high p_T collider probe for any generic Z' scenario wherein it is possible to extract the pattern of the Wilson coefficients.

The note is organized as follows: In Section 2 we compute the fits for the anomalies in $b \to sll$ processes for the two different scenarios. In Section 3 we develop a strategy to extract the Wilson coefficients at collider and we finally conclude.

2 $b \to sll$ processes: B anomalies

In bulk custodial models in RS , neutral currents at tree level receive contribution to the four fermions operators due to the exchange of $X \in Z_{SM, Z_X, Z_H, \gamma^{(1)}}$. The expression for the coupling of the SM fermions to these NP states can be written as:

$$\mathcal{L}_{NP} \subset X_\mu \left[\alpha_L^{bs}(X)(\bar{s}_L\gamma^\mu b_L) + \alpha_R^{bs}(X)(\bar{s}_R\gamma^\mu b_R) + \bar{\mu}\left(\alpha_V^l(X)\gamma^\mu - \alpha_A^l(X)\gamma^\mu\gamma^5\right)\mu\right] \tag{7}$$

where $\alpha_{V,A}^l(X) = \frac{\alpha_L^l(X) \pm \alpha_R^l(X)}{2}$. α gives the coupling in the zero mode fermions to the gauge bosons in the mass basis of the former. The Wilson co-efficients for each gauge field X can then be written as:

$$\Delta C_9 = -\frac{\sqrt{2}\pi}{M_X^2 G_F \alpha} \alpha_L^{bs}(X) \alpha_V^l(X), \quad \Delta C_{10} = \frac{\sqrt{2}\pi}{M_X^2 G_F \alpha} \alpha_L^{bs}(X) \alpha_A^l(X), \tag{8}$$

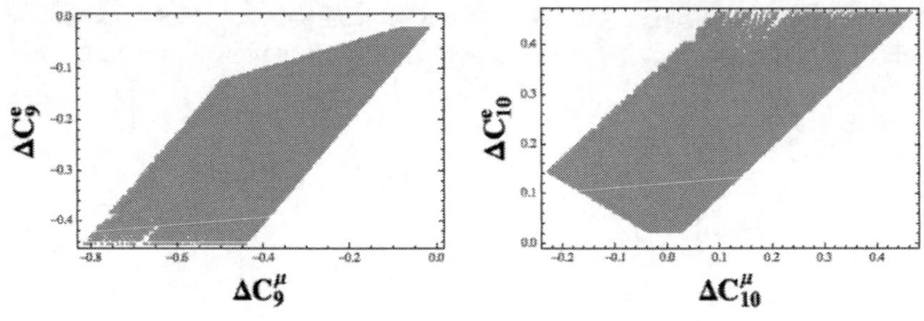

Figure 1. Scenario **A**: Plot show the correlation in the C_9 and C_{10} parameter plane for both the electron and the muon. The bottom row gives the range of c_{b_L} if $-1.4 < C_{10}^{\mu} < -0.7$. We use $M_{KK} = 3\ TeV$

The primed operators vanish due the presence of an accidental $U(3)$ in the coupling of the down quark singlets to the gauge KK states. Assuming the the up-sector quark to be in the mass-diagonal basis, down quark rotation matrices are simply $D_L \sim V_{CKM}$. We now discuss the following two scenarios:

1) Scenario **A**: In this scenario the lepton singlets have relatively larger coupling to the NP than the doublets. The doublets have universal bulk wave function with $c > 0.5$. The scanning range for the doublets and the muon singlets is chosen to be $0.45 < c < 0.55$. Thus even though this scenario can admit a relatively larger coupling of muon singlets to NP than the corresponding doublets, the other possibility still exists. The ranges chosen for c parameter scan is: $c_{Q_3} \in [0.4, 0.5]$, $c_{\mu_L} = c_L \in [0.45, 0.55]$ and $c_{\mu_R} \in [0.45, 0.55]$. The scan in this scenario has a couple of unique features: The $\Delta C_9^{\mu} - \Delta C_{10}^{\mu}$ plot in Fig. 1 has a feature where ΔC_{10}^{μ} may vanish while the contribution to ΔC_9^{μ} is considerable. This is an artifact of the similar scanning range for the lepton doublets and the muon singlets. This corresponds to the explicit case where $\alpha_L^l(X) = \alpha_R^l(X)$. Corresponding to this, we discuss two minor possibilities:
a) ΔC_{10}^{μ} vanishes and the contributions to $\Delta C_{9,10}^{e}$ are very small, thereby reducing the analysis to a one-dimensional fit with contributions mainly from ΔC_9^{μ}. The best fit for this case is $\Delta C_9^{\mu} = -1.5$ with a 2σ region $[-2.9, -1.73]$ [4, 17]. From the bottom plot of Fig.1, we find that we can obtain a fit only in the 3σ region where the electron contribution is negligible.
b) The tight condition $1-D$ scenario can be further relaxed, once the electrons also contribute considerably. This opens up the possibility of a four dimensional fit in the plane of Wilson coefficients and the results for this case are shown in Fig.1. Lower row of Fig. 1 gives the different correlations: $C_9^{\mu} - C_9^{e}$ (top) and $C_{10}^{\mu} - C_{10}^{e}$ (bottom). The 2-σ regions for a 4D fit to the data is [12]

$$\frac{C_9^{\mu}}{C_9^{SM}} \in [-0.33, 0.06]; \quad \frac{C_9^{e}}{C_9^{SM}} \in [-2.23, 0.74]; \quad \frac{C_{10}^{\mu}}{C_{10}^{SM}} \in [-0.29, 0.14]; \quad \frac{C_{10}^{e}}{C_{10}^{SM}} \in [-2.60, 0.60]. \tag{9}$$

With a four dimensional fit, it is relatively easier to find solutions which satisfy the above regions.

The non-negligible values of the ΔC^e is due to left doublets having $c \sim 0.5$ thereby resulting in a mildly larger coupling to the NP states than would be expected of states having $c \geq 0.55$. For these choices of c parameters corresponding to the values in Fig.1, fitting the muon mass requires choosing the $O(1)$ Yukawa ~ 0.03. Though slightly fine tuned

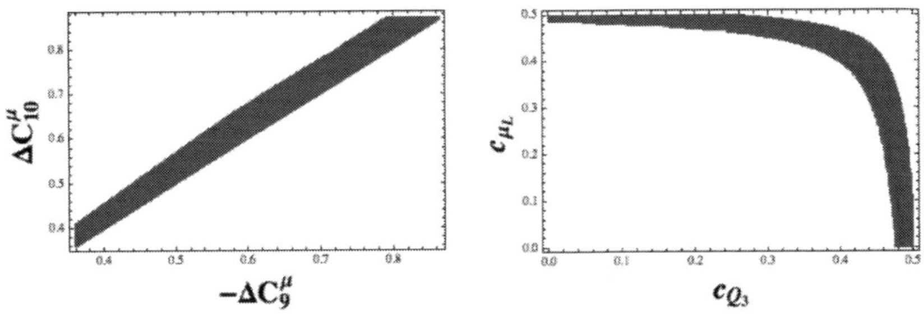

Figure 2. Scenario **B**: Left plot gives the distribution for ΔC_9 and ΔC_{10}. The corresponding c parameters ranges are given in the right plot.

with regards to the fit to the muon mass, this scenario is more favorable with regards to an anarchic neutrino mixing matrix but also in suppressing FCNC in the lepton sector through the implementation of 5D MFV.

2) Scenario **B**: The non-universality in this case exists in the coupling of the lepton doublets to NP . Further, we assume that the τ doublets to be relatively closer to the IR brane than the μ doublets ($c_{\tau_L} < c_{\mu_L}$). All the lepton singlets and the electron doublet satisfy $c > 0.55$. This results in $\Delta C^e_{9,10}$ much smaller than $\Delta C^\mu_{9,10}$, with its magnitude being at most ~ 0.2. For most of the region, it effectively reduces this to a 2-D fit where the value of $C^e_{9,10}$ is an order of magnitude less,

Top plot of Fig.2 gives the correlation in the ΔC^μ_9-ΔC^μ_{10} plane which correspond to points which satisfy $0.36 < |\Delta C^\mu_{9,10}| < 0.87$. The correlation also shows that there exists solutions for which $\Delta C^\mu_9 = -\Delta C^\mu_{10}$ reducing it to a 1-D fit as discussed in [16, 17]. The two relevant parameters for the fits to the B-anomalies in this case are $c_{Q_3} - c_{\mu_L}$ and the correlation is shown in the bottom plot of Fig. 2.

3 Correlating the pattern of solutions to high p_T searches at colliders

In the previous section we demonstrated two different scenarios in the same model which can explain the B anomalies. Scenario A in particular included contributions due to both the electron and the muon. They were associated with braod spread in the ranges for the Wilson coefficients. We now develop a correlation wherein the pattern involving electrons and/or muons can be probed at high energy p-p collisions. To begin with, we consider the following neutral current Lagrangian:

$$\mathcal{L}_{eff} = \frac{\lambda_{bs}}{M^2}(\bar{s}\gamma_\mu b)\left[\lambda_e(\bar{e}\gamma^\mu e) + \lambda_\mu(\bar{\mu}\gamma^\mu\mu) + \lambda_\tau(\bar{\tau}\gamma^\mu\tau)\right] + \ldots \tag{10}$$

Given the Lagrangian in Eq. 9, the Wilson-coefficients for the $R(K^*)$ anomalies are identified as:

$$C^e_9 = -\frac{\sqrt{2}\pi}{G_F\alpha}\frac{\lambda_{bs}\lambda^V_e}{M^2} \quad C^\mu_9 = -\frac{\sqrt{2}\pi}{G_F\alpha}\frac{\lambda_{bs}\lambda^V_\mu}{M^2} \quad C^e_{10} = -\frac{\sqrt{2}\pi}{G_F\alpha}\frac{\lambda_{bs}\lambda^{AV}_e}{M^2} \quad C^\mu_{10} = -\frac{\sqrt{2}\pi}{G_F\alpha}\frac{\lambda_{bs}\lambda^{AV}_\mu}{M^2} \tag{11}$$

where $\lambda^V = \frac{L+R}{2}$ and $\lambda^{AV} = \frac{L-R}{2}$ are the vector and axial vector coupling respectively of the leptons to Z'. $L(R)$ is the coupling strength of doublets (singlets) to Z'. In the first instance,

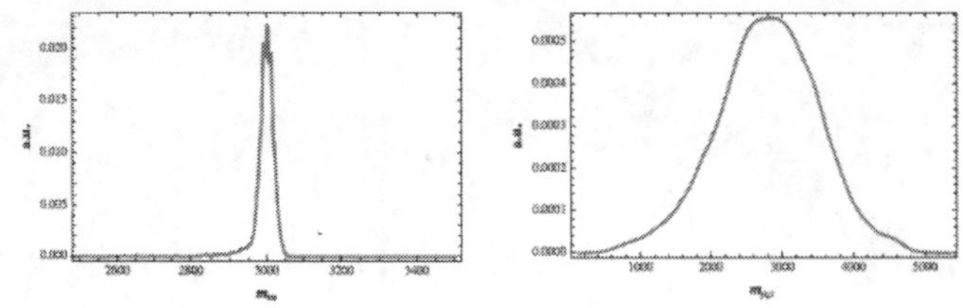

Figure 3. Mass reconstruction for the electron (left) and muon (right) for $M_{Z'} = 3$ TeV

we assume $L = R$, implying the axial-vector couplings of the leptons to Z' vanish. In this case the ratio of C_9 for muons and electrons is simply $\frac{C^9_\mu}{C^9_e} = \frac{\lambda^V_\mu}{\lambda^V_e}$ where the quark dependence cancels out.

In a direct production of Z' in pp collisions decaying into electrons and muons. The ratio of number di-muons to di-electrons at luminosity $\mathcal{L}$ is simply given as:

$$\delta = \frac{\sigma_{Z'}(\lambda^V_\mu)^2 \mathcal{L} \epsilon_\mu}{\sigma_{Z'}(\lambda^V_e)^2 \mathcal{L} \epsilon_e} = \frac{N_\mu}{N_e}. \tag{12}$$

In the event, the acceptance efficiency of the electrons and the muons are similar, then the ratio in Eq. 11 would exactly mirror the ratio of the Wilson coefficients. However, as shown in Table. 1, electrons and muons are characterized by different acceptance efficiencies. In evaluating the numbers in Table 1, the first row gives the number of events with atleast a single leptons and bottom row computes the number of leptons over a given threshold.

	ϵ_μ	ϵ_e
Simple Isolation(> 1 leptons)	59.33	39.79
Mass cuts (> 1000GeV)	58.79	39.61

Table 1. Comparison of acceptance efficiencies for electrons and muons for $m_{Z'} = 3000$ GeV

Note that the methodology of counting the number of di-leptons over a given threshold is rather unconventional. Traditional methods involve counting events in a given bin around a possible resonant mass. The electron and the muon are characterized by different reconstruction efficiencies. The latter is associated with the large smearing of the four moments especially at high p_T while the electrons have a sharp reconstruction independent of the mass of the resonance. This is illustrated in Fig. 3. Correspondingly Fig. 4 gives the corresponding sensitivity for two different mass bins around the pole mass: $M_{Z'} \pm 200$ GeV (left) and $M_{Z'} \pm 500$ GeV (right). Note that the efficiency changes depending on the choice of bin due to different reconstruction efficiencies. As a result in order to overcome the difference in the reconstruction efficiency, we propose to simply count the di-lepton events over a given threshold.

Further to ensure best implementation of the ration in Eq. 11 at colliders, we must ensure it is due to difference in couplings of the Z' to the leptons and not due to different acceptance efficiency. Increasing the acceptance of the electrons can be achieved using the electron jet techniques and using the following substructure variables to identify the electron [22]:
A) Hadronic energy fraction: As electron deposit their energy in the $E-cal$, the corresponding jets are associated with very small $H-cal$ component of the energy. This helps in limiting

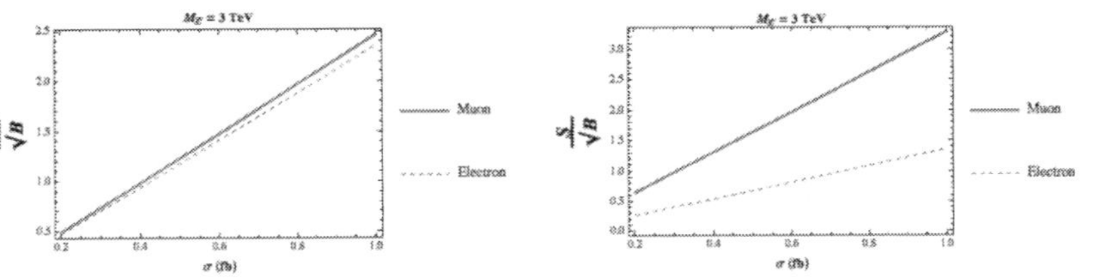

Figure 4. The plot gives the comparison of corresponding sensitivity as a function of $\sigma_{Z'}$ at $\mathcal{L}_{int}$ = 3000 fb^{-1} for two different bin choices: The left column corresonds to a choice $M_{Z'} \pm 200$ GeV and the right column corresponds to $M_{Z'} \pm 500$ GeV. Note that for all cases we assume B.R($Z' \to \mu\mu$) = 4 B.R($Z' \to ee$)

$m_{Z'}$ (GeV)	ϵ_μ	ϵ_e (Electron jets)	ϵ_τ (tau jets)
2000	71.45	64.75	31.25
2500	66.35	63.06	37.28
3000	58.79	60.37	40.88
3500	51.68	59.50	43.98

Table 2. Comparison of efficiencies of electron jets and muons for different $m_{Z'}$ masses. A selectrion criteria of the $m_{\mu\mu}$ or $m_{j_0 j_1}$ (> 1000) GeV is imposed. For the electron jets the QCD fake rate is < 1 in 3×10^5 events. For τ jets the QCD fake rate is 0.2%

QCD to a great extent and B) Tracks: We demand that the leading jet is associated with exactly one track, while the subleading jet may have 0 or 1 track. The latter condition facilitates in accepting events which would otherwise have been rejected by the standard electron identification criteria.

Table 2 gives a comparison of the muon and 'electron' efficiencies using electron jets. We find that the acceptance efficiencies are within 5 – 10% of each other. Thus any discrepancy in the observation of the ratio can be attributed to the difference in the couplings of the NP to electrons and muons. The third column of Table 2 gives the acceptance efficiency for tau jets. This technique can be used to extract information on C_9^τ, though with a lesser accuracy than the leptons and have implications for processes like $B \to K^*\tau\tau$

4 Conclusions

We discussed a warped framework to explain the observations of lepton flavour universality violation in the semileptonic decay of B mesons. One particular solution led to a case where both electron and muon contribute. This provided a motivation to extract the nature of Wilson coefficients at collider by probing the resonant decay of Z' into leptons. We propose a strategy to ensure that any observed deviations in the $N_{ee}/N_{\mu\mu}$ is primarily due to differential coupling of NP to leptons.

Acknowledgements

I would like to thank the organizers of QCD@work 2018 for the opportunity to present my work in the workshop.

References

[1] R. Aaij *et al.* [LHCb Collaboration], Phys. Rev. Lett. **113**, 151601 (2014) doi:10.1103/PhysRevLett.113.151601 [arXiv:1406.6482 [hep-ex]].

[2] M. Bordone, G. Isidori and A. Pattori, Eur. Phys. J. C **76**, no. 8, 440 (2016) doi:10.1140/epjc/s10052-016-4274-7 [arXiv:1605.07633 [hep-ph]].

[3] G. Buchalla, A. J. Buras and M. E. Lautenbacher, Rev. Mod. Phys. **68**, 1125 (1996) doi:10.1103/RevModPhys.68.1125 [hep-ph/9512380].

[4] L. S. Geng, B. Grinstein, S. Jäger, J. Martin Camalich, X. L. Ren and R. X. Shi, Phys. Rev. D **96**, no. 9, 093006 (2017) doi:10.1103/PhysRevD.96.093006 [arXiv:1704.05446 [hep-ph]].

[5] L. Randall and R. Sundrum, Phys. Rev. Lett. **83**, 3370 (1999) doi:10.1103/PhysRevLett.83.3370 [hep-ph/9905221].

[6] K. Agashe, A. Delgado, M. J. May and R. Sundrum, JHEP **0308**, 050 (2003) doi:10.1088/1126-6708/2003/08/050 [hep-ph/0308036].

[7] M. Blanke, A. J. Buras, B. Duling, S. Gori and A. Weiler, JHEP **0903**, 001 (2009) doi:10.1088/1126-6708/2009/03/001 [arXiv:0809.1073 [hep-ph]].

[8] M. Blanke, A. J. Buras, B. Duling, K. Gemmler and S. Gori, JHEP **0903**, 108 (2009) doi:10.1088/1126-6708/2009/03/108 [arXiv:0812.3803 [hep-ph]].

[9] M. Bauer, S. Casagrande, U. Haisch and M. Neubert, JHEP **1009**, 017 (2010) doi:10.1007/JHEP09(2010)017 [arXiv:0912.1625 [hep-ph]].

[10] S. Casagrande, F. Goertz, U. Haisch, M. Neubert and T. Pfoh, JHEP **1009**, 014 (2010) doi:10.1007/JHEP09(2010)014 [arXiv:1005.4315 [hep-ph]].

[11] G. D'Ambrosio and A. M. Iyer, arXiv:1712.08122 [hep-ph].

[12] T. Hurth, F. Mahmoudi and S. Neshatpour, Nucl. Phys. B **909**, 737 (2016) doi:10.1016/j.nuclphysb.2016.05.022 [arXiv:1603.00865 [hep-ph]].

[13] A. J. Buras, D. Buttazzo, J. Girrbach-Noe and R. Knegjens, JHEP **1511**, 033 (2015) doi:10.1007/JHEP11(2015)033 [arXiv:1503.02693 [hep-ph]].

[14] T. K. Komatsubara, Prog. Part. Nucl. Phys. **67**, 995 (2012) doi:10.1016/j.ppnp.2012.04.001 [arXiv:1203.6437 [hep-ex]].

[15] K. Shiomi [KOTO Collaboration], arXiv:1411.4250 [hep-ex].

[16] B. Capdevila, A. Crivellin, S. Descotes-Genon, J. Matias and J. Virto, arXiv:1704.05340 [hep-ph].

[17] W. Altmannshofer, P. Stangl and D. M. Straub, Phys. Rev. D **96**, no. 5, 055008 (2017) doi:10.1103/PhysRevD.96.055008 [arXiv:1704.05435 [hep-ph]].

[18] G. Buchalla and A. J. Buras, Nucl. Phys. B **548**, 309 (1999) doi:10.1016/S0550-3213(99)00149-2 [hep-ph/9901288].

[19] C. Patrignani *et al.* [Particle Data Group], Chin. Phys. C **40**, no. 10, 100001 (2016). doi:10.1088/1674-1137/40/10/100001

[20] F. Mescia and C. Smith, Phys. Rev. D **76**, 034017 (2007) doi:10.1103/PhysRevD.76.034017 [arXiv:0705.2025 [hep-ph]].

[21] G. Buchalla and A. J. Buras, Nucl. Phys. B **412**, 106 (1994) doi:10.1016/0550-3213(94)90496-0 [hep-ph/9308272].

[22] F. Conventi, G. D 'Ambrosio, A. M. Iyer and E. Rossi et al., To appear [arXiv:1810.xxxx].

[23] A. Chakraborty, A. M. Iyer and T. S. Roy, Nucl. Phys. B **932**, 439 (2018) doi:10.1016/j.nuclphysb.2018.05.019 [arXiv:1707.07084 [hep-ph]].

Pseudoscalar pole contribution to the hadronic light-by-light piece of a_μ

Adolfo Guevara[1,*], *Pablo* Roig[2,**], and *Juan José* Sanz Cillero[1,***]

[1]Departamento de Física Teórica and UPARCOS, Universidad Complutense de Madrid, Plaza de las Ciencias 1, 28040 Madrid, Spain
[2]Centro de Investigación y de Estudios Avanzados del IPN, Apartado Postal 14-740, 07000, Ciudad de México, Mexico

Abstract. We have studied the $P \to \gamma^\star\gamma^\star$ form factor in Resonance Chiral Theory, with $P = \pi^0, \eta, \eta'$, to compute the contribution of the pseudoscalar pole to the hadronic light-by-light piece of the anomalous magnetic moment of the muon. In this work we allow the leading $U(3)$ chiral symmetry breaking terms, obtaining the most general expression for the form factor of order $O(m_P^2)$. The parameters of the Effective Field Theory are obtained by means of short distance constraints on the form factor and matching with the expected behavior from QCD. Those parameters that cannot be fixed in this way are fitted to experimental determinations of the form factor within the spacelike momentum region of the virtual photon. Chiral symmetry relations among the transition form factors for π^0, η and η' allow for a simultaneous fit to experimental data for the three mesons. This shows an inconsistency between the BaBar π^0 data and the rest of the experimental inputs. Thus, we find a total pseudoscalar pole contribution of $a_\mu^{P,HLbL} = (8.47 \pm 0.16) \cdot 10^{-10}$ for our best fit (neglecting the BaBar π^0 data). Also, a preliminary rough estimate of the impact of NLO in $1/N_C$ corrections and higher vector multiplets (asym) enlarges the uncertainty up to $a_\mu^{P,HLbL} = (8.47 \pm 0.16_{\rm stat} \pm 0.09_{N_C}{}^{+0.5}_{-0.0}{}_{\rm asym})$.

1 Introduction

Nowadays, the anomalous magnetic moment of the muon, a_μ, has been predicted to an outstanding precision of $O\left[(\alpha/\pi)^5\right]$ for purely electromagnetic effects ($a_\mu^{\rm QED}$)[1] and to two loops[1] precision in electroweak corrections ($a_\mu^{\rm EW}$)[2]. This property, which is measured through the decay of muons [3], has been measured to an extraordinary precision of ten significant figures. An interesting point is that the experimental error is orders of magnitude larger than those estimated for $a_\mu^{\rm QED}$ and $a_\mu^{\rm EW}$, however, the uncertainty of the hadronic contributions is of the same order of the experimental error[2]. There is an incompatibility among the sum of all contributions with respect to the experiment of $\sim 3.5\sigma$, however, there are new experiments that will measure a_μ with an experimental error reduced at least by

*e-mail: adguevar@ucm.es
**e-mail: proig@fis.cinvestav.mx
***e-mail: jjsanzcillero@ucm.es

[1]The three loop Leading Logarithms are found to give negligible contribution of $O(10^{-12})$ [2].

© The Authors, published by EDP Sciences. This is an open access article distributed under the terms of the Creative Commons Attribution License 4.0 (http://creativecommons.org/licenses/by/4.0/).

a factor 4, namely E34 at J-PARC [4] and muon g-2 at Fermilab [5]. If one is to assume the Standard Model (SM) is all that is needed to understand this difference, then there is something about the SM that is not understood; if the assumption is that the difference stems from Beyond Standard Model (BSM) effects, the SM must be further improved to reduce the uncertainties in order to have a more controlled SM background for the search of BSM effects. Whatever the case is, the fact that the experimental precision in a_μ will be further reduced forces theoreticians to give a more precise determination of a_μ.

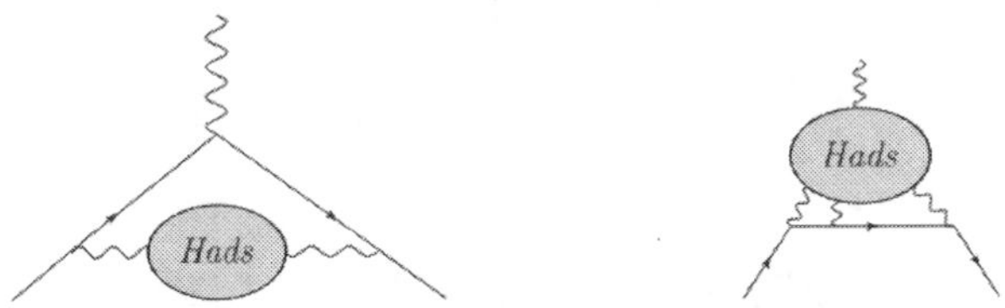

Figure 1. Hadronic contributions to a_μ. The diagram on the left-hand-side represents all contributions from the hadronic vacuum to the self energy of the virtual photon, called Hadronic Vacuum Polarization (HVP). The diagram on the right-hand-side represents all contributions from elastic scattering of two photons, called Hadronic Light-by-Light scattering (HLbL).

As said above, the main uncertainty comes from hadronic parts. These are contributions involving quarks and their interactions. This can be separated into two pieces, shown in the left-hand-side diagram of Fig. 1. The largest of both, a_μ^{HVP}, is the piece stemming from the self energy of virtual photons from the polarization of the hadronic vacuum (HVP) which can, however, be determined through dispersion relations using data available on $R_{\rm had} = \sigma(e^+e^- \to \text{hadrons})/\sigma(e^+e^- \to \mu^+\mu^-)$. The remaining hadronic piece, a_μ^{HLbL} depends on $\gamma\gamma \to \gamma\gamma$ scattering, which involves strong interactions (HLbL), as shown in the right-hand-side diagram of Fig. 1. The latter cannot be obtained in the same manner[2] as a_μ^{HVP}, and so, has to be obtained either numerically or in a model dependent way. The approach we follow is the latter.

The HLbL can be divided in three parts, which are the leading order parts of the HLbL piece [3] shown in Fig. 2. The sum of diagrams 2 (b) and 2 (c) are one order of magnitude suppressed as compared to the pseudoscalar exchange shown in diagram 2 (a). We focus on the pseudoscalar pole contribution to the HLbL piece, $a_\mu^{P,HLbL}$, which in order to be fully described needs only the Transition Form Factor (TFF), $\mathcal{F}_{P\gamma^*\gamma^*}(q^2, p^2)$. As it has been shown in ref [9], this TFF gives most of its contribution to a_μ at Euclidian squared photon momenta $\lesssim 1$ GeV^2. Therefore, it will be saturated mainly with the lightest resonant part of the (TFF) and higher energies effects will be suppressed. To compute the TFF we rely on χPT [10, 11] extended to include the lightest multiplet of meson resonances [12] (Resonance Chiral Theory, RχT). Instead of using the complete basis of VVP and VJP operators for resonances [13], we will rely on the simpler basis given in [14] since, for describing vertices involving only one pseudo-Goldstone both are equivalent [15]. We will use the Lagrangian in ref. [13] to include effects due to pseudoscalar resonances. The novelty in our approach is that we account for all the leading order terms that break explicitly chiral symmetry, which enter as corrections in powers of the squared pseudo-Goldstone bosons, m_P^2. Here we will only present those operators necessary for a consistent description of the NLO terms in m_P^2.

[2] See, however, the outstanding effort done in this direction from [6, 8].

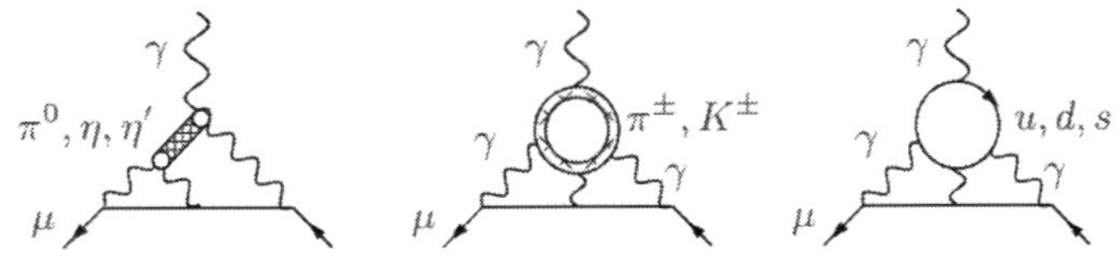

(a) [L.D.] (b) [L.D.] (c) [S.D.]

Figure 2. Leading order contributions in α from HLbL scattering to a_μ. The left-hand-side diagram and its photon momenta permutations give the pseudoscalar pole contribution, $a_\mu^{P,HLbL}$, to a_μ.

2 Flavor $U(3)$ breaking

For the complete description of the full base of operators used [11–14, 16, 17], we refer to [18]. To consistently include all terms breaking $U(3)$, an $O(p^6)$ odd-intrinsic Lagrangian without resonances must be considered for RχT [16]. In addition to the Wess-Zumino-Witten term [11], the relevant operators here are

$$\begin{aligned} O_7^W &= i\epsilon_{\mu\nu\alpha\beta}\langle \chi_- f_+^{\mu\nu} f_+^{\alpha\beta}\rangle, \\ O_8^W &= i\epsilon_{\mu\nu\alpha\beta}\langle \chi_-\rangle \langle f_+^{\mu\nu} f_+^{\alpha\beta}\rangle, \\ O_{22}^W &= i\epsilon_{\mu\nu\alpha\beta}\langle u^\mu \{\nabla_\rho f_+^{\rho\nu}, f_+^{\alpha\beta}\}\rangle. \end{aligned} \tag{1}$$

A correction to the vector resonance-photon coupling will be given by the interaction[3]

$$\mathcal{L}_{VJ} = \frac{\lambda_V}{\sqrt{2}}\langle V_{\mu\nu}\{f_+^{\mu\nu}, \chi_+\}\rangle. \tag{2}$$

There is also a correction to the mass of the vector resonances from V-V interactions[4]

$$\mathcal{L}_{VV} = -e_m^V\{V_{\mu\nu}V^{\mu\nu}\chi_+\}. \tag{3}$$

As a result, the masses of the vector resonances are given by

$$M_\rho^2 = M_\omega^2 = M_V^2 - 4e_m^V m_\pi^2, \qquad M_\phi^2 = M_V^2 - 4e_m^V \Delta_{2K\pi}^2, \tag{4}$$

where $\Delta_{2K\pi}^2 = 2m_K^2 - m_\pi^2$ and M_V is the mass associated with the vector nonet in the chiral and large N_C limits. The TFF is defined in ref. [18] along with the short distance constraints relating parameters of the model. Its complete expression before imposing such constraints can be found therein.

3 Transition Form Factor

By imposing the short-distance behavior

$$\lim_{q^2\to\infty} \mathcal{F}_{P\gamma^\star\gamma^\star}(q^2, q^2) = O(q^{-2}) \qquad \text{and} \qquad \lim_{q^2\to\infty} \mathcal{F}_{P\gamma\gamma^\star}(0, q^2) = O(q^{-2}), \tag{5}$$

we find some relations among the parameters of the theory that simplify the expression of the TFF. For π^0 we find[5]

$$\mathcal{F}_{\pi\gamma^\star\gamma^\star}(q_1^2, q_2^2) = \frac{32\pi^2 m_\pi^2 F_V^2 d_{123}^\star - N_C M_V^2 M_\rho^2}{12\pi^2 F_\pi D_\rho(q_1^2) D_\rho(q_2^2)}, \tag{6}$$

[3]This interaction term is the only single-trace operator $O(m_P^2)$ from those given in [17].
[4]This term generates a mass splitting effect in the nonet of resonances, inducing an explicit $U(3)$ breaking effect.
[5] $d_2^\star$ and $d_{123}^\star$ have been defined to account for pseudoscalar resonance effects.

where $D_R(s) = M_R^2 - s$ with the resonance masses given by (4), $d_{123}^\star$ is a free parameter and F_π is the π decay constant. Analogously, the simplified expression for the TFF of the η after imposing eq. (5) is given by

$$\mathcal{F}_{\eta\gamma^\star\gamma^\star}(q_1^2,q_2^2) = \frac{1}{12\pi^2 F D_\rho(q_1^2)D_\rho(q_2^2)D_\phi(q_1^2)D_\phi(q_2^2)} \times \tag{7}$$
$$\left\{ -\frac{N_C M_V^2}{3}\left[5C_q M_\rho^2 D_\phi(q_1^2)D_\phi(q_2^2) - \sqrt{2}C_s M_\phi^2 D_\rho(q_1^2)D_\rho(q_2^2)\right]\right.$$
$$+\frac{32\pi^2 F_V^2 d_{123}^\star m_\eta^2}{3}\left[(5C_q D_\phi(q_1^2)D_\phi(q_2^2) - \sqrt{2}C_s D_\rho(q_1^2)D_\rho(q_2^2)\right]$$
$$\left. -\frac{256\pi^2 F_V^2 d_2^\star}{3}\left[(5C_q \Delta_{\eta\pi}^2 D_\phi(q_1^2)D_\phi(q_2^2) + \sqrt{2}C_s \Delta_{2K\pi\eta}^2 D_\rho(q_1^2)D_\rho(q_2^2)\right]\right\},$$

where $d_2^\star$ is another free parameter and $C_{q/s}$ are the $\eta - \eta'$ mixing parameters. One can get the η'-TFF from eq. (7) by making $C_q \to C_q'$, $C_s \to -C_s'$ and $m_\eta \to m_{\eta'}$.

As it can be seen from the previous expressions, we were not able to match the precise QCD behavior [19, 20] $\lim_{q^2\to\infty} \mathcal{F}_{P\gamma^\star\gamma^\star}(q^2,q^2) \propto 1/q^2$, which will be taken into account as a systematic uncertainty. This implies an underestimation of the form factor, and thus will enter as an asymmetric error. However, it is worth to mention that the correct behavior in q^2 with approximately the correct coefficient is achieved for the singly virtual TFF, $\lim_{q^2\to\infty} \mathcal{F}_{P\gamma^\star\gamma}(q^2,0) \approx 2F/q^2$, after the fit to experimental data.

4 Pole contribution to a_μ^{HLbL}

The parameters $d_2^\star$, $d_{123}^\star$, M_V, e_m^V and the $\eta-\eta'$ mixing parameters are all fitted to experimental data from BaBar ([21] for π-TFF and [22] for $\eta^{(\prime)}$-TFF), CLEO [23] and CELLO [24], LEP [25] and Belle [26]. The complete details of the fits can be seen on ref. [18]. To compute $a_\mu^{P,HLbL}$ we rely on the loop integral representation of ref. [9]. The $a_\mu^{P,HLbL}$ contribution is obtained using a MonteCarlo run based on a normal distribution for the fit parameters, including correlations[6],

$$a_\mu^{P,HLbL} = (8.47 \pm 0.16)\cdot 10^{-10}. \tag{8}$$

Also, the separate parts of the three pseudo-Goldstone mesons are computed in the same way, giving

$$a_\mu^{\pi^0,HLbL} = (5.81 \pm 0.09)\cdot 10^{-10}, \tag{9a}$$
$$a_\mu^{\eta,HLbL} = (1.51 \pm 0.06)\cdot 10^{-10}, \tag{9b}$$
$$a_\mu^{\eta',HLbL} = (1.15 \pm 0.07)\cdot 10^{-10}. \tag{9c}$$

Notice that our result for the π^0 contribution is in agreement within 1.8σ with that obtained using dispersion relations [27].

Note that by taking the uncorrelated error as a sum in quadratures from the three previous values one gets a smaller uncertainty ($\sim \pm 0.13 \cdot 10^{-10}$) than that given in eq. (8). This is due to correlations among the three results, since the total result is obtained performing the simultaneous integral of the three contributions. In Table 1 we compare our result with some previous determinations of $a_\mu^{P,HLbL}$.

[6] By making the same calculation for the TFF in the chiral limit, we obtain $a_\mu^{P,HLbL} = 8.27 \cdot 10^{-10}$, where the change is essentially given by the η contribution. We note that, however, we keep the physical masses for the integration kernels.

Table 1. Comparison of different predictions for the pseudoscalar-pole contributions to a_μ^{HLbL}. The 'sta' error is due to fit uncertainties. The details on the other errors are given in section 5

Reference	$10^{10} \cdot a_\mu^{P,HLbL}$
Knecht and Nyffeler (2002) [9]	8.3 ± 1.2
Hayakawa and Kinoshita (2002) [28]	8.3 ± 0.6
Bijnens, Pallante and Prades (2002) [29]	8.5 ± 1.3
Erler and Toledo Sánchez (2006) [30]	$13.7\ ^{+2.7}_{-1.5}$
Roig, Guevara and López Castro (2014) [31]	8.60 ± 0.25
Masjuan and Sánchez-Puertas (2017) [32]	9.4 ± 0.5
Czyż, Kisza and Tracz (2018) [33]	8.28 ± 0.34
This work [18]	$8.47 \pm 0.16_{\text{sta}} \pm 0.09_{N_C}\ {}^{+0.5}_{-0.0}{}_{\text{asym}})$

5 Further error analysis

Our previous result does not take into account NLO effects in the $1/N_C$ expansion in which RχT relies. However, these effects can be estimated by taking into account how the main contribution at this order to the ρ meson propagator, namely the $\pi\pi$ and $K\overline{K}$ loops, affect the result for $a_\mu^{P,HLbL}$. This loop effects are included by changing the denominator of the propagator [34]

$$M_\rho^2 - q^2 \longrightarrow M_\rho^2 - q^2 + \frac{q^2 M_\rho^2}{96\pi^2 F_\pi^2}\left(A_\pi(q^2) + \frac{1}{2}A_K(q^2)\right), \tag{10}$$

where

$$A_P(q^2) = \ln\frac{m_P^2}{M_\rho^2} + 8\frac{m_P^2}{q^2} - \frac{5}{3} + \sigma_P^3(q^2)\ln\left(\frac{\sigma_P(q^2)+1}{\sigma_P(q^2)-1}\right), \tag{11}$$

being $\sigma_P(q^2) = \sqrt{1 - \frac{4m_P^2}{q^2}}$. Note that $A_P(q^2)$ is real for $q^2 < 4m_P^2$. Thus, the ρ propagator provided by eq. (10) is real in the whole space-like region $q^2 < 0$. This gives a total contribution to the uncertainty

$$\Delta a_\mu^{P,HLbL}{}_{1/N_C} = \pm 0.09 \cdot 10^{-10} \tag{12}$$

As said in section 3, the form factor is underestimated due to the incorrect asymptotic behavior of the doubly off-shell form factor. This was previously noted by [9], where it was pointed out that the correct asymptotic behavior can be recovered considering two vector meson resonance multiplets instead of just one. Therefore, in order to give an estimation of the uncertainty stemming from this incorrect limit we compute $a_\mu^{P,HLbL}$ considering two multiplets and compare the result with that using one multiplet, both in the chiral limit. The values for the fit parameters are kept identical for both estimations of $a_\mu^{P,HLbL}$. For a more detailed description of the form factor and the short distance constraints obtained among parameters we refer to [18]. This uncertainty is estimated to be

$$\Delta a_\mu^{P,HLbL}{}_{\text{asym}} = {}^{+0.5}_{-0.0} \cdot 10^{-10}. \tag{13}$$

6 Conclusions

We have given a more accurate description of the TFF within the framework for RχT, including terms up to order m_P^2 for the first time in a chiral invariant Lagrangian approach. This led to computing a more theoretically precise contribution from the P-pole to a_μ, with an improved precision. By looking at the difference of our result with that using the TFF in the chiral limit ($0.20 \cdot 10^{-10}$) it is evident that further orders in m_P^2 will be negligible. Future works will be directed towards reducing the uncertainty from the asymptotic limit, which is the largest one. A way to reduce such uncertainty could be by taking into account data form doubly off-shell TFF such as that given by BaBar for the η'-TFF [35]. Considering all possible contributions to the error we get

$$a_\mu^{P,HLbL} = (8.47 \pm 0.16_{\text{sta}} \pm 0.09_{1/N_C}\,{}^{+0.5}_{-0}{}_{\text{asym}}) \cdot 10^{-10}, \qquad (14)$$

where the first error (sta) comes from the fit of the TFF, the second from possible $1/N_C$ corrections and the last from the wrong asymptotic behavior estimated through the effects of heavier resonances in the TFF.

Acknowledgements

This work was supported by CONACYT Projects No. FOINS-296-2016 ('Fronteras de la Ciencia') and 250628 ('Ciencia Básica'), by the Spanish MINECO Project FPA2016-75654-C2-1-P and by the Spanish Consolider-Ingenio 2010 Programme CPAN (CSD2007-00042). A. G. acknowledges CONACYT for the support 'Estancia Posdoctoral en el Extranjero'. J.J.S.C. would like to thank Z.H. Guo for discussions on the $\eta-\eta'$ mixing. P.R. acknowledges discussions on the short distance constraints with Bastian Kubis, Andreas Nyffeler, Hans Bijnens and Gilberto Colangelo during the 'Muon g-2 Theory Initiative Hadronic Light-by-Light working group workshop' held at Univ. of Connecticut, 12-14 March 2018.

References

[1] T. Kinoshita and M. Nio, Phys. Rev. D **73** (2006) 053007.

[2] C. Patrignani *et al.*, Particle Data Group collab., Chin. Phys. **C 40** (2016) 100001.

[3] F. Jegerlehner and A. Nyffeler, Phys. Rept. **477** (2009) 1.

[4] H. Iinuma, H. Nakayama, K. Oide, K. i. Sasaki, N. Saito, T. Mibe and M. Abe, Nucl. Instrum. Meth. A **832** (2016) 51.

[5] W. Gohn, [Muon g-2 Collaboration], arXiv:1801.00084 [hep-ex], [FERMILAB-CONF-17-602-PPD].

[6] G. Colangelo, M. Hoferichter, M. Procura and P. Stoffer, JHEP **1409** (2014) 091; G. Colangelo, M. Hoferichter, B. Kubis, M. Procura and P. Stoffer, Phys. Lett. B **738** (2014) 6; G. Colangelo, M. Hoferichter, M. Procura and P. Stoffer, JHEP **1509** (2015) 074; Phys. Rev. Lett. **118** (2017) no.23, 232001; JHEP **1704** (2017) 161.

[7] M. Hoferichter, B. L. Hoid, B. Kubis, S. Leupold and S. P. Schneider, Phys. Rev. Lett. **121** (2018) 112002.

[8] V. Pauk and M. Vanderhaeghen, Phys. Rev. D **90** (2014) no.11, 113012; A. Nyffeler, Phys. Rev. D **94** (2016) no.5, 053006; I. Danilkin and M. Vanderhaeghen, Phys. Rev. D **95** (2017) no.1, 014019; F. Hagelstein and V. Pascalutsa, Phys. Rev. Lett. **120** (2018) no.7, 072002.

[9] M. Knecht and A. Nyffeler, Phys. Rev. D **65** (2002) 073034.

[10] S. Weinberg, Physica A **96** (1979) 327. J. Gasser and H. Leutwyler, Annals Phys. **158** (1984) 142; Nucl. Phys. B **250** (1985) 465.

[11] J. Wess and B. Zumino, Phys. Lett. **37B** (1971) 95; E. Witten, Nucl. Phys. B **223** (1983) 422.

[12] G. Ecker, J. Gasser, A. Pich, E. De Rafael, Nucl. Phys. B**321** (1989) 311; G. Ecker, J. Gasser, H. Leutwyler, A. Pich, E. De Rafael, Phys. Lett. B**223** (1989) 425.
[13] K. Kampf and J. Novotny, Phys. Rev. D **84** (2011) 014036.
[14] P. D. Ruiz-Femenia, A. Pich and J. Portoles, JHEP **0307** (2003) 003.
[15] P. Roig and J. J. Sanz Cillero, Phys. Lett. B **733** (2014) 158.
[16] J. Bijnens, L. Girlanda and P. Talavera, Eur. Phys. J. C **23** (2002) 539.
[17] V. Cirigliano, *et al.*,Nucl. Phys. B **753** (2006) 139.
[18] A. Guevara, P. Roig and J. J. Sanz-Cillero, JHEP **1806** (2018) 160.
[19] S. J. Brodsky and G. R. Farrar, Phys. Rev. Lett. **31** (1973) 1153; G. P. Lepage and S. J. Brodsky, Phys. Rev. D **22** (1980) 2157.
[20] V. A. Nesterenko and A. V. Radyushkin, Phys. Lett. B**128** (1983) 439; V. A. Novikov, Mikhail A. Shifman, A. I. Vainshtein, M. B. Voloshin, V. I. Zakharov, Nucl. Phys. B**237** (1984) 525.
[21] B. Aubert *et al.* [BaBar Collaboration], Phys. Rev. D **80** (2009) 052002.
[22] P. del Amo Sanchez *et al.* [BaBar Collaboration], Phys. Rev. D **84** (2011) 052001.
[23] J. Gronberg *et al.* [CLEO Collaboration], Phys. Rev. D **57** (1998) 33.
[24] H. J. Behrend *et al.* [CELLO Collaboration], Z. Phys. C **49** (1991) 401.
[25] M. Acciarri *et al.* [L3 Collaboration], Phys. Lett. B **418** (1998) 399.
[26] S. Uehara, *et al.* (Belle Collaboration), Phys. Rev. D **86** (2012) 092007.
[27] M. Hoferichter, B. L. Hoid, B. Kubis, S. Leupold and S. P. Schneider, arXiv:1808.04823 [hep-ph].
[28] M. Hayakawa and T. Kinoshita, Phys. Rev. D **57** (1998) 465 Erratum: [Phys. Rev. D **66** (2002) 019902].
[29] J. Bijnens, E. Pallante and J. Prades, Nucl. Phys. B **626** (2002) 410.
[30] J. Erler and G. Toledo Sanchez, Phys. Rev. Lett. **97** (2006) 161801.
[31] P. Roig, A. Guevara and G. López Castro, Phys. Rev. D **89** (2014) no.7, 073016.
[32] P. Masjuan and P. Sánchez-Puertas, Phys. Rev. D **95** (2017) no.5, 054026.
[33] H. Czyz, S. Ivashyn, A. Korchin and O. Shekhovtsova, Phys. Rev. D **85** (2012) 094010.
[34] D. Gómez Dumm, A. Pich and J. Portolés, Phys. Rev. D **62** (2000) 054014.
[35] J. P. Lees *et al.* [BaBar Collaboration], [arXiv:1808.08038 [hep-ex]].

The Belle II experiment: Status and prospects

Alberto Martini[1,2,*]

[1]Dipartimento di Matematica e Fisica, Università degli Studi Roma Tre, Via della Vasca Navale, 84 Rome 00146, Italy
[2]INFN Sezione di Roma Tre, Via della Vasca Navale, 84 Rome 00146, Italy

Abstract. This article aims to describe the Belle II experiment, its status and physics prospects for the next several years. Belle II is situated in Japan, at the KEK laboratory and it is the upgraded version of the Belle experiment. It uses a new collider named SuperKEKB, a second generation of B-factory based on the innovative Nano-Beam scheme technique, which is expected to collect an integrated luminosity of 50 ab^{-1}. Using this amount of data, together with improved detector performances, it will be possible to provide important contributions about several flavour physics topics (*i.e.* CKM matrix elements, FCNC processes, quarkonium states etc..) through high precision measurements. The main aim of Belle II is to investigate new physics scenarios and validate highly suppressed SM predictions. The experiment is almost completely assembled; it already took the first data without the vertex detector installed while the data taking will start in February 2019.

1 Introduction

An essential role in the understanding of the Standard Model (SM) and its mechanisms is played by the heavy flavour physics. In the past, experiments at B-factories investigated deeply this field (BaBar using PEP II and Belle using KEKB) producing important discoveries concerning the B meson physics and more. So far many SM predictions have still to be verified and the investigation of New Physics (NP) processes is crucial. For those reasons a second generation of B-factories is rising together with a corresponding experiment. That experiment is Belle II which relies on an international collaboration composed of more than 700 researchers from 25 countries and it takes advantage of the largest Japanese laboratory for particle physics KEK, in Tsukuba. The new SuperKEKB collider [1] (Fig. 1 left), together with an improved detector system with respect to its predecessor Belle, have been assembled and tested at KEK, and it plays a key role in the investigation of NP.

Thanks to several upgrades, Belle II will collect data up to a total integrated luminosity of 50 ab^{-1} and will take under control the increased background coming from the higher instant luminosity $\mathcal{L}_{max} = 8 \cdot 10^{35}$ cm^{-2} s^{-1} (Fig. 1 right). The whole dataset will allow to investigate several highly suppressed processes and some of them will be introduced later (see Ref. [2] for more details). Furthermore, the low-background environment provided by the B-factory, *w.r.t.* an hadronic machine, will produce large data samples of B, D, τ giving exclusive advantages to the experiment.

The time-line of Belle II can be divided into three main phases (Fig. 2):

*e-mail: alberto.martini@roma3.infn.it,alberto.martini@uniroma3.it

© The Authors, published by EDP Sciences. This is an open access article distributed under the terms of the Creative Commons Attribution License 4.0 (http://creativecommons.org/licenses/by/4.0/).

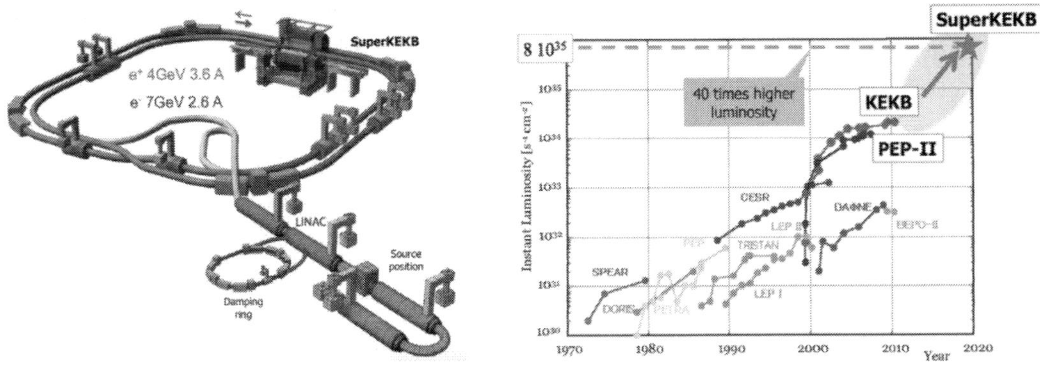

Figure 1. SupeKEKB schematic view (left) and luminosity comparison between SuperKEKB and colliders of the past (right).

- Phase 1: it has been successfully completed in 2016. It concerned the commissioning of SuperKEKB;
- Phase 2: it ended in July 2018 and some physics data have been collected without the silicon vertex detector installed. Those data will provide some first physics results, mainly in the dark sector field, even with partial detector installed.
- Phase 3: it will start in February 2019 and it consists in the data taking with the whole detector installed.

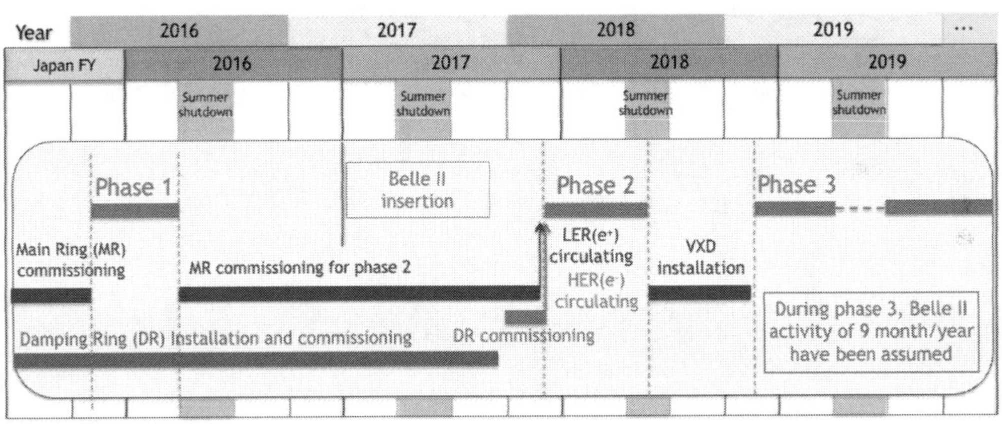

Figure 2. Status and future plan scheme of the experiment.

2 Collider and detector

SuperKEKB uses asymmetric electron and positron beams resulting in a boost of the center of mass (CM) system corresponding to a factor $\beta\gamma = 0.28$ in order to enable time-dependent CP violation measurements. The CM energy will be just above the threshold of the $\Upsilon(4S)$

resonance production which predominantly decays into $B\bar{B}$ couples. In order to increase the statistics collected by the experiment, SuperKEKB uses an innovative technique named Nano-Beam scheme, originally proposed for the Italian SuperB project by P. Raimondi [3], which will allow to reach 40 times higher luminosity *w.r.t.* KEKB. The luminosity of the machine can be written as follows:

$$\mathcal{L} = \frac{\gamma_\pm}{2er_e}(1 + \frac{\sigma_y^*}{\sigma_x^*})\frac{I_\pm \xi_{y\pm} R_L}{\beta_{y_\pm} R_{\xi_y}} \tag{1}$$

where r_e, e and γ are respectively the electron classical radius, the elementary electric charge and the Lorentz factor. The $\pm$ signs distinguish the positron (+) from the electron (-) while the ratio between the parameters R_L and R_{ξ_y} represents a geometrical reduction factor. The overall value of the parameters just mentioned is around 1 so the luminosity depends on the remaining parameters: the total beam current $I_\pm$ (increased by a factor ~2), the vertical beam-beam parameter $\xi_{y\pm}$ (slightly increased) and the vertical beta function at the IP $\beta^*_{y\pm}$ (decreased by a factor ~20). The Nano-Beam scheme allows to squeeze the beams at the interaction point up to ~50 nm among y and ~5 μm along x, as shown in Fig. 3. As a consequence the machine background increased thus, in order to compensate this effect, the Belle II detector has to be improved.

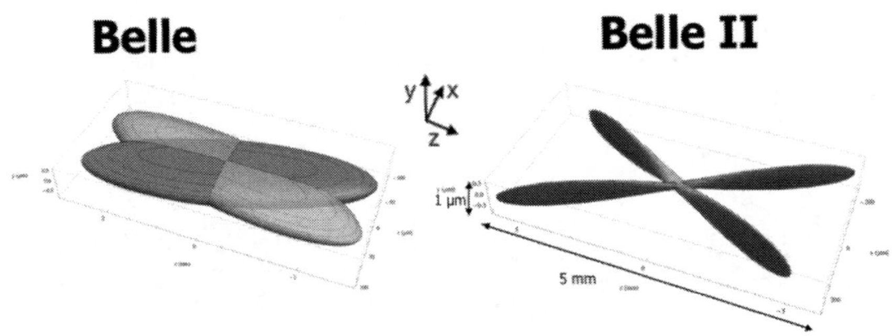

Figure 3. The Nano-Beam scheme improvements *w.r.t.* Belle.

The detector (Fig. 4), starting from the interaction point, is made of [4]:

- Vertex detector (VXD): it is composed by 2 layers of completely new Pixel vertex Detector (PXD) and 4 layers of fast Silicon-strip Vertex Detector (SVD). With respect to the vertex detector used by Belle, the VXD is faster and will increase the vertex resolution of a factor ~2;
- Central Drift Chamber (CDC): it is the main tracking detector with an improved resolution thanks to smaller cell size;
- Particles identification system: it is made of the Time-Of-Propagation counter (TOP) placed in the barrel region and the Aerogel Ring Imaging Cherenkov detector (ARICH) placed in the forward end-cap region. Their main aim is to distinguish pions from kaons.
- Electromagnetic Calorimeter (ECL): it is substantially the same of the one used for the Belle detector, with a faster read-out electronics;

- K_L and μ detector (KLM): it has been improved by substituting all the Resistive Plate Chamber (RPC) layers with scintillators in the end-caps region and the first 2 layers in the barrel region.

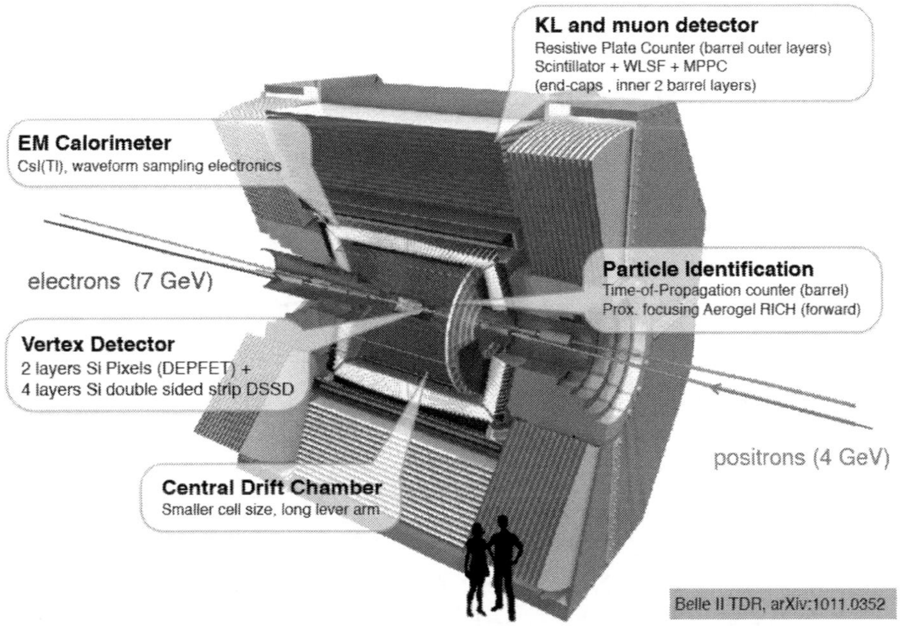

Figure 4. Schematic view of the Belle II detector.

2.1 Physics program

The experiment aims to investigate with high precision several heavy flavour physics fields as a B-factory [2] . The detector together with the collider will provide important advantages in the context of a physics analysis. SuperKEKB will produce coherent couples of B mesons from Y(4S) resonance in a clean environment *w.r.t.* experiments at hadronic machines and large data samples of B, D and τ with low background will be collected. This is the substantial difference between Belle II and its main competitor LHCb whose results are very important to make as much comparisons as possible. A valuable aspect is that, combining the results of both experiments it is possible to cover almost every heavy flavour physics field. Another important characteristic of the Belle II detector is a good hermeticity (around 90% of the whole solid angle) which, together with good reconstruction efficiency and resolution for neutral particles as η^0, K^0 and π^0, will provide important advantages for decays with missing energy. Moreover, Belle II takes advantage of a powerful tool called Full Event Interpretation (FEI) [5]. It consists in the complete reconstruction of the B-tag side in both semi-leptonic (more efficient but less precise) and hadronic (more precise but less efficient) ways in order to obtain the flavour of the B-tag. Thus it is possible to infer the flavour of the B-signal and isolate particles coming from it, having large advantages in analysis with missing energy and missing mass (Fig. 5). The FEI algorithm can be exploited because of the background-clean

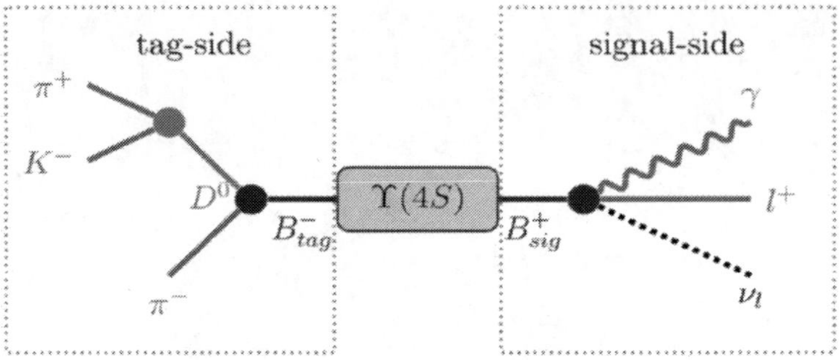

Figure 5. Schematic example of the FEI algorithm mechanism in a particular case.

environment and the very good hermeticity of the detector. It will provide a greater efficiency, *w.r.t.* previous versions of the algorithm, yielding a larger effective sample size usable in the measurement.

Taking advantages of the exclusive features just mentioned, part of the Belle II physics program can be summarized as below:

- Unitarity Triangle (UT) angles and CKM matrix elements (Fig. 6 left): CP violating measurements (time-dependent and time-integrated) allow to discover new possible CP-violating phases that indicates the existence of SM extensions;
- Dark sector: search for dark matter candidates *i.e.* dark photon (Fig. 6 right). Belle II will use a dedicated single photon trigger in order to be able to reconstruct its decay into an invisible final state (the signature is the presence of a single photon and missing energy). Some results can be obtained before the beginning of Phase 3.

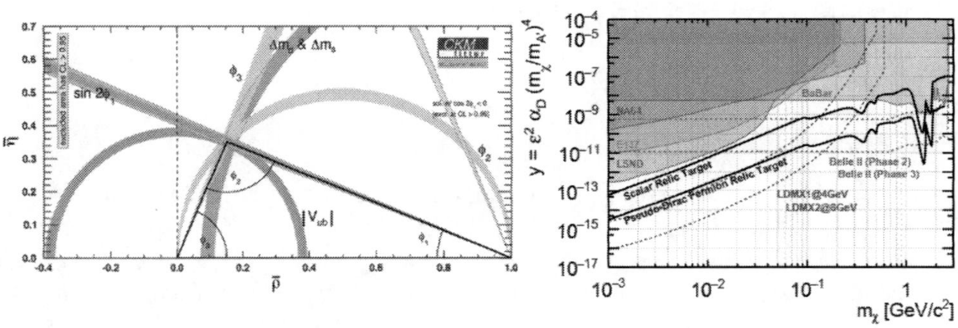

Figure 6. Expected UT constraints that can be obtained with the whole statistic of Belle II (left) and limits on the coupling vs the mass of the dark candidate (right).

- Flavour Changing Neutral Current (FCNC): penguin processes described by quark transitions like $b \to s$ and mixing processes of neutral meson states allow to search for NP in loops;
- Leptonic decays and Lepton Flavour Violation (LFV): study of τ and leptonic B decays in order to probe NP scenarios which take into account NP models *i.e.* extended Higgs sector or right-handed neutrino couplings. LFV can be investigated mainly thanks to the clean environment provided by a B-factory which makes Belle II highly competitive (Fig. 7);

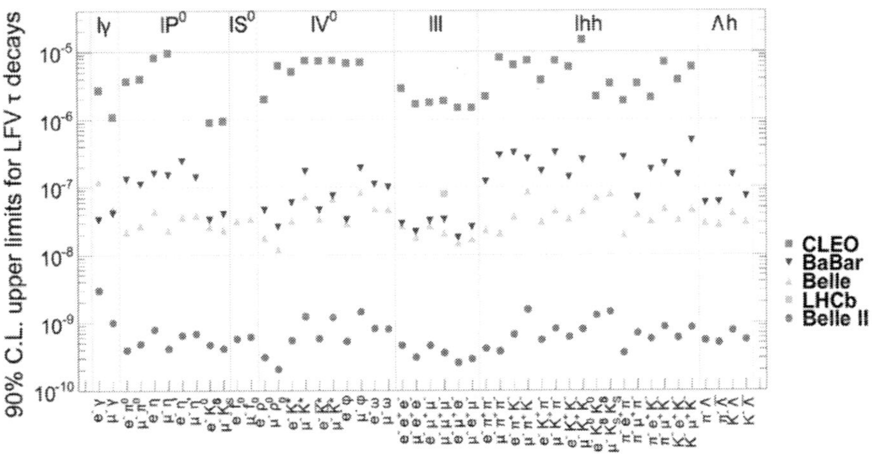

Figure 7. Expected upper limits on the LFV in several τ decays [2].

- Hadronic spectroscopy and quarkonium: a different center of mass energy of the collider is needed in order to produce resonances like $\Upsilon(3S)$, $\Upsilon(5S)$ and $\Upsilon(6S)$ allowing to study several intermediate bounded states and their properties. Quarkonium can be produced in different ways, some of them accessible only by a B-factory like the two-photon interaction and the double charmonium production. Currently there are many predicted intermediate states to be investigated and Belle II can provide the needed sensitivity to do that.

Using a part of the Phase 2 data, Belle II has been able to obtain several known results like the K_S mass reconstruction as shown in Fig. 8 left. It is also important to underline that the instant luminosity reached during Phase 2 was of $\sim 10^{33} cm^{-2} s^{-1}$, obtained by squeezing the beams up to a value of β of 6 mm along y direction at the interaction point (Fig. 8 right).

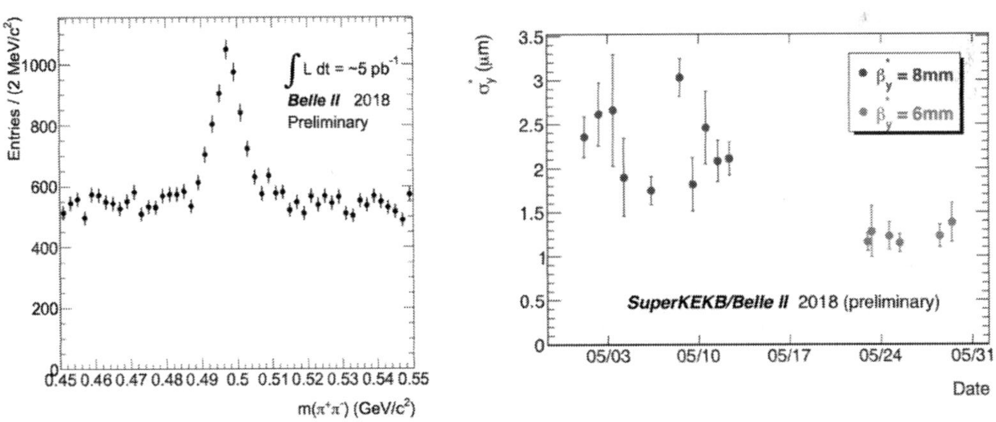

Figure 8. K_S mass distribution (left) and vertical beam size at the interaction point vs time (right).

2.2 Conclusions

The Belle II experiment expected to collect an huge dataset of 50 ab^{-1} which will largely extend our knowledge of flavour physics. The experiment successfully completed the accelerator commissioning and it already took some data with partial detector installed while the data taking will begin on February 2019. With the first collected data, Belle II already produced some known results and first physics studies are ongoing, mainly in the dark sector field.
The physics program of the experiment is wide. Belle II will be competitive not only in the B-physics field but also in the D and τ physics, as it will provide clean datasets *w.r.t.* experiments at hadron colliders. Belle II aims to search for NP signals and most of the sensitivity studies have already been performed by the collaboration [2].

References

[1] Y.Ohnishietal et al. 2013, Accelerator design at SuperKEKB, Prog. Theor. Exp. Phys. **03A011**, 17

[2] E. Kou *et al.*, The Belle II Physics book, 2018, **arXiv:1808.10567 [hep-ex]**.

[3] P. Raimondi, M. Zobov and D. Shatilov, Proceedings of EPAC08, Genoa, Italy, 2620 (2008).

[4] Abe T et al. 2010, Belle II Technical Design Report, **arXiv:1011.0352**, 480

[5] T. Keck et al., The Full Event Interpretation – An exclusive tagging algorithm for the Belle II experiment, **arXiv:1807.08680 [hep-ex]**.

Theory of rare K decays

Giancarlo D'Ambrosio[1,*]
[1] *INFN-Sezione di Napoli, Via Cintia, 80126 Napoli, Italia*

Abstract. I review rare kaon decays. I introduce the flavor problem and possible solutions. Very rare kaon decays like $K \to \pi\nu\bar{\nu}$ are very important to this purpose: we study also $K \to \pi l^+ l^-$, $K \to \pi\pi ee$ where chiral dynamics is important to disentangle short distance effects. We discuss also the decays $K^0 \to \mu^+\mu^-$, which have received recently some attention due to the measurement by LHCB.

1 Introduction and $K \to \pi\nu\bar{\nu}$

Rare kaon decays furnish challenging MFV probes and will severely constrain additional flavor physics motivated by NP [1]. SM predicts the $V - A \otimes V - A$ effective hamiltonian

$$\mathcal{H} = \frac{G_F}{\sqrt{2}} \frac{\alpha}{2\pi \sin^2\theta_W} (\underbrace{V_{cs}^* V_{cd} X_{NL}}_{\lambda x_c} + \underbrace{V_{ts}^* V_{td} X(x_t)}_{A^2\lambda^5 (1-\rho-i\eta) x_t}) \bar{s}_L \gamma_\mu d_L \, \bar{\nu}_L \gamma^\mu \nu_L, \quad (1)$$

$x_q = m_q^2/M_W^2$, θ_W the Weak angle and X's are the Inami-Lin functions with Wilson coefficients known at two-loop electroweak corrections and the main uncertainties is due to the strong corrections in the charm loop contribution. The structure in (1) leads to a pure CP violating contribution to $K_L \to \pi^0\nu\bar{\nu}$, induced only from the top loop contribution and thus proportional to $\Im m(\lambda_t)$ ($\lambda_t = V_{ts}^* V_{td}$) and free of hadronic uncertainties. This leads to the prediction

$$\mathcal{B}(K_L \to \pi^0\nu\bar{\nu})_{\rm SM} = (2.9 \pm 0.2) \times 10^{-11} \quad \mathcal{B}(K^+ \to \pi^+\nu\bar{\nu})_{\rm SM} = (8.3 \pm 0.9) \times 10^{-11},$$

where the parametric uncertainty due to the error on $|V_{cb}|$, ρ and η is shown.

Typical BSM predict new flavor structures that might affect $K \to \pi\nu\bar{\nu}$ that now can be tested at NA62 and KOTO [2]; we describe two different BSM effects i) new flavor structures for ϵ' avoiding $\Delta S = 2$ constraints (Fig. 1) [1, 3] and ii) attempts to describe B-anomalies [4], typically induce large flavor effects at O(1) TeV [5]. i) the recent lattice results for $K \to 2\pi$ leave open the possibility of BSM for ϵ'; to isospin breaking terms in $\Im(A_2)$ have been studied [3] in Fig.1. We expect effects at most 10% in $K^+ \to \pi^+\nu\bar{\nu}$ while are more sizable for $K_L \to \pi^0\nu\bar{\nu}$. Theoretically addressing flavor in Randall Sundrum models is more challenging: we have studied the so called flavor anarchy scenario with 5D MFV and custodial symmetry; the only sources of flavor breaking are two 5D anarchic Yukawa matrices. These matrices also generate also the bulk masses, which are responsible for the

*e-mail: gdambros@na.infn.it

© The Authors, published by EDP Sciences. This is an open access article distributed under the terms of the Creative Commons Attribution License 4.0 (http://creativecommons.org/licenses/by/4.0/).

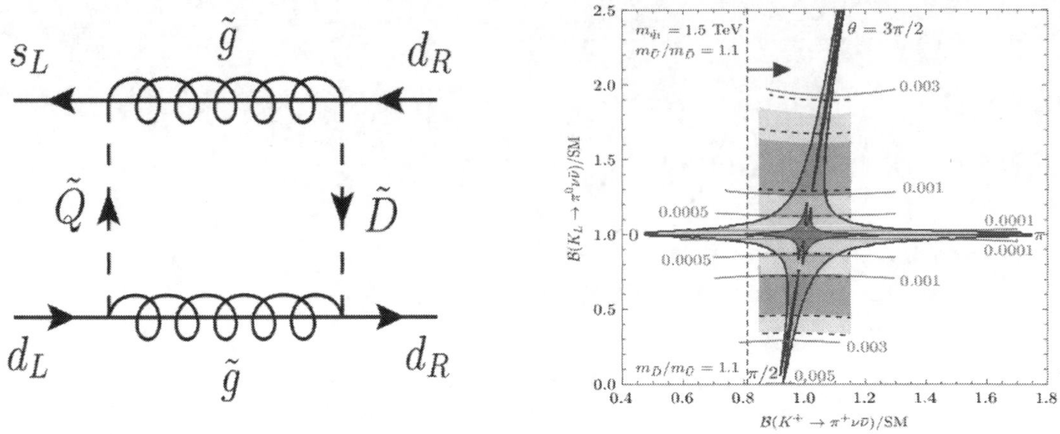

Figure 1. Impact of $K \to 2\pi$ isospin breaking terms ($\Im(A_2)$) (left) on $K \to \pi\nu\bar{\nu}$, typical effects, (right)

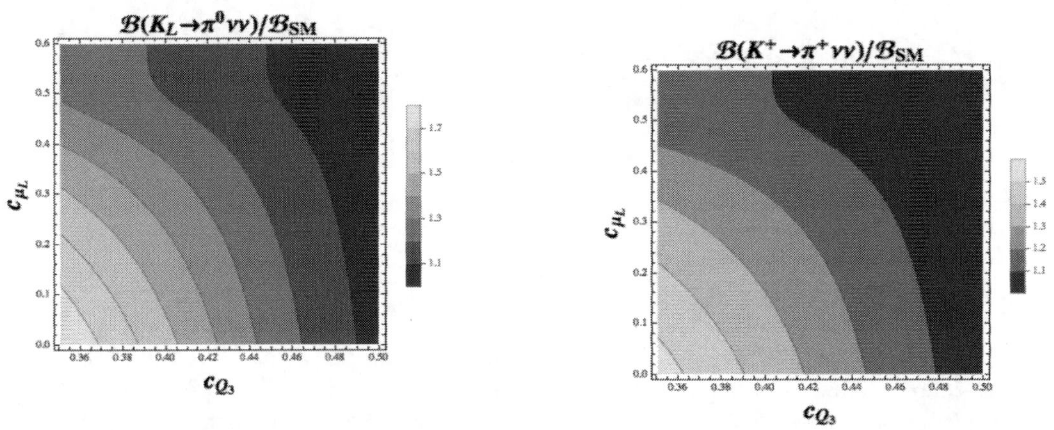

Figure 2. RS scenario to explain B-anomalies: $B(K \to \pi\nu\bar{\nu})$ ranges as a function of fermion profiles (c_i's)

resulting flavor hierarchy. The theory flows to a next to minimal flavor violation model where flavor violation is dominantly coming from the 3rd generation. We show that it is possible to find a range of parameters for bulk masses satisfying experimental flavor constraints, but also we explain the neutral B-anomalies, requiring NP flavor scale at O(1) TeV. Then we address $K \to \pi\nu\bar{\nu}$–decays: we show the TH predictions as a function f the bulk fermion masses in Fig.2 [5]. A natural issue is to test O(1) TeV physics at LHC; we are trying to apply the technique of Ref. [6] to this purpose.

2 $K_{L,S} \to \mu^+\mu^-$

Recent $K_S \to \mu\bar{\mu}$ LHCB measurement is very interesting and unexpected

$$B(K_S \to \mu\bar{\mu})_{LHCB} < 9 \times 10^{-9} \text{ at } 90\ \% \text{ CL} \qquad B(K_S \to \mu\bar{\mu})_{SM} = (5.0 \pm 1.5) \times 10^{-12}. \tag{2}$$

It represents an important milestone since it has improved the previous limit, $< 3.2\times 10^{-7}$ at 90 % CL, lasted 40 years. It is based on a production of 10^{13} K_S per fb^{-1} inside the LHCB acceptance and it is obtained using 1.0 fb^{-1} of pp collisions at $\sqrt{s} = 7$ TeV collected in 2011.

Two photon exchange generates the dominant contribution for both K_L and K_S decays to two muons [7]. The structure of weak and electromagnetic interactions entails a vanishing CP conserving short distance contribution to $K_S \to \mu^+\mu^-$. Indeed the SM short diagrams (similar to $K \to \pi\nu\bar{\nu}$ in Fig. 2) lead to the SM effective hamiltonian similar to eq. (1).

The LD contributions to $K_S \to \mu^+\mu^-$ Fig. (4) have been computed reliably in CHPT ($B = (5.0 \pm 1.5) \times 10^{-12}$). The relevant short distance contributions are

$$B(K_S \to \mu\bar{\mu})^{SD}_{SM} = 1 \times 10^{-5} |\Im(V^*_{ts} V_{td})|^2 \sim 10^{-13} \qquad \text{vs} \qquad B(K_S \to \mu\bar{\mu})_{NP} \leq 10^{-11} \tag{3}$$

We have shown that in some appealing susy scenario in Fig. (3) [8] large allowed new physics contributions (NP) can be substantially larger that SM SD contributions.

The short distance hamiltonian will contribute also to $K_L \to \mu\bar{\mu}$, through a CP conserving amplitude, $\Re(A_{\text{short}})$, that has to be disentangled from the large LD two-photon exchange contributions, $A_{\gamma\gamma}$: the absorptive LD contribution is much larger than SD, in the rate respectively 25 times larger than dispersive; total ($B_{\text{expt}} = (6.84 \pm 0.11) \times 10^{-9}$) To extract SD info the situation would be better if we would know the sign of $A_{\gamma\gamma}$, theoretically and experimentally unknown. While K_L−decays outside the LHCB fiducial volume the interference $A(K_L \to \mu\bar{\mu})^* A(K_S \to \mu\bar{\mu})$ may affect the LHCB K_S−rates: we can study the time interference $K_{S,L} \to \mu\mu$; this can be done by flavor tagging $K^0\bar{K}^0$, specifically by detecting the associated $\pi^\pm$ and (or) $K^\mp$, determining the impurity parameter $D = \frac{K^0 - \bar{K}^0}{K^0 + \bar{K}^0}$. Then interference term will affect the measured branching [7]:

$$\begin{aligned}\mathcal{B}(K^0_S \to \mu^+\mu^-)_{\text{eff}} = \tau_S \left(\int_{t_{\min}}^{t_{\max}} dt e^{-\Gamma_S t} \varepsilon(t) \right)^{-1} \Bigg[\int_{t_{\min}}^{t_{\max}} dt \Bigg\{ \Gamma(K^0_S \to \mu^+\mu^-) e^{-\Gamma_S t} \\ + \frac{D f_K^2 M_K^3 \beta_\mu}{8\pi} \text{Re} \left[i \left(A_S A_L - \beta_\mu^2 B_S^* B_L \right) e^{-i\Delta M_K t} \right] e^{-\frac{\Gamma_S + \Gamma_L}{2} t} \Bigg\} \varepsilon(t) \Bigg],\end{aligned} \tag{4}$$

Then we are i) increasing the sensitivity to short distance and ii) possibly determining the sign $A_{L\gamma\gamma}$

$$\sum_{\text{spin}} \mathcal{A}(K_1 \to \mu^+\mu^-)^* \mathcal{A}(K_2 \to \mu^+\mu^-) \sim \underbrace{\text{Im}[\lambda_t] y'_{7A}}_{SD} \left\{ \underbrace{A^\mu_{L\gamma\gamma}}_{LD} - 2\pi \sin^2\theta_W \left(\text{Re}[\lambda_t] y'_{7A} + \text{Re}[\lambda_c] y_c \right) \right\} \tag{5}$$

Experimentally, one can also access an *effective* branching ratio of $K^0_S \to \mu^+\mu^-$ [7] which includes an interference contribution with $K^0_L \to \mu^+\mu^-$ in the neutral kaon sample.

LHCB has also a beautiful kaon physics program [10–12]

	PDG	Prospects
$K_S \to \mu\mu$	$< 9 \times 10^{-9}$ at 90% CL	(LD)$(5.0 \pm 1.5) \cdot 10^{-12}$ NP $< 10^{-11}$
$K_L \to \mu\mu$	$(6.84 \pm 0.11) \times 10^{-9}$	difficult : SD << LD
$K_S \to \mu\mu\mu\mu$	−	SM LD $\sim 2 \times 10^{-14}$
$K_S \to ee\mu\mu$	−	$\sim 10^{-11}$
$K_S \to eeee$	−	$\sim 10^{-10}$
$K_S \to \pi^+\pi^-\mu^+\mu^-$	−	SM LD $\sim 10^{-14}$

(6)

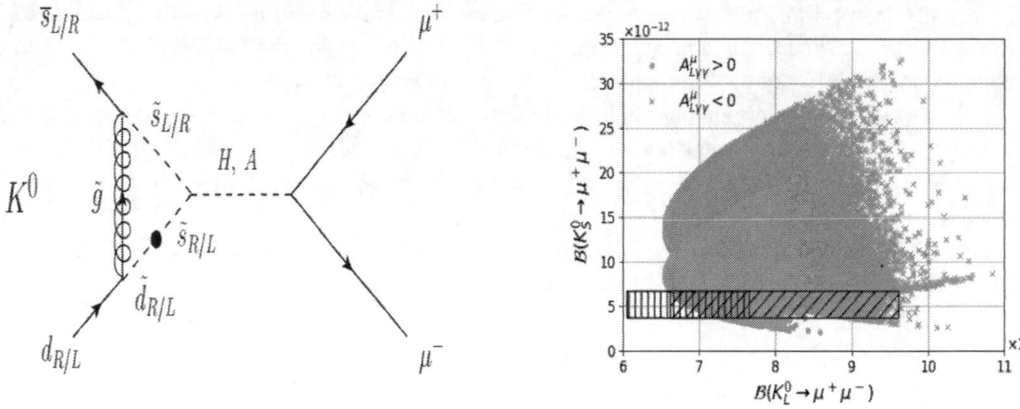

Figure 3. Susy scenario: $K_S \to \mu\mu$ diagram (left), theory predictions: in dashed area no interference effects are considered (right)

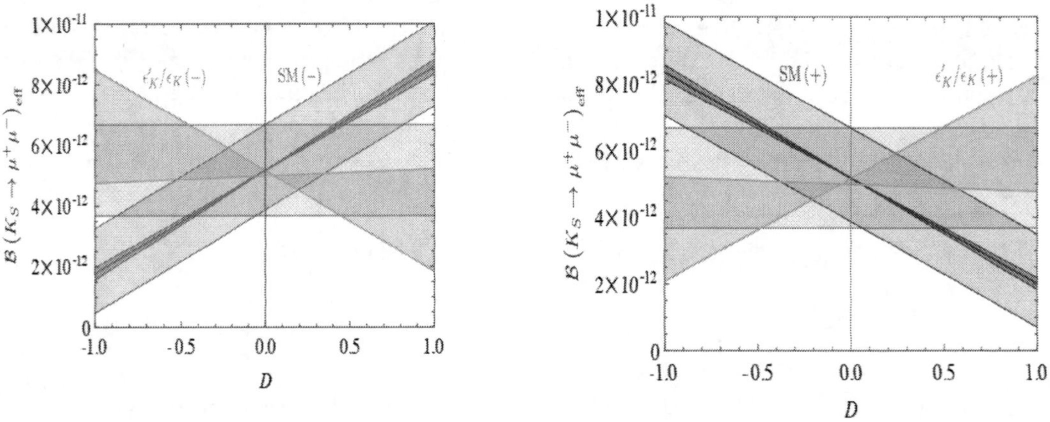

Figure 4. K_S LD diagram (left), interference effect from eq. 5 on $\mathcal{B}(K_S \to \mu^+\mu^-)$ depending on the $A_{L\gamma\gamma}$ sign: negative (center and in red SM while in green NP contributions) and positive (right and in blue SM while in green NP contributions).

3 The weak chiral lagrangian

In Ref. [9] we have studied how to determine the weak $O(p^4)$ chiral countertems in

$$\mathcal{L}_{\Delta S=1} = \mathcal{L}^2_{\Delta S=1} + \mathcal{L}^4_{\Delta S=1} + \cdots = G_8 F^4 \underbrace{\langle \lambda_6 D_\mu U^\dagger D^\mu U \rangle}_{K \to 2\pi/3\pi} + \underbrace{G_8 F^2 \sum_i N_i W_i}_{K^+ \to \pi^+\gamma\gamma, K \to \pi l^+ l^-} + \cdots .$$

Due to the accurate NA48/2 study of the decays $K^\pm \to \pi^\pm\pi^0\gamma$ and $K^\pm \to \pi^\pm\pi^0 e^+e^-$ the subset of CT's in the table can be determined

$$\begin{array}{llll} K^\pm \to \pi^\pm\gamma^* & N_{14}^r - N_{15}^r & a_+ = -0.578 \pm 0.016 & \text{NA48/2} \\ K_S \to \pi^0\gamma^* & 2N_{14}^r + N_{15}^r & a_S = (1.06^{+0.26}_{-0.21} \pm 0.07) & \text{NA48/1} \\ K^\pm \to \pi^\pm\pi^0\gamma & N_{14}^r - N_{15}^r - N_{16}^r - N_{17} & X_E = (-24 \pm 4 \pm 4)\ \text{GeV}^{-4} & \text{NA48/2} \\ K^+ \to \pi^+\gamma\gamma & N_{14}^r - N_{15}^r - 2N_{18}^r & \hat{c} = 1.56 \pm 0.23 \pm 0.11 & \text{NA48/2} \end{array} \quad (7)$$

4 $K_{S,L} \to l^+l^-$, $K_{S,L} \to l^+l^-l^+l^-$ and $K_S \to \pi^+\pi^-l^+l^-$

Recent LHCB limit on $K_S \to \mu\overline{\mu}$ in the table is close to test interesting New Physics (NP) models. A high precision measurement can test the short distance (SD) SM but it requires to improve the long distance (LD) prediction with auxiliary channels [10]. $K_L \to \mu\mu$: the small ratio SD/LD $\sim \frac{1}{30}$ may obscure an experimental improvement on the rate. The situation would be a bit ameliorated if the sign for $A(K_L \to \gamma\gamma)$ would be known. Help to this ambiguity could come from the experimental study of $K_{S,L} \to l^+l^-l^+l^-$ [10] As shown in table these channels are at reach in a high intensity machine and they may also give LD distance info needed for a better control of $K_L \to \mu\mu$. These four body decays have also a peculiar feature, similarly to $K_{S,L} \to \pi^+\pi^-e^+e^-$, the two different helicity amplitudes interfere; then one can measure the sign $K_L \to\to \gamma^*\gamma^* \to l^+l^-l^+l^-$ by studying the time interference K_S K_L which it has a decay length $2\Gamma_S$ [10].

The interplay between LHCB and NA62 program is nicely shown in ref [12].

$K_S \to \pi^+\pi^-l^+l^-$ Following the study of $K^+ \to \pi^+\pi^0 l^+l^-$in Ref. [13] we have studied the decay $K_S \to \pi^+\pi^-l^+l^-$ [9], that it has been studied by NA48/2 and it is a target of LHCB. One finds that the long-distance contributions to $K_S \to \pi^+\pi^-\gamma^*$ can be determined with remarkable accuracy, namely

$$BR(K_S \to \pi^+\pi^-e^+e^-) = \underbrace{4.74 \cdot 10^{-5}}_{\text{Brems.}} + \underbrace{4.39 \cdot 10^{-8}}_{\text{Int.}} + \underbrace{1.33 \cdot 10^{-10}}_{\text{DE}} . \quad (8)$$

This number is in excellent agreement with the PDG average [14]:

$$BR(K_S \to \pi^+\pi^-e^+e^-)_{exp} = (4.79 \pm 0.15) \times 10^{-5} . \quad (9)$$

Similarly, one can predict that

$$BR(K_S \to \pi^+\pi^-\mu^+\mu^-) = \underbrace{4.17 \cdot 10^{-14}}_{\text{Brems.}} + \underbrace{4.98 \cdot 10^{-15}}_{\text{Int.}} + \underbrace{2.17 \cdot 10^{-16}}_{\text{DE}} . \quad (10)$$

Acknowledgements

I would like to thank the organizers of the workshop QCD@work, in particular Fulvia De Fazio and Pietro Colangelo. I also acknowledge collaboration with L- Cappiello, O. Cata, A. Iyer, T. Kitahara, D: Maritinez Santos and K. Yamamoto.

References

[1] A. Crivellin, G. D'Ambrosio, T. Kitahara and U. Nierste, Phys. Rev. D **96**, no. 1, 015023 (2017) doi:10.1103/PhysRevD.96.015023 [arXiv:1703.05786 [hep-ph]].

[2] Riccardo Fantechi, these Proceedings

[3] T. Kitahara, U. Nierste and P. Tremper, Phys. Rev. Lett. **117**, no. 9, 091802 (2016) doi:10.1103/PhysRevLett.117.091802 [arXiv:1604.07400 [hep-ph]].

[4] T. Hurth, C. Langenbruch and F. Mahmoudi, JHEP **1711**, 176 (2017) doi:10.1007/JHEP11(2017)176 [arXiv:1708.04474 [hep-ph]].
[5] G. D'Ambrosio and A. M. Iyer, arXiv:1712.08122 [hep-ph].
[6] A. Chakraborty, A. M. Iyer and T. S. Roy, arXiv:1707.07084 [hep-ph].
[7] G. D'Ambrosio and T. Kitahara, Phys. Rev. Lett. **119**, no. 20, 201802 (2017) doi:10.1103/PhysRevLett.119.201802 [arXiv:1707.06999 [hep-ph]].
[8] V. Chobanova, G. D'Ambrosio, T. Kitahara, M. Lucio Martinez, D. Martinez Santos, I. S. Fernandez and K. Yamamoto, JHEP **1805**, 024 (2018) doi.org/10.1007/JHEP05(2018)024 [arXiv:1708.04474 [hep-ph]] arXiv:1711.11030 [hep-ph].
[9] L. Cappiello, O. Catà and G. D'Ambrosio, Eur. Phys. J. C **78**, no. 3, 265 (2018) doi:10.1140/epjc/s10052-018-5748-6 [arXiv:1712.10270 [hep-ph]].
[10] G. D'Ambrosio, D. Greynat and G. Vulvert, Eur. Phys. J. C **73**, 2678 (2013) [arXiv:1309.5736 [hep-ph]].
[11] V. Cirigliano, G. Ecker, H. Neufeld, A. Pich and J. Portoles, Rev. Mod. Phys. **84**, 399 (2012) doi:10.1103/RevModPhys.84.399 [arXiv:1107.6001 [hep-ph]].
[12] A. A. Alves Junior *et al.*, arXiv:1808.03477 [hep-ex].
[13] L. Cappiello, O. Cata, G. D'Ambrosio and D. N. Gao, Eur. Phys. J. C **72**, 1872 (2012) Erratum: [Eur. Phys. J. C **72**, 2208 (2012)] doi:10.1140/epjc/s10052-012-2208-6, 10.1140/epjc/s10052-012-1872-x [arXiv:1112.5184 [hep-ph]].
[14] M. Tanabashi *et al.* [ParticleDataGroup], Phys. Rev. D **98** (2018) no.3, 030001. doi:10.1103/PhysRevD.98.030001
[15] F. Feruglio, P. Paradisi and A. Pattori, Phys. Rev. Lett. **118** (2017) no.1, 011801 doi:10.1103/PhysRevLett.118.011801 [arXiv:1606.00524 [hep-ph]].
[16] A. Crivellin, G. D'Ambrosio, M. Hoferichter and L. C. Tunstall, Phys. Rev. D **93**, no. 7, 074038 (2016) doi:10.1103/PhysRevD.93.074038 [arXiv:1601.00970 [hep-ph]].
[17] E. Coluccio Leskow, G. D'Ambrosio, D. Greynat and A. Nath, Phys. Rev. D **93**, no. 9, 094031 (2016) doi:10.1103/PhysRevD.93.094031 [arXiv:1603.09721 [hep-ph]].

Recent results from NA62 and NA48

Riccardo Fantechi[1]; e-mail: fantechi@cern.ch

[1]INFN - Sezione di Pisa

Abstract. This paper reviews recent results form the NA62 and NA48 experiments at CERN. NA62 has searched for the $K \to \pi\nu\bar{\nu}$ with an in-flight decay technique. The analysis of $1.2 \cdot 10^{11}$ decays has found one event compatible with the Standard Model prediction, giving an upper limit for the branching ratio of $1.4 \cdot 10^{-9}$ (95% CL) in the background assumption. Prospects for improvements in the measurement are given. Results on the K_{l3} decay form factors from NA48/2 are presented as well as the first observation of the decay $K^{\pm} \to \pi^{\pm}\pi^{0}e^{+}e^{-}$ and the measurement of its branching ratio from the same experiment.

1 Introduction

1.1 The quest for $K^{+} \to \pi^{+}\nu\bar{\nu}$ and $K_{L} \to \pi^{0}\nu\bar{\nu}$

The ultra rare decays $K \to \pi\nu\bar{\nu}$ are, among the many rare flavour changing neutral current K and B decays, a powerful tool to search for new physics through underlying mechanisms of flavour mixing. Several circumstances allow to compute the SM branching ratio to very high precision:

- the $O(G_F^2)$ electroweak amplitudes exhibit a power-like GIM mechanism

*for the NA62 Collaboration: R. Aliberti, F. Ambrosino, R. Ammendola, B. Angelucci, A. Antonelli, G. Anzivino, R. Arcidiacono, M. Barbanera, A. Biagioni, L. Bician, C. Biino, A. Bizzeti, T. Blazek, B. Bloch-Devaux, V. Bonaiuto, M. Boretto, M. Bragadireanu, D. Britton, F. Brizioli, M.B. Brunetti, D. Bryman, F. Bucci, T. Capussela, A. Ceccucci, P. Cenci, V. Cerny, C. Cerri, B. Checcucci, A. Conovaloff, P. Cooper, E. Cortina Gil, M. Corvino, F. Costantini, A. Cotta Ramusino, D. Coward, G. D'Agostini, J. Dainton, P. Dalpiaz, H. Danielsson, N. De Simone, D. Di Filippo, L. Di Lella, N. Doble, B. Dobrich, F. Duval, V. Duk, J. Engelfried, T. Enik, N. Estrada-Tristan, V. Falaleev, R. Fantechi, V. Fascianelli, L. Federici, S. Fedotov, A. Filippi, M. Fiorini, J. Fry, J. Fu, A. Fucci, L. Fulton, E. Gamberini, L. Gatignon, G. Georgiev, S. Ghinescu, A. Gianoli, M. Giorgi, S. Giudici, F. Gonnella, E. Goudzovski, C. Graham, R. Guida, E. Gushchin, F. Hahn, H. Heath, T. Husek, O. Hutanu, D. Hutchcroft, L. Iacobuzio, E. Iacopini, E. Imbergamo, B. Jenninger, K. Kampf, V. Kekelidze, S. Kholodenko, G. Khoriauli, A. Khotyantsev, A. Kleimenova, A. Korotkova, M. Koval, V. Kozhuharov, Z. Kucerova, Y. Kudenko, J. Kunze, V. Kurochka, V.Kurshetsov, G. Lanfranchi, G. Lamanna, G. Latino, P. Laycock, C. Lazzeroni, M. Lenti, G. Lehmann Miotto, E. Leonardi, P. Lichard, L. Litov, R. Lollini, D. Lomidze, A. Lonardo, P. Lubrano, M. Lupi, N. Lurkin, D. Madigozhin, I. Mannelli, G. Mannocchi, A. Mapelli, F. Marchetto, R. Marchevski, S. Martellotti, P. Massarotti, K. Massri, E. Maurice, M. Medvedeva, A. Mefodev, E. Menichetti, E. Migliore, E. Minucci, M. Mirra, M. Misheva, N. Molokanova, M. Moulson, S. Movchan, M. Napolitano, I. Neri, F. Newson, A. Norton, M. Noy, T. Numao, V. Obraztsov, A. Ostankov, S. Padolski, R. Page, V. Palladino, C. Parkinson, E. Pedreschi, M. Pepe, M. Perrin-Terrin, L. Peruzzo, P. Petrov, F. Petrucci, R. Piandani, M. Piccini, J. Pinzino, I. Polenkevich, L. Pontisso, Yu. Potrebenikov, D. Protopopescu, M. Raggi, A. Romano, P. Rubin, G. Ruggiero, V. Ryjov, A. Salamon, C. Santoni, G. Saracino, F. Sargeni, V. Semenov, A. Sergi, A. Shaikhiev, S. Shkarovskiy, D. Soldi, V. Sougonyaev, M. Sozzi, T. Spadaro, F. Spinella, A. Sturgess, J. Swallow, S. Trilov, P. Valente, B. Velghe, S. Venditti, P. Vicini, R. Volpe, M. Vormstein, H. Wahl, R. Wanke, B. Wrona, O. Yushchenko, M. Zamkovsky, A. Zinchenko.

© The Authors, published by EDP Sciences. This is an open access article distributed under the terms of the Creative Commons Attribution License 4.0 (http://creativecommons.org/licenses/by/4.0/).

- the top-quark loops largely dominate the matrix element
- the sub-leading charm-quark contributions have been computed at NNLO order [1]
- the hadronic matrix element can be extracted from the branching ratio of the $K^+ \to \pi^0 e^+ \nu$ decay, well known experimentally [2].

The current predictions for the branching ratios are [3]:

$$BR(K^+ \to \pi^+ \nu\bar{\nu}) = (8.4 \pm 1.0) \times 10^{-11}$$
$$BR(K_L \to \pi^0 \nu\bar{\nu}) = (3.4 \pm 0.6) \times 10^{-11}$$

The uncertanties are dominated by the experimental knowledge of the external inputs. These decays are sensitive to extensions of the SM, like for instance:

- MSSM analyses [6, 7]
- Littlest Higgs with T-parity [8]
- Simplified Z and Z' [9]
- Randall-Sundrum models with protective custodial symmetry [10]
- LFU violation models [11]

Thanks to the SM suppression and existing constraints from K physics, variations of the $K \to \pi\nu\bar{\nu}$ BRs from the SM predictions induced by new physics at mass scales up to 100 TeV/c^2 can be observed with a 10% precision.

The decay $K^+ \to \pi^+\nu\bar{\nu}$ has been observed by the experiments E787 and E949 at the Brookhaven National Laboratory and the measured branching ratio is $1.73^{+1.15}_{-1.05} \times 10^{-10}$ [12]. A limit $BR(K_L \to \pi^0\nu\bar{\nu}) < 2.6 \times 10^{-8}$ at 90%CL has been published by the E391a Collaboration[13].

1.2 Semileptonic kaon decays

The K_{l3} decay width in the absence of radiative corrections can be represented by the Dalitz plot density depending on E_l and E_π [14] and the vector and scalar form factors, $f_+(t)$ and $f_0(t)$. For K_{e3} decays only $f_+(t)$ will contribute. The following form factor parametrizations have been used in the analysis: the quadratic parametrization [15], the pole parametrization [16] and the dispersive parametrization [17]. An improvement on the precision of the form factors measurement contributes to reduce the uncertainty of $|V_{us}|$.

1.3 The decay $K^\pm \to \pi^\pm\pi^0 e^+ e^-$

Radiative kaon decays, dominated by long-distance effects, could bring information on short-distance processes considering specific modes in specific kinematical areas. The case of $K^\pm \to \pi^\pm\pi^0 e^+ e^-$ is relevant as the virtual photon giving the e^+e^- pair can be emitted either by Inner Bremsstrahlung (IB) or by Direct Emission (DE). These components, as well as their interference, offer interesting short and long-distance observables which can be studied in a kinematical space including the non-zero $q^2 = m^2_{ee}$ variable. Only few theoretical publications are available [18–20] and the decay has not yet been observed. The NA48/2 measurements of the magnetic and electric terms in the $K^\pm \to \pi^\pm\pi^0\gamma$ decay has been used in [18] to predict the branching ratio of the single components of the $K^\pm \to \pi^\pm\pi^0 e^+ e^-$ decay.

2 The NA62 experiment at CERN

NA62 aims to collect about 100 $K^+ \to \pi^+ \nu\bar{\nu}$ decays in two years of running, to measure the branching ratio with 10% precision, with an acceptance of ~ 10%. This measurement requires a beam line providing at least 10^{13} kaon decays, integrated over the data taking period. In addition a background rejection factor O(10^{12}) is required, to extract the signal in the overwhelming abundance of the main K^+ decay modes. Indeed $K^+ \to \pi^+\pi^0$ (21%) and $K^+ \to \mu^+\nu$ (63%) should be rejected using respectively photon veto detectors and muon identifiers. The use of a high energy kaon beam (75 GeV/c) has an advantage in the first case, because for $P_{\pi^+} < 35$ GeV/c the energy of the photons from the π^0 is larger than 40 GeV and the efficiency for photon rejection is very high. Additional particle ID is needed to eliminate other backgrounds and to identify the pa rent kaon inside the unseparated beam.

The beam line can provide the required intensity of about 3×10^{12} protons/pulse at an energy of 75 GeV with 1% momentum bite. The beam contains π^+ (70%), protons (23%) and K^+ (6%). The rate seen by the detectors along the beam line, integrated over a surface of 12.5 cm^2, is about 750 MHz. The decay region is defined to be a 80 m volume downstream of the last beam line element. The decay volume is evacuated down to 10^{-6} mbar, to minimize the multiple scattering of the decay products and the number of interactions of the beam with the residual gas. The rate downstream is given mainly by kaon decay products and is about 10 MHz.

The layout of the detector is shown in Fig.1; the description of the apparatus can be found in [21].

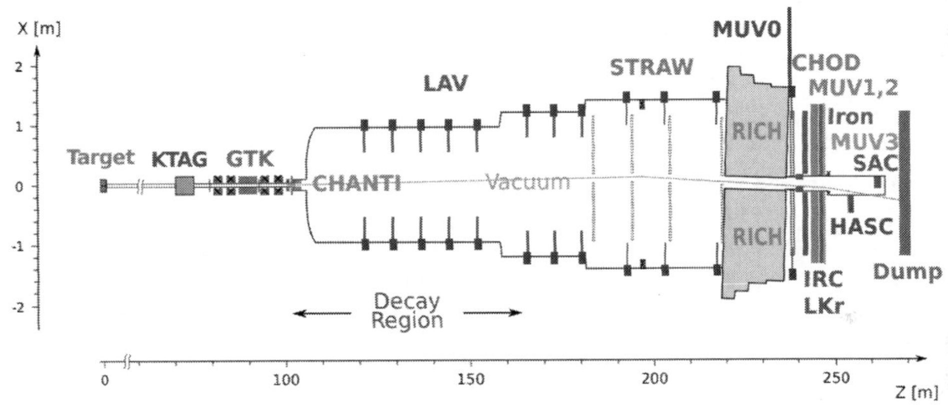

Figure 1. Horizontal view of the NA62 detector. The various labels refers to the detector elements.

3 $K^+ \to \pi^+\nu\bar{\nu}$ analysis strategy

The analysis is based for the time being on a cut and count method, defining signal regions using the squared missing mass, $m^2_{miss} = (p_K - p_\pi)^2$ (Fig. 2), assuming the output particle is a π^+ and computed for events with a single Straw track and for events with more than one track if no combination gives a vertex. Events with Straw track momentum larger than 35 GeV/c are rejected to better suppress the $K^+ \to \pi^+\pi^0$, as lower track momentum implies higher π^0 energy, better detectable. A lower cut in track momentum of 15 GeV/c is applied, as in this case the pion is below the threshold in the RICH.

Events with hits in either the LAV, IRC, SAC and HASC in time with the π^+ are rejected, as well as those with extra clusters in the LKr if they are more than 10 cm away from the

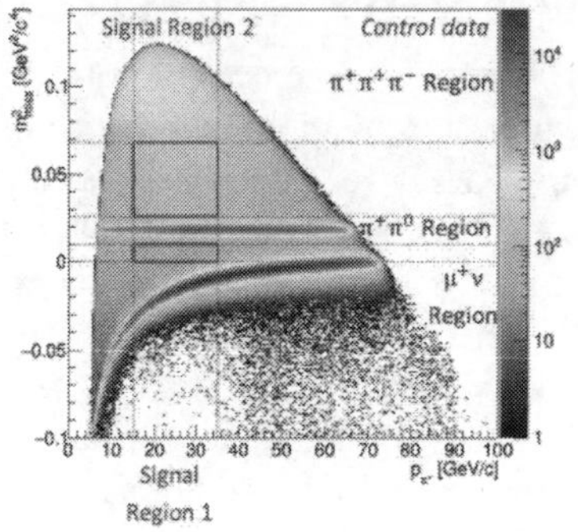

Figure 2. Kaon decays squared missing mass versus Straw track momentum.

Table 1. Expected signal and background events for the 2016 data

Process	Expected events in signal regions
$K^+ \to \pi^+ \nu\bar{\nu}$	$0.267 \pm 0.001_{stat} \pm 0.02_{syst} \pm 0.032_{ext}$
$K^+ \to \pi^+\pi^0(\gamma)IB$	$0.064 \pm 0.007_{stat} \pm 0.006_{syst}$
$K^+ \to \pi^+\nu(\gamma)IB$	$0.020 \pm 0.003_{stat} \pm 0.003_{syst}$
$K^+ \to \pi^+\pi^-e^+\nu$	$0.018^{+0.024}_{-0.017}\vert_{stat} \pm 0.009_{syst}$
$K^+ \to \pi^+\pi^+\pi^-$	$0.002 \pm 0.001_{stat} \pm 0.002_{syst}$
Upstream Background	$0.050^{+0.090}_{-0.030}\vert_{stat}$
Total Background	$0.15 \pm 0.09_{stat} \pm 0.01_{syst}$

impact point of the π^+. Muon background is rejected using the combined information from the calorimeters and the RICH.

The single event sensitivity (SES) is $SES = (3.15 \pm 0.01_{stat} \pm 0.24_{syst}) \cdot 10^{-10}$, where the major contribution to the systematic error is due to the estimation of the loss due to random veto, which will improve in the future. Two signal regions are defined: Region I between the $K^+ \to \mu^+\nu$ and $K^+ \to \pi^+\pi^0$ peaks and Region II between the $K^+ \to \pi^+\pi^0$ peak and the area populated by $K^+ \to \pi^+\pi^+\pi^-$. The signal acceptance is 4%, of which 1% in Region I and 3% in Region II.

Table 1 gives the number of expected signal and background events. One event is found in Region II (Fig. 3): the ring of the decay track in the RICH is shown in Fig. 4, consistent with a pion track.

In the assumption that the event is background, a 95% upper limit could be placed

$$BR(K^+ \to \pi^+\nu\bar{\nu}) < 14 \times 10^{-10}$$

If the candidate is taken to be signal, the corresponding branching ratio at 68% CL is

$$BR(K^+ \to \pi^+\nu\bar{\nu}) < 28^{+44}_{-23} \times 10^{-11}$$

This result is not competitive with the previous ones, but it shows that the in-flight technique works. NA62 has accumulated during 2017 20 times more statistics and it is continuing to take data in 2018, accumulating at the end of run about 20 SM events. The collaboration is seeking approval to continue the data taking after LS2 to complete the measurement with a precision of 10%.

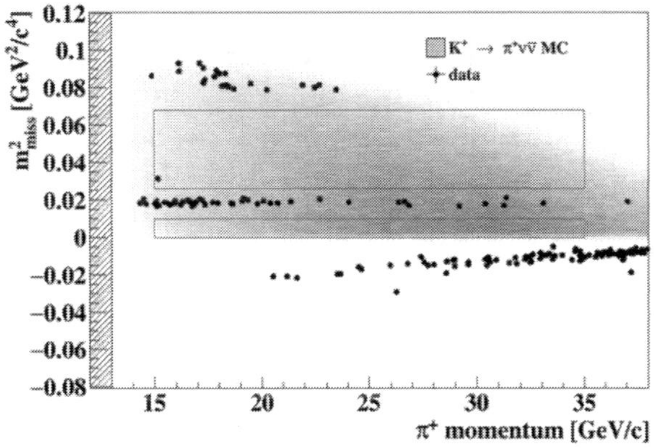

Figure 3. The events passing the analysis cuts, with the candidate in Region II

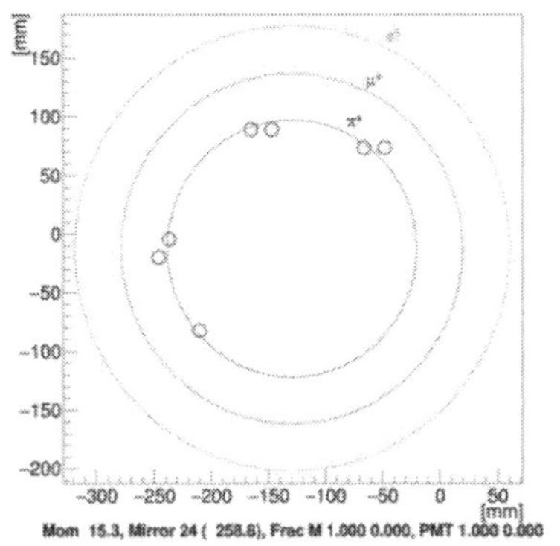

Figure 4. RICH hits for the candidate event

4 Form factors of $K^{\pm} \to \pi^0 l^{\pm} \nu$ decays

The total number of events on the final samples of data taken with the NA48 detector [22] are 4.278×10^6 K_{e3} and 2.907×10^6 $K_{\mu 3}$.

The measurements of the form factors are obtained by minimizing the difference between data (background subtracted) and simulated Dalitz plots in 5×5 MeV^2 bins. The semileptonic radiative Monte Carlo samples have been simulated with a specific generator [27].

NA48/2 is the first experiment measuring the form factors using both K^+ and K^-. The $K_{\mu 3}$ result is dominated by the statistical error, the K_{e3} by the uncertainty on background. The $K_{\mu 3}$ and K_{e3} are in agreement within each other and the combined results are competitive with the current world average. To avoid partially correlated systematic uncertainties in the averaging of the $K_{\mu 3}$ and K_{e3} results, the complete analysis has been repeated considering the two decay modes as a unique data set, containing two Dalitz plots simultaneously fitted with a common set of form factor parameters. Fig. 5 shows the 39% confidence contours for the joint form factors using the quadratic parametrization, together with earlier measurements.

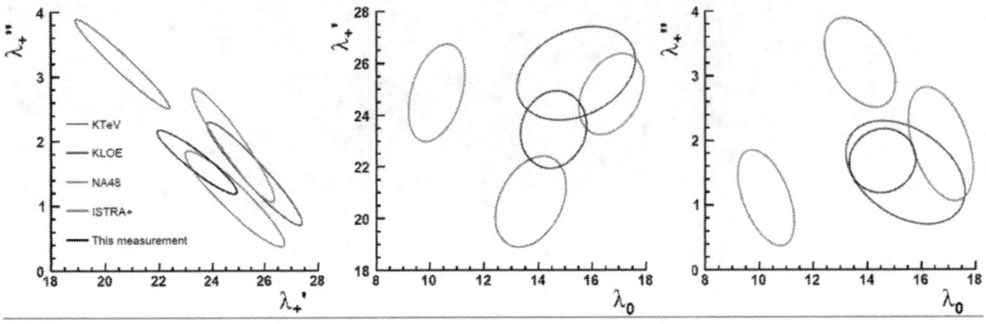

Figure 5. Confidence levels for the quadratic parametrization of the joint form factors. The ellipses correspond to 39%CL (one sigma contour).

5 First observation of the decay $K^\pm \to \pi^\pm\pi^0 e^+ e^-$

The main background contributions to the $K^\pm \to \pi^\pm\pi^0 e^+ e^-$ signal are $K_{3\pi D}$ where one photon is lost or merged with another particle and $K_{2\pi D}$ where an extra photon combines with the Dalitz photon to mimic a $\pi^0 \to \gamma\gamma$ decay. The first background is suppressed with a cut on the squared invariant mass of the $\pi^+\pi^0$ system to be greater than 0.12 GeV/c^2. The second one with the condition that the two possible invariant masses $m_{ee\gamma}$ were more than 7 MeV/c^2 away from the mass of the π^0. A total number of 5076 candidates has been selected. $\pi^+\pi^0_D$ is used as normalization with a total of 16.8 million candidates.

A detailed simulation based on GEANT3[24] has been used to compute the acceptances for the signal and the background, as well as for the determination of the background. Following the approach in [18], separate acceptances were computed for each component of the decay and then combined with the expected weights [18][23]. Several generators [25][26][27] have been used.

The most relevant background contributions to the signal are $K_{3\pi D}$ with 159 ± 8 events and $K_{2\pi D}$ with 130 ± 24 events (Fig. 6). For the normalization, $K_{\mu 3D}$ contributes with 15182 ± 173 events, while K_{e3D} with 10334 ± 140 events (Fig. 7).

The branching ratio is computed with

$$BR(K^\pm \to \pi^\pm\pi^0 e^+ e^-) = BR_{norm} \cdot \frac{(N_S - N_{BS}) \cdot A_N \cdot \epsilon_{L1N} \cdot \epsilon_{L2N}}{(N_N - N_{BN}) \cdot A_S \cdot \epsilon_{L1S} \cdot \epsilon_{L2S}}$$

The preliminary result is

$$BR(K^\pm \to \pi^\pm\pi^0 e^+ e^-) = (4.22 \pm 0.06_{stat} \pm 0.04_{syst} \pm 0.13_{ext}) \times 10^{-6}$$

Predictions from [18][23] are 4.183×10^{-6} for IB only, 4.229×10^{-6} including DE and INT terms and 4.145×10^{-6} including also isospin breaking corrections, compatible with our result within the experimental errors.

6 Conclusions

NA62 has demonstrated the goodness of the in-flight technique to search for $K^+ \to \pi^+\nu\bar{\nu}$ decays. With the analysis of the 2016 data, one candidate has been found with an expected background of 0.15 events. A 95% CL limit could be placed on the branching ratio as $BR(K^+ \to \pi^+\nu\bar{\nu}) < 14 \times 10^{-10}$.

Precise form factors for the semileptonic decays of $K^\pm$ have been obtained from the analysis of the NA48/2 data.

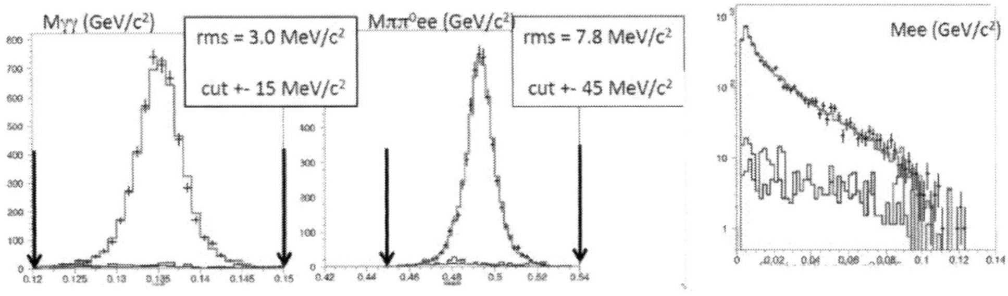

Figure 6. Data-MC comparison for the signal: reconstructed invariant masses for π^0 (left), for the kaon (center) and for m_{ee} in log scale.

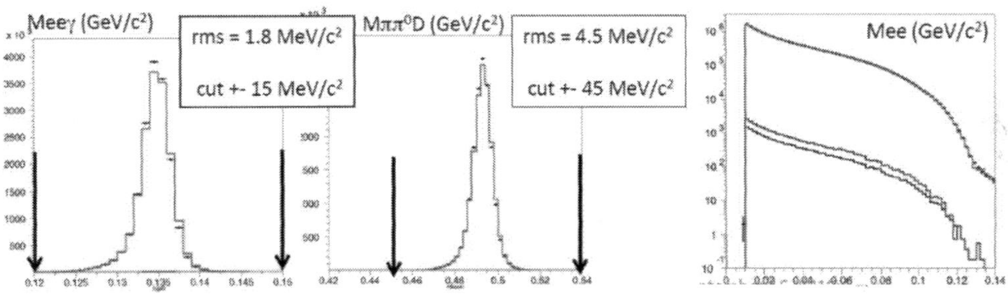

Figure 7. Normalization mode. Left: $M_{\pi^0 D}$. Center: $M_{\pi\pi^0_D}$. Right: m_{ee} spectrum in log scale.

NA48/2 reported the first observation of the decay $K^{\pm} \rightarrow \pi^{\pm}\pi^0 e^+ e^-$ and measured a preliminary branching ratio, in agreement with theoretical predictions.

References

[1] A. J. Buras, M. Gorbahn, U. Haisch and U. Nierste, JHEP **0611** (2006) 002.
[2] C. Amsler et al., Phys. Lett. **B667** (2008) 1.
[3] A.J. Buras, D. Buttazzo, J. Girrbach-Noe and R. Knegjens, JHEP **1511**, 33 (2015).
[4] M. Blanke, A.J. Buras and S. Recksiegel, Eur. Phys. J. **C76** (2016) 182.
[5] M. Blanke, A.J. Buras, B. Duiling, K. Gemmler and S. Gori, JHEP **0903** (2009) 108.
[6] T. Blazek and P. Matak, Int. J. Mod. Phys. A **29** (2014) no. 27, 1450162.
[7] G. Isidori, F. Mescia, P. Paradisi, C. Smith and S. Trine, JHEP **0608** (2006) 064.
[8] M. Blanke, A.J. Buras and S. Recksiegel, Eur. Phys. J. **C76** (2016) 182.
[9] A.J. Buras, D. Buttazzo and R. Knegjens, JHEP **1511**(2015) 166.

[10] M. Blanke, A.J. Buras, B. Duiling, K. Gemmler and S. Gori, JHEP **0903** (2009) 108.
[11] M. Bordone, D. Buttazzo, G. Isidori and J. Monnard, Eur. Phys. J. **C77** (2017) 618.
[12] S. Adler et al., Phys. Rev. **D 79** (2009) 092004.
[13] J.K. Ahn et al., Phys. Rev. **D 81** (2010) 072004.
[14] L.M. Chounet, J.M. Gaillard, M. Gaillard, Phys.Rept. **4**, (1972) 199.
[15] C. Patrignani et al. (Particle Data Group), Chin. Phys. **C 40**, (2016) 100001.
[16] P. Lichard, Phys. Rev. D **D 55**, (1997) 5385.
[17] V. Bernard, M. Oertel, E. Passemar, J. Stern, Phys. Rev. **D 80**, (2009) 034034.
[18] L. Cappiello, O. Cata, G. D'Ambrosio and D. N. Gao, Eur. Phys. J. C **72** (2012) 1872.
[19] H. Pichl, Eur. Phys. J. C **20** (2001) 371.
[20] S. R. Gevorkyan and M. H. Misheva, Eur. Phys. J. C **74** (2014) 2860.
[21] E. Cortina Gil et al. [NA62 Collaboration], JINST **12** (2017) P05025.
[22] V. Fanti et al., Nucl.Instrum. Meth. **A574**, (2007) 433.
[23] L. Cappiello, O. Cata and G. D'Ambrosio, [arXiv:17012.10270 [hep-ph]].
[24] GEANT3 Detector Description & Simulation Tool CERN Program Library **W5013** (1994).
[25] E. Barberio and Z. Was, Comput. Phys. Commun. **79** (1994) 291.
[26] T. Husek, K .Kampf and J. Novotny, Phys. Rev. D **92** (2015) 054027.
[27] C. Gatti, Eur. Phys. J. **C45** (2006) 417.

Long-distance QCD effects in FCNC $B \to \gamma l^+ l^-$ decays

Anastasiia Kozachuk[1,2,*], *Dmitri* Melikhov[2,3,4,**], and *Nikolai* Nikitin[1,2,5,***]

[1]M. V. Lomonosov Moscow State University, Faculty of Physics, 119991 Moscow, Russia
[2]D. V. Skobeltsyn Institute of Nuclear Physics, M. V. Lomonosov Moscow State University, 119991 Moscow, Russia
[3]Institute for High Energy Physics, Austrian Academy of Sciences, Nikolsdorfergasse 18, A-1050 Vienna, Austria
[4]Faculty of Physics, University of Vienna, Boltzmanngasse 5, A-1090 Vienna, Austria
[5]A. I. Alikhanov Institute for Theoretical and Experimental Physics, 117218 Moscow, Russia

Abstract. This presentation reviews the main results of our recent work [1] on rare radiative leptonic decays $B_{d,s} \to \gamma\mu^+\mu^-$ and $B_{d,s} \to \gamma e^+e^-$ induced by flavour-changing neutral currents (FCNC) in the Standard Model.

Rare FCNC decays of B-mesons are forbidden at the tree level in the Standard Model and occur only via loop diagrams. Respectively, their branching ratios are very small, of order $10^{-8} - 10^{-10}$ [2]. New particles can propagate in the loops and hence such processes are expected to be sensitive to the possible effects of New Physics. Several FCNC decays have been observed experimentally, exhibiting a few deviations from the Standard Model at the level of $2 - 3\,\sigma$ (see discussion in [3, 4]).

An appropriate framework for the theoretical description of FCNC B-decays is the effective field theory: virtual heavy particles of the SM with the masses much greater than m_b (i.e., W, Z, and t-quark) are integrated out yielding the $b \to q$ effective Hamiltonian [5–7]

$$H_{\rm eff}^{b\to q} = \frac{G_F}{\sqrt{2}} V_{tq}^* V_{tb} \sum_i C_i(\mu) O_i^{b\to q}(\mu). \tag{1}$$

The basis operators $O_i^{b\to q}(\mu)$ contain only dynamical light degrees of freedom (u, d, s, c, and b-quarks, leptons, photons and gluons). These light particles appear as dynamical degrees of freedom in the diagrams for the amplitudes of rare FCNC B-decays. The Wilson coefficients $C_i(\mu)$ absorb the contributions of the heavy particls (W, Z, and t in the SM) given by the box and the penguin digrams; taking account of hard gluon exchanges in the Feynman diagrams leads to the dependence of C_i on the scale μ.

Finally, the amplitude of the radiative leptonic B-decay is given by

$$A(B_q \to \gamma ll) = \langle \gamma ll | H_{\rm eff}^{b\to q} | B_q \rangle. \tag{2}$$

The presence of the B-meson in the initial state leads to the necessity of treating nonperturbative QCD effects; the amplitudes of FCNC rare leptonic B-decays involve a great variety of such nonperturbative QCD contributions.

*e-mail: anastasiia.kozachuk@cern.ch
**e-mail: dmitri_melikhov@gmx.de
***e-mail: nikolai.nikitine@cern.ch

© The Authors, published by EDP Sciences. This is an open access article distributed under the terms of the Creative Commons Attribution License 4.0 (http://creativecommons.org/licenses/by/4.0/).

Top-quark contributions

The diagrams generated by the t-quarks (as well as other heavy particles of the SM) in the loops are presented in Fig. 1 and Fig. 2. The B-decay amplitudes corresponding to these diagrams are described in terms of $B \to \gamma$ transition form factors:

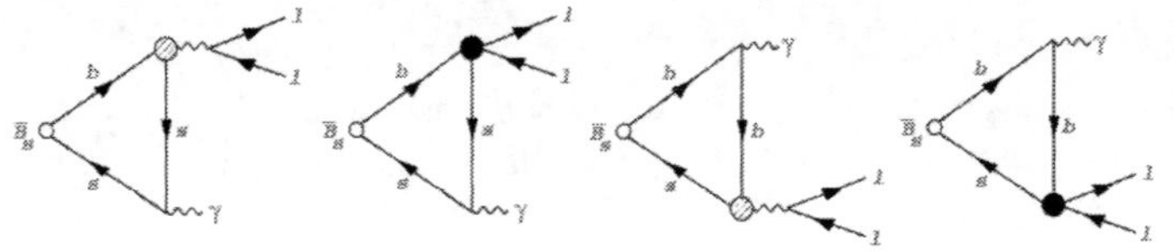

Figure 1. Real photon emission from the valence quark.

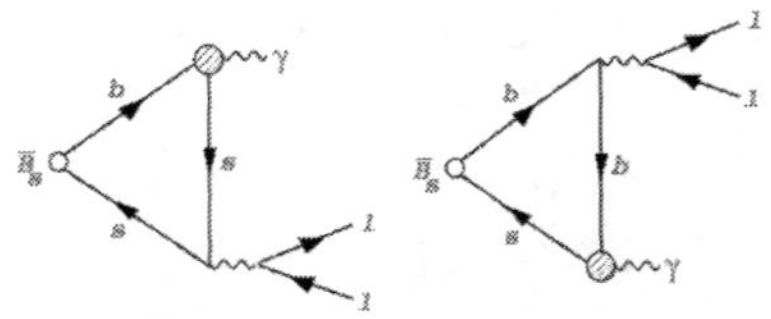

Figure 2. Real photon emission from the penguin FCNC vertex.

The diagrams of Fig. 1 yield contributions that have no singularities in the physical decay region. For these contributions, we made use of the dispersion approach based on the relativistic constituent quark picture [8–12]: within this approach, the form factors are given by relativistic spectral representations via the meson relativistic wave functions. The transition form factors from the dispersion approach satisfy all rigorous constraints known from QCD for these quantities [9, 13]. The model parameters (i.e., the constituent quark masses and the hadron wave functions) have been fixed by the requirement to reproduce the known values of meson weak decay constants [1]. The left diagram of Fig. 2 leads to the contributions to the form factors that do have singularities (resonances and cuts) in the physical decay region. To calculate these contributions, we combined the results from the dispersion approach at $q^2 < 0$ with the gauge-invariant version of vector meson dominance [14].

Finally, Fig. 3 displays the Bremsstrahlung contribution to the $B_{s,d} \to \gamma\ell^+\ell^-$ amplitude; it is given in terms of the B-meson decay constant f_B and is proportional to the mass of the lepton in the final state [15].

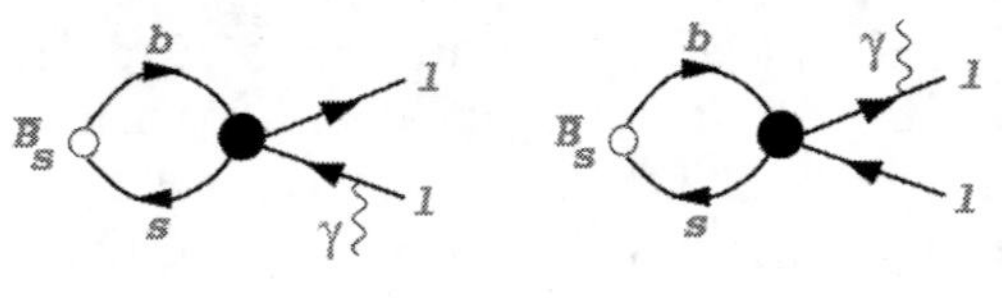

Figure 3. The Bremsstrahlung contribution.

Charm-quark contributions

Whereas heavy degrees of freedom (t, W, Z) are described by the effective Hamiltonian, light degrees of freedom, in particular c and u quarks, remain dynamical and their contributions in the loops should be taken into account. The diagrams for the charm-loop contributions are presented in Fig. 4: the numerically dominant penguin diagram (left) and the weak-annihilation diagram (right). When the W-boson line is shrinked to a point, and the soft-gluon exchanges between the charm-loop and the B-meson loop are taken into account, the charming penguin is reduced to two classes of diagrams shown in Fig. 5.

The structure of the charm-loop contributions to the $B_s \to \gamma l^+ l^-$ amplitude has the form:

$$H_{\mu\alpha}(k',k) = -\frac{G_F}{\sqrt{2}} V_{cb} V^*_{cs} e \left[\epsilon_{\mu\alpha k' k} H_V - i \left(g_{\alpha\mu}\, kk' - k'_\alpha k_\mu \right) H_A \right], \tag{3}$$

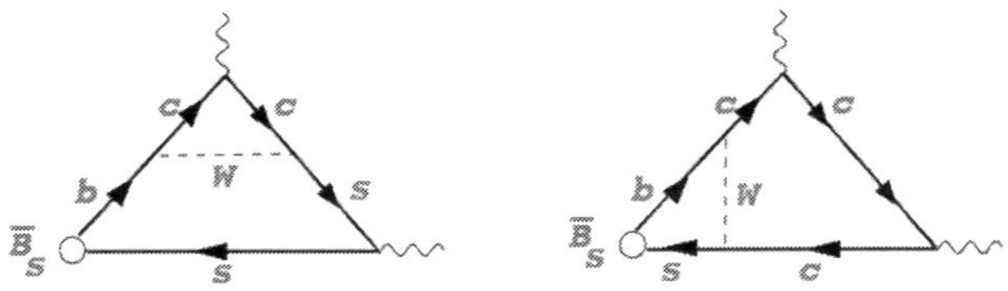

Figure 4. Diagrams with c-quarks in the loops: penguin diagram (left), weak annihilation diagram (right)

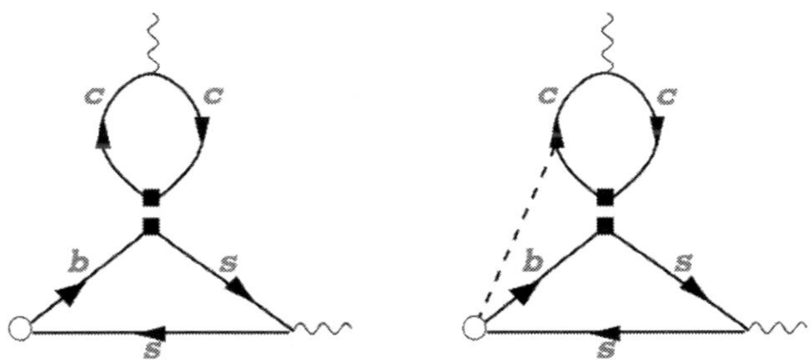

Figure 5. Factorizable (left) and nonfactorizable (right) charming loops. Dashed line in the right figure stands for the soft gluon.

with $0 < q^2 < M_B^2$, including the region of the charmonium resonances. Perturbative QCD cannot be applied here and non-perturbative approaches based on hadron degrees of freedom are necessary, see discussion in [16–23]. For $H_i(q^2,0)$ one may write dispersion representation in q^2 with two subtractions [18]:

$$H_i(q^2,0) = a_i + b_i q^2 + (q^2)^2 \left\{ \sum_{\psi=J/\psi,\psi'} \frac{f_\psi \mathcal{A}^i_{B\psi\gamma}}{m_\psi^3 (m_\psi^2 - q^2 - i m_\psi \Gamma_\psi)} + h_i(q^2) \right\}, \quad i = V, A, \quad (4)$$

where the functions $h_i(q^2)$ describe the hadron continuum including the broad charmonium states lying above the DD threshold. At $q^2 > 4M_D^2$, we take into account the known broad vector ψ_n ($n = 3, \ldots, 6$) resonances. The subtraction constants a_i and b_i are determined by matching to the known results from light-cone sum rules at $q^2 \le 4m_c^2$, including nonfactorizable corrections calculated in [18]. The absolute values of the amplitudes $\mathcal{A}^i_{B\psi\gamma}$ may be measured in $B \to \psi_i \gamma$ decays. It should be understood, however, that nonfactorizable gluons may generate complex relative phases between the contributions of different charmonia [19]. These possible non-universal (i.e., process-dependent and thus in general different in $B \to \gamma ll$ and $B \to Vll$) relative phases cannot be determined by pQCD-based calculations and provide one of the main sources of the theoretical uncertainties for rare radiative leptonic decays. Further details and subtleties see [1, 24].

Results

The differential branching ratios are shown in Fig. 6. The plots in Fig. 6 correspond to the description of the charm-loop effects according to Eq. (4), and further assuming that all charmonia contribute with the same positive sign (coinciding with the sign of the factorizable contribution).

In the region $q^2 \le 6$ GeV^2, the charming loops provide a small contribution at the level of a few percent, so the distributions in this region may be predicted with a good accuracy, mainly limited by the form-factor uncertainty. Our estimates [1] read

$$\begin{aligned} \mathcal{B}(\bar{B}_s \to \gamma l^+ l^-)|_{q^2 \in [1,6]\,\mathrm{GeV}^2} &= (6.01 \pm 0.08) 10^{-9} \\ \mathcal{B}(\bar{B}_d \to \gamma l^+ l^-)|_{q^2 \in [1,6]\,\mathrm{GeV}^2} &= (1.02 \pm 0.15) 10^{-11}. \end{aligned} \quad (5)$$

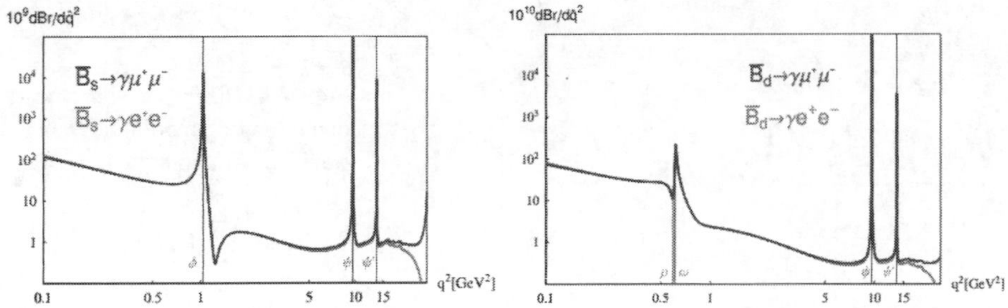

Figure 6. Differential branching fractions for $B_s \to \gamma l^+l^-$ (left) and $B_d \to \gamma l^+l^-$ (right) decays. Blue lines - $\mu^+\mu^-$ final state, red lines - e^+e^- final state.

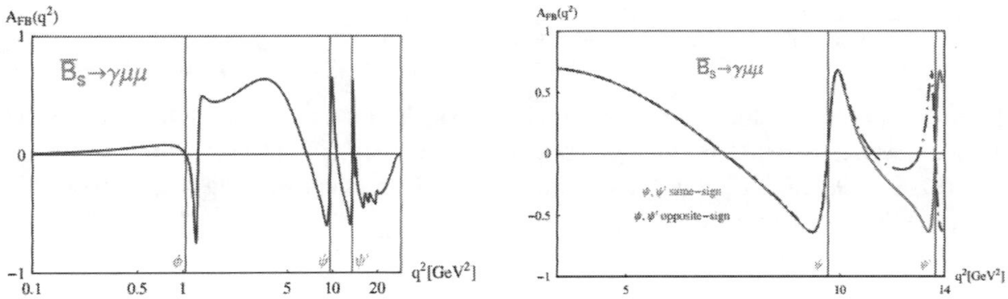

Figure 7. Forward-backward asymmetry for $B_s \to \gamma\mu\mu$ decays. The left plot shows the asymmetry for all q^2 for the contributions of all charmonia taken of the same positive sign, equal to that of the factorizable contribution; the right plot shows the sensitivity of A_{FB} to the relative signs of ψ and ψ': solid (red) line - both signs positive, dashed (blue) line - the ψ contribution taken positive, whereas the ψ' contribution negative.

For the $B_s \to \gamma l^+l^-$ transition, the dominant contribution is given by the ϕ-meson. Its parameters are known well, leading to the small uncertainty in the $B_s \to \gamma l^+l^-$ decay rate integrated over the range of $q^2 = [1, 6]$ GeV^2. For the $B_d \to \gamma l^+l^-$ transition, the known contribution of the vector resonances is less important, and the branching ratio uncertainty reflects to a large extent the 10% uncertainty in the form factor contributions given by Fig. 1.

The A_{FB} for $\bar{B}_s \to \gamma\mu\mu$ is shown in Fig. 7. The right plot exhibits the sensitivity of A_{FB} in the region between ψ and ψ' to the relative signs of the contributions of these states.

Further results and details see our paper [1].

Acknowledgments

D. M. thanks the Organizers of QCD@work for their warm hospitality in Matera and FWF for finacial support under project P29028. The results given in Eq. (4) were obtained with support of grant 16-12-10280 of the Russian Science Foundation (A. K. and N. N.).

References

[1] A. Kozachuk, D. Melikhov and N. Nikitin, *Rare FCNC radiative leptonic $B_{s,d} \to \gamma l^+l^-$ decays in the standard model*, Phys. Rev. D **97**, no. 5, 053007 (2018) [arXiv:1712.07926]

[2] A. Ali, *B decays, flavor mixings and CP violation in the SM*, [hep-ph/9606324].

[3] D. Guadagnoli, D. Melikhov and M. Reboud, *More Lepton Flavor Violating Observables for LHCb's Run 2*, Phys. Lett. B **760**, 442 (2016) [arXiv:1605.05718].

[4] D. Guadagnoli, *Flavor anomalies on the eve of the Run-2 verdict*, Mod. Phys. Lett. A**32**, 1730006 (2017) [arXiv:1703.02804].

[5] B. Grinstein, M. J. Savage and M. B. Wise, *$B \to X(s)e^+e^-$ in the Six Quark Model*, Nucl. Phys. B **319** 271 (1989).

[6] A. J. Buras and M. Munz, *Effective Hamiltonian for $B \to X(s)e^+e^-$ beyond leading logarithms in the NDR and HV schemes*, Phys. Rev. D **52**, 186 (1995) [hep-ph/9501281].

[7] G. Buchalla, A. J. Buras, and M. E. Lautenbacher, *Weak decays beyond leading logarithms*, Rev. Mod. Phys. **68**, 1125 (1996) [hep-ph/9512380].

[8] D. Melikhov, *Form-factors of meson decays in the relativistic constituent quark model*, Phys. Rev. D **53**, 2460 (1996) [hep-ph/9509268].

[9] D. Melikhov, *Heavy quark expansion and universal form-factors in quark model*, Phys. Rev. D **56**, 7089 (1997) [hep-ph/9706417].

[10] D. Melikhov, *Dispersion approach to quark binding effects in weak decays of heavy mesons*, Eur. Phys. J. direct **C4**, 2 (2002) [hep-ph/0110087].

[11] M. Beyer and D. Melikhov, *Form-factors of exclusive $b \to u$ transitions*, Phys. Lett. B **436**, 344 (1999) [hep-ph/9807223].

[12] D. Melikhov and B. Stech, *Weak form-factors for heavy meson decays: An Update*, Phys. Rev. D **62**, 014006 (2000) [hep-ph/0001113].

[13] F. Kruger and D. Melikhov, *Gauge invariance and form-factors for the decay $B \to \gamma l^+l^-$*, Phys. Rev. D **67**, 034002 (2003) [hep-ph/0208256].

[14] D. Melikhov, O. Nachtmann, V. Nikonov and T. Paulus, *Masses and couplings of vector mesons from the pion electromagnetic, weak, and pi gamma transition form-factors*, Eur. Phys. J. **C34**, 345 (2004) [hep-ph/0311213].

[15] D. Melikhov and N. Nikitin, *Rare radiative leptonic decays $B(d,s) \to \gamma l^+l^-$*, Phys. Rev. D **70**, 114028 (2004) [hep-ph/0410146].

[16] A. Ali, T. Mannel, T. Morozumi, *Forward backward asymmetry of dilepton angular distribution in the decay $b \to sl^+l^-$*, Phys. Lett. B **273**, 505 (1991).

[17] M. Beneke, G. Buchalla, M. Neubert, and C. T. Sachrajda, *Penguins with Charm and Quark-Hadron Duality*, Eur. Phys. J. **C61**, 439 (2009) [arXiv:0902.4446].

[18] A. Khodjamirian, T. Mannel, A. Pivovarov, and Y.-M. Wang, *Charm-loop effect in $B \to K^{(*)}l^+l^-$ and $B \to K^*\gamma$*, JHEP **09**, 089 (2010) [arXiv:1006.4945].

[19] J. Lyon and R. Zwicky, *Resonances gone topsy turvy - the charm of QCD or new physics in $b \to sl^+l^-$?*, [arXiv:1406.0566].

[20] M. Ciuchini, M. Fedele, E. Franco, S. Mishima, A. Paul, L. Silvestrini, and M. Valli, *$B \to K^*l^+l^-$ decays at large recoil in the Standard Model: a theoretical reappraisal*, JHEP **1606**, 116 (2016) [arXiv:1512.07157].

[21] S. Brass, G. Hiller, and I. Nisandzic, *Zooming in on $B \to K^{(*)}l^+l^-$ decays at low recoil*, Eur. Phys. J. C**77**, 16 (2017) [arXiv:1606.00775].

[22] LHCb Collaboration, R. Aaij *et al.*, *Measurement of the phase difference between short- and long-distance amplitudes in the $B^+ \to K^+\mu^+\mu^-$ decay*, Eur. Phys. J. C**77**, 161 (2017) [arXiv:1612.06764].

[23] T. Blake, U. Egede, P. Owen, G. Pomery, K. Petridis, *An empirical model of the long-distance contributions to $\bar{B}^0 \to \bar{K}^{*0}\mu^+\mu^-$* [arXiv:1709.03921].

[24] A. Kozachuk and D. Melikhov, *Nonfactorizable charm-loop effects in rare FCNC B-decays*, [arXiv:1805.05720].

EPJ Web of Conferences **192**, 00032 (2018) https://doi.org/10.1051/epjconf/201819200032
QCD@Work 2018

Prospects for Higgs Boson Measurements and Beyond Standard Model Physics at the High-Luminosity LHC with CMS

Rosamaria Venditti[1,*], for the CMS Collaboration

[1]INFN, Sezione di Bari, Via Orabona 4, Bari

Abstract. The High-Luminosity Large Hadron Collider (HL-LHC) is a major upgrade of the LHC, expected to deliver an integrated luminosity of up to 3000/fb over one decade. The very high instantaneous luminosity will lead to about 200 proton-proton collisions per bunch crossing (pileup) superimposed to each event of interest, therefore providing extremely challenging experimental conditions. The scientific goals of the HL-LHC physics program include precise measurement of the properties of the recently discovered standard model Higgs boson and searches for beyond the standard model physics (heavy vector bosons, SUSY, dark matter and exotic long-lived signatures, to name a few). In this contribution we will present the strategy of the CMS experiment to investigate the feasibility of such search and quantify the increase of sensitivity in the HL-LHC scenario.

1 Introduction

This paper presents physics studies and motivations that lead the development of the strategy of the future operation of the LHC collider with high luminosity (HL-LHC), also called LHC Phase-2. After the third long shutdown of the machine, that will take place around 2023, the instantaneous luminosity will be increased up to five times the design value, with 40 MHz operation (one bunch crossing every 25 ns) at a center of mass energy of 14 TeV. This yields challenging conditions for the CMS detector: up to 200 overlaying (pileup) events, high rates and high radiation levels, especially in the forward regions. An upgrade of the detectors is mandatory, in order to fully exploit the augmented luminosity and efficiently cope the new data taking conditions. It is foreseen that the total accumulated luminosity will reach 3000/fb of data, corresponding to ten times the data expected at the end of Phase-1. This environment offers great opportunities to shade some light on some unexplored or not fully explored physics searches. Test of the Standard Model can be performed with sufficient precision in the Higgs sector, by accurately measuring the couplings to fermions and bosons. Searches for the associated production of two Higgs boson and measurement of the trilinear Higgs boson coupling can be a unique opportunity to test the Higgs potential and is sensitive to the presence of new physics that can be explored in both resonant and non-resonant searches. The measurement of the differential cross-section of the Higgs boson is also an interesting test-bench of the Standard Model, providing that a percent precision can be reached.

*e-mail: rosamaria.venditti@cern.ch

© The Authors, published by EDP Sciences. This is an open access article distributed under the terms of the Creative Commons Attribution License 4.0 (http://creativecommons.org/licenses/by/4.0/).

The SM does not provide answers to the remaining questions. Those require new physics. In fact, the scalar nature of the particle, presents theoretical challenges. Radiative corrections to the Higgs should cause the mass to increase to very high values. New physics must appear at masses not too far from 1 TeV to cancel this growth. Deviations from perfect SM behavior because of its interaction with other forms of matter, including dark matter, could answer some very fundamental questions, such as the origin of the matter-antimatter asymmetry of the universe. The Supersymmetric (SUSY) particles, foreseen by the homonymous theory and still unobserved by date up to the TeV scale, could offer the advantage to cancel the the growth of the Higgs mass from radiative corrections. The rate of production and the characteristics of decays of SUSY particles depend on their mass spectrum, which is not predicted, so the search has to investigate many possibilities. SUSY also predicts several more Higgs-type particles. Searches for these have also been undertaken but so far no additional Higgs bosons have been found. In what follows, the study of some selected benchmark physics processes, both in the Higgs sector and in the search for New Physics at the HL-LHC will be presented.

As mentioned above, a general upgrade of the detectors is crucial to maximize physics potential and maintain a good particle reconstruction in the harsh environment of HL-LHC. The planned upgrades of the CMS experiment in view of the HL-LHC phase will take place around 2023 and are fully described in [1], [2], [3], [4].

2 Higgs Boson Signal Strength

The present status of the Higgs strength and couplings precision measurement performed by the CMS experiment is summarized in [5], where the the main decay channels of the Higgs bosons are exploited (WW, ZZ, $\gamma\gamma$, $\mu\mu$, $\tau\tau$, $\bar{b}b$). Using the 2016 CMS dataset (36/fb) 10-20% uncertainties are reached on the main parameters, and the signal strength relative to the standard model is close to unity [5].

The HL-LHC will be a Higgs factory: over 170 million Higgs bosons will be produced during the whole data taking period in 3000/fb. The increased size of this dataset will allow to achieve high precision measurements, down to the level of a few percents. The results summarized in this report are based on CMS public measurements performed using the 2015 and early 2016 proton-proton datasets, projected to larger datasets of 3000/fb assuming a center of mass energy of 13 TeV, thus using a smaller dataset with respect to the one used for extracting the measurement in [5]. The projections on the Higgs coupling are presented under different scenarios assumed for the size of systematic uncertainties, which are expected to bracket a realistic extrapolation. The incorporation of the performance of the upgraded detectors and the effect of higher pileup conditions are included in both scenarios:

- S1+ : All systematic uncertainties are kept constant with integrated luminosity.
- S2+ : Theoretical uncertainties scaled down by a factor 1/2, while experimental systematic uncertainties are scaled down by the square root of the integrated luminosity until they reach a defined lower limit based on estimates of the achievable accuracy with the upgraded detector.

The rates of Higgs boson production and decay into $ZZ \rightarrow 4l$ and $\gamma\gamma$ final states, parametrized using strength parameters μ (defined as the ratios between the observed rates and the expected ones in the SM) are showed in Fig. 1. The extrapolations show that all the couplings precisions are statistically limited, so theoretical and and experimental systematics will play a central role in the future measurements. Further improvements are expected by incorporating features of recent Run 2 analyses (e.g. ttH) and the full Run 2 dataset. The benefit given by the increased dataset is clearly shown in the Higgs to top quarks coupling (accessible via

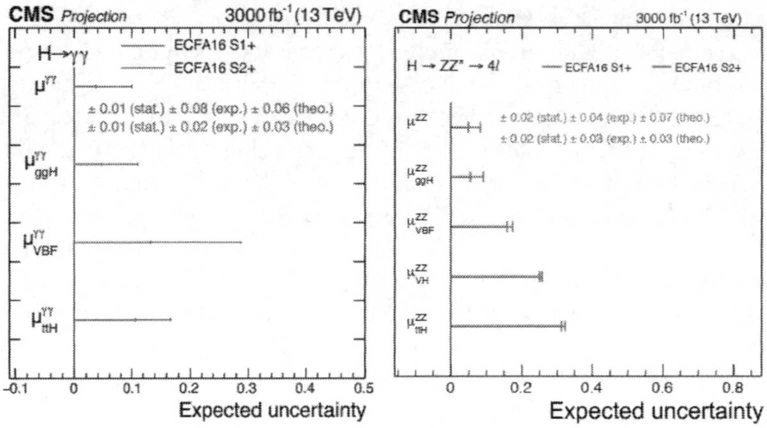

Figure 1. The projected 68% CL uncertainties in the Higgs boson signal strength for different production modes at 3000/fb, in the $\gamma\gamma$ (top plot) and $ZZ \to 4l$ (bottom plot) channels with S1(+) in green and S2(+) in red

gluon fusion production mechanism) although systematics also important. One of the main goals of HL-LHC physics program is to bring uncertainty on the rate down to 5% level

3 Higgs differential cross-sections

Differential cross sections provide an interesting portal to a number of physical observables. The shape of the differential cross section distribution can be tested with respect to SM expectation: it has been shown in [6] that small variation of the Higgs boson couplings with fermions, introduced by BSM processes, can lead to significant shape distortions. The differential cross section as a function of the Higgs transverse momentum is sensitive to the modifications introduced by the of effective Higgs Yukawa couplings at low p_T and to finite top mass effects at high p_T. The present status of this measurement performed by CMS quotes a 30% to 40% uncertainties depending on the considered p_T range [7].

The projection for the HL-LHC phase of the differential cross section measurement as a function of the transverse momentum of the Higgs boson is shown in Fig. 2. The theoretical

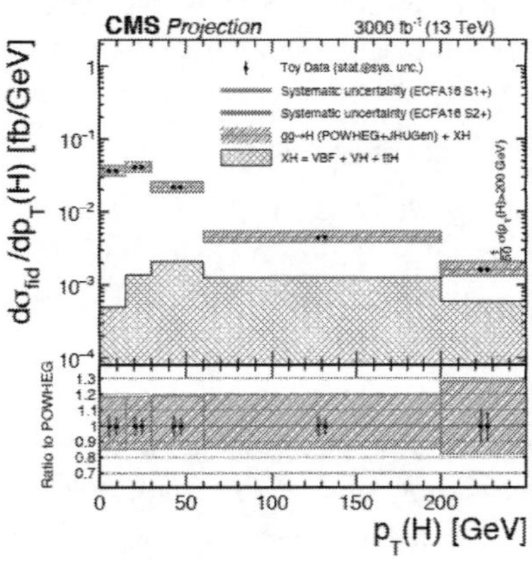

Figure 2. Projections for the differential fiducial cross section measurement of the Higgs boson transverse momentum at 3000/fb. The theoretical uncertainty in the differential gluon fusion cross section, which does not affect the measurement, is taken at NLO and shown in magenta. The last bin represents the integrated cross section for $p_T \geq 200$ GeV and is scaled by 50 for presentation.

uncertainties in the total signal cross section are not relevant and the cross section is measured in a fiducial phase space closely matching the experimental acceptance. The statistical uncertainty of the measurements ranges from 4 to 9% depending on the p_T bin. The las bin ($p_T \geq 200$ GeV) is still dominated by the statistical uncertainty even at 3000/fb [10].

4 Projection on Double Higgs Boson Searches

The measurement of the Higgs boson self coupling is a fundamental test of the SM. In this context, a extremely small cross section is predicted for the main double Higgs bosons (HH) production mechanism, (gluon fusion) σ_{SM} =33 fb at NNLO of the perturbative QCD expansion [8]. Because of such small cross section, arising from the interference of the two loop induced processes involved at the tree level (associated double Higgs production and self coupling), LHC experiments are not yet sensitive to SM HH production with the current data. The best present upper limit based on in Run 2 dataset is given by ATLAS that excluded at 95% confidence level a signal strength $\mu \geq 7$ [12]. However, BSM effects can strongly modify the HH production cross section and kinematics and be observed at the LHC and, at higher scale, at HL-LHC.

The direct production of new resonant states (called X in the following) decaying to HH, or resonant production, is predicted in many models such as scalars sectors extended with a doublet or a singlet, or extra dimension theories. Although different in motivation, they result in a similar experimental signature, but require experimentally the study of a broad m_X range to be sensitive to a large variety of models. This results effectively in anomalous Higgs boson couplings that affect the non-resonant production, and may manifest with large modifications of both the cross section and the kinematic properties of the double Higgs boson final state. The exploration of many HH decay channels is crucial in this context to ensure a broad coverage of the possible BSM effects thanks to the complementarity of the different final states.

The extrapolations from Run II to HL-LHC luminosity are based on 2.3-2.7/fb of 2015 CMS dataset and shown in Fig. 3 [10], with the projections for the four final states currently under scrutiny by the CMS collaboration: $HH \to \gamma\gamma bb$, $HH \to \tau\tau bb$, $HH \to 4b$ as well as $HH \to VVbb$, the latter looking at the $ll\nu\nu$ final state with $l = e, \mu$ and where b is a jet induced by a b-quak [1]. The most sensitive channel comes from the decay mode of $HH \to \gamma\gamma bb$ with

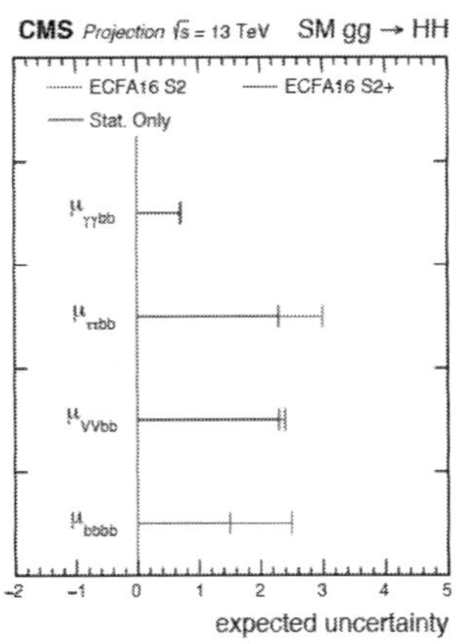

Figure 3. Projection of the sensitivity to the SM $gg \to HH$ production at 3000/fb, based on 13 TeV preliminary analyses performed with data collected in 2015. The uncertainty in the signal modifier $\mu = \sigma/\sigma_{SM}$ is provided assuming different scenarios on the systematic uncertainties.

a significance of 1.47 standard deviations when only considering statistical uncertainty. The

[1]The extrapolations presented in this document assume $\sqrt{s}$ = 13 TeV. However, the nominal center-of-mass energy at the HL-LHC is $\sqrt{s}$=14 TeV with a predicted cross section of 39.51 fb, corresponding to an increase in cross section by 18%. Assuming a background scaling with the center of mass energies (14/13 ~ 1.08), the projected results are expected to underperform by 1.18/ $\sqrt{(1.08)}$ ~ 15% with respect to the HL-LHC energy conditions.

scenario assumes uncertainty on the reconstructed b-tagged jet of 1% level and upgraded detector performance with an average of 200 pileup events. As shown, each of the channel alone is insufficient to observe the Higgs boson pair production. However, combining with these channels including other decay modes is promising.

In the $HH \to \tau\tau bb$ channel two complementary approaches have been pursued by extrapolating the projection using the full 2016 CMS dataset (36/fb) and running the full analysis on the signal and background Monte Carlo samples that incorporates the HL-LHC and Phase 2 CMS detector upgrades and operation conditions. Both the procedures are found to obtain compatible results, namely a cross section ratio σ_{HH}/σ_{SM} ~1.6 [3]. The distribution of the final discriminant used to extract the the results, the stransverse mass that is defined in [9], is showed in Fig.4 normalized to the full Phase 2 expected luminosity. Similar studies are ongoing in the other channels.

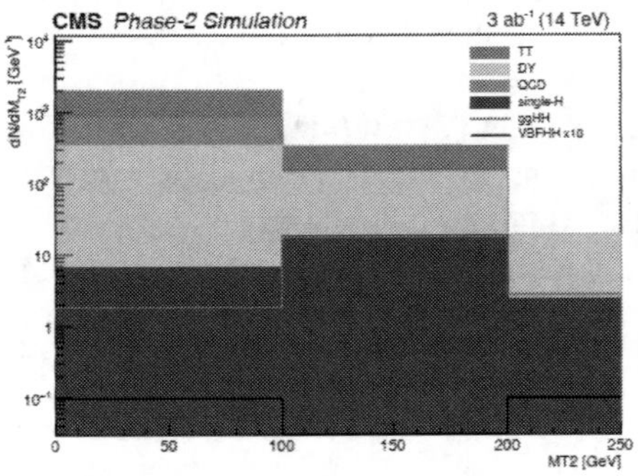

Figure 4. Distribution of the stransverse mass, defined as the largest mass of the parent particle that is compatible with the kinematic constraints of the event. In the case of the bb$\tau\tau$ decay, where the dominant background is $t\bar{t}$ production, the parent particle is interpreted as the top quark that decays into a bottom quark and a W boson.

5 Rare process

The measurement of the Yukawa couplings of the Higgs boson to fermions is one of the important studies to check the consistency with the SM and to search for possible deviations. The $H \to \mu\mu$ process can serve as a probe for coupling to fermion of second generation, provided that sufficient luminosity is collected to account for the small branching fraction (2.2×10^{-4} in the SM) and reach a sizeable sensitivity. With the large datasets of the HL-LHC, searching for rare decays of the Higgs boson becomes accessible. Moreover this channel has been exploited as a benchmark for the CMS tracker upgrade. Thanks to the reduced material budget and improved spatial resolution of the upgraded tracker, about 65% better invariant mass resolution can be achieved. This leads to an uncertainty of 5% for couplings to muons and 10% for the cross section measurement with 3000/fb [2]. Figure 5 shows the dimuon invariant mass distribution of $H \to \mu\mu$ decays for muons in the central region of the CMS detector as taken from simulated events that incorporates the CMS Phase II upgrade conditions.

6 Search for Supersymmetric particles

The present LHC has excluded large parts of the natural SUSY parameter space; limits for strong SUSY production are above 1 TeV and top and bottom squarks are already highly constrained. There are still many opportunities in the electroweak sector though which may be the dominating sector if squarks and gluinos are heavy. In most SUSY breaking scenarios the supersymmetric partners of the electroweak gauge bosons (EWK-inos) are expected to have a mass of the order 100 GeV based on naturalness. Given that EWK-inos are produced via

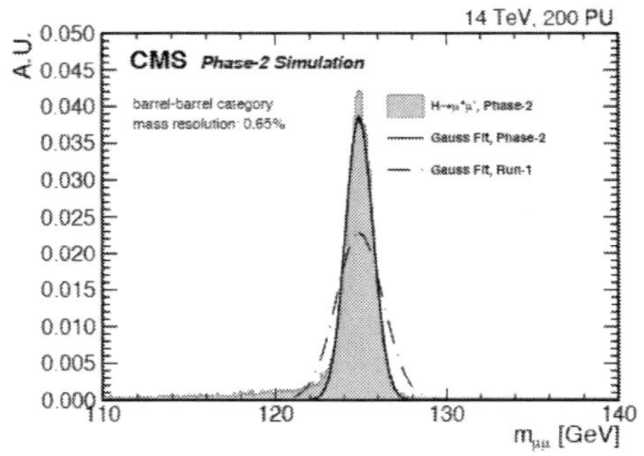

Figure 5. The di-muon invariant mass distribution for $H \to \mu\mu$ decays for muons in the central region, simulated with the Phase-2 detector.

EWK production the cross sections are small and HL-LHC has a large potential to increase sensitivity.

A brand-new CMS study for EWK-inos exploits the striking signature of same-sign leptons. It searches for mass degenerate $\tilde{\chi}^0_4, \tilde{\chi}^\pm_2$ are then expected to decay into Higgsinos (which are taken to be the lightest SUSY particles) emitting same charge W bosons with a total branching ratio close to 25% (Fig. 6).

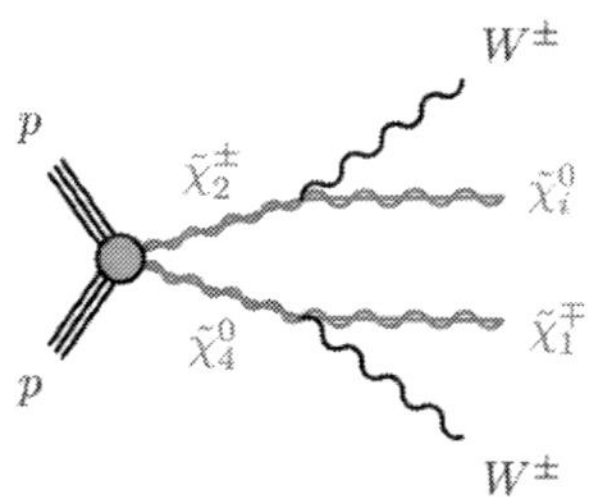

Figure 6. Example signal process which yields two same-sign leptons and large missing energy in the final state.

Given the small mass difference, the W bosons are soft, thus decaying into leptons with very low p_T that are quite challanging to detect both at the trigger and offline reconstruction level. The full analysis has been performed using the fast simulation tool Delphes [13], that accounts for the performance of the upgraded CMS detector on the reconstructed final state objects. Main benefits come from the upgraded high-granularity calorimeters that improved the missing energy measurement and the extended Muon system acceptance, that yields factor 3 reduction of the irreducible WZ background. Two Higgisno mass scenarios considered:

- m($\tilde{\chi}^1$) = 150 GeV is representative of the region of the parameter space outside the reach of the Run 2
- m($\tilde{\chi}^1$) =250 GeV is outside the sensitivity reach of the same search when extrapolated to the HL-LHC

The spectrum of the transverse mass of the lepton candidate (built by combining the lepton p_T and missing energy), is shown in Fig.7 top panel. Wino-like mass degenerate $\tilde{\chi}^0_4, \tilde{\chi}^\pm_2$ are excluded at 95% CL up to 900 GeV in the tested mass scenarios as shown in Fig. 7 (bottom panel).

7 Exotic Searches

It may happen that the only hints of new physics are rather exotic signatures that cannot be detected with conventional analyses New physics could manifest itself with a non-standard

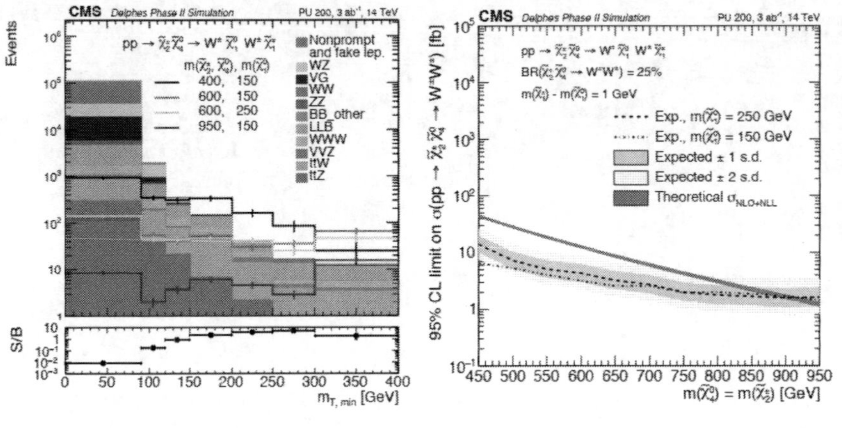

Figure 7. Top panel: the discriminating variable mT,min reconstructed from the leptons and missing transverse energy. The seven bins would allow an extraction of the signal from the background. Botton panel: exclusion limit in case of no observation of a signal.

signature, requiring dedicated trigger and reconstruction algorithms and specific detector features. One example are slow moving particles, also known as Heavy Stable Charged Particles (HSCP) foreseen by Split SUSY scenarios. These particles feature very high mass and small velocity. One way to detect such particle is their anomalous energy loss (dE/dx) [11] provided they are triggered. Muons from the decay of slow moving particles can already be identified at the trigger level based on the difference in arrival time with respect to relativistic SM particles. The muons from such decays are delayed w.r.t. to the time-zero bunch crossing (BX) and will spill over in neighbouring BXs. Fig. 8 (left) illustrates such a time difference as seen by CMS RPC chambers instrumenting the barrel and the forward muon system. The ns time resolution of the upgraded muon trigger allows to detect slowly moving charged particles by their hits. This will significantly extend the trigger capability to very slow particles with beta ≤ 0.5 (red) compared to present conditions, as displayed in Fig. 8 right [4].

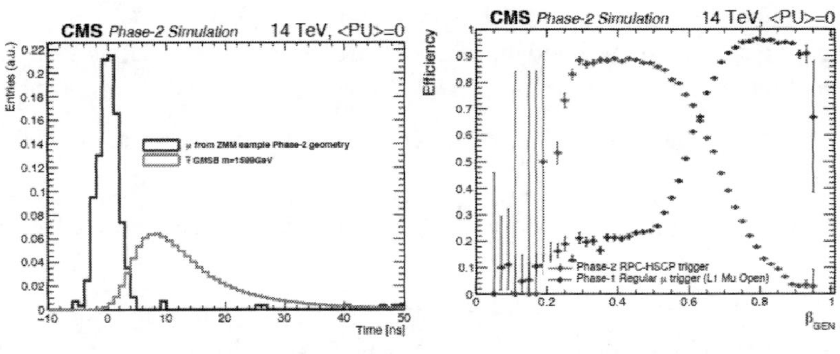

Figure 8. Left: measured time for a particle to traverse CMS with respect to the transit time of a particle traveling at the speed of light. Muons from the decay of Z bosons are relativistic ($\beta = 1$) and HSCPs are slower. Right: the efficiency as a function of b of the standard L1 muon trigger without any pT threshold, and the RPC-HSCP Phase-2 trigger with 1.56 ns sampling time

An important class of Gauge Mediated SUSY breaking models predict long-lived particles that would potentially lead to displaced signatures. For particles of a few hundred GeV mass, impact parameters can reach up to approximately one meter (or longer) for sufficiently large lifetimes In this context, an interesting benchmark channel studied in the context of the CMS Phase II upgrade is the search for the supersymmetric partners of the muons (s-muons) that can be the co-NLSP (i.e. they are the next to lightest LSP) and almost degenerate in mass. This search has low sensitivity in the present LHC runs, given the extremely small cross section, around 10^{-2} fb for 1 TeV s-muon masses but became accessible in the HL-LHC scenario. The search sensitivity depends on the s-muons decay lengths: larger displacement (driven by longer s-muon lifetimes) allow to test a phase space region that is free from the SM background. In the proposed benchmark channel, the s-muon decay to a SM muon and a gravitino (whose signature in the detector is the missing energy) with lifetimes which can be very long such that the displacement is of the order of meter. Muon triggering and reconstruction can be performed by the muon system in stand-alone, without the tracker. Discovery sensitivity of 3σ significance can be reached with 3000b of data (Fig. 9), providing full trigger and offline reconstruction efficiency even in the pile-up 200 environment.

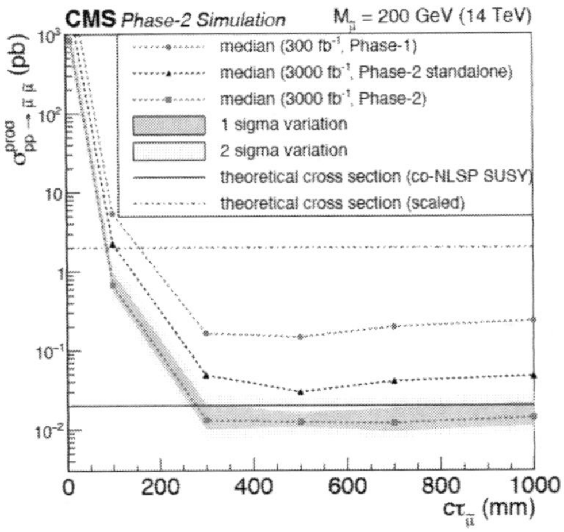

Figure 9. The 95% CL upper limits of $qq \to \tilde{\mu}\tilde{\mu}$ with $\tilde{\mu}$ to a muon and a gravitino as a function of the decay length for s-muon mass of 200 GeV. The theoretical cross section for the specific model is represented by the blue solid line. For different SUSY breaking scales, $\tan\beta$ or otherwise modified parameters, the cross sections may be 100 times larger, reflected by the blue dash-dotted line. Green (yellow) shaded bands show the one (two) sigma range of variation of the expected 95% CL limits. Phase II results with an average 200 pileup events and an integrated luminosity of 3000/fb are compared to results obtained with 300/fb. The black line shows the sensitivity without the muon reconstruction algorithm dedicated to the displaced muon and mainly relying on the muon system chambers. Using the standard muon reconstruction algorithms reduces the reconstruction efficiency by a factor three.

8 Conclusions

In the middle of 2026, LHC will reach a peak instantaneous luminosity of 7.5×10^{34}cm−2s−1. The HL-LHC will be the first Higgs factory, opening up the possibility to perform high precision measurement thanks to the to large amounts of events that is planned to collect, corresponding to an integrated luminosity of 3000/fb, more than 10 times larger than data by the end of Run 3. Thanks to the high statistics, rare processes in the Higgs sector will become accessible as well as the possibility to study the any possible deviation from the Standard Model. Towards the high precision era, a key role will be thus played by the theoretical and experimental systematics. The large Phase 2 dataset will enrich the potential to test physics beyond the SM, providing that the CMS experiment will be able to cope the challenging operation conditions. An extensive detector upgrade program has been planned and will start around 2023. Long-term exploration of Higgs sector and New Physics searches at HL-LHC will have crucial impact on our understanding of nature. There is a coordinated effort by ATLAS, CMS and theorists to update and combine results which will be documented in a

CERN Yellow Report at the end of the year. Better precisions should come from the optimized Phase-2 detectors documented in TDRs and from the better analysis tools developed for Run 2.

References

[1] CMS Collaboration, CERN-LHCC-2015-10, https://cds.cern.ch/record/2020886
[2] CMS Collaboration, CMS-TDR-17-001, https://cds.cern.ch/record/2272264.
[3] CMS Collaboration, CMS-TDR-019, https://cds.cern.ch/record/2293646
[4] CMS Collaboration, CMS-TDR-17-003
[5] CMS Collaboration, CMS-PAS-HIG-17-031, 2018, http://cds.cern.ch/record/2308127
[6] F. Bishara et al., arXiv:1606.09253v3
[7] CMS Collaboration, CMS-HIG-17-028, https://cds.cern.ch/record/2628757
[8] D. de Florian et. al., 10.23731/CYRM-2017-002, arXiv:1610.07922
[9] A. J. Barr, M. J. Dolan, C. Englert, and M. Spannowsky, Phys. Lett. B 728 (2014) 308.
[10] CMS Collaboration, CMS-PAS-FTR-16-002, http://cds.cern.ch/record/2266165.
[11] CMS Collaboration, CMS-PAS-EXO-14-007
[12] ATLAS Collaboration, ATLAS-CONF-2018-043
[13] The DELPHES 3 Collaboration, de Favereau, J., Delaere, C. et al. J. High Energ. Phys. (2014) 2014: 57, https://doi.org/10.1007/JHEP02(2014)057

The Light CP-Even Higgs Boson Mass of the MSSM at Three-Loop Accuracy

Edilson A. Reyes Rojas[1,*] and *A. Raffaele* Fazio[1,**]

[1]Departamento de Fisica, UNAL, Bogotá, Colombia

Abstract. In this paper we are going to discuss the relevant aspects related to the calculation of the higher order corrections to the lightest CP-even Higgs boson mass of the minimal supersymmetric standard model with real parameters (rMSSM). We have computed these corrections using the Feynman diagrammatic approach with a precision of three-loop in the SUSYQCD sector. The renormalization scheme adopted in this work is based on a variant of dimensional regularization where the so called dimensional reduction is performed in order to preserve supersymmetry to all perturbative orders. The calculation will extend the region of validity of previous studies to the whole parameters space. The emerging loop integrals have been computed by exploiting new proposed approaches based on the dispersion relation techniques for the numerical calculation of three-loop vacuum integrals with arbitrary mass scales.

1 Introduction

The analysis of the scale dependent properties of the Higgs sector in the Standard Model (SM) and beyond the SM scenarios (BSM) have been discussed by several authors for the last decade. The currently accepted analysis in the SM was presented in a completely way up to two-loop order in the references [1] and [2]. In these works the triviality, the vacuum stability problem and the hierarchy problem were studied by assuming that no new physics appears up to the Planck scale (Λ_P) and therefore by assuming an unnatural amount of fine-tuning (10^{34}) for the squared Higgs boson mass at the electroweak (EW) scale. In this approach the near-criticality of the Higgs quartic coupling (λ) and its beta-function is used as a guideline to go beyond the SM. However, there are strong physical arguments to believe that the SM is an effective field theory valid in a limited energy range below Λ_P. These include the dark matter observations, the baryon asymmetry in the universe, the neutrino oscillations, the Yukawa hierarchies in flavour physics, etc. If one insists that the naturalness argument is valid and would represent a real problem of the theory, then one has to conclude that physics beyond the SM has to appear at the TeV scale in the next runs of the LHC or in the coming experiments at future colliders like the ILC. One way to test that is through the comparison between the experimental observations and the theoretical predictions in order to confirm whether the Higgs boson is actually the predicted particle by the SM or is part of an extended model. Any deviation of the experimental results from the SM predictions would suppose a signal of new physics beyond the SM.

*e-mail: eareyesro@unal.edu.co
**e-mail: arfazio@unal.edu.co

© The Authors, published by EDP Sciences. This is an open access article distributed under the terms of the Creative Commons Attribution License 4.0 (http://creativecommons.org/licenses/by/4.0/).

The most popular BSM scenario that may describe these new phenomena is formulated as a supersymmetric extension of the SM, the minimal supersymmetric standard model (MSSM). The MSSM can provide for a solution to the fine-tuning and the hierarchy problem by the existence of new bosonic degrees of freedom that cancel the loop contributions with the unwanted quadratic divergences. Moreover, similar contributions render the effective potential stable, thus the MSSM simultaneously cures the vacuum stability problem. In most of the benchmark scenarios of the MSSM the Higgs boson found at LHC corresponds to the lightest CP-even Higgs boson with a mass M_h which is not a free input parameter but a prediction coming from the parameters of the theory. The upper bound on its predicted mass at leading order (LO) is given by the Z gauge boson mass, $M_Z = 91.2$ GeV, leading to the exclusion of the MSSM at current collider experiments. However, higher order quantum corrections to the MSSM Higgs boson mass lead to a considerably large shift in the upper limit towards $M_h \leq 135$ GeV. Additionally, the expected experimental uncertainty associated to the Higgs boson mass is of the order of 100-200 MeV at the LHC [3] and this value could go even down at the future ILC reaching a roughly value of 50 MeV [4]. At two-loop order the theoretical uncertainty has been estimated to 3-5 GeV [5] and as a consequence, it is mandatory to perform calculations at next-to-next-to-next to leading order (NNNLO) to reduce the theoretical uncertainty to the same order as the experimental accuracy expected at the LHC and ILC colliders. Our goal in this article is take you through the details of the computation of the radiative corrections to M_h in the rMSSM at this level of accuracy.

2 Radiative Corrections to $\mathrm{M_h}$ in the MSSM

The Higgs sector of the MSSM have five physical Higgs bosons, three of them are neutral, the others two are charged. In lowest order these are the lightest (h) and heavy (H) CP-even Higgs bosons, the CP-odd Higgs boson (A), and the two charged Higgs bosons ($H^{\pm}$). At tree level, the Higgs sector can be parametrized in terms of the gauge boson masses (M_W and M_Z), the mass of the CP-odd Higgs boson (M_A) and $tan\beta = v_2/v_1$, which is the ratio of the two vacuum expectation values. The masses of the five Higgs boson particles follow as predictions. For the MSSM the status of higher-order corrections to the masses and mixing angles in the neutral Higgs sector is quite advanced and widely studied due to radiative corrections can give large contributions to the values of these parameters. Regarding the corrections to the lightest Higgs boson mass the state of the art is as follows. At one-loop level the full corrections are found in reference [6] with real parameters. The detailed results of a Feynman diagrammatic (FD) calculation of the leading two-loop QCD corrections at order $O(\alpha\alpha_s)$ can be found in [7], in particular the $O(\alpha_b\alpha_s)$ [8] and $O(\alpha_t\alpha_s)$ [9] contributions using the FD approach are known in the limit where the external momentum vanishes and in the MSSM version with complex parameters. In this limit there is an alternative procedure to compute the above corrections, the Effective Potential (EP) approach. The analytical expressions for the two-loop leading-log neutral Higgs boson masses using the MSSM effective potential can be found in [10] and the references therein. A comparison of the corresponding two-loop results in the FD and EP approaches of the mass of the lightest CP-even Higgs boson in the MSSM at $O(\alpha\alpha_s)$ can be found in [11]. The FD approach has the advantage that it can allow for non-vanishing external momentum, in contrast to the EP method where all contributions are evaluated at zero external momentum. An evaluation of the momentum dependence of the two-loop corrections, including all the terms involving the QCD couplings, in the modified dimensional reduction scheme ($\overline{DR}$) was presented in [12]. The renormalization scheme $\overline{DR}$ is based on a variant of the dimensional reduction scheme (DRED) where the UV divergences are minimally subtracted [13]. In this scheme all gamma matrices lives in the quasi-four-dimensional space (Q4S) [14] while loop momenta remain in the D-dimensional

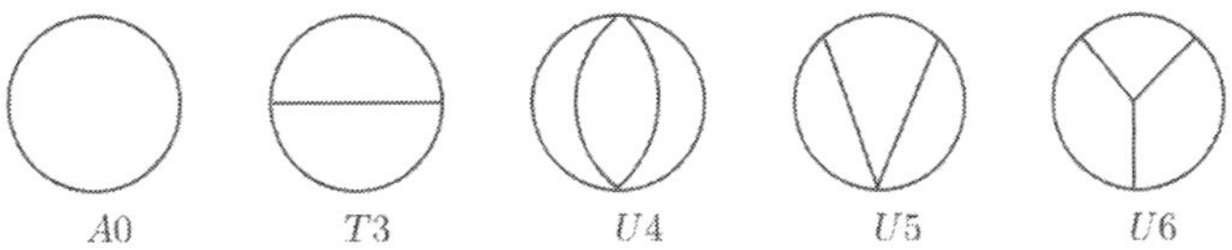

Figure 1: Basis of three-loop vacuum integrals obtained from the IBP reduction with Reduze.

space. This regularization procedure is performed in order to preserve supersymmetry to all perturbative orders. The latest status of the momentum-dependent two-loop corrections was discusses recently in [15, 16] using a hybrid on-shell-$\overline{DR}$ scheme, where M_A and the tadpoles are renormalized on-shell, whereas the Higgs boson fields and $tan\beta$ are renormalized $\overline{DR}$ and including corrections of $O(p^2\alpha_t\alpha_s)$.

3 Higgs Boson Mass at Three-loop Accuracy

In this section we are going to discuss our diagrammatic computation of the corrections to M_h at three-loop accuracy. We restrict the calculation to the supersymmetric quantum chromodynamics (SUSYQCD) sector where the Higgs couples to the top quark or its super-partner. In the perturbative expansion of the Higgs mass, these are the terms of order $\alpha_s^2\alpha_t$, where α_s is the square of the strong coupling constant and α_t denotes the coupling of the Higgs to the top quark. This is not the first computation of M_h at this order, there is already the diagrammatic computation performed by Harlander et al. [17], where M_h was computed at $O(\alpha_t\alpha_s^2)$. Their results were expressed in terms of $\overline{DR}$ parameters for the quark and squark masses and implemented in the computer program H3m for Mathematica. However, the contributions obtained in this work have exploited the methods of asymptotic expansion in order to provide precise approximations in some relevant mass hierarchies. We have gone beyond these approximations and therefore we have avoided perform asymptotic expansions at the integral level in order to obtain a general expression valid for any election of the input masses.
Taking in mind the above statement we have organized our computation as follows. The three-loop radiative correction to M_h is obtained by evaluating the neutral Higgs boson self-energies ($\sum$) and the tadpoles (T) for the fields h, H and A according with the equation

$$\hat{\sum}_{hh}^{(3)} = \sum_{hh}^{(3)} - Re\left(\sum_{AA}^{(3)}\right)\left(s_\alpha s_\beta + c_\alpha c_\beta\right)^2 - \left(\frac{e}{2s_W M_W}\right)\left(\frac{T_h^{(3)}}{\sqrt{2}}C_1 + \frac{T_H^{(3)}}{\sqrt{2}}C_2\right), \quad (1)$$

where the letter sigma with a hat represents the three-loop renormalized correction to M_h, while $\sum$ and T are three-loop unrenormalized quantities evaluated at zero external momentum. The coefficients C_1 and C_2 are functions of the mixing angles α and β which are responsible for the diagonalization of the bilinear part of the Higgs potential. All the necessary Feynman diagrams and their corresponding amplitudes are generated with FeynArts [18]. In total there are 7738 self-energies and 7180 tadpoles to be evaluated. The generated amplitudes with FeynArts are not regularized. With the help of the Mathematica package FeynCalc [19] we have written a routine that implements the regularization of Feynman integrals by dimensional reduction. Thus, we performed the Dirac and Lorentz algebra over numerators of the integrals in the Q4S space. It is worth to mention that the Q4S algebra for diagrams which contain the cubic vertex quark-squark-gluino requires a special treatment. This vertex introduces the γ_5 matrix in the Dirac traces. We have dealt γ_5 as an anti-commuting object, which is allowed in our computation because traces with an odd number of γ_5 always contain less than four gamma matrices. By other side, sum over Color indices and the Color algebra was done with the help of the package SUNSimplify of FeynCalc. After performing the

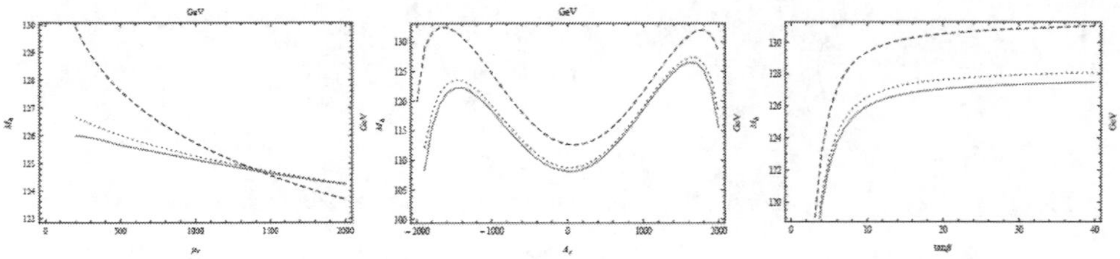

(a) A_t = 1500 GeV
$tan\beta$ = 10

(b) μ_r = 200 GeV
$tan\beta$ = 10

(c) A_t = 1500 GeV
μ_r = 250 GeV

Figure 2: Dependence of M_h on (a) μ_r; (b) A_t and (c) $tan\beta$. Black dashed and blue dotted lines are the one- and two-loop predictions of FeynHiggs. The red solid line depicts our three-loop predictions.

algebra of numerators, each amplitude can be expressed as a superposition of a set of scalar integrals which can contain irreducible numerators. We have classified and listed all different scalar integrals that we need in our calculation, in total there are 3525 scalar integrals that we must evaluate. This set of scalar integrals are not independent of each other but related by the integration by parts (IBP) and Lorentz invariance (LI) identities. We have used the IBPs to generate a homogeneous system of linear equations where the scalar integrals are the unknowns. The system can be reduced to a small set of irreducible integrals, the so called Master Integrals. This is something that cannot be done by hand, because there are thousand of equations. Thus, we have used the program Reduze [20], an implementation of the Laporta algorithm, in order to reduce the set of scalar integrals to a superposition of a small set of master integrals. We have found the basis of six master integrals depicted in Figure 1, where each topology can contain at most four different mass scales. The functions A0 and T3 have very well known analytical solutions for any configuration of masses, while the three-loop functions U4, U5 and U6 have an analytical solution just for the cases where there are one or two independent mass scales but when there are three or four independent scales a numerical evaluation of the three-loop vacuum integrals is required.
The analysis of convergence and the numerical evaluation of the integrals which are unknown analytically was performed with the help of the program TVID developed by A. Freitas [21]. TVID uses the discontinuities coming from the one-loop self-energy and the one-loop vertex in four dimensions to produce dispersion relations that are useful in the evaluation of the three-loop vacuum integrals in terms of one- and two-dimensional integral representations. In all cases we have been able to get analytical expressions for the divergent part of the master integrals while the finite part could be numerically evaluated with up to 10 digits of accuracy. It is possible to reach this precision because the numerical integrations are at most 2-dimensional and therefore there is a controlled treatment of any singularities. It is also necessary an additional sub-renormalization procedure due to the presence of non-local divergences in the unrenormalized Higgs self-energies as well as in the Higgs tadpoles. As a consequence, we have to include also diagrams with counter-term insertions in order to remove sub-divergences. At three-loop level we need the renormalization of the gluino mass, the squark masses, the stop mixing angle and the top quark mass at the one-loop order. In addition we need the two-loop counterterms for the stop mixing angle, the stop quark masses and the top quark mass. The necessary expressions for these counterterms can be consulted in the review [22] and references therein.
Once the UV divergences and sub-divergences are renormalized we get a finite correction to M_h. Using this contribution we have studied the dependence of the Higgs boson mass M_h on the renormalization scale μ_r (Fig. 2a), the soft-breaking parameter A_t (Fig. 2b) and the input parameter $tan\beta$ (Fig. 2c). In Figure 2 we show these dependences for a benchmark scenario where all the squark masses, except the stop mass, have the same value defined to

be M_{SUSY} = 1000 GeV, while we have chosen a value of $M_{\tilde{g}}$ = 1500 GeV for the gluino mass and M_t = 173.3 GeV for the top quark mass. To generate the MSSM Higgs predictions at one and two -loop level we have used the code FeynHiggs [23]. These contributions are represented in the plots of Fig. 2 with the black dashed and blue dotted curves, respectively. Besides, the red solid line represents our three-loop prediction. The three-loop corrections are quite sizeable, amounting a size of about 1 TeV and with negative sign for our election of input parameters. It leads to a more stable dependence of M_h with the renormalization scale μ_r and have a strong dependence on $tan\beta$ for small values close to $tan\beta$ = 5, while for large values above $tan\beta$ = 20 the variation of M_h is marginal.

4 Conclusion

Following the Feynman diagrammatic procedure described in the above section, we have obtained a finite expression for the renormalized three-loop correction to the lightest Higgs boson mass. Taking in mind that nowadays in the MSSM is no clear at priori what the hierarchies of the masses will turn out to be, we have avoided the application of asymptotic expansions and we have obtained the correction in terms of a set of three-loop master integrals whose numerical evaluation is possible thanks to the dispersion relation techniques for any arbitrary mass hierarchy. A complete phenomenological analysis of our numerical results will be presented in future works.

Acknowledgements. This work is partially financially supported by the research grant N. 34729, Minimal Supersymmetric Standard Model Higgs Boson Masses to High Accuracy of the call CONVOCATORIA NACIONAL DE PROYECTOS PARA EL FORTALECIMIENTO DE LA INVESTIGACIÓN, CREACIÓN E INNOVACIÓN DE LA UNIVERSIDAD NACIONAL DE COLOMBIA 2016-2018.

References

[1] Buttazzo D., Degrassi G., Giardino P.P., Giudice G., Sala F., Salvio A. and Strumia, A., JHEP, **1312**, 089 (2013).
[2] B. A. Kniehl, A. F. Pikelner and O. L. Veretin, Nucl. Phys. B **896**, 19 (2015).
[3] CERN Report No. CERN-LHCC-2006-021; CMS-TDR- 008-2, (2006), http://cmsdoc.cern.ch/cms/cpt/tdr/.
[4] J. A. Aguilar-Saavedra et al. [ECFA/DESY LC Physics Working Group] (2001) [arXiv:hep-ph/0106315].
[5] B. C. Allanach, A. Djouadi, J. L. Kneur, W. Porod and P. Slavich, JHEP **0409**, 044 (2004).
[6] A. Dabelstein, Nucl. Phys. B **456**, 25 (1995).
[7] S. Heinemeyer, W. Hollik and G. Weiglein, Eur. Phys. J. C **9**, 343 (1999).
[8] S. Heinemeyer, W. Hollik, H. Rzehak and G. Weiglein, Eur. Phys. J. C **39**, 465 (2005).
[9] S. Heinemeyer, W. Hollik, H. Rzehak, G. Weiglein, Phys. Lett. B **652**, 300 (2007).
[10] M. Carena, M. Quiros and C. Wagner, Nucl. Phys. B **461**, 407 (1996).
[11] M. Carena, H. Haber, S. Heinemeyer, W. Hollik, C. Wagner, and G. Weiglein, Phys. B **580**, 29 (2000).
[12] S. Martin, Phys. Rev. D **71**, 016012 (2005).
[13] I. Jack et al., Phys. Rev. D **50**, 5481 (1994). W. Siegel, Phys. Lett. B **84**, 193 (1979).
[14] D. Stockinger, JHEP **0503**, 076 (2005).
[15] S. Borowka, T. Hahn, S. Heinemeyer, G. Heinrich and W. Hollik, Eur. Phys. J. C **74**, 8, 2994 (2014).
[16] G. Degrassi, S. Di Vita and P. Slavich, Eur. Phys. J. C **75**, 2, 61 (2015).
[17] R. Harlander, P. Kant, L. Mihaila and M. Steinhauser, JHEP **1008**, 104 (2010).
[18] T. Hahn, Comput. Phys. Commun. **140**, 418 (2001).
[19] V. Shtabovenko, R. Mertig, F. Orellana, Comput. Phys. Commun. **207**, 432 (2016).
[20] C. Studerus, Comput. Phys. Commun. **181**, 1293 (2010).
[21] A. Freitas, JHEP **1611**, 145 (2016) [arXiv:1609.09159 [hep-ph]].
[22] L. Mihaila, Adv. High Energy Phys. **2013**, 607807 (2013).
[23] T. Hahn, S. Heinemeyer, W. Hollik, H. Rzehak and G. Weiglein, Comput. Phys. Commun. **180**, 1426 (2009).

EPJ Web of Conferences **192**, 00034 (2018) https://doi.org/10.1051/epjconf/201819200034
QCD@Work 2018

Extra Quarks and Bileptons in BSM Physics in a 331 Model

Claudio Corianò[1,*], *Antonio* Costantini[1,**]

[1] *Dipartimento di Matematica e Fisica "Ennio De Giorgi", Università del Salento and INFN-Lecce, Via Arnesano, 73100 Lecce, Italy*

Abstract. We describe some salient features of the 331_F (Frampton-Pisano-Pleitez) bilepton model, in which the constraints of anomaly cancelation require the number of generations to be three. In a class of six models, four of which characterised by a β parameter describing the embedding of the hypercharge in the $SU(3)_L$ symmetry, a specific choice for β allows bileptons in the spectrum, i.e. vectors and scalars of lepton numbers ± 2. At the same time the model allows exotic quarks, with the third quark generation treated asymmetrically respect to the other two. Bileptons generate specific signatures in the form of multilepton final states in Drell-Yan like processes, with and without associated jets, which can be searched for at the LHC.

1 Introduction

At the beginning of June 2018 the ATLAS collaboration reported the measurement of the coupling of the top quark to the Higgs boson [1]. At the same time the CMS collaboration published the results of a similar analysis [2]. From their common results it is established that the coupling of the top quark to the Higgs boson is compatible with the prediction of the Standard Model (SM), confirming that the mechanism of mass generation of the heaviest known quark is similar to that of the lighter ones [3, 4]. These findings secure the mechanism of spontaneous symmetry breaking (SSB) - induced by a scalar sector and associated Yukawa couplings - as a general feature of the SM. If we consider BSM (beyond the SM) models, in particular models with multiple SSB steps, by the same token, it is natural to expect that extra fermionic states will be generated, widely distributed in their mass. Their charges are directly linked to the underlying patterns of anomaly cancellation, which, in some cases, take to electric charges which differ by those typical of the three generations of the SM, and are classified as *exotics*.

2 GUT's and 331's

The idea that the gauge symmetry ($SU(3)_C \times SU(2)_L \times U(1)_Y \equiv 321$) of the SM is part of a larger one has been pursued since the 70's, after that it was noticed that the unification of the gauge couplings could be achieved, provided that the underlying gauge group was *simple* or a product of *simple* factors

* e-mail: claudio.coriano@le.infn.it
** e-mail: antonio.costantini@le.infn.it

© The Authors, published by EDP Sciences. This is an open access article distributed under the terms of the Creative Commons Attribution License 4.0 (http://creativecommons.org/licenses/by/4.0/).

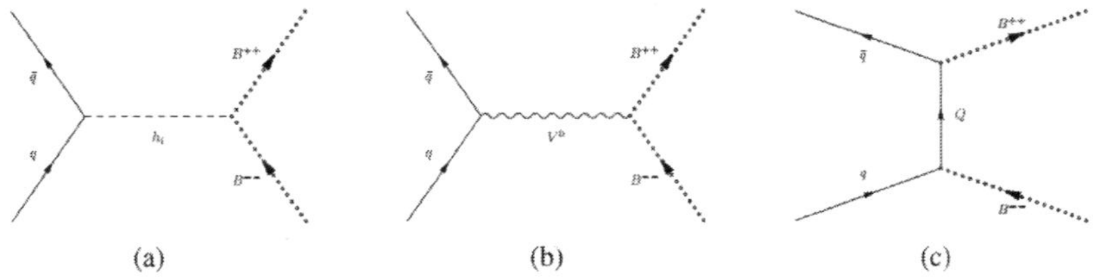

Figure 1. Typical contributions to events with two doubly-charged bosons in the final state and no extra jets. (a) and (b): contributions due to the mediation of a scalar (a) and a vector (b) boson. (c): t-channel exchange of exotic quarks Q. $B^{\pm}$ denotes either a vector ($Y^{\pm}$) or a scalar ($H^{\pm}$) bilepton.

(see [7] for an overview). Such symmetries with multiple *simple* groups can be obtained by the sequential breakings of simple groups of higher rank, most notably within E_6 and $SO(10)$ scenarios realized at a large scale, the grand unification scale or GUT. Among these, we mention the left-right symmetric model ($SU(3)_C \times SU(2)_L \times SU(2)_R \times U(1)_{B-L}$) which can be derived from the breaking of $SO(10)$. Other examples are models based on $SU(3)_c \times SU(3)_L$, later extended by an extra $U(1)_X$ in order to get the correct fermion masses. Tree level considerations on the structure of the couplings on the 331 models suggest that the breaking 331 $\rightarrow$ 321 could take place at the TeV scale, rather than at the usual GUT scale of $10^{13} - 10^{15}$ GeV, which has been the motivation for studying such class of models, especially in the flavour sector (see [12, 13]). There are various realization of such models since the embeddings of 321 into 331 can be quite different.

3 $SU(3)_c \times SU(3)_L \times U(1)_X$: exotics in the Frampton-Pisano-Pleitez model

The gauge structure of the 331_F model of [5, 6] is $SU(3)_c \times SU(3)_L \times U(1)_X$, with the fermions in the fundamental of $SU(3)_c$ arranged into triplets of $SU(3)_L$. The three families of quarks are treated asymmetrically with respect to the weak $SU(3)$ ($SU(3)_L$), with the first two families given by

$$Q_1 = \begin{pmatrix} u_L \\ d_L \\ D_L \end{pmatrix}, \quad Q_2 = \begin{pmatrix} c_L \\ s_L \\ S_L \end{pmatrix}, \quad Q_{1,2} \in (\mathbf{3}, \mathbf{3}, -1/3) \tag{1}$$

under $SU(3)_c \times SU(3)_L \times U(1)_X$, whereas the third family is

$$Q_3 = \begin{pmatrix} b_L \\ t_L \\ T_L \end{pmatrix}, \quad Q_3 \in (\mathbf{3}, \bar{\mathbf{3}}, 2/3). \tag{2}$$

D, S and T are extra (exotic) quarks. In this respect, the number of generations is underwritten by the contraints of anomaly cancellation, which single out this model as unique respect to the general class of BSM scenarios. The lepton sector is assigned to the representation $\bar{3}$ of the same gauge group. Universality is instead respected in the lepton sector, where the three lepton generations are ordinarily arranged into triplets of $SU(3)_L$,

$$l = \begin{pmatrix} e_L \\ \nu_L \\ e_R^c \end{pmatrix}, \quad l \in (\mathbf{1}, \bar{\mathbf{3}}, 0), \tag{3}$$

with $e_R^c = i\sigma_2 e_R^*$. The presence of exotic particles in the spectrum of the 331 model is due to the specific embedding of the hypercharge Y in the $SU(3)_L \times U(1)_X$ gauge symmetry, which is defined by

$$Y_{\mathbf{3}} = \beta T_8 + X\mathbf{1} \qquad Y_{\bar{\mathbf{3}}} = -\beta T_8 + X\mathbf{1} \tag{4}$$

for the $\mathbf{3}$ and the $\bar{\mathbf{3}}$ representations of $SU(3)_L$, respectively, with generators $T_i = \lambda_i/2$ $(i = 1, \ldots 8)$, corresponding to the Gell-Mann matrices, and $T_8 = \mathrm{diag}\left[\frac{1}{2\sqrt{3}}(1, 1, -2)\right]$. The charge operator is given by

$$Q_{em,\mathbf{3}} = Y_{\mathbf{3}} + T_3 \qquad Q_{em,\bar{\mathbf{3}}} = Y_{\bar{\mathbf{3}}} - T_3 \tag{5}$$

in the fundamental and anti-fundamental representations of of $SU(3)_L$, respectively, where $T_3 = \mathrm{diag}\left[\frac{1}{2}(1, -1, 0)\right]$.

4 Family embedding with exotics

If we choose the $SU(2)_L \times U(1)_Y$ hypercharge assignments of the SM as

$$Y(Q_L) = 1/6, \qquad Y(L) = -1/2, \qquad Y(u_R) = 2/3, \qquad Y(d_R) = -1/3, \qquad Y(e_R) = -1 \tag{6}$$

and denote by q_X the $U(1)_X$ charges, the breaking of the symmetry $SU(3)_L \times U(1)_X \to SU(2)_L \times U(1)_Y$, for the $\mathbf{3}$ of $SU(3)$ is

$$(\mathbf{3}, q_X) \to \left(2, \frac{\beta}{2\sqrt{3}} + q_X\right) + \left(1, -\frac{\beta}{\sqrt{3}} + q_X\right), \tag{7}$$

and for the $\bar{\mathbf{3}}$

$$(\bar{\mathbf{3}}, q_X') \to \left(2, -\frac{\beta}{2\sqrt{3}} + q_X'\right) + \left(1, +\frac{\beta}{\sqrt{3}} + q_X'\right). \tag{8}$$

The X-charge is fixed by the condition that the first two components of the Q_1 and Q_2 triplets carry the same hypercharge as the quark doublets $Q_L = (u, d)_L$ of the Standard Model, yielding

$$q_X = \frac{1}{6} - \frac{\beta}{2\sqrt{3}}. \tag{9}$$

The minimal version of the 331_F model, in our conventions, is obtained by $\beta = \sqrt{3}$, resulting in

$$Q_{em}(Q_{1,2}) = \mathrm{diag}(2/3, -1/3, -4/3), \tag{10}$$

with two exotic quarks D and S of charge -4/3 whereas

$$Q_{em\,\bar{\mathbf{3}}}(Q_3) = \mathrm{diag}\left(-\frac{1}{3}, \frac{2}{3}, \frac{5}{3}\right). \tag{11}$$

The exotic quark T carries electric charge 5/3. For the leptons we have the ususal charge assignments and the three generations are treated symmetrically with

$$Q_{em}(l) = \mathrm{diag}(-1, 0, 1). \tag{12}$$

Both left- and right-handed components of the SM leptons are fitted into the same $SU(3)_L$ multiplet. The condition of absence of colorless fractionally charged fermions fixes the value of β to 4 possible values $(\pm\sqrt{3}, \pm 1/\sqrt{3})$, of which the FPP model is one, plus two additional models. The first is a flipped 331 version [8], with all the quarks in the same representation and the leptons in different ones; while the second is embedded in E_6 and there is family replication for both leptons and quarks. Among these models states with bilepton charges are present only in 331_F and in the E_6 version [9].

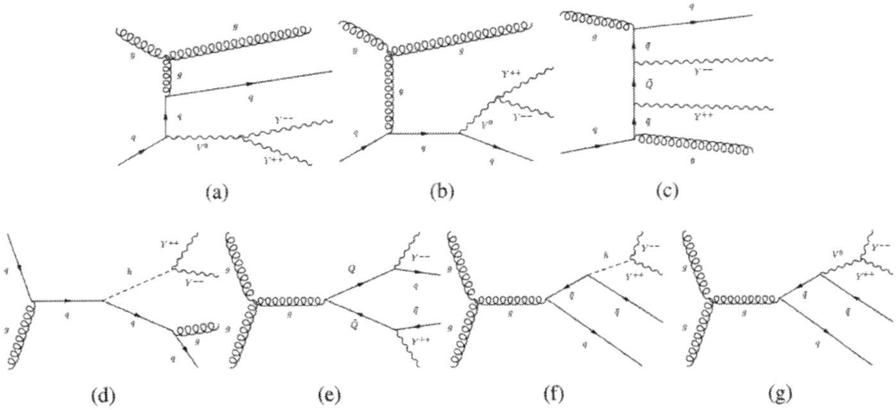

Figure 2. Bilepton signal from quark-gluon fusion and gluon-gluon fusion.

5 The scalars of 331_F

The scalars of the 331_F model, responsible for the electroweak symmetry breaking (EWSB), come in three triplets of $SU(3)_L$

$$\rho = \begin{pmatrix} \rho^{++} \\ \rho^{+} \\ \rho^{0} \end{pmatrix} \in (1,3,1), \quad \eta = \begin{pmatrix} \eta^{+} \\ \eta^{0} \\ \eta^{-} \end{pmatrix} \in (1,3,0), \quad \chi = \begin{pmatrix} \chi^{0} \\ \chi^{-} \\ \chi^{--} \end{pmatrix} \in (1,3,-1). \tag{13}$$

The breaking $SU(3)_L \times U(1)_X \to U(1)_{em}$ is obtained in two steps. The vacuum expectation value of the neutral component of ρ causes the breaking from $SU(3)_L \times U(1)_X$ to $SU(2)_L \times U(1)_Y$. Then the usual spontaneous symmetry breaking mechanism from $SU(2)_L \times U(1)_Y$ to $U(1)_{em}$ is obtained through the vevs of the neutral components of η and χ. Exotic quarks play an important role in the production of vector bileptons ($B^{\pm}$) in $q\bar{q}$ annihilaton channels, as shown in Fig. 1. Compared to an ordinary Drell-Yan process where the coupling of the final state leptons to the intermediate gauge bosons is direct, in this case it is mediated by the bilepton pairs, each decaying to same sign leptons with a distinct signature of multi-lepton final states. At the same time, respect to an ordinary Drell-Yan process, an extra contribution is generated by a t-channel exchange of an exotic quark (Q) at Born level (Fig. 1 c). Amplitudes with no associated jets can be generated also via the sextet scalars, by replacing $Y^{\pm}$ with $H^{\pm}$.

5.1 The triplet sector

Coming to the electroweak SSB mechanism, this is realised in the 331_F model by giving a vev to the neutral component of the triplets ρ, η and χ, with Yukawa interactions

$$\begin{aligned} \mathcal{L}^{Yuk.}_{q,triplet} = & (y_d^1\, Q_1\eta^* d_R + y_d^2\, Q_2\eta^* s_R + y_d^3\, Q_3\chi\, b_R^* \\ & + y_u^1\, Q_1\chi^* u_R^* + y_u^2\, Q_2\chi^* c_R^* + y_u^3\, Q_3\eta\, t_R^* \\ & + y_E^1\, Q_1\rho^* D_R^* + y_E^2\, Q_2\rho^* S_R^* + y_E^3\, Q_3\rho\, T_R^*) + \text{h.c.}, \end{aligned} \tag{14}$$

where y_d^i, y_u^i and y_E^i are the Yukawa couplings for down-, up-type and exotic quarks, respectively. The masses of the exotic quarks are related to the vev of the neutral component of $\rho = (0, 0, v_\rho)$. In this case the invariants are, in their representation content,

$$\begin{aligned} Q_1\,\rho^* D_R^*, Q_1\,\rho^* S_R^* &\sim (3,3,-1/3)\times(1,\bar{3},-1)\times(\bar{3},1,4/3) \\ Q_3\,\rho T_R^* &\sim (3,\bar{3},2/3)\times(1,3,1)\times(\bar{3},1,-5/3), \end{aligned} \tag{15}$$

which induce the breaking $SU(3)_c \times SU(3)_L \times U(1)_X \to SU(3)_c \times SU(2)_L \times U(1)_Y$. If the relation $y_E^i \sim 1$ is satisfied, using $v_\rho \gg v_{\eta,\chi}$, it is clear that the masses of the exotics are $\mathcal{O}$(TeV), which is of particular interest for searches of such states at the LHC. Notice that the model has the appealing feature that the condition of reality of the gauge couplings sets a constraint on the location of the extra vacua, which is limited to the TeV region.

5.2 The sextet sector

The triplet sector is not sufficient to account for the masses of the leptons of the SM. This is the reason for introducing a scalar sextet. A typical Dirac mass term for the leptons in the SM is associated with the operator $\bar{l}_L H e_R$, with $l_L = (\nu_{eL}, e_L)$ being the $SU(2)_L$ doublet. The representation content is $(\bar{2}, 1/2) \times (2, 1/2) \times (1, -1)$ (for l, H and e_R, respectively) in $SU(2)_L \times U(1)_Y$. In the 331 the L and R components of the lepton (e) are in the same multiplet and therefore the identification of an $SO(1,3) \times SU(3)_L$ singlet needs two leptons in the same representation. The flavour structure of the Yukawa terms is given by the trilinear terms

$$\mathcal{L}^{Yuk}_{l,triplet} = G^{\eta}_{ab}\, l^i_a \cdot l^j_b\, \eta^{*k} \epsilon^{ijk} + \text{h.c.} \tag{16}$$

where a and b are flavour indices while i, j, k belong to the 3 of $SU(3)_L$. Some considerations on the gauge charges and fermion statistics of this interaction imply that G^{η}_{ab} is antisymmetric in flavour space and as such it is not sufficient to account for the masses of all the leptons. This problem is overcome by the introduction of a scalar sector, sextet of $SU(3)_L$, denoted as σ

$$\sigma = \begin{pmatrix} \sigma_1^{++} & \sigma_1^{+}/\sqrt{2} & \sigma^0/\sqrt{2} \\ \sigma_1^{+}/\sqrt{2} & \sigma_1^0 & \sigma_2^{-}/\sqrt{2} \\ \sigma^0/\sqrt{2} & \sigma_2^{-}/\sqrt{2} & \sigma_2^{--} \end{pmatrix} \in (1, 6, 0), \tag{17}$$

leading to the Yukawa term

$$\mathcal{L}^{Yuk.}_{l,sextet} = G^{\sigma}_{ab} l^i_a \cdot l^j_b \sigma^*_{i,j}, \tag{18}$$

with G^{σ}_{ab} which is symmetric in flavour space. This allows to singlet out the representation 6 of $SU(3)_L$, contained in $l^i_a \cdot l^j_b$, by combining it with the flavour-symmetric σ^*, i.e. $\bar{6}$. Multi-lepton final states are a specific feature of the model generated by the decay of doubly charged vectors and scalars.

6 Quark exotics in Drell-Yan like distributions

We show in Fig. 2 the amplitudes involved in the analysis of bileptons with associated jets in the final state. Each exotic (Q) and non exotic (q) quark contributions interfere in the production of the intermediate $Y^{\pm}$. In Drell-Yan, as usual, the p_T (transverse momentum) distributions of a lepton pair refer to processes containing 2 leptons of opposite signs plus one (quark or gluon) extra jet in the final state, at leading order in the QCD coupling. The contributions in Fig. 2 can be viewed as a

(more complex) generalization of that process, with the total p_T of the 4 leptons from the decay of $Y^\pm$ vectors balanced by those of the two extra jets. The number of amplitudes which contribute to such final states is quite large and results have been presented in [10], to which we refer for more details. Such processes can equally be generated by a pair of $H^\pm$ states if the contribution of the scalar sextet is included.
The inclusion of the sextet representation in the potential $V(\rho, \eta, \chi, \sigma)$ enlarges the number of physical states in the spectrum, allowing a richer scenario. In fact we now have, after electroweak SSB five scalar Higgses, three pseudoscalar Higgses, four charged Higgses and three doubly-charged Higgses. For instance, the physical doubly-charged Higgs states are expressed in terms of the gauge eigenstates and the elements of the rotation matrix R^C as [11]

$$H_i^{++} = \mathrm{R}_{i1}^{2C}\rho^{++} + \mathrm{R}_{i2}^{2C}(\chi^{--})^* + \mathrm{R}_{i3}^{2C}\sigma_1^{++} + \mathrm{R}_{i4}^{2C}(\sigma_2^{--})^*. \tag{19}$$

In [10] the study has been focused only on processes involving intermediate vectors rather than scalars, by considering benchmark points where the scalar bileptons are much heavier than the vector ones. In [11], on the other hand, the study has been focused on the Drell-Yan like contributions shown in Fig. 1, which provide the golden plated mode for the study of bileptons at the LHC. In this second study the $H^\pm$ and $Y^\pm$ states have been chosen to be close in mass. In this case the analysis has been focused on observables of the final state leptons which may allow to distinguish between processes mediated either by the $Y^\pm$ or by the $H^\pm$. Setting a distinction between the scalar ($H^\pm$) and vector ($Y^\pm$) contributions to the 4-lepton channel is an important experimental challenge, since searches for doubly-charged scalars are performed in a model-independent way, which differ from the scenario offered by this specific model, due to the presence of vector bileptons. This and the presence of exotics are peculiarities of the model which we hope will be addressed in direct experimental searches at the LHC.

7 Conclusions

In the class of models models with a 331 symmetry, the FPP model has a special status. It carries specific signatures in the form of scalar and vector bileptons which can be searched for at the LHC. Drell-Yan like processes are the golden plated modes for the discovery of such realizations. We have briefly overviewed their spectra and the the specific embedding of the hypercharge in the $SU(3)_L$ symmetry which take to exotic quarks in the spectrum. The asymmetric treatment of the 3 generations is a crucial signature of many 331 models, in which the number of families is underwritten by the cancellation of the gauge anomalies. At the same time, they are constrained to lay at the TeV scale.

Acknowledgements
We thank Paul Frampton and Gennaro Corcella for collaborating to this analysis. This work is partially supported by INFN of Italy under Iniziativa Specifica QFT-HEP.

References

[1] M. Aaboud *et al.* [ATLAS Collaboration], Phys. Lett. B **784** (2018) 173.
[2] A. M. Sirunyan *et al.* [CMS Collaboration], Phys. Rev. Lett. **120** (2018) no.23, 231801.
[3] M. Aaboud *et al.* [ATLAS Collaboration], arXiv:1808.08238 [hep-ex].
[4] A. M. Sirunyan *et al.* [CMS Collaboration], arXiv:1808.08242 [hep-ex].
[5] P. H. Frampton, Phys. Rev. Lett. **69** (1992) 2889.
[6] F. Pisano and V. Pleitez, Phys. Rev. D **46** (1992) 410

[7] R.M. Fonseca, M. Hirsch, Phys. Rev. **D94**, 115003 (2016), 1607.06328
[8] R. M. Fonseca and M. Hirsch, JHEP **08**, 003 (2016)
[9] L. A. Sánchez, W. A. Ponce and R. Martinez, PhysRev **D64**, 075013 (2001)
[10] G. Corcella, C. Corianò, A. Costantini, P.H. Frampton, Phys. Lett. **B773**, 544 (2017)
[11] G. Corcella, C. Corianò, A. Costantini, P.H. Frampton, Phys. Lett. **B785**, 73 (2018)
[12] A.J. Buras, F. De Fazio, J. Girrbach, M.V. Carlucci, JHEP **02**, 023 (2013)
[13] S. Descotes-Genon, M. Moscati, G. Ricciardi (2017), [arXiv:1711.03101 [hep-ph]]

Study of $\Upsilon(1S)$ radiative decays to $\gamma\pi^+\pi^-$ and γK^+K^-

Antimo Palano[1,*]

[1]*INFN and University of Bari, Via Orabona 4, 70125 Bari, Italy*

Abstract. We study the $\Upsilon(1S)$ radiative decays to $\gamma\pi^+\pi^-$ and γK^+K^- using data recorded with the *BABAR* detector operating at the SLAC PEP-II asymmetric-energy e^+e^- collider at center-of-mass energies at the $\Upsilon(2S)$ and $\Upsilon(3S)$ resonances. The $\Upsilon(1S)$ resonance is reconstructed from the decay $\Upsilon(nS) \to \pi^+\pi^-\Upsilon(1S)$, $n = 2, 3$.

1 Introduction

The existence of gluonium states is still an open issue for Quantum Chromodynamics (QCD). Lattice QCD calculations predict the lightest gluonium states to have quantum numbers $J^{PC} = 0^{++}$ and 2^{++} and to be in the mass region below 2.5 GeV/c^2 [1]. In particular, the $J^{PC} = 0^{++}$ glueball is predicted to have a mass around 1.7 GeV/c^2. Radiative decays of heavy quarkonia, in which a photon replaces one of the three gluons from the strong decay of J/ψ or $\Upsilon(1S)$, can probe color-singlet two-gluon systems that produce gluonic resonances. J/ψ decays have been extensively studied in the past [2]. In this analysis [3] we study $\Upsilon(1S)$ decays, taking into account that the experimental observation of radiative $\Upsilon(1S)$ decays is challenging because their rate is suppressed by a factor of ≈ 0.025 compared to J/ψ radiative decays, which are of order 10^{-3} [4].

2 Events reconstruction

We reconstruct the decay chains

$$\Upsilon(2S)/\Upsilon(3S) \to (\pi_s^+\pi_s^-)\Upsilon(1S) \to (\pi_s^+\pi_s^-)(\gamma\pi^+\pi^-) \tag{1}$$

and

$$\Upsilon(2S)/\Upsilon(3S) \to (\pi_s^+\pi_s^-)\Upsilon(1S) \to (\pi_s^+\pi_s^-)(\gamma K^+K^-), \tag{2}$$

where we label with the subscript s the slow pions from the direct $\Upsilon(2S)$ and $\Upsilon(3S)$ decays.

Events with balanced momentum are required to satisfy energy balance requirements. For each combination of $\pi_s^+\pi_s^-$ candidates, we first require both particles to be identified loosely as pions and compute the recoiling mass

$$M^2_{\rm rec}(\pi_s^+\pi_s^-) = |p_{e^+} + p_{e^-} - p_{\pi_s^+} - p_{\pi_s^-}|^2, \tag{3}$$

where p is the particle four-momentum. The distribution of $M^2_{\rm rec}(\pi_s^+\pi_s^-)$ is expected to peak at the squared $\Upsilon(1S)$ mass for signal events. Figure 1 shows the combinatorial recoiling mass $M_{\rm rec}(\pi_s^+\pi_s^-)$ for $\Upsilon(2S)$ and $\Upsilon(3S)$ data, where narrow peaks at the $\Upsilon(1S)$ mass can be observed.

* e-mail: antimo.palano@ba.infn.it

© The Authors, published by EDP Sciences. This is an open access article distributed under the terms of the Creative Commons Attribution License 4.0 (http://creativecommons.org/licenses/by/4.0/).

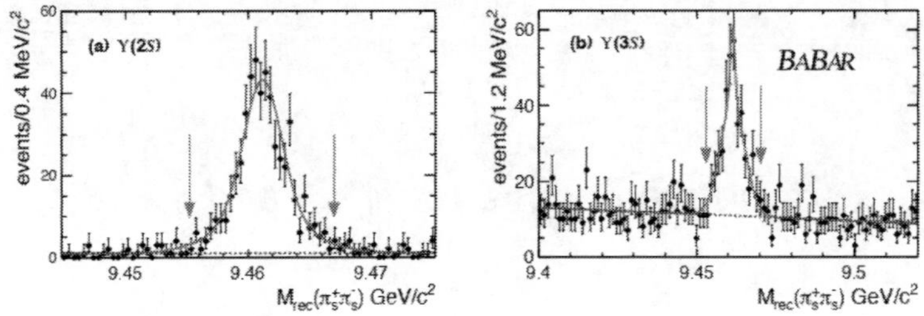

Figure 1. Combinatorial recoiling mass $M_{\rm rec}$ to $\pi_s^+\pi_s^-$ candidates for (a) $\Upsilon(2S)$ and (b) $\Upsilon(3S)$ data.

We select signal event candidates by requiring

$$|M_{\rm rec}(\pi_s^+\pi_s^-) - m(\Upsilon(1S))_f| < 2.5\sigma, \tag{4}$$

where $m(\Upsilon(1S))_f$ indicates the fitted $\Upsilon(1S)$ mass value. To reconstruct $\Upsilon(1S) \to \gamma\pi^+\pi^-$ or $\Upsilon(1S) \to \gamma K^+K^-$ decays, we require a loose identification of both pions or kaons and isolate the decays by requiring

$$9.1 \text{ GeV}/c^2 < m(\gamma h^+h^-) < 9.6 \text{ GeV}/c^2, \tag{5}$$

where $h = \pi, K$.

3 Study of the $\pi^+\pi^-$ and K^+K^- mass spectra

The $\pi^+\pi^-$ mass spectrum, for $m(\pi^+\pi^-) < 3.0$ GeV/c^2 and summed over the $\Upsilon(2S)$ and $\Upsilon(3S)$ datasets with 507 and 277 events, respectively, is shown in Fig. 2(Left). The spectrum shows $I = 0$, $J^P =$ even^{++} resonance production, with low backgrounds above 1 GeV/c^2. We observe a rapid drop around 1 GeV/c^2 characteristic of the presence of the $f_0(980)$, and a strong $f_2(1270)$ signal. The data also suggest the presence of weaker resonant contributions.

The K^+K^- mass spectrum, summed over the $\Upsilon(2S)$ and $\Upsilon(3S)$ datasets with 164 and 63 events, respectively, is shown in Fig. 2(Right) and also shows resonant production, with low background. Signals at the positions of $f_2'(1525)$ and $f_0(1710)$ can be observed.

We make use of a phenomenological model to extract the different $\Upsilon(1S) \to \gamma R$ branching fractions, where R is an intermediate resonance. We perform a simultaneous binned fit to the $\pi^+\pi^-$ mass spectra from the $\Upsilon(2S)$ and $\Upsilon(3S)$ datasets. We describe the low-mass region (around the $f_0(500)$) using a relativistic S-wave Breit-Wigner lineshape having free parameters. We describe the $f_0(980)$ using the Flatté [6] formalism. The $f_2(1270)$ and $f_0(1710)$ resonances are represented by relativistic Breit-Wigner functions with parameters fixed to PDG values [7]. In the high $\pi^+\pi^-$ mass region we include a single resonance $f_0(2100)$ having a width fixed to the PDG value (224 ± 22) and unconstrained mass. For the $\Upsilon(3S)$ data we also include $\rho(770)^0$ background with parameters fixed to the PDG values. The fit is shown in Fig. 2. It has 16 free parameters and $\chi^2 = 182$ for ndf=152, corresponding to a p-value of 5%. We note the observation of a significant S-wave in $\Upsilon(1S)$ radiative decays. This observation was not possible in the study of J/ψ radiative decay to $\pi^+\pi^-$ because of the presence of a strong, irreducible background from $J/\psi \to \pi^+\pi^-\pi^0$ [8]. No evidence is found for a $\Upsilon(1S) \to \pi^+\pi^-\pi^0$ decay in present data. We obtain the following $f_0(500)$ parameters:

$$m(f_0(500)) = 0.856 \pm 0.086 \text{ GeV}/c^2, \Gamma(f_0(500)) = 1.279 \pm 0.324 \text{ GeV}, \tag{6}$$

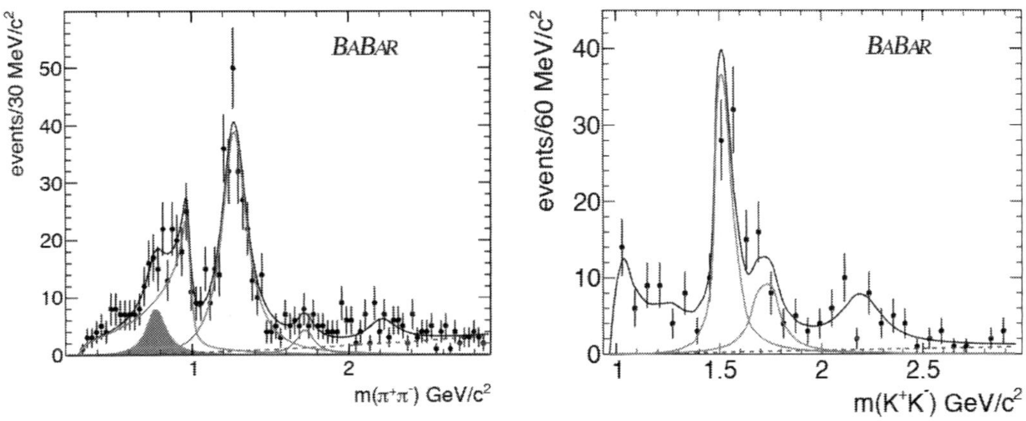

Figure 2. (Left) $\pi^+\pi^-$ mass distribution from $\Upsilon(1S) \to \pi^+\pi^-\gamma$ for the combined $\Upsilon(2S)$ and $\Upsilon(3S)$ datasets. The full (red) curves indicate the S-wave, $f_2(1270)$, and $f_0(1710)$ contributions. The shaded (gray) area represents the fitted $\rho(770)^0$ background. (Right) K^+K^- mass distribution from $\Upsilon(1S) \to K^+K^-\gamma$ for the combined $\Upsilon(2S)$ and $\Upsilon(3S)$ datasets. The (red) curves show the contributions from $f_2'(1525)$ and $f_0(1710)$. Dashed (blue) lines indicate the background contributions.

and $\phi = 2.41 \pm 0.43$ rad. The fraction of S-wave events associated with the $f_0(500)$ is $(27.7 \pm 3.1)\%$.

We perform a binned fit to the combined K^+K^- mass spectrum using the following model. The $f_0(980)$ is parameterized according to the Flatté formalism. The $f_2(1270)$, $f_2'(1525)$, $f_0(1500)$, and $f_0(1710)$ resonances are represented by relativistic Breit-Wigner functions with parameters fixed to PDG values. We include an $f_0(2200)$ contribution having parameters fixed to the PDG values. The fit shown in Fig. 2(Right). It has six free parameters and $\chi^2 = 35$ for ndf=29, corresponding to a p-value of 20%. The resonances yields and significances are given in Table 1. Systematic uncertainties are dominated by the PDG uncertainties on resonances parameters.

Table 1. Resonances yields and statistical significances from the fits to the $\pi^+\pi^-$ and K^+K^- mass spectra for the $\Upsilon(2S)$ and $\Upsilon(3S)$ datasets. The symbol $f_J(1500)$ indicates the signal in the 1500 MeV/c^2 mass region.

Resonances ($\pi^+\pi^-$)	Yield $\Upsilon(2S)$	Yield $\Upsilon(3S)$	Significance
S-wave	$133 \pm 16 \pm 13$	87 ± 13	12.8σ
$f_2(1270)$	$255 \pm 19 \pm 8$	$77 \pm 7 \pm 4$	15.9σ
$f_0(1710)$	$24 \pm 8 \pm 6$	$6 \pm 8 \pm 3$	2.5σ
Resonances (K^+K^-)	**Yield $\Upsilon(2S) + \Upsilon(3S)$**		**Significance**
$f_0(980)$	47 ± 9		5.6σ
$f_J(1500)$	$77 \pm 10 \pm 10$		8.9σ
$f_0(1710)$	$36 \pm 9 \pm 6$		4.7σ

4 Efficiency correction

The efficiency distributions as functions of mass, for the $\Upsilon(2S)/\Upsilon(3S)$ data and for the $\pi^+\pi^-\gamma$ and $K^+K^-\gamma$ final states, are found to have an almost uniform behavior for all the final states. We define the helicity angle θ_H as the angle formed by the h^+, in the h^+h^- rest frame, and the γ in the $h^+h^-\gamma$ rest frame. We also define θ_γ as the angle formed by the radiative photon in the $h^+h^-\gamma$ rest frame with respect to the $\Upsilon(1S)$ direction in the $\Upsilon(2S)/\Upsilon(3S)$ rest frame. We label with $\epsilon(m, \cos\theta_H)$ the efficiency computed as a function of the h^+h^- effective mass and the helicity angle $\cos\theta_H$. This is used only to obtain efficiency-corrected mass spectra. We label with $\epsilon(\cos\theta_H, \cos\theta_\gamma)$ the efficiency computed, for each resonance mass window, as a function of $\cos\theta_H$ and $\cos\theta_\gamma$. This is used to obtain the efficiency-corrected angular distributions and branching fractions of the different resonances. To obtain the efficiency correction weight w_R for the resonance R we divide each event by the efficiency $\epsilon(\cos\theta_H, \cos\theta_\gamma)$

$$w_R = \frac{\sum_{i=1}^{N_R} 1/\epsilon_i(\cos\theta_H, \cos\theta_\gamma)}{N_R}, \tag{7}$$

where N_R is the number of events in the resonance mass range.

5 Angular analysis

To obtain information on the angular momentum structure of the $\pi^+\pi^-$ and K^+K^- systems in $\Upsilon(1S) \to \gamma h^+h^-$ we study the dependence of the $m(h^+h^-)$ mass on the helicity angle θ_H. A better way to observe angular effects is to plot the $\pi^+\pi^-$ mass spectrum weighted by the Legendre polynomial moments, corrected for efficiency and shown in Fig. 3. In a simplified environment, the moments are related to the spin 0 (S) and spin 2 (D) amplitudes by the equations [9]:

$$\begin{aligned}
\sqrt{4\pi}\langle Y_0^0\rangle &= S^2 + D^2, \\
\sqrt{4\pi}\langle Y_2^0\rangle &= 2SD\cos\phi_{SD} + 0.639D^2, \\
\sqrt{4\pi}\langle Y_4^0\rangle &= 0.857D^2,
\end{aligned} \tag{8}$$

where ϕ_{SD} is the relative phase. Therefore we expect to observe spin 2 resonances in $\langle Y_4^0\rangle$ and S/D interference in $\langle Y_2^0\rangle$. The results are shown in Fig. 3(Top). We clearly observe the $f_2(1270)$ resonance in $\langle Y_4^0\rangle$ and a sharp drop in $\langle Y_2^0\rangle$ at the $f_2(1270)$ mass, indicating the interference effect. The distribution of $\langle Y_0^0\rangle$ is just the scaled $\pi^+\pi^-$ mass distribution, corrected for efficiency. Similarly, we plot in Fig. 3(Bottom) the K^+K^- mass spectrum weighted by the Legendre polynomial moments, corrected for efficiency. We observe signals of the $f_2'(1525)$ and $f_0(1710)$ in $\langle Y_4^0\rangle$ and activity due to S/D interference effects in the $\langle Y_2^0\rangle$ moment.

Resonance angular distributions in radiative $\Upsilon(1S)$ decays from $\Upsilon(2S)/\Upsilon(3S)$ decays are rather complex. In this section we perform a simplified Partial Wave Analysis (PWA) solving directly the system of Eq. (8). Figure 4 shows the resulting S-wave and D-wave contributions to the $\pi^+\pi^-$ and K^+K^- mass spectra, respectively. Due to the presence of background in the threshold region, the $\pi^+\pi^-$ analysis is performed only on the $\Upsilon(2S)$ data.

We note that in the case of the $\pi^+\pi^-$ mass spectrum we obtain a good separation between S and D-waves, with the presence of an $f_0(980)$ resonance on top of a broad $f_0(500)$ resonance in the S-wave and a clean $f_2(1270)$ in the D-wave distribution. Integrating the S-wave amplitude from threshold up to a mass of 1.5 GeV/c^2, we obtain an integrated, efficiency corrected yield $N(S\text{-wave}) = 629 \pm 128$.

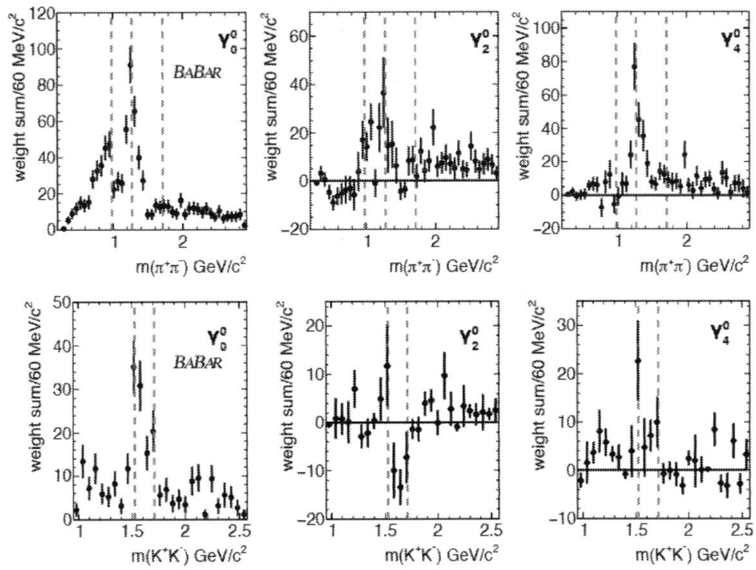

Figure 3. The distributions of the most relevant unnormalized Y_L^0 moments for $\Upsilon(1S) \to \gamma\pi^+\pi^-$ (Top) and $\Upsilon(1S) \to \gamma K^+K^-$ (Bottom) as functions of mass. The lines indicate the positions of $f_0(980)$, $f_2(1270)$, and $f_0(1710)$ for $\pi^+\pi^-$ and $f_2'(1525)$ and $f_0(1710)$ for K^+K^-.

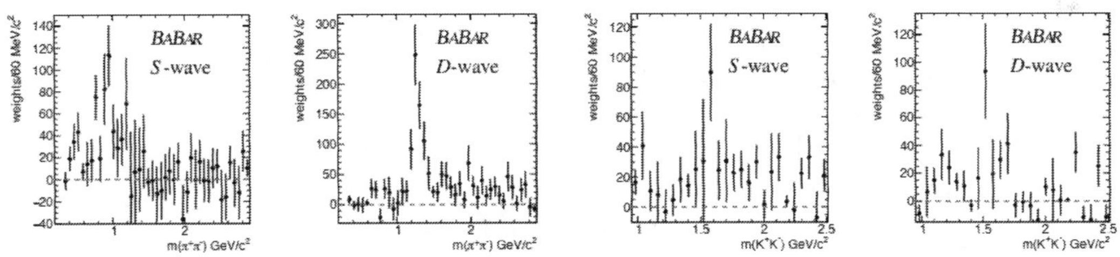

Figure 4. S and D-wave contributions from the simple PWA of the $\pi^+\pi^-$ mass spectrum for the $\Upsilon(2S)$ data (Left) and of the K^+K^- mass spectrum for the combined $\Upsilon(2S)/\Upsilon(3S)$ data (Right).

In the case of the K^+K^- PWA the structure peaking around 1500 MeV/c^2 appears in both S and D-waves suggesting the presence of $f_0(1500)$ and $f_2'(1525)$. In the $f_0(1710)$ mass region there is not enough data to discriminate between the two different spin assignments. This pattern is similar to that observed in the Dalitz plot analysis of charmless $B \to 3K$ decays [10]. Integrating the S and D-wave contributions in the $f_2'(1525)/f_0(1500)$ mass region, we obtain a fraction of S-wave contribution $f_S(K^+K^-) = 0.53 \pm 0.10$.

The full $\Upsilon(nS)$ angular distributions for decays Eq. (1) and Eq. (2) are expressed in terms of θ_γ and θ_H where the $\Upsilon(1S)$ can be produced with helicity 0 or 1. A spin 2 resonance, on the other hand, can have three helicity states. We make use of the helicity formalism [11, 12] to derive the angular distributions for a spin 2 and spin 0 resonance as reported in Ref. [3]. We perform a 2D unbinned maximum likelihood fit for each resonance using the spin 2 and spin 0 full angular distributions. We first fit the $f_2(1270)$ angular distributions and allow a background contribution of 16% from the S-wave having fixed parameters. Figure 5 shows the uncorrected fit projections on $\cos\theta_H$ and $\cos\theta_\gamma$. The $\cos\theta_\gamma$ spectrum is approximately uniform, while $\cos\theta_H$ shows structures well-fitted by the spin 2 hypothesis.

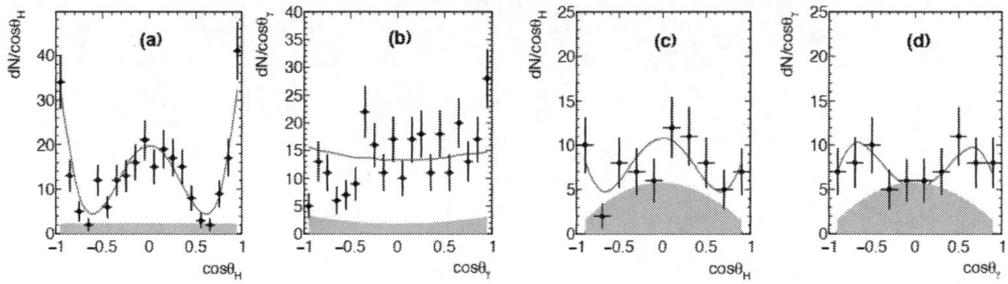

Figure 5. Uncorrected (a) $\cos\theta_H$ and (b) $\cos\theta_\gamma$ distributions in the $f_2(1270) \to \pi^+\pi^-$ mass region. Uncorrected (c) $\cos\theta_H$ and (d) $\cos\theta_\gamma$ distributions in the $f_J(1500) \to K^+K^-$ mass region. The full (red) lines are the projections from the fit with the spin 2 hypothesis. The shaded (gray) area represents the S-wave contribution.

We fit the K^+K^- data in the $f_J(1500)$ mass region, where many resonances can contribute: $f_2'(1525)$, $f_0(1500)$ [10], and $f_0(1710)$. We fit the data using a superposition of S and D waves, having helicity contributions as free parameters, and free S-wave contribution. We obtain an S-wave contribution of $f_S(K^+K^-) = 0.52 \pm 0.14$, in agreement with the estimate obtained from the simple PWA. Fit projections are shown in Fig. 5, giving an adequate description of the data.

6 Measurement of branching fractions

We determine the branching fraction $\mathcal{B}(R)$ for the decay of $\Upsilon(1S)$ to photon and resonance R using the expression

$$\mathcal{B}(R) = \frac{N_R(\Upsilon(nS) \to \pi_s^+\pi_s^-\Upsilon(1S)(\to R\gamma))}{N(\Upsilon(nS) \to \pi_s^+\pi_s^-\Upsilon(1S)(\to \mu^+\mu^-))} \times \mathcal{B}(\Upsilon(1S) \to \mu^+\mu^-), \tag{9}$$

where N_R indicates the efficiency-corrected yield for the given resonance. To reduce systematic uncertainties, we first compute the relative branching fraction to the reference channel $\Upsilon(nS) \to \pi^+\pi^-\Upsilon(1S)(\to \mu^+\mu^-)$, which has the same number of charged particles as the final states under study. We then multiply the relative branching fraction by the well-measured branching fraction $\mathcal{B}(\Upsilon(1S) \to \mu^+\mu^-) = 2.48 \pm 0.05\%$ [7].

We determine the reference channel corrected yield using the method of "B-counting", also used to obtain the number of produced $\Upsilon(2S)$ and $\Upsilon(3S)$ [5]. Taking into account the known branching fractions of $\Upsilon(2S)/\Upsilon(3S) \to \pi_s^+\pi_s^-\Upsilon(1S)$ we obtain

$$N(\Upsilon(2S) \to \pi_s^+\pi_s^-\Upsilon(1S)(\to \mu^+\mu^-)) = (4.35 \pm 0.12_{\text{sys}}) \times 10^5 \tag{10}$$

and

$$N(\Upsilon(3S) \to \pi_s^+\pi_s^-\Upsilon(1S)(\to \mu^+\mu^-)) = (1.32 \pm 0.04_{\text{sys}}) \times 10^5 \tag{11}$$

events. Table 2 gives the measured branching fractions. In all cases we correct the efficiency corrected yields for isospin and for PDG measured branching fractions [7]. We report the first observation of

Table 2. Measured $\Upsilon(1S) \to \gamma R$ branching fractions.

Resonance	$\mathcal{B}(10^{-5})$
$\pi\pi$ S-wave	$4.63 \pm 0.56 \pm 0.48$
$f_2(1270)$	$10.15 \pm 0.59\ ^{+0.54}_{-0.43}$
$f_0(1710) \to \pi\pi$	$0.79 \pm 0.26 \pm 0.17$
$f_J(1500) \to K\bar{K}$	$3.97 \pm 0.52 \pm 0.55$
$f_2'(1525)$	$2.13 \pm 0.28 \pm 0.72$
$f_0(1500) \to K\bar{K}$	$2.08 \pm 0.27 \pm 0.65$
$f_0(1710) \to K\bar{K}$	$2.02 \pm 0.51 \pm 0.35$

$f_0(1710)$ in $\Upsilon(1S)$ radiative decay with a significance of 5.7σ, combining $\pi^+\pi^-$ and K^+K^- data. To determine the branching ratio of the $f_0(1710)$ decays to $\pi\pi$ and $K\bar{K}$, we remove all the systematic uncertainties related to the reference channels and of the γ reconstruction. Labelling with N the efficiency-corrected yields for the two $f_0(1710)$ decay modes, we obtain

$$\frac{\mathcal{B}(f_0(1710) \to \pi\pi)}{\mathcal{B}(f_0(1710) \to K\bar{K})} = \frac{N(f_0(1710) \to \pi\pi)}{N(f_0(1710) \to K\bar{K})} = 0.64 \pm 0.27_{\text{stat}} \pm 0.18_{\text{sys}}, \tag{12}$$

in agreement with the world average value of $0.41^{+0.11}_{-0.17}$ [7].

These results may contribute to the long-standing issue of the identification of a scalar glueball.

References

[1] Y. Chen *et al.* Phys. Rev. D **73**, 014516 (2006).
[2] L. Köpke and N. Wermes, Phys. Rept. **174**, 67 (1989).
[3] J. P. Lees *et al.* [BaBar Collaboration], Phys. Rev. D **97** (2018) no.11, 112006
[4] S.B. Athar *et al.* (CLEO Collaboration), Phys. Rev. D **73**, 032001 (2006).
[5] Ed. A.J. Bevan, B. Golob, Th. Mannel, S. Prell, and B.D. Yabsley, "*The Physics of the B Factories*", Eur. Phys. J. C **74**, 302 (2014).
[6] S.M. Flatté, Phys. Lett. B **63**, 224 (1976).
[7] C. Patrignani *et al.* (Particle Data Group), Chin. Phys. C **40**, 100001 (2016).
[8] J. Becker *et al.* (Mark III Collaboration), Phys. Rev. D **35**, 2077 (1987).
[9] G. Costa *et al.* Nucl. Phys. B**175** 402 (1980).
[10] J.P. Lees *et al.* (*BABAR* Collaboration), Phys.Rev. D **85**, 112010 (2012).
[11] L. Breva-Newell, "Decays of the $\Upsilon(1S)$ into a photon and two charged hadrons", PhD thesis, hep-ex/0412075 (2004).
[12] J.D. Richman, "An Experimenter's Guide to the Helicity Formalism", CALT-68-1148 (1984).
[13] W. M. Yao *et al.* (Particle Data Group), J. Phys. G **33**, 1 (2006).

Decays of the $f_0(1370)$ scalar glueball candidate in pp Central Exclusive Production (CEP) and in antiproton annihilations at rest

Ugo Gastaldi

INFN-Ferrara, Via Saragat 1, 44122 Ferrara, Italy

Abstract. Decays into $\pi^+\pi^-$ of the $f_0(1370)$ are the main source of an isolated structure localized between 1.2 and 1.5 GeV in the $\pi^+\pi^-$ mass spectrum measured in pp Central Exclusive Production (CEP) at $\sqrt{s}$=200 GeV at very low four momentum transfer |t| by the STAR experiment. These data confirm in the $\pi^+\pi^-$ decay channel the existence of the $f_0(1370)$ as an isolated well identified structure previously observed in K^+K^-, K_sK_s, $\pi^+\pi^-\pi^+\pi^-$, $2\pi^0\pi^+\pi^-$ and $4\pi^0$ decays measured in pbar annihilations at rest. The ensemble of these data point at a high gg content of the $f_0(1370)$. CEP interactions at higher energies favour production of 0^{++} and 2^{++} mesons. Selection of events with lower |t| at both proton vertices suppresses 2^{++} structures. LHC runs dedicated to pp CEP measurements at low |t| could then provide a unique source of all the low energy scalars. This would make it clear if and where scalar gluonium is resident and the nature (composition in terms of qqbar, qqqbarqbar, qqbar-qqbar and gg) of $f_0(500)$, $f_0(980)$, $f_0(1370)$, $f_0(1500)$ and $f_0(1710)$.

1 Introduction

The possibility of the existence of glueballs is a basic qualitative prediction of quantum chromodynamics (QCD)[1,2]. The lowest lying glueball states are expected to have the same 0^{++} J^{PC} quantum numbers as the vacuum. While for the 0^{-+} pseudoscalar, 1^{--} vector and 2^{++} tensor meson nonets there are two observed isoscalar partners for two places in each nonet (respectively η and η', ω and φ, $f_2(1270)$ and $f'_2(1525)$, there are 4 or 5 observed isoscalar mesons (respectively σ, $f_0(980)$, $f_0(1370)$, $f_0(1500)$ and $f_0(1710)$ for the two places of the isoscalar members of the 0^{++} ground state nonet. The candidates for the two places are 4 or 5 depending whether σ and $f_0(1370)$ are considered as distinct separated objects or they are part of a single continuum. The existence of the σ meson is considered established since some years (see [3] for a review) and σ is currently called $f_0(500)$. Doubts instead concern the existence of the $f_0(1370)$ as an individual isolated structure (see [4] for a review). Concerning the nature of the σ and $f_0(980)$ 0^{++} mesons, various scenarios envisaged in the literature include (because of mixing) the possibility of a qqbar, qqqbarqbar, qqbar-qqbar and gg content in their wave functions[3-7]. If the 0^{++} isoscalars are 5 it is more difficult to exclude the hypothesis of the presence (or even dominance) of a gg content in some of them. Production in glue rich processes (central production mediated by double pomeron exchange, pbar annihilations, J/ψ and ψ' radiative decays, heavy meson decays) and absence in $\gamma\gamma$ production and meson exchange are criteria useful to characterize the gg content of the scalars [8]. Relative decay branching ratios into $\sigma\sigma$, $\rho\rho$, $\pi\pi$, KKbar, $\eta\eta$ for the three heavier scalars, and also into $\eta\eta'$ for $f_0(1500)$ and $f_0(1710)$, give other selection criteria to identify their gg content [8-11]. Currently there is no consensus concerning the experimental observation of a

© The Authors, published by EDP Sciences. This is an open access article distributed under the terms of the Creative Commons Attribution License 4.0 (http://creativecommons.org/licenses/by/4.0/).

scalar glueball nor on the possible gg content in the scalars experimentally observed (for reviews see [4,12-17]).

An isolated $\pi^+\pi^-$ peak practically without background and well separated from the $f_0(980)$ signal by a valley nearly devoid of events present in pp CEP data of the STAR experiment [18] has been interpreted as a clear manifestation of the $f_0(1370)$ meson as an isolated structure [19,20]. This interpretation is based on STAR data [18, 21-25], on the scaling laws of CEP [8,26] and on results of the AFS experiment [27,28] at the CERN ISR. This points are discussed in section 2.

In section 3 we recall previous data where the $f_0(1370)$ is present as an isolated clearly visible structure in K^+K^-, K_sK_s, $\pi^+\pi^-\pi^+\pi^-$, $2\pi^0\pi^+\pi^-$ and $4\pi^0$ decays measured in pbar annihilations at rest. Some of these data have been ignored in the PDG compilations of the last 20 years concerning $f_0(1370)$ and $f_0(1500)$, some are used in data averages to give mass, width and decay branching ratios of $f_0(1500)$, and are discarded in the $f_0(1370)$ section. Besides establishing –together with the STAR data- the existence of the $f_0(1370)$ as an isolated structure, these data are essential to extract the production rates and decay branching ratios of both $f_0(1370)$ and $f_0(1500)$, in order to identify their gg content. The orders of magnitude of the decay branching ratios of $f_0(1370)$ into $\sigma\sigma$, $\rho\rho$, $\pi\pi$, KKbar, $\eta\eta$ relative to $\pi\pi$ decays are 6, 3, 1, 1, 0.02. These data point at a large gg content of the $f_0(1370)$ meson and need a definitive confirmation by CEP measurements at LHC, also because there are conflicting results (based on low energy CEP data) of the WA102 experiment.

Section 4 gives conclusions, stresses the importance of the assessment of the properties of the $f_0(1370)$ for the observation of a scalar glueball and highlights the substantial improvements achievable by pp CEP experiments at LHC in the medium and longer term for the spectroscopy of all low energy scalars.

2 $f_0(1370)$ from $\pi^+\pi^-$ decays in pp Central Exclusive Production

Fig.1 [18] shows row data of $\pi^+\pi^-$ pairs in pp CEP data at $\sqrt{s}$=200 GeV produced by the STAR experiment at RICH in a 2009 run with β^*=20 m setting of the storage rings and precision forward detectors inserted in two sets of roman pots positioned at 55.5 and 58.5 m distance from the central detector at its two sides . This configuration of machine and forward detectors gives access to the kinematical region for the four momentum transfer square $|t_1|$ and $|t_2|$ at both proton vertices with $0.003 < |t_1|, |t_2| < 0.03$ GeV^2. This $|t_1|$, $|t_2|$ acceptance window of the 2009 STAR run -shown in fig. 2- is the $|t|$ window with lowest values so far exploited in pp CEP experiments.

Fig. 3 [23-25] shows row data of $\pi^+\pi^-$ pairs in pp CEP data at $\sqrt{s}$=200 GeV produced by the STAR experiment at RICH in a 2015 run with β^*=0.85 m setting of the storage rings and the two sets of roman pots positioned at the two sides of the central detector at 15.8 and 17.6 m distance. This second configuration of the machine and forward detectors gives access to the kinematical region with four momentum transfer square $|t_1|$ and $|t_2|$ at both proton vertices in the window $0.03 < |t_1|, |t_2| < 0.3$ GeV^2 (also shown in fig.2). The sharp drop in fig. 3 at 1 GeV is due to the $f_0(980)$ scalar meson; the strong peak near 1.3 GeV is due to the $f_2(1270)$ tensor meson. These two signals permit to check the energy calibration of the experiment. The sharp drop near 1 GeV is also present in fig.1, where the energy region comprised between 1 and 1.2 GeV is nearly empty of events, while a peak with about 60 events is present in the energy window 1.2-1.5 GeV. The peak in fig.1 between 1.2 and

1.5 GeV has been interpreted in [19,20] as due essentially to the $f_0(1370)$.

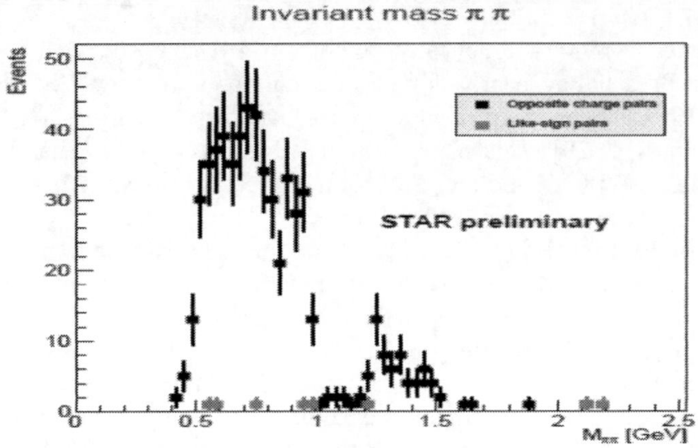

Fig. 1. $\pi^+\pi^-$ mass plot of STAR pp CEP raw data of the 200 GeV 2009 run with $0.003<|t_1|$, $|t_2|<0.03$ GeV^2 kinematic coverage of both scattered protons (Fig. from [18]).

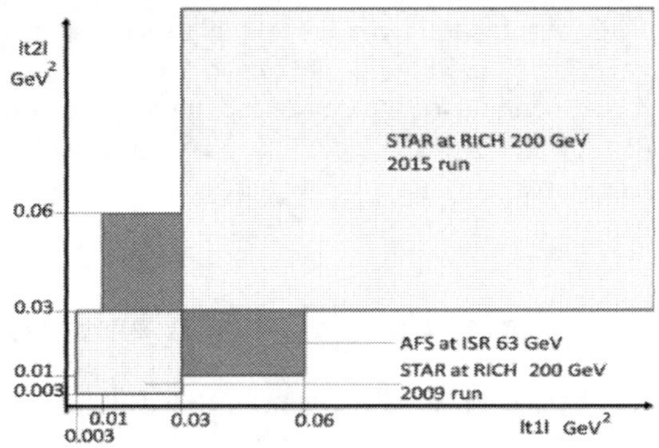

Fig. 2. Kinematical acceptances in the $|t_1|$, $|t_2|$ plane of the AFS [27,28] and STAR [18,23] pp CEP experiments

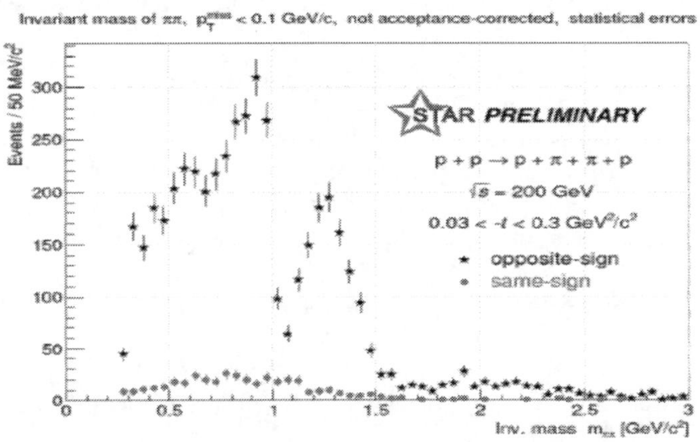

Fig. 3. $\pi^+\pi^-$ mass plot of STAR pp CEP raw data of the 200 GeV 2012 run with $0.03<|t_1|$, $|t_2|<0.3$ GeV^2 kinematic coverage of both scattered protons (Fig. from [23]).

The angular distribution of the $\pi^+\pi^-$ events in the peak is not available, but the S-wave nature of the signal can be deduced by the fact that already at $\sqrt{s}$= 63 GeV the $\pi^+\pi^-$ signal of the AFS experiment was dominantly S-wave up to 1.7 GeV [27,28,16] with a higher ltl window than STAR and because, as shown by comparing the 2009 and 2012 STAR data, lowering the ltl window suppresses dramatically the 2^{++} $f_2(1270)$ signal. The structure in fig.1 between 1.2 and 1.5 GeV centered around 1.35 GeV is too large to be associated to the $f_0(1500)$, which has a width of the order of 100 MeV, and is expected at 1.5 GeV. The statistics is too low to extract meaningfully a ratio between the $f_0(1370)$ and the $f_0(1500)$ contributions. Since most of the events in the peak occur at masses below 1.5 GeV, the $f_0(1500)$ may contribute to the spectrum via a destructive interference with the $f_0(1370)$ amplitude, as suggested in the analysis of the AFS data [28]. Independently of the interpretation of the structure in the 1.2-1.5 GeV energy region, quite noticeable is the fact that the STAR $\pi^+\pi^-$ spectrum of fig.1 drops nearly to zero at 1 GeV. This may be the result of the interference of the amplitudes of the low energy tail of the $f_0(1370)$ with the high energy part of the $f_0(980)$ plus the effect of the KKbar threshold, but very likely it might be due to the vanishing of the S.wave continuum for events selected in the low ltl kinematical region. The S-wave continuum, which is usually invoked with its destructive interference with the $f_0(980)$ amplitude to generate the drop at 1 GeV, is drastically reduced in comparison to the AFS data. It looks like the σ meson, which generates the broad structure above 0.5 GeV, is confined below 1 GeV. In other words, under this hypothesis the "red dragon" glueball proposed by Minkowski and Ochs [29], which features a low energy body centered at about 0.6 GeV and a head extending below the $f_0(1500)$, would be split into two separate parts, the σ and the relatively narrow $f_0(1370)$.

3 $f_0(1370)$ from $K\bar{K}$ and 4 pion decays in $p\bar{p}$ annihilations at rest

Decays of $f_0(1370)$ into K^+K^- and into K_sK_s are directly observable in $\pi^0K^+K^-$, $\pi^-K_sK_s$ and $\pi^0K_sK_s$ Dalitz plots and in the respective K^+K^- and K_sK_s mass plots of $p\bar{p}$ annihilations at rest in liquid H_2 and $n\bar{p}$ annihilations at rest in liquid D_2 targets (see fig. 4 from ref. [30]. K^+K^- decays of $f_0(1370)$ are the main source of the diagonal K^+K^- band comprised between the K^{*+} and K^{*-} bands in the $\pi^0K^+K^-$ Dalitz plot of the top left plate of fig.4 and of the peak at 1.4 GeV in the K^+K^- mass plot of the top right plate of fig. 4. Decays into K^+K^- of the $f_0(1500)$ are not easily visible in ppbar annihilations in liquid H_2 targets [31-33] (neither in the $\pi^0K^+K^-$ Dalitz plot, nor in the K^+K^- mass plot) because the $f_0(1500)$ signal is weak in liquid H_2 targets, since it is only partially produced from S-wave initial atomic states (while the $f_0(1370)$ signal is dominantly produced from 0^{-+} S-wave initial atomic states), and S-wave annihilations dominate in liquid H_2 targets (while P-wave annihilations dominate in low density H_2 targets) [32-34]. Moreover the weak $f_0(1500)$ signal may be masked by the nearly overlapping $f_2'(1525)$ signal.

Destructive interference of the $f_0(1500)$ amplitude with the $f_0(1370)$ amplitude generates the deep narrow flat valley at $M(K_sK_s)$=1.5 GeV in the $\pi^0K_sK_s$ Dalitz plot and in the K_sK_s mass plot obtained [30] using CERN [35,36] and BNL [37] bubble chamber data (see the bottom left and right plates in fig. 5). This K_sK_s valley in the $\pi^0K_sK_s$ bubble chamber Dalitz plot and in the K_sK_s mass plot represents probably the best evidence of KKbar decays of $f_0(1500)$. The same effect is visible in the high statistics $\pi^0K_lK_l$ Dalitz plot and in the associated K_lK_l mass plot produced by Crystal Barrel [38] (see figs.1 and 2 in [38]), where the valley is half as deep because of background and less mass resolution (notice that ref.[38] uses the frequency of ppbar $\rightarrow \pi^0K_sK_s$ annihilations in liquid H_2 as frequency of the their ppbar $\rightarrow \pi^0K_lK_l$ annihilations).

In the $\pi^- K_s K_s$ Dalitz plot produced [30] using CERN [39] and BNL [40] bubble chamber data (see the central left plate in fig. 4) the effect of the interference of the $f_0(1370)$ band with the K^{*-} bands is clearly noticeable . The $K_s K_s$ decays of $f_0(1370)$ are the main source of the peak at about 1.4 GeV in the $K_s K_s$ mass plot of the central right plate of fig.4. These data represent probably the best direct evidence of $K\bar{K}$ $f_0(1370)$ decays. A direct quantitative comparison of $\pi^- K_s K_s$ and $\pi^- 2\pi^0$ annihilation data in liquid D_2 would help providing relative decay branching ratios of $K\bar{K}$ and $\pi\pi$ decays of $f_0(1370)$, but it is not available.

Data of $p\bar{p}$ annihilations at rest into three final states ($\pi^+\pi^-\pi^0$, $K^+K^-\pi^0$ and $K_sK^+\pi^-$/ $K_sK^-\pi^+$) at 3 densities of the H_2 target (liquid H_2, NTP gaseous H_2 and 5 mbar NT gaseous H_2) have been collected and analyzed by the Obelix experiment at LEAR in a coupled channel analysis [32,33]. The spectra at the 3 target densities have markedly different shapes because the relative production of intermediate resonances ($\pi\pi$, KKbar, πK) depends dramatically on the J^{PC} initial atomic states of annihilation, and the J^{PC} fractions of annihilation depend substantially on the target density [34]. The $K^+K^-\pi^0$ data in liquid H_2 constrain the width of the $f_0(1370)$ so that the hypothesis of a broad $f_0(1370)$ extending below 1 GeV is naturally discarded, and the $\pi^+\pi^-\pi^0$ data are fit with sensible priors for the $f_0(1370)$ and the $f_0(1500)$ width and mass. The data show that the $f_0(1370)$ is essentially produced only from $J^{PC} = 0^{-+}$ S-wave initial states and that for $f_0(1500)$ there is a relevant production from $J^{PC} = 1^{++}$ P-wave initial states (this feature may explain why $f_0(1500)$ and not $f_0(1370)$ are observed in $p\bar{p}$ annihilations in flight at 900 and at 1640 MeV/c [41]). The coupled channel analysis shows that the ratio between the $K\bar{K}$ and $\pi\pi$ couplings of $f_0(1370)$ is of the order of 1 [32,33].

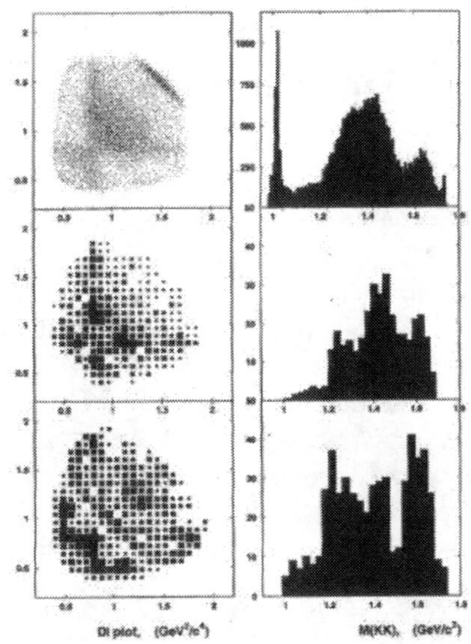

Fig. 4 $\pi^0K^+K^-$, $\pi^-K_sK_s$, $\pi^0K_sK_s$ Dalitz plots (top, middle and bottom left plates) and respective K^+K^- and K_sK_s mass plots (right plates) of pbar annihilations at rest in liquid H_2 and D_2 targets (picture derived from figs. 1, 2 and 5 of ref. [30]).

Decays of $f_0(1370)$ into $\pi^+\pi^-\pi^+\pi^-$ and into $4\pi^0$ measured in $\pi^-\pi^+\pi^-\pi^+\pi^-$ [42,43] and into $\pi^- 4\pi^0$ [44-47] pbar annihilations at rest in liquid D_2 are the dominant feature (well distinct from phase space) in the respective mass plots (see fig.5 left plate from ref. [43] and fig. 5 right plate from ref.[44-46]). The $4\pi^0$ data are particularly interesting because they do not suffer from any combinatorial background and there cannot be a $\rho\rho$ decay. These data show that $f_0(1370)$ decays dominantly into $\sigma\sigma$ (with a frequency twice that of the $\rho\rho$ decays and 6 times larger than $\pi\pi$ decays) and that 4 pion decays of $f_0(1370)$ are 10 times more frequent than 4 pion decays of $f_0(1500)$ [13].

The scenario which emerges is compatible with the expectation for a glueball to decay with equal rates into $\pi\pi$ and $K\bar{K}$ deriving from the equal couplings of gluons to $n\bar{n}$ and $s\bar{s}$ quarks [1,8] and expectations from QCD sum rules and low energy theorems for a scalar glueball to decay dominantly into σσ [48]. There is however the caveat that in the analysis of WA102 pp CEP data at 29 GeV the ρρ decay branching ratio of $f_0(1370)$ is found to be more than 4 times larger than the σσ decay branching ratio [49] and the ratios of the decay branching ratios of $f_0(1370)$ into $\pi\pi$, $K\bar{K}$ and ηη to the decay branching ratios of $f_0(1370)$ into $\pi\pi$ are of the order of 1, 0,5 and 0,2 [50,51].

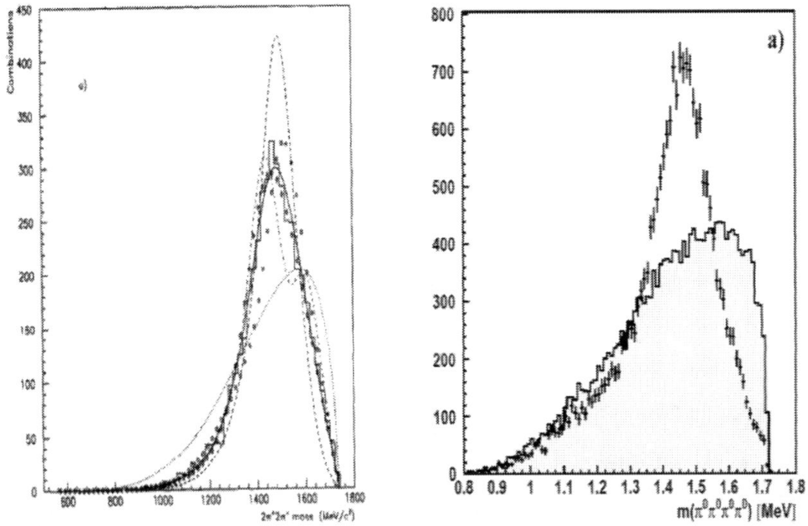

Fig. 5. $\pi^+\pi^-\pi^+\pi^-$ invariant mass in npbar → $\pi^-\pi^+\pi^-\pi^+\pi^-$ annihilations in liquid D_2; the histogram shows data, the dotted line shows phase space (left plate, from ref.[43]). $4\pi^0$ invariant mass in npbar → π^- $4\pi^0$ annihilations at rest in liquid D_2; the peak –data points with errors- is mainly due to $f_0(1370)$ decays to $4\pi^0$, the colored histogram shows phase space (right plate, from ref. [44-46]).

4 Conclusions and prospects

The STAR data confirm indications of earlier CEP experiment at lower energies that moving to higher energies and lower |t| windows singes out pomeron-pomeron interactions, suppresses 2^{++} production and therefore produces spectra where all low mass 0^{++} mesons can appear with little background.

The clear STAR signal of the $f_0(1370)$ decays into $\pi^+\pi^-$ pairs confirms experimentally the data of pbar annihilations at rest and shows that the σ meson and $f_0(1370)$ are distinct separate objects. The limited width of $f_0(1370)$ and its mass centered around 1370 MeV validate the analysis of data of pbar annihilations at rest into 3 pseudoscalars that used the hypothesis of the existence of $f_0(1370)$ as an individual object [32,33,43-47,52,53]. The ratios of the decay branching ratios of $f_0(1370)$ into σσ, ρρ, $\pi\pi$, KKbar and ηη to the decay branching ratios of $f_0(1370)$ into $\pi\pi$ are then of the order of 6, 3, 1, 1 and 0.02. There is however conflict between these decay branching ratios and the values obtained by WA102 in CEP measurements at 29 GeV [49-51]. The assessment of the values of the branching ratios of decays of the $f_0(1370)$ scalar meson into σσ and ρρ pairs and into $\pi\pi$, $K\bar{K}$, ηη pairs of pseudoscalar mesons derived from pbar annihilations at rest is discussed in more detail in ref. [54]. The decay properties of $f_0(1370)$ measured in pbar annihilations at rest and its production properties match the characteristics expected from an object that has a large gluon content [8-10,48].

CEP experiments at LHC at low |t| look extremely promising. The very large cms energy selects dominant pomeron-pomeron production. Low |t| select 0^{++} production within pomeron-pomeron production. Experiments equipped with precision detectors inside roman pots can approach the circulating beams and measure events with low |t| at both proton vertices. This is the case of ALFA-ATLAS and CMS-TOTEM, that have taken CEP data in 2015 with $\beta^* = 90$ m optics of LHC and a |t| coverage which could go down to 0.03 GeV^2 [55]. With CEP data at LHC it should be possible in the medium term to establish with confidence the relative decay branching ratios of $f_0(1370)$ and $f_0(1500)$ (and also $f_0(1710)$) into $\pi^+\pi^-$, K^+K^-, K_SK_S and compare with the values measured in antiproton annihilations at rest. Other important tasks will be to study of the narrow signal at 1450 MeV observed in CEP experiments at the Omega spectrometer in the $\pi^+\pi^-\pi^+\pi^-$ decay channel [56-59] and interpreted as possibly due to the interfering amplitudes of $f_0(1370)$ and $f_0(1500)$[58,59], to study the other 4 pion decay channels, and to compare the $\sigma\sigma$ and $\rho\rho$ decays of $f_0(1370)$ and $f_0(1500)$. In the longer term, measurements with larger β^* and central detectors tuned to measure both low energy charged and neutral prongs the |t| window could be extended down to 0.003 GeV^2 and CEP could be measured in kinematical regimes where 0^{++} production should be definitively overwhelming and permit to do a complete study of all low energy scalars.

References

1. H. Fritzsch and M. Gell-Mann, Current algebra: quarks and what else ?, Proceedings of the XVI Int. Conf. on High Energy Physics, Chicago, Vol.2, 135, J. D. Jackson, A. Roberts, Eds., eConfC720906V2 (1972), arXiv:hep-ph/0208010(2002).
2. H. Fritzsch and P. Minkowski, Nuovo Cimento **30** A, 393 (1975)
3. J. R. Pelaez, Phys. Rep. **658**, 1 (2016)
4. W. Ochs, J. Phys. G: Nucl. Part. Phys. **40**, 043001,1 pp. 1-68 (2013)
5. R. L. Jaffe, Phys. Rev. D **15**, 267 (1977); R. L. Jaffe, Phys. Rev. D **15**, 281 (1977)
6. J. Weinstein and N. Isgur, Phys. Rev. D **41**, 2236 (1990)
7. M. Pennington, arXiv:1411.7902 [nucl-th] (2014)
8. D. Robson, Nucl. Phys. B **130**, 328 (1977)
9. S. Narison and G. Veneziano, Int. J. Mod. Phys. A **4**, .2751 (1989)
10. M. Strohmeier-Presicek et al., Phys. Rev. D **60**, 054010 (1999)
11. M. S. Chanowitz, Phys. Rev. Lett. **95**, 172001 (2005)
12. U. Gastaldi, Nucl. Phys. B (Proc. Suppl.) **96**, 234 (2001).
13. U. Thoma, Eur. Phys. J. A **18**, 135, (2003)
14. E. Klempt and A. Zeitsev, Phys. Rep. **454**, 1 (2007) .
15. V. Crede and C. A. Meyer, Prog. Particle and Nucl. Phys. **63**, 74 (2009)
16. M. G. Albrow, T. D. Coughlin and J. R. Forshaw, arXiv:1006.1289 [hep-ph] (2010); M. Albrow, Int. J. Mod. Phys. A **29** (Nr.28), 1 (2014); M. Albrow, arXiv:1701.09092, (2017).
17. H. Cheng, C. Chua and K. Liu, Phys. Rev. D **92**, 094006, 1 (2015)
18. J. Turnau for the STAR Collaboration, EPJ Web of Conferences **37** 06010, 1 (2012)
19. U. Gastaldi and M. Berretti, arXiv:1804.09121 [hep-ph] (2018).
20. U. Gastaldi and M. Berretti, arXiv:1809.04496 [hep-ph] (2018).

21. J. Turnau, XXII Int. Workshop on Deep-Inelastic Scattering and Related Subjects, 28 April-2 May, POS (DIS2014) 098, 1 (2014)
22. L. Adamczuk, W. Guryn and J. Turnau, Int. J. Mod. Phys. A **29** (28), 1446010, pp. 1 (2014)
23. W. Guryn, EPJ Web of Conferences **120** 02008, 1 (2016)
24. R. Sikora for the STAR Collaboration, arXiv:1611.07823 [nucl-ex], (2016).
25. K. Yip, Central Exclusive Production at the STAR experiment, presented at the 3rd Elba Workshop on Forward Physics, La Biodola, Isola d'Elba, (2016).
26. S. N. Ganguli and D. P. Roy, Phys. Rep. **67**, 201 (1980)
27. AFS Collaboration, T. Akesson et al., Phys. Lett. **133** B, 268 (1983)
28. AFS Collaboration, T. Akesson et al., Nucl. Phys. B **264**, 154 (1986)
29. P. Minkowski and W. Ochs, Eur. Phys. J. C **9**, 283 (1999); P. Minkowski and W. Ochs, Eur. Phys. J. C **9**, 283 (1999)
30. V. Ableev et al., Nucl. Phys. B (Proc. Suppl.) **86,** 351 (2000); A. Ableev et al., ppbar → $K_sK_s\pi^0$ annihilations at rest in liquid H_2 and $f_0(1500)\rightarrow K_sK_s$ decays, LNL-INFN(Rep)142/99 (1999)
31. A. Abele et al. (Crystal barrel Collaboration), Phys. Lett B **468**, 178 (1999)
32. M. Bargiotti et al. (Obelix Collaboration), Eur. Phys. J. C **26**, 371 (2003)
33. M. Bargiotti et al. (Obelix Collaboration), Phys. Lett. **B 561**, 233 (2003)
34. U. Gastaldi et al., Phys. Lett. B **320**, 193 (1994)
35. R. Armenteros et al., Phys. Lett. **17**, 170 (1965)
36. B. Conforto et al., Nucl. Phys. B **3**, 469 (1967)
37. N. Barash et al., Phys. Rev. **139**, 1659 (1965)
38. A. Abele et al. (Crystal barrel Collaboration), Phys. Lett B **385**, 425 (1996)
39. A. Bettini et al., Nuoco Cimento **63** A, 1199 (1969)
40. L. Gray et al., Phys. Rev. D **27**, 307 (1983)
41. C. Amsler et al. (Crystal Barrel Collaboration), Phys. Lett. B **639**, 165 (2006)
42. A. Bettini et al., Nuovo Cimento **42** A, 695 (1966)
43. M. Gaspero, Nucl. Phys. A **562**, 407 (1993)
44. U. Thoma (Crystal barrel Collaboration), Nucl. Phys. B (Proc. Suppl.) **56** A, 216 (1997)
45. U. Thoma, Nucl. Phys. A **675**, 76c (2000)
46. A. Abele et al. (Crystal barrel Collaboration), Eur. Phys. J. C **19**, 667 (2001)
47. A. Abele et al. (Crystal barrel Collaboration), Eur. Phys. J. C **21**, 261 (2001)
48. S. Narison, Nucl. Phys. B **509**, 1998, pp. 312 (1998); S. Narison, Phys. Rev. D **73**, 2006, 114024 (2006); S. Narison, Nucl. Phys. B (Proc. Suppl.)186, 306 (2009)
49. WA102 Collaboration, D. Barberis et al., Phys. Lett. B **471**, 440 (2000)
50. WA102 Collaboration, D. Barberis et al., Phys. Lett. B **462**, 462 (1999).
51. WA102 Collaboration, D. Barberis et al., Phys. Lett. B **479**, 59 (2000)
52. A. Abele et al., Nucl. Phys. A **609**, 562 (1996)
53. D. V. Bugg, Eur. Phys. J. C **52**, 55 (2007)
54. U. Gastaldi, to be published in proceedings of QCD 18, Montpellier, France, 2-6 July (2018)
55. The TOTEM Collaboration, CERN-LHCC D 013-009/LHCC-P-007 (2013)
56. WA76 Collaboration, T. A. Armstrong et al., Phys. Lett. B **228**, 536 (1989)
57. WA91 Collaboration, S. Abatsis et al., Phys. Lett. B **324**, 509 (1994)
58. WA91 Collaboration, F. Antinori et al., Phys. Lett. B **353,** 589 (1995)
59. WA102 Collaboration, D. Barberis et al., Phys. Lett. B **413**, 217 (1997)

Heavy flavour spectroscopy and hadron properties from LHCb

Antimo Palano[1,*]

[1] *INFN and University of Bari, Via Orabona 4, 70125 Bari, Italy*

Abstract. The LHCb experiment is designed to study the properties and decays of heavy flavored hadrons produced in pp collisions at the LHC. The data collected enables precision spectroscopy studies of beauty and charm hadrons. The latest results on spectroscopy of conventional and exotic hadrons are reviewed.

1 Introduction

In the simple quark model, only two types of quark combinations are required to account for the existing hadrons, i.e. $q\bar{q}$ combinations form mesons, while baryons are made up of three quarks. However, in the quark model proposed by Gell-Mann and Zweig in 1960s [1] other SU(3) color-neutral combinations of quarks and gluons such as gg glueballs, $q\bar{q}g$ hybrids, $q\bar{q}q\bar{q}$ tetraquarks, $q\bar{q}qqq$ pentaquarks etc. are allowed. The world's largest data sample of beauty and charm hadrons collected by LHCb provides great opportunities for studying the production and properties of heavy hadrons. The present document describes some recent results from LHCb in exotic and standard hadron spectroscopy.

2 The observation of pentaquarks in $\Lambda_b^0 \to J/\psi pK^-$

Evidences/observations of pentaquarks candidates were reported around 2005 by several experiments, always with low statistics and approximate statistical treatment of the data. For a summary and references see ref. [2]. High statistics searches gave negative results (see for example ref. [3]).

LHCb experiment studied the decay of $\Lambda_b^0 \to J/\psi pK^-$ [4]. A sample of $26,007 \pm 166$ signal candidates, on 5.4% background, within ±15 MeV of the $J/\psi pK^-$ mass peak has been selected using the full Run I dataset. An anomalous peaking structure in the $J/\psi p$ invariant mass spectrum was observed on the Dalitz plot. This structure can also be seen in the invariant mass projections shown in Fig. 1. If the peak structure observed in Fig. 1(b) represents a resonance strongly decaying into $J/\psi p$, the minimal valence quarks should be $c\bar{c}uud$, a charmonium pentaquark state. A full six-dimensional amplitude fit with resonance invariant mass, three helicity angles and two differences between decay planes has been applied to describe the data. It allowed interference between the decays $\Lambda_b^0 \to J/\psi\Lambda^*$ and $\Lambda_b^0 \to P_c^+K^-$, with $\Lambda^* \to pK^-$ and $P_c^+ \to J/\psi p$, where P_c^+ indicates the pentaquark candidate. The amplitude model in the fit contains 14 well-defined Λ^* states listed by Particle Data Group [5] and two P_c^+ states: $P_c(4380)^+$ and $P_c(4450)^+$. With a full amplitude analysis, the masses of these two states are measured to be $4380 \pm 8(\text{stat}) \pm 29(\text{syst})$ MeV and $4449.8 \pm 1.7(\text{stat}) \pm 2.5(\text{syst})$ MeV, with the widths

* e-mail: antimo.palano@ba.infn.it

© The Authors, published by EDP Sciences. This is an open access article distributed under the terms of the Creative Commons Attribution License 4.0 (http://creativecommons.org/licenses/by/4.0/).

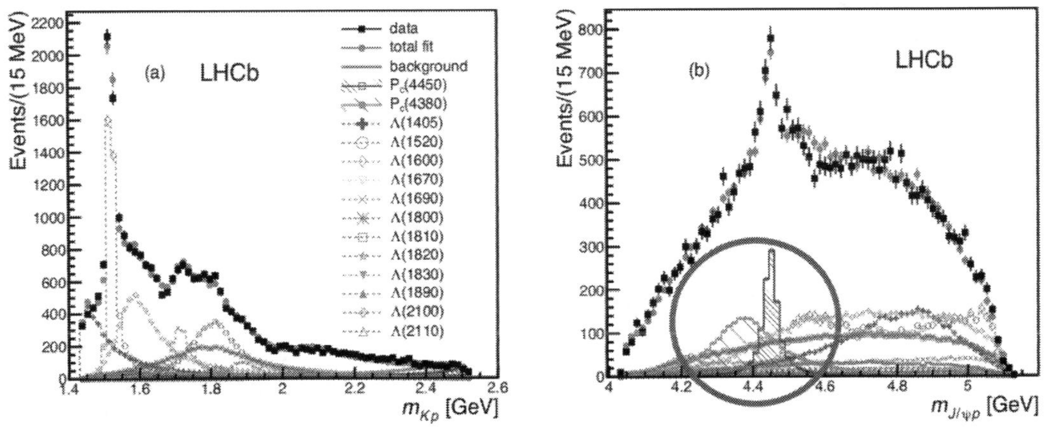

Figure 1. Dalitz plot projections of $\Lambda_b^0 \to J/\psi pK^-$ with superimposed the results from the amplitude analysis.

$205 \pm 18(\text{stat}) \pm 86(\text{syst})$ MeV and $39 \pm 5(\text{stat}) \pm 19(\text{syst})$ MeV, respectively. The preferred spin-parity values are $(3/2^-, 5/2^+)$, $(3/2^+, 5/2^-)$ and $(5/2^+, 3/2^-)$, where the first one is the J^P assignment given by the best fit. This analysis has been extended with a model-independent approach [6] which gave consistent results. Results consistent with expectations have also been obtained from the amplitude analysis of the Cabibbo-suppressed decay $\Lambda_b^0 \to J/\psi p\pi^-$ [7].

3 Confirmation of resonant nature of $Z_c(4430)^-$

The $Z_c(4430)^-$ state was first reported by the Belle collaboration [8] as a charged resonance structure in the $\psi'\pi^-$ invariant mass distribution in the decay $B^0 \to \psi' K^+\pi^-$. The BaBar collaboration could explain the enhancement as a reflection of the known K^* states [9].

The LHCb experiment collected about 25,000 B^0 signal events, a yield one order of magnitude larger than the BaBar or Belle datasets, and observed the $Z_c(4430)^-$ with a significance larger than 13.9σ [10]. The LHCb collaboration performed both a model independent analysis as well as a full amplitude analysis. The projection of the fit including the $Z_c(4430)^-$ state is shown in Fig. 2 and evidences a good description of the data. The measured parameters are $M(Z_c) = 4475 \pm 7^{+15}_{-25}$ MeV, $\Gamma(Z_c) = 172 \pm 13^{+37}_{34}$ MeV. The Argand diagram of the $Z_c(4430)^-$ amplitude (Fig. 2) shows the resonance behavior for the first time. The spin-parity was measured to be 1^+, by excluding other hypotheses by at least 9.7σ. For a charged charmonium state, $Z_c(4430)^-$ has a minimum quark content of $c\bar{c}u\bar{d}$ which clearly does not fit into the traditional quark model.

4 Study of $B \to J/\psi\phi K$

There has been a great deal of experimental and theoretical interest in $J/\psi\phi$ mass structures in $B^+ \to J/\psi\phi K^+$ decays since the CDF collaboration presented 3.8σ evidence for a near-threshold $X(4140)$ mass peak, with width $\Gamma = 11.7$ MeV [11]. Subsequent measurements resulted in a confusing experimental situation. LHCb experiment has reconstructed 4289 ± 151 $B^+ \to J/\psi\phi K^+$ decays, with roughly uniform efficiency and a relatively low background across the entire $J/\psi\phi$ mass range [12]. A full amplitude analysis has been performed, including resonant contributions from K^* resonances

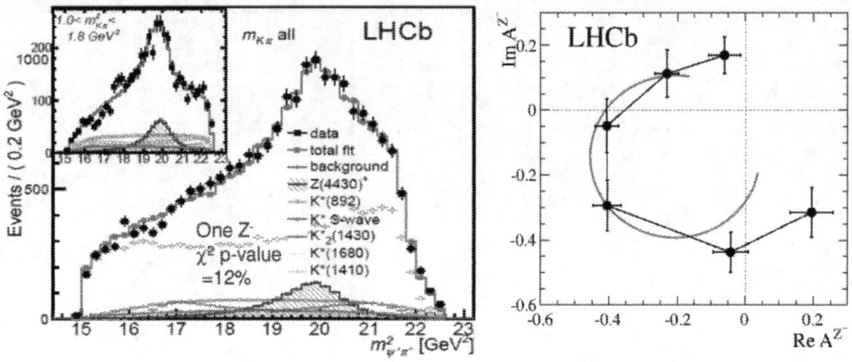

Figure 2. (Left) $\psi'\pi^-$ mass distribution from $\bar{B}^0 \to \psi'\pi^- K^+$ with superimposed the results from the amplitude analysis. (Right) Argand diagram for $Z_c(4430)^-$.

Table 1. Parameters of the resonances decaying to $J/\psi\phi$ resulting from the amplitude analysis of $B^+ \to J/\psi\phi K^+$

Resonance	σ	J^{PC}	M (MeV)	$\Gamma(MeV)$
X(4140)	8.4	1^{++}	$4160 \pm 4^{+5}_{-3}$	$83 \pm 21^{+21}_{-14}$
X(4274)	5.8	1^{++}	$4273 \pm 8^{+17}_{-4}$	$56 \pm 11^{+8}_{-11}$
X(4500)	6.1	0^{++}	$4506 \pm 11^{+12}_{-15}$	$92 \pm 21^{+21}_{-20}$
X(4700)	5.6	0^{++}	$4704 \pm 10^{+14}_{-24}$	$120 \pm 31^{+42}_{-33}$

decaying to ϕK^+. Four resonant contributions in the $J/\psi\phi$ final state are found to contribute, whose parameters and quantum numbers are given in Table 1. The fit results are superimposed to the $J/\psi\phi$ mass spectrum in fig. 3. It can be noted that the $X(4140)$ width is substantially larger than previously

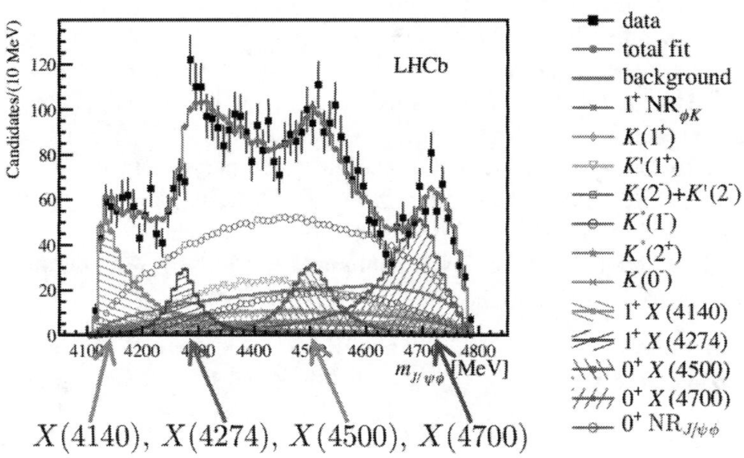

Figure 3. $J/\psi\phi$ mass projection with superimposed the fit result from the amplitude analysis of $B^+ \to J/\psi\phi K^+$

determined. The interpretation of these states is still an open issue.

5 Observation of five new Ω_c^0 states

The spectroscopy of singly charmed baryons cqq' is intricate. With three quarks and numerous degrees of freedom, many states are expected. Among the expected charmed baryon states the spectrum of Ω_c baryons, which have quark content css and isospin zero, was largely unknown: only the Ω_c and $\Omega_c(2770)^0$, presumed to be the $J^P = 1/2^+$ and $3/2^+$ ground states, were observed.

The LHCb experiment has reconstructed a large ($\approx 10^6$ events) and high-purity (83%) sample of Ξ_c^+ baryons, reconstructed in the Cabibbo-suppressed decay mode $pK^+\pi^-$ [13]. Surprisingly, in the $\Xi_c^+K^-$ mass spectrum, shown in Fig. 4, five new, narrow excited Ω_c states are observed: the $\Omega_c(3000)^0$, $\Omega_c(3050)^0$, $\Omega_c(3066)^0$, $\Omega_c(3090)^0$, and $\Omega_c(3119)^0$. The measured masses and widths are reported in Table 2. The data indicate also the presence of a broad structure around 3188 MeV that is fitted as a single resonance but could be produced in other ways, for example as a superposition of several states. In addition, the partially reconstructed decay $\Omega_c(3066)^0$, $\Omega_c(3090)^0$ and $\Omega_c(3119)^0$ in their possible decay to $\Xi_c'^+K^-$ are observed via their feed-down in the low mass region. This observation has generated large interest, especially for the presence of very narrow states, and many phenomenological models have been proposed, including the possibility of the presence of pentaquark states.

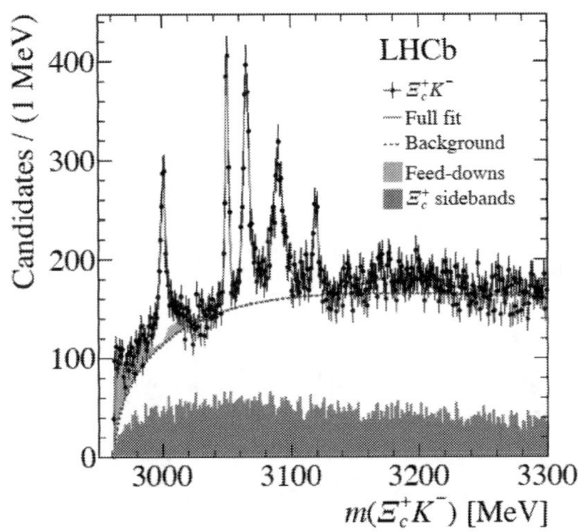

Figure 4. $\Xi_c^+K^-$ mass spectrum with the observation of five new Ω_c states.

6 Observation of the double charmed baryon Ξ_{cc}^{++}

The constituent quark model of hadrons predicts the existence of three weakly decaying ground state $J^P = 1/2^+$ baryons with two charmed valence quarks: Ξ_{cc}^+ (ccd) and Ξ_{cc}^{++} (ccu), which form an isospin doublet, and Ω_{cc}^+ (ccs). The only previously published evidence of their existence comes from the observations of the Ξ_{cc}^+ baryon claimed by the SELEX collaboration [14]. They observed narrow peaking low statistics structures at a mean mass of 3518.7 ± 1.7 MeV in the mass spectra of $\Lambda_c^+K^-\pi^+$ and pD^+K^-.

LHCb experiment has searched for double charmed baryon states using the decay mode $\Xi_{cc}^{++} \to \Lambda_cK^-\pi^+\pi^+$, with $\Lambda_c \to pK^-\pi^+$ (which has a 10% branching fraction) [15]. The $\Lambda_cK^-\pi^+\pi^+$ mass

Table 2. Measured parameters of the five new Ω_c states.

Resonance	Mass (MeV)	Γ (MeV)
$\Omega_c(3000)^0$	$3000.4 \pm 0.2 \pm 0.1^{+0.3}_{-0.5}$	$4.5 \pm 0.6 \pm 0.3$
$\Omega_c(3050)^0$	$3050.2 \pm 0.1 \pm 0.1^{+0.3}_{-0.5}$	$0.8 \pm 0.2 \pm 0.1$
		$< 1.2 MeV$, 95% CL
$\Omega_c(3066)^0$	$3065.6 \pm 0.1 \pm 0.3^{+0.3}_{-0.5}$	$3.5 \pm 0.4 \pm 0.2$
$\Omega_c(3090)^0$	$3090.2 \pm 0.3 \pm 0.5^{+0.3}_{-0.5}$	$8.7 \pm 1.0 \pm 0.8$
$\Omega_c(3119)^0$	$3119.1 \pm 0.3 \pm 0.9^{+0.3}_{-0.5}$	$1.1 \pm 0.8 \pm 0.4$
		$< 2.6 MeV$, 95% CL
$\Omega_c(3188)^0$	$3188.1 \pm 4.8 \pm 12.7$	$60 \pm 15 \pm 11$

spectrum is shown in Fig. 5(a) where a prominent signal can be observed. LHCb observes 313 ± 33

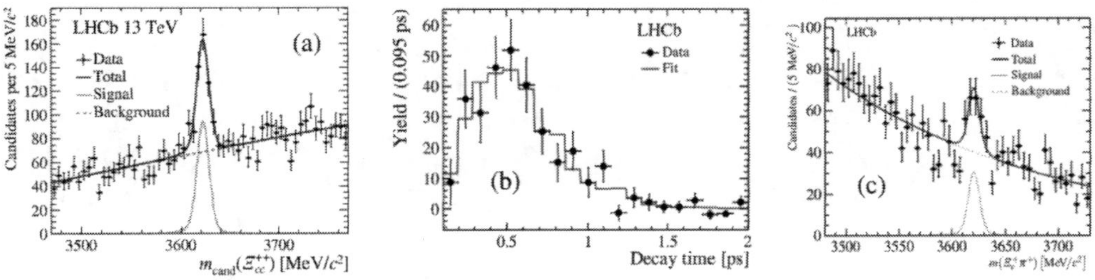

Figure 5. (a) $\Lambda_c K^- \pi^+ \pi^+$ mass spectrum with the observation of Ξ_{cc}^{++}. (b) Background-subtracted decay-time distribution of selected Ξ_{cc}^{++} candidates with superimposed fit results. (c) Observation of $\Xi_{cc}^{++} \to \Xi_c^+ \pi^+$.

decays with a significance of $> 12\sigma$ of a Ξ_{cc}^{++} state that has a mass of $m(\Xi_{cc}^{++}) = 3621.40 \pm 0.72(\text{stat}) \pm 0.27(\text{syst}) \pm 0.14(\Lambda_c)$ MeV and a width that is much smaller than experimental resolution and that is consistent with a particle that flies a significant distance before decaying. This is the first published observation of the Ξ_{cc}^{++} baryon. The published SELEX Ξ_{cc}^{+} state has a mass that is 103 ± 2 MeV smaller than the Ξ_{cc}^{++} state, so the two observations are inconsistent with being isospin partners. A consistent interpretation of the two observations may prove challenging.

7 Lifetime measurement of Ξ_{cc}^{++}

LHCb experiment has performed a Ξ_{cc}^{++} lifetime measurement relative to that of the $\Lambda_b^0 \to \Lambda_c^+ \pi^- \pi^+ \pi^-$ which has the same topology [16]. An unbinned maximum likelihood fit to the background subtracted (using the sFit method) Ξ_{cc}^{++} decay time distribution has been performed using the function

$$R(t) = \frac{f_{\Xi_{cc}^{++}}}{f_{\Lambda_b^0}} \times \frac{\epsilon_{\Lambda_b^0}(t)}{\epsilon_{\Xi_{cc}^{++}}(t)} = R(0) \times e^{-(\frac{1}{\tau_{\Xi_{cc}^{++}}} - \frac{1}{\tau_{\Lambda_b^0}})t},$$

where ϵ indicate the time dependent efficiencies. The fitted and background subtracted Ξ_{cc}^{++} lifetime distribution is shown in Fig. 5(b). The measured lifetime is

$$\tau_{\Xi_{cc}^{++}} = 0.256^{+0.024}_{-0.022} \pm 0.014\ ps,$$

in the lower end but compatible with the expected theoretical predictions. This confirms the weakly decay nature of the newly discovered Ξ_{cc}^{++} state.

8 Observation of a new $\Xi_{cc}^{++} \to \Xi_c^+ \pi^+$ decay mode

Searching for new decay modes is critical to understand the dynamics of weak decays of doubly heavy baryons. The present LHCb analysis makes use of a data sample of 1.7 fb^{-1} collected at 13 TeV [17]. The normalization channel $\Xi_{cc}^{++} \to \Lambda_c K^- \pi^+ \pi^+$ is used and a significant peak (5.9σ) of 91 ± 20 events in the $\Xi_c^+ \pi^+$ mass spectrum (shown in Fig. 5(c)), where $\Xi_c^+ \to pK^-\pi^+$ is observed. The resulting mass measurement is $m(\Xi_{cc}^{++}) = 3620 \pm 1.5(\text{stat}) \pm 0.4(\text{sys}) \pm 0.3(\Xi_c^+)$ MeV/c^2, consistent with the previous measurement. The ratio of branching fractions is also measured as

$$\frac{\mathcal{B}(\Xi_{cc}^{++} \to \Xi_c^+ \pi^+) \times \mathcal{B}(\Xi_c^+ \to pK^-\pi^+)}{\mathcal{B}(\Xi_{cc}^{++} \to \Lambda_c K^-\pi^+\pi^+) \times \mathcal{B}(\Lambda_c \to pK^-\pi^+)} = 0.035 \pm 0.009(\text{stat}) \pm 0.003(\text{syst}).$$

9 Measurement of the Ω_c baryon lifetime

Charm baryon lifetimes are known much less precisely than charm meson ones. The expected lifetime hierarchy should be: $\tau_{\Xi_c^+} > \tau_{\Lambda_c^+} > \tau_{\Xi_c^0} > \tau_{\Omega_c^0}$ and current measurements are consistent with this. LHCb experiment makes use a sample of approximately 10^3 semileptonic $\Omega_b^- \to \Omega_c^0 \mu^- \bar{\nu}_\mu X$, with $\Omega_c^0 \to pK^-K^-\pi^+$ [18] and a dataset of 3 fb^{-1}. To reduce uncertainties the ratio with respect to the $D^+ \to K^-\pi^+\pi^+$ lifetime is measured. The Ω_c^0 signal is shown in Fig. 6 together with the decay time distribution. The fitted Ω_c^0 lifetime is $\tau_{\Omega_c^0} = 268 \pm 21_{\text{stat}} \pm 10_{\text{syst}} \pm 2_{D^+}\ fs$, four times larger than and

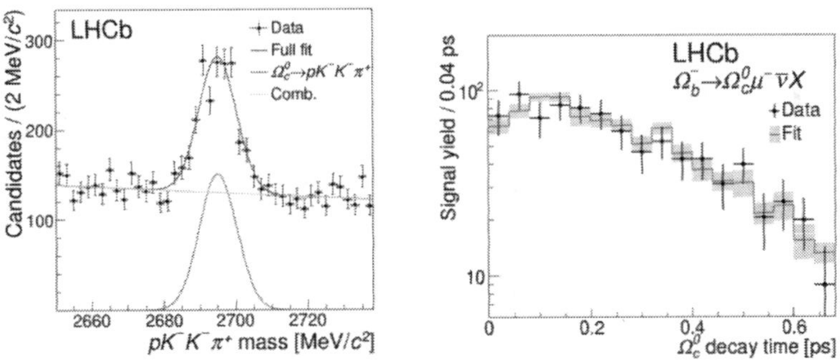

Figure 6. (Left) $\Omega_c^0 \to pK^-K^-\pi^+$ candidates and (Right) Ω_c^0 decay time distribution with superimposed fit result.

inconsistent with the world average value of (69 ± 12) fs. The present measurement produces the modified lifetimes hierarchy as $\tau_{\Xi_c^+} > \tau_{\Omega_c^0} > \tau_{\Lambda_c^+} > \tau_{\Xi_c^0}$.

10 Observation of a new Ξ_b^- resonance

A new $\Xi_b(6227)^-$ state is observed by LHCb decaying to both $\Xi_b^0\pi^-$ and $\Lambda_b^0 K^-$ [19]. The analysis makes use of a dataset of 1.0 fb^{-1} at 7 TeV, 2.0 fb^{-1} at 8 TeV and 1.5 fb^{-1} at 13 TeV. Three independent reconstructed decay chains are used: $\Xi_b(6227)^- \to \Lambda_b^0(\to \Lambda_c^+\pi^-)K^-$, $\Xi_b(6227)^- \to \Xi_b^0(\to \Xi_c^+\mu^- X)\pi^-$, and $\Xi_b(6227)^- \to \Lambda_b^0(\to \Lambda_c^+\mu^- X)K^-$. In the case of semileptonic decays the missing p_ν

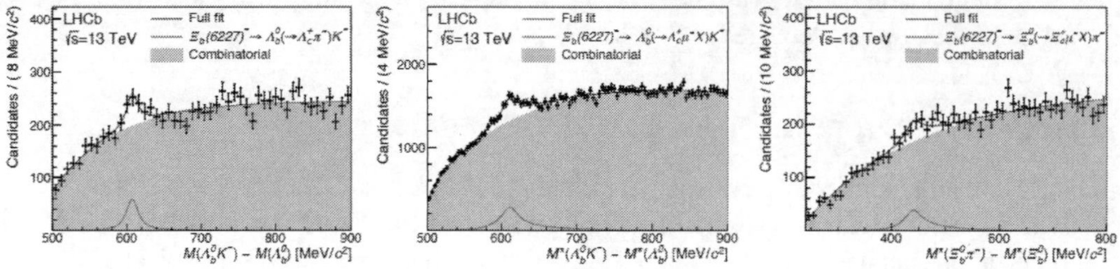

Figure 7. Observation of the new $\Xi_b(6227)^-$ state in the three reconstructed decay modes for the 13 TeV data.

momentum is reconstructed assuming a zero-mass particle that balances the momentum transverse to the direction of Λ_b^0/Ξ_b^0. The total invariant mass is then constrained to have the known Λ_b^0/Ξ_b^0 mass. The new resonance is observed in three different mass spectra as shown in Fig. 7 for the 13 TeV data. Similar signals are observed in the 7/8 TeV data. The new state could be identified as $\Xi_b(1P)$ or $\Xi_b(2S)$ states. Resonance resonance parameters are

$$m_{\Xi_b(6227)^-} = 6226.9 \pm 2.0\text{stat} \pm 0.3\text{syst} \pm 0.2(\Lambda_b^0)\,\text{MeV}/c^2, \Gamma_{\Xi_b(6227)^-} = 18.1 \pm 5.4\text{stat} \pm 1.8\text{syst}\,\text{MeV}/c^2.$$

References

[1] M. Gell-Mann, Phys. Lett. **8**, 214 (1964).
[2] A. Dzierba, C. Mayer and A. Szczepaniak, hep-ex/04120 (2005).
[3] B. Aubert *et al.* [BaBar Collaboration], Phys. Rev. Lett. **95**, 042002 (2005). J. M. Link *et al.* [FOCUS Collaboration], Phys. Lett. B **639**, 604 (2006).
[4] R. Aaij *et al.* [LHCb Collaboration], Phys. Rev. Lett. **115**, 072001 (2015).
[5] C. Patrignani et al. (Particle Data Group), Chin. Phys. C, **40**, 100001 (2016) and 2017 update.
[6] R. Aaij *et al.* [LHCb Collaboration], Phys. Rev. Lett. **117**, no. 8, 082002 (2016).
[7] R. Aaij *et al.* [LHCb Collaboration], Phys. Rev. Lett. **117**, no. 8, 082003 (2016).
[8] S. K. Choi *et al.* [Belle Collaboration], Phys. Rev. Lett. **100**, 142001 (2008).
[9] B. Aubert *et al.* [BaBar Collaboration], Phys. Rev. D **79**, 112001 (2009).
[10] R. Aaij *et al.* [LHCb Collaboration], Phys. Rev. Lett. **112**, no. 22, 222002 (2014).
[11] T. Aaltonen *et al.*, [CDF Collaboration], Phys. Rev. Lett. **102** , 242002 (2009).
[12] R. Aaij *et al.* [LHCb Collaboration], Phys. Rev. Lett. **118**, 022003 (2017).
[13] R. Aaij *et al.* [LHCb Collaboration], Phys. Rev. Lett. **118**, 182001 (2017).
[14] A. Ocherashvili *et al.* [SELEX Collaboration], Phys. Lett. B **628**, 18 (2005).
[15] R. Aaij *et al.* [LHCb Collaboration], Phys. Rev. Lett. **119**, 112001 (2017).
[16] R. Aaij *et al.* [LHCb Collaboration], Phys. Rev. Lett. **121** (2018) no.5, 052002.
[17] R. Aaij *et al.* [LHCb Collaboration], arXiv:1807.01919 [hep-ex].
[18] R. Aaij *et al.* [LHCb Collaboration], Phys. Rev. Lett. **121** (2018) 092003.
[19] R. Aaij *et al.* [LHCb Collaboration], Phys. Rev. Lett. **121** (2018) no.7, 072002.

Global Dyson–Schwinger–Bethe–Salpeter Approach to Mesons with Open Flavour

Thomas Hilger[1,2,*], María Gómez-Rocha[3,**], Andreas Krassnigg[2,***], and Wolfgang Lucha[1,****]

[1]Institute for High Energy Physics, Austrian Academy of Sciences, Nikolsdorfergasse 18, A-1050 Vienna, Austria
[2]Institute of Physics, University of Graz, NAWI Graz, A-8010 Graz, Austria
[3]European Centre for Theoretical Studies in Nuclear Physics and Related Areas, Villa Tambosi, 38123 Villazzano (Trento), Italy

Abstract. Exploiting an interplay of the Bethe–Salpeter equation enabling us to regard mesons as bound states of quark and antiquark and the Dyson–Schwinger equation controlling the dressed quark propagator, we amend existing studies of quarkonia by a comprehensive description of open-flavour mesons composed of all conceivable combinations of quark flavour. Employing throughout a fixed set of model parameters, we predict some basic characteristics of these mesons, *i.e.*, their masses, leptonic decay constants and corresponding in-hadron condensates entering in a generalized formulation of the Gell-Mann–Oakes–Renner relation.

1 Poincaré-Covariant Analysis of Bound States: the Physics Case

Within the framework of quantum field theories, the Bethe–Salpeter equation, underpinned by that Dyson–Schwinger equation that governs the dressed quark propagator, offers an approach to quark–antiquark bound states in Poincaré-covariant manner. This quark Dyson–Schwinger equation is part of an *infinite* tower of coupled Dyson–Schwinger equations which forces us to truncate this tower somehow to a *finite* set of coupled relations. The merits of such a covariant approach are evident: ⋆ Needless to say, quark models constitute a very convenient formalism for comprehensive investigations of hadronic states by comparatively simple technical means. ⋆ Nonperturbative approaches get rather limited or constrained by technical or computational boundary conditions. ⋆ Covariant treatments use QCD input and modelling to bridge this gap. Understandably, the first target of a covariant approach usually is the case of quarkonia, bound states of a quark and its antiquark, and hence flavourless. In order to arrive at a comprehensive picture, we complete this kind of investigations by applying a single common formalism to *all* conceivable flavour combinations and fathom its implications for the predicted meson masses, decay constants, and in-meson condensates, by comparison with experiment or other findings:

★ Clearly, covariant descriptions of open-flavour mesons have been, and will remain, limited.

★ Nevertheless, such covariant studies produce utmost comprehensive sets of predictions [1].

*e-mail: thomas.hilger@uni-graz.at
**e-mail: mariagomezrocha@gmail.com
***e-mail: andreas.krassnigg@uni-graz.at
****e-mail: Wolfgang.Lucha@oeaw.ac.at

© The Authors, published by EDP Sciences. This is an open access article distributed under the terms of the Creative Commons Attribution License 4.0 (http://creativecommons.org/licenses/by/4.0/).

2 Dyson–Schwinger–Bethe–Salpeter Liaison

Within the Bethe–Salpeter formalism for the Poincaré-covariant description of bound states, a bound state, of total momentum P, composed of quark and antiquark of relative momentum p, is represented by this bound state's Bethe–Salpeter *amplitude* $\Gamma(p;P)$ or Bethe–Salpeter *wave function* $\chi(p;P)$, being related by the dressed propagators $S_{1,2}$ of the bound-state constituents:

$$\chi(p;P) \equiv S_1(p+\eta P)\,\Gamma(p;P)\,S_2(p-(1-\eta)P)\,, \qquad \eta \in [0,1]\,.$$

These dressed propagators can be found as solutions of the Dyson–Schwinger equation for the quark two-point function. In rainbow truncation, this Dyson–Schwinger equation involves the quark wave-function renormalization constant, Z_2, and bare mass, m_b, the transverse projector

$$T_{\mu\nu}(k) \equiv \delta_{\mu\nu} - \frac{k_\mu k_\nu}{k^2}$$

in the free gluon propagator in Landau gauge, its translationally invariant integration measure, Pauli–Villars regularized at a scale Λ and indicated by $\int_q^\Lambda$, and an effective coupling, $k^2\,\mathcal{G}(k^2)$, subsuming or mimicking all effects of both full gluon propagator and full quark–gluon vertex:

$$S^{-1}(p) = Z_2\,(\mathrm{i}\,\gamma\cdot p + m_b) + \frac{4}{3}\,Z_2^2 \int_q^\Lambda \mathcal{G}\big((p-q)^2\big)\,T_{\mu\nu}(p-q)\,\gamma_\mu\,S(q)\,\gamma_\nu\,.$$

A mass renormalization Z_m relates bare, m_b, and at a scale μ renormalized, $m_q(\mu)$, quark mass:

$$m_b = Z_m\,m_q(\mu)\,.$$

On an equal footing, Bethe–Salpeter amplitude or Bethe–Salpeter wave function of the bound state are governed by the homogeneous Bethe–Salpeter equation. For mesons, the latter reads, in order to preserve the axial-vector Ward–Takahashi identity of QCD, in its ladder truncation,

$$\Gamma(p;P) = -\frac{4}{3}\,Z_2^2 \int_q^\Lambda \mathcal{G}\big((p-q)^2\big)\,T_{\mu\nu}(p-q)\,\gamma_\mu\,\chi(q;P)\,\gamma_\nu\,.$$

Expansion of $\Gamma(p;P)$ in Lorentz covariants reformulates this bound-state equation as a system of four (for all bound states of spin zero) or eight (for all bound states of nonzero spin) coupled equations. An estimate of part of the systematic uncertainties inherent to this treatment can be acquired by adopting for the effective couplings $k^2\,\mathcal{G}(k^2)$ at least two different (rather popular) models [2, 3]. From the solution for some bound state's Bethe–Salpeter amplitude $\Gamma(p;P)$ and mass M, we infer its decay constant f and in-hadron condensate [4], *i.e.*, the matrix element of the chiral density $\bar{q}_1\,\gamma_5\,q_2$ between vacuum and bound state, below quoted in form of $|\langle\bar{q}\,q\rangle|^{1/3}$.

3 It's a Long Way to ~~Tipperary~~ Bound States, It's a Long Way to Go

Within a covariant formalism, the classification of predicted bound states in terms of quantum numbers is not as straightforward as in a nonrelativistic treatment. Spectra of quark–antiquark mesons, with total spin $s = 0, 1$ and orbital angular momentum ℓ, can be classified, in terms of their angular momentum J, parity $P = (-1)^{\ell+1}$, and charge-conjugation parity $C = (-1)^{\ell+s}$, as

$$\text{ordinary:}\quad J^{PC} \in \{0^{++},0^{-+},1^{++},1^{+-},1^{--},2^{++},2^{-+},2^{--},3^{++},3^{+-},3^{--},4^{++},4^{-+},4^{--},\dots\}\,,$$
$$\text{exotic:}\quad J^{PC} \in \{0^{+-},0^{--},1^{-+},2^{+-},3^{-+},4^{+-},5^{-+},\dots\}\,.$$

This situation carries over to open-flavour mesons where one encounters *quasi-exotic* mesons, mirroring exotic mesons of the equal-mass case [5]: the Poincaré-covariant setup utilized here predicts more meson states than corresponding nonrelativistic quark models. That insight can be visualized via the contributions to the norm of $\chi(p;P)$ of the different Lorentz covariants of various values of ℓ attributable to them in the given meson's rest frame, as illustrated by Fig. 1.

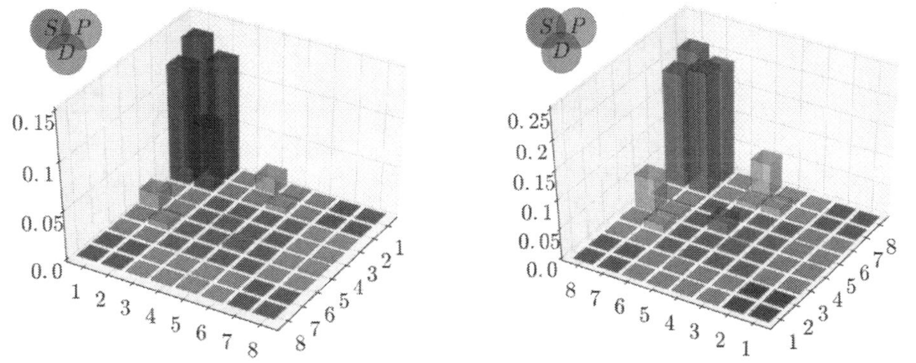

Figure 1. Lorentz covariants: 8×8 relative contributions to ρ (left) or $\rho(1450)$ (right) [6, Figs. 7 and 12].

4 Meson Masses, Decay Constants, and In-Hadron Condensates [1]

For a given meson composed of antiquark $\bar{q}$ and quark q', we collect our predictions [1] for the masses $M_{\bar{q}\,q'}$, leptonic decay constants $f_{\bar{q}\,q'}$, and in-meson condensates $\langle\bar{q}\,q\rangle$ of its ground state and lowest-lying radial excitations in shared plots, separately for all flavour combinations $\bar{q}\,q'$ for $M_{\bar{q}\,q'}$ and $f_{\bar{q}\,q'}$, but combined to a single plot for $\langle\bar{q}\,q\rangle$, as exemplified by Figs. 2 through 4. For comparison with experiment, we merge our results [1] for the effective-interaction models adopted [2, 3] for two fits of the involved quark masses to single predictions (*cf.* Figs. 5 and 6).

It goes without saying that the example outcomes shown here merely serve as a teaser: the complete sets of predictions can be found in Ref. [1]. A raw guess of the size of the systematic uncertainties inherent to the adopted approach can be gained by variation of the model details.

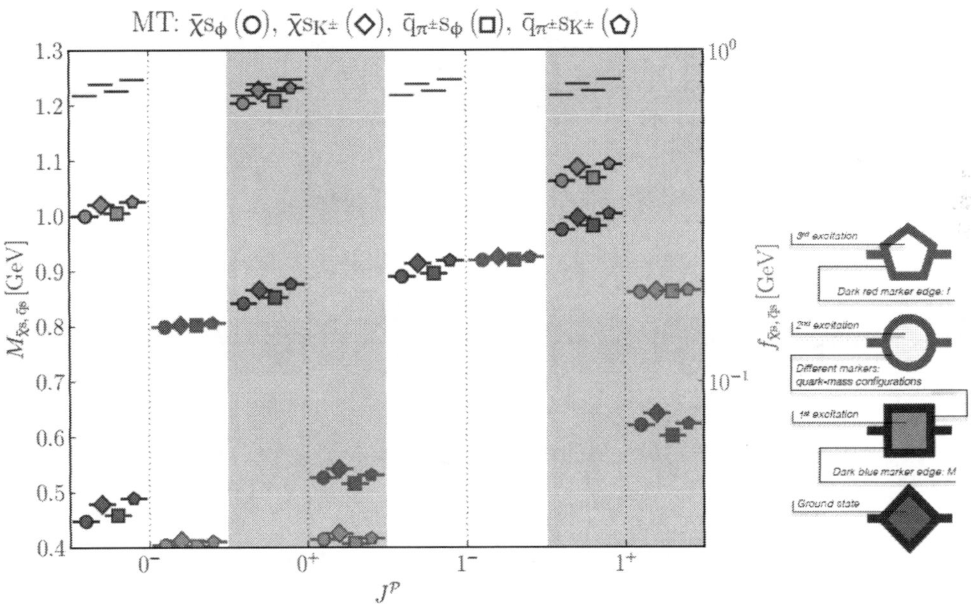

Figure 2. Masses and decay constants [1, Fig. 14]: mesons built up by s and chiral (χ) or light (q) quark.

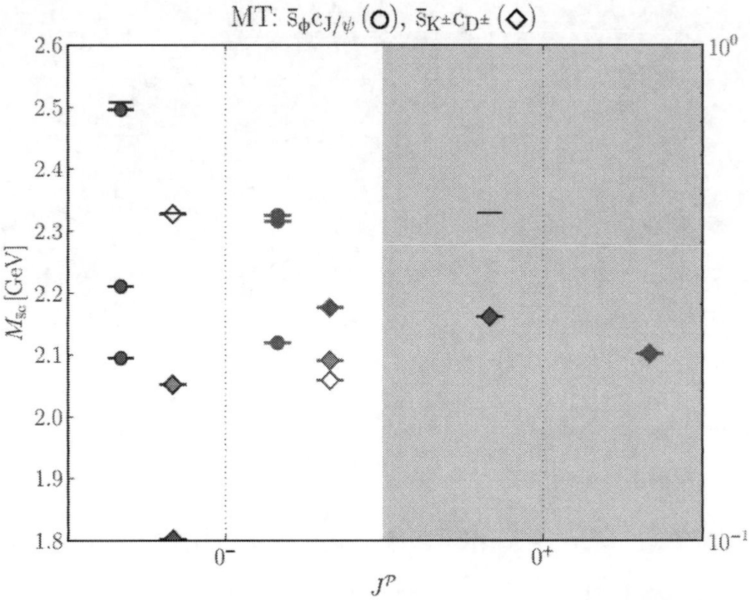

Figure 3. Charmed, strange mesons' masses and decay constants [1, Fig. 17] from the model of Ref. [2].

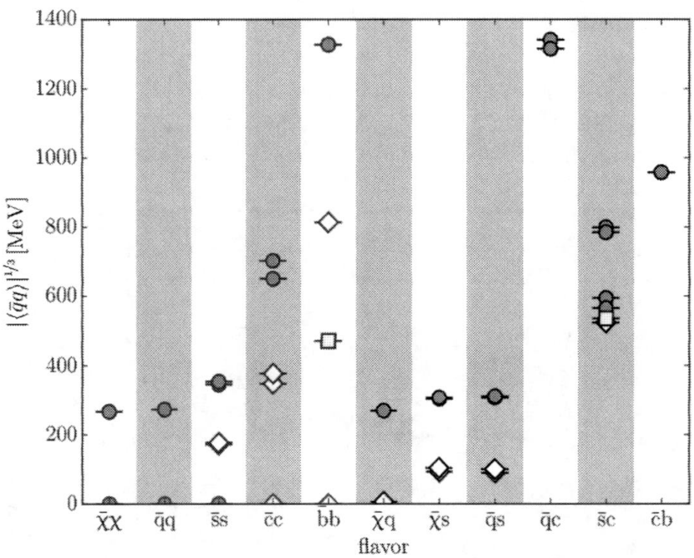

Figure 4. In-hadron condensates $\langle \bar{q}\, q \rangle$ of all mesons under study [1, Fig. 21], from the model of Ref. [2].

References

[1] T. Hilger, M. Gómez-Rocha, A. Krassnigg, and W. Lucha, Eur. Phys. J. A **53** (2017) 213, arXiv:1702.06262 [hep-ph].

[2] P. Maris and P. C. Tandy, Phys. Rev. C **60** (1999) 055214, arXiv:nucl-th/9905056.

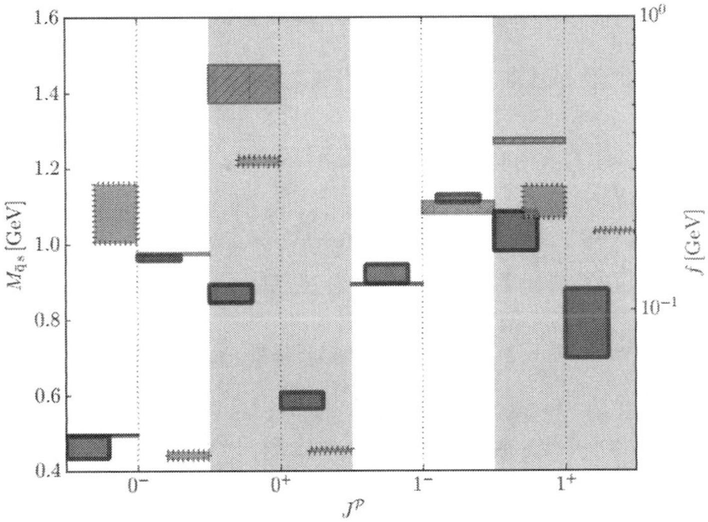

Figure 5. Masses and decay constants of ($\bar{q}\,s$) mesons: overall result (boxes) vs. experiment [1, Fig. 24].

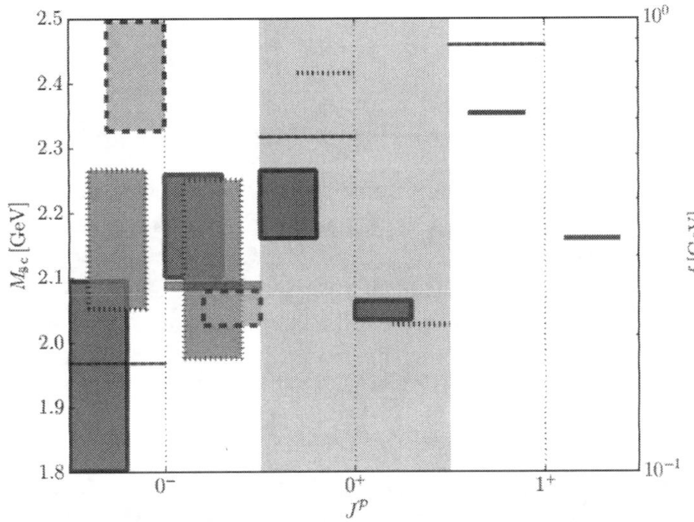

Figure 6. Masses and decay constants of ($\bar{s}\,c$) mesons: overall result (boxes) vs. experiment [1, Fig. 25].

[3] R. Alkofer, P. Watson, and H. Weigel, Phys. Rev. D **65** (2002) 094026, arXiv:hep-ph/0202053.

[4] P. Maris, C. D. Roberts, and P. C. Tandy, Phys. Lett. B **420** (1998) 267, arXiv:nucl-th/9707003.

[5] T. Hilger and A. Krassnigg, Eur. Phys. J. A **53** (2017) 142, arXiv:1605.03464 [hep-ph].

[6] T. Hilger, M. Gómez-Rocha, and A. Krassnigg, Eur. Phys. J. C **77** (2017) 625, arXiv:1508.07183 [hep-ph].

Heavy-light mesons in a contact interaction

Marco A. Bedolla[1,2,*] and *Elena* Santopinto[1,**]

[1]INFN, Sezione di Genova, via Dodecaneso 33, 16146 Genova, Italy.
[2]Instituto de Física y Matemáticas, Universidad Michoacana de San Nicolás de Hidalgo, Edificio C-3, Ciudad Universitaria, Morelia, Michoacán 58040, México.

Abstract. We present a unified formalism for the analysis of mesons. It is provided by a symmetry preserving Schwinger-Dyson-Bethe-Salpeter-Equation (SDBSE) treatment of a vector-vector contact interaction. The contact interaction (CI) model offers a simple-to-implement alternative to perform exploratory studies of QCD, within the SDBSE framework. Within the limitation of this model, we calculate the spectrum of D and B mesons. Our results are in agreement within 2% when compare with experimental data, lattice QCD and other SDBSE calculations involving sophisticated interaction kernels.

1 Introduction

The Schwinger-Dyson-Bethe-Salpeter Equation (SDBSE) approach has proven to be a reliable instrument to explain a wide range of QCD phenomena [1–3]. However, brute force numerical evaluations depict problems when evaluating at large momentum transfer regions [4, 5], and the community endeavor consists in solving these problems from other perspectives [6–9].

The CI model appeared as an alternative to full QCD studies. In this model, quarks interact not via mass-less vector-boson exchanges, but instead through a symmetry preserving vector-vector contact interaction [10–14]. This interaction is embedded within the SDBSE approach in the rainbow-ladder approximation, implement confinement through a proper time regularization scheme. A fully consistent treatment of the CI model is simple to execute, and it produces useful results that can be compared and contrasted with full QCD calculations and experimental data.

In this work, we present a study of heavy-light systems with the contact interaction. Our results are a direct application of this model, that we developed in our previous papers [15–17]. We present our heavy-light mesons results and compare them experimental data and other covariant model setups [18–21]. For a complete model revision refer to Ref. [15].

2 SDBSE approach and CI model

Since this work is a direct application of the unified model presented in Ref. [15], we only sketch the basic formulae. The complete description of the CI model is found in Refs. [14, 17].

*e-mail: bedolla@ifm.umich.mx
**e-mail: santopinto@ge.infn.it

© The Authors, published by EDP Sciences. This is an open access article distributed under the terms of the Creative Commons Attribution License 4.0 (http://creativecommons.org/licenses/by/4.0/).

2.1 Contact Interaction Model

The DSBSE formalism solves the bound-state problem in terms of their building blocks (quarks) and their interactions with gluons. In order to solve the meson bound state equation, we need to know the quark propagator, the gluon propagator and the quark-gluon interaction. In the contact interaction model, we assume that the quark-gluon interaction is led by symmetry-preserving vector×vector contact interaction; here, quarks are attached through the interaction defined as

$$g^2 D_{\mu\nu}(k) = \frac{4\pi\alpha_{\rm IR}}{m_g^2}\delta_{\mu\nu} \equiv \frac{1}{m_G^2}\delta_{\mu\nu}, \tag{1}$$

$$\Gamma^a_\mu(p,q) = \frac{\lambda^a}{2}\gamma_\mu, \tag{2}$$

where $m_g = 800\,\text{MeV}$ is a gluon mass scale which is in fact generated dynamically in QCD [22], and $\alpha_{\rm IR}$ is the CI model parameter, which can be interpreted as the interaction strength in the infrared [23, 24].

With this interaction, we obtain a constant mass function. Because the integrals we need to solve are divergent, we must adopt a regularization procedure. We employ the proper time regularization scheme [25] an get the mass function expression

$$M_f = m_f + \frac{M_f^3}{3\pi^2 m_G^2}\Gamma(-1, \tau_{\rm UV} M_f^2, \tau_{\rm IR} M_f^2), \tag{3}$$

where $\Gamma(a, z_1, z_2)$ is the generalized incomplete Gamma function:

$$\Gamma(a, z_1, z_2) = \Gamma(a, z_1) - \Gamma(a, z_2). \tag{4}$$

The parameters $\tau_{\rm IR}$ and $\tau_{\rm UV}$ are infrared and ultraviolet regulators, respectively. A nonzero value for $\tau_{\rm IR} \equiv 1/\Lambda_{\rm IR}$ implements confinement [26]. Since the CI is nonrenormalizable theory, $\tau_{\rm UV} \equiv 1/\Lambda_{\rm UV}$ becomes part of the model and therefore sets the scale for all dimensional quantities.

2.2 CI model running coupling

quark	$\hat{\alpha}_{IR}$ [GeV^{-2}]	Λ_{UV} [GeV]	α	Ratio
u, d, s	4.565	0.905	3.739	1
c	0.228	2.400	1.547	0.414
b	0.035	6.400	1.496	0.400

Table 1. Dimensionless coupling constant $\alpha = \hat{\alpha}_{\rm IR}\Lambda_{\rm UV}^2$, where $\hat{\alpha}_{IR} = \alpha_{\rm IR}/m_g^2$, for the contact interaction, extracted from a best-fit to data, as explained in Ref. [15]. Fixed parameters are $m_g = 0.8\,\text{GeV}$ and $\Lambda_{IR} = 0.24\,\text{GeV}$.

In a previous paper [15], we explained how the CI can be used to study light and heavy mesons. When studying the heavy sector, a change in the model parameters has to be done: an increase in the ultraviolet regulator, and a reduction in the coupling strength. Subsequently, we figured out that different set of parameters are needed in order to study each sector: light, charm and bottom, as displayed in Table 1. With these parameters, we defined a dimensionless coupling α guided by [27, 28]

$$\alpha = \frac{\alpha_{IR}}{m_g^2}\Lambda_{\rm UV}^2. \tag{5}$$

The drop in α, in relation to its value in the light-quarks sector, can be read off from the last column of Table 1. Indeed, α is reduced by a factor of $2.1 - 2.3$ on going from the light to the heavy sector, instead of the apparent large factors listed in Refs. [15, 17].

Moreover, as a reminiscent of the running coupling QCD with the momentum scale at which it is measured, an inverse logarithmic curve can fit reasonably well the functional dependence of $\alpha(\Lambda_{UV})$. The fit reads

$$\alpha(\Lambda_{\rm UV}) = a \ln^{-1}(\Lambda_{\rm UV}/\Lambda_0) \,, \tag{6}$$

where $a = 0.923$ and $\Lambda_0 = 0.357$ [15]. With this fit, we can recover the value of the strength coupling α once given a value of Λ_{UV}.

3 Results

In order to calculate mass spectra and decay constants, we follow the expressions found in Ref. [17]. Additionally, we include two distinct spin-orbit(SO) parameters,

$$g_{\rm SO}^{1^+} = 0.312/3, \quad g_{\rm SO}^{0^+} = 0.4/3\,. \tag{7}$$

to calculate the spectrum of scalar and axial-vector mesons, respectively. With similar values, it is possible to match the mass splitting $m_{a_1} - m_\rho = 0.45$GeV and $m_\sigma - \rho = 0.29$GeV [29].

In our approach, to solve these unequal mass systems, we find $\Lambda_{\rm IR}$ and α through Eq. (6) to obtain the pseudoscalar meson experimental mass value. Then, with these parameters, we calculate the other mesons masses, the pseudoscalar and vector decay constants.

We present plots and tables comparing our results on D and B mesons, We compare with experimental data and other models results.

3.1 *D* Mesons

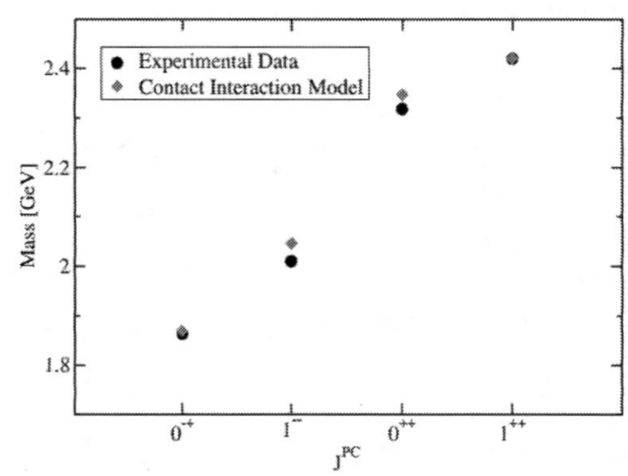

Figure 1. Contact interaction results for $c\bar{u}$ ($\bar{c}u$) mass spectrum using model parameters found in Table 2. PDG data are from Ref. [31].

Table 2. Ground-state of D $u\bar{c}(\bar{u}c)$ mass spectrum. The CI results were obtained with the best-fit parameter set: $m_g = 0.8\,\text{GeV}$, $\alpha_{IR} = 0.93\pi/8.047$, $\Lambda_{\text{IR}} = 0.24\,\text{GeV}$, and $\Lambda_{\text{UV}} = 1.905\,\text{GeV}$. The current-quark mass are $m_n = 0.007\,\text{GeV}$ and $m_c = 1.09\,\text{GeV}$ and the dynamically generated constituent-like mass are $M_n = 0.049\,\text{GeV}$ and $M_c = 1.451\,\text{GeV}$. *The decay constants are from a Lattice calculation [32].

	masses and decay constants [MeV]			
	$(m, f)_{D(1S)}$	$(m, f)_{D^*(1S)}$	$m_{D_0(1P)}$	$m_{D_1(1P)}$
Experiment [31]*	(1864, 149)	(2010, 196)	2318	2420
CI-model	(1869, 425)	(2046, 182)	2347	2422
CI-subtr [30]	(1869, 146)	(2011, 169)	···	···
NST1 [18]	(1850, 108)	(2040, 113)	···	···
NST2 [18]	(1880, 183)	···	···	···
HGKL1 [21]	(1868, 228)	···	···	···
HGKL2 [21]	(1869, 678)	···	···	···
	amplitudes			
E_H	4.292	0.592	0.073	0.039
F_H	0.064	···	···	···

The spectrum of D mesons is displayed in Fig. 1. In Table 2, we compare the spectrum with experimental data and other models predictions. Our mass spectrum is in excellent agreement with experimental results and other models predictions. However, the decay constants predictions with different SDBSE methods still need more improvement.

3.2 B mesons

Table 3. Ground-state of B $n\bar{b}(\bar{n}b)$ mass spectrum. The CI results were obtained with the best-fit parameter set: $m_g = 0.8\,\text{GeV}$, $\alpha_{IR} = 0.93\pi/168.569$, $\Lambda_{\text{IR}} = 0.24\,\text{GeV}$, and $\Lambda_{\text{UV}} = 6.605\,\text{GeV}$. The current-quark mass are $m_n = 0.007\,\text{GeV}$ and $m_b = 3.8\,\text{GeV}$ and the dynamically generated constituent-like mass are $M_n = 0.0140\,\text{GeV}$ and $M_b = 4.575\,\text{GeV}$. *The decay constants are from a recent Lattice calculation [33].

	masses and decay constants [GeV]			
	$(m, f)_{B(1S)}$	$(m, f)_{B^*(1S)}$	$m_{B_0(1P)}$	$m_{B_1(1P)}$
Experiment [31]*	(5279, 186)	(5325, 175)	···	5725
CI-model	(5273, 129)	(5314, 371)	5660	5704
NST1 [18]	(5270, 74)	(5150, 187)	···	···
NST2 [18]	(5150, 187)	···	···	···
	amplitudes			
E_H	0.143	0.149	0.002	0.001
F_H	0.000	···	···	···

Table 3 and Fig. 2 provide the CI model predicted B-mesons mass spectrum. Similar to our previous results, the mass spectrum agrees perfectly with experimental data and other models results. Additionally, thanks to the model simplicity, we present a B_0 mass value. We expect that our value will be in good agreement once it is an experimental value.

It is important to mention that the light quark gets a really small dressing when it is bounded to the bottom quark. This means that in a heavy-light systems, the light quark plays

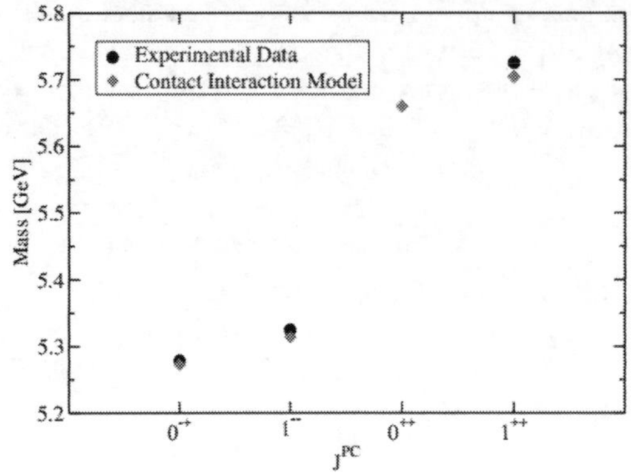

Figure 2. Contact interaction results for $b\bar{n}$ ($\bar{b}n$) mass spectrum using model parameters found in Table 3. PDG data are from Ref. [31].

a secondary role while the heavy quark has the most significant effects. Speaking about the decay constants, due to the mass difference between the bottom and light quarks, we aim to improve the results in future works.

4 Conclusions

We calculated the spectrum of D and B mesons. Our mass spectrum is in a good agreement with experimental data and other model results. However, the decay constants still need more improvement in these unequal mass systems. On the other hand, in D mesons we reduce the light coupling by a factor of ≈ 8, while in B mesons this factor is ≈ 168. This indicates that our model coupling with high sensitivity in the heavy sector

Thus, we found that the reduction in the coupling model is dependent on the quark masses involved in the bound-state. We fitted the contact interaction coupling with an inverse logarithmic curve. In future works, we will exploit this feature to study baryons and exotics with a variety of quarks components.

References

[1] P. Jain and H. J. Munczek, Phys. Rev. D **48**, 5403 (1993)
[2] C. D. Roberts and A. G. Williams, Prog. Part. Nucl. Phys. **33**, 477 (1994)
[3] A. Bashir, L. Chang, I. C. Cloet, B. El-Bennich, Y. X. Liu, C. D. Roberts and P. C. Tandy, Commun. Theor. Phys. **58**, 79 (2012)
[4] P. Maris and P. C. Tandy, Phys. Rev. C **62**, 055204 (2000)
[5] J. Chen, M. Ding, L. Chang and Y. x. Liu, Phys. Rev. D **95**, no. 1, 016010 (2017)
[6] N. Nakanishi, Phys. Rev. **130**, 1230 (1963).
[7] L. Chang, I. C. Cloët, C. D. Roberts, S. M. Schmidt and P. C. Tandy, Phys. Rev. Lett. **111**, no. 14, 141802 (2013)
[8] K. Raya, L. Chang, A. Bashir, J. J. Cobos-Martinez, L. X. Gutiérrez-Guerrero, C. D. Roberts and P. C. Tandy, Phys. Rev. D **93**, no. 7, 074017 (2016)
[9] K. Raya, M. Ding, A. Bashir, L. Chang and C. D. Roberts, Phys. Rev. D **95**, no. 7, 074014 (2017)

[10] L. X. Gutierrez-Guerrero, A. Bashir, I. C. Cloet and C. D. Roberts, Phys. Rev. C **81**, 065202 (2010)
[11] H. L. L. Roberts, C. D. Roberts, A. Bashir, L. X. Gutierrez-Guerrero and P. C. Tandy, Phys. Rev. C **82**, 065202 (2010)
[12] H. L. L. Roberts, L. Chang, I. C. Cloet and C. D. Roberts, Few Body Syst. **51**, 1 (2011)
[13] H. L. L. Roberts, A. Bashir, L. X. Gutierrez-Guerrero, C. D. Roberts and D. J. Wilson, Phys. Rev. C **83**, 065206 (2011)
[14] C. Chen, L. Chang, C. D. Roberts, S. Wan and D. J. Wilson, Few Body Syst. **53**, 293 (2012)
[15] K. Raya, M. A. Bedolla, J. J. Cobos-Martínez and A. Bashir,
[16] M. A. Bedolla, K. Raya, J. J. Cobos-Martínez and A. Bashir, Phys. Rev. D **93**, no. 9, 094025 (2016)
[17] M. A. Bedolla, J. J. Cobos-Martínez and A. Bashir, Phys. Rev. D **92**, no. 5, 054031 (2015)
[18] T. Nguyen, N. A. Souchlas and P. C. Tandy, AIP Conf. Proc. **1361**, 142 (2011)
[19] C. S. Fischer, S. Kubrak and R. Williams, Eur. Phys. J. A **51**, 10 (2015)
[20] M. Gómez-Rocha, T. Hilger and A. Krassnigg, Phys. Rev. D **93**, no. 7, 074010 (2016)
[21] T. Hilger, M. Gómez-Rocha, A. Krassnigg and W. Lucha, Eur. Phys. J. A **53**, no. 10, 213 (2017)
[22] P. Boucaud, J. P. Leroy, A. L. Yaouanc, J. Micheli, O. Pene and J. Rodriguez-Quintero, Few Body Syst. **53**, 387 (2012)
[23] D. Binosi, C. Mezrag, J. Papavassiliou, C. D. Roberts and J. Rodriguez-Quintero, Phys. Rev. D **96**, no. 5, 054026 (2017)
[24] A. Deur, S. J. Brodsky and G. F. de Teramond, Prog. Part. Nucl. Phys. **90**, 1 (2016)
[25] D. Ebert, T. Feldmann and H. Reinhardt, Phys. Lett. B **388**, 154 (1996)
[26] C. D. Roberts, Prog. Part. Nucl. Phys. **61**, 50 (2008)
[27] R. L. S. Farias, G. Dallabona, G. Krein and O. A. Battistel, Phys. Rev. C **73**, 018201 (2006)
[28] R. L. S. Farias, G. Dallabona, G. Krein and O. A. Battistel, Phys. Rev. C **77**, 065201 (2008)
[29] Y. Lu, C. Chen, C. D. Roberts, J. Segovia, S. S. Xu and H. S. Zong, Phys. Rev. C **96**, no. 1, 015208 (2017)
[30] F. E. Serna, B. El-Bennich and G. Krein, Phys. Rev. D **96**, no. 1, 014013 (2017)
[31] C. Patrignani *et al.* [Particle Data Group], Chin. Phys. C **40**, no. 10, 100001 (2016).
[32] D. Becirevic, V. Lubicz, F. Sanfilippo, S. Simula and C. Tarantino, JHEP **1202**, 042 (2012)
[33] B. Colquhoun *et al.* [HPQCD Collaboration], Phys. Rev. D **91**, no. 11, 114509 (2015)

Simplified Bethe–Salpeter Description of Basic Pseudoscalar-Meson Features

Wolfgang Lucha[1,*]

[1]Institute for High Energy Physics, Austrian Academy of Sciences, Nikolsdorfergasse 18, A-1050 Vienna, Austria

Abstract. We assess a description of pseudoscalar mesons as pseudo-Goldstone bosons by its compatibility with some Gell-Mann–Oakes–Renner-type relation.

1 Pseudoscalar Mesons of Goldstone Nature

Goldstone's theorem necessitates the presence of a massless boson in the spectrum of physical particles for any spontaneously broken chiral symmetry of quantum chromodynamics (QCD); these hypothetical Goldstone bosons are identified with the ground-state pseudoscalar mesons (pions, kaons, η). Their finite (but comparatively small) masses are attributed to the additional explicit breakdown of the chiral symmetries of QCD enforced by nonvanishing quark masses.

We analyze the Goldstone-boson nature of the lightest pseudoscalar mesons by means of a formalism [1–4] situated somewhere between the fully relativistic Bethe–Salpeter description of bound states [5], with several still to be resolved inherent obstacles, and the latter's extreme instantaneous limit, represented by its three-dimensional reduction devised by Salpeter [6]. A rather promising tool to judge the merits of such kind of intermediate framework proves to be, among others, the fulfilment of (generalized) Gell-Mann–Oakes–Renner-type relations [7] by characteristic features (*viz.*, decay constants or condensates) of light pseudoscalar mesons [8].

2 Quark–Antiquark Bound-State Formalism

The homogeneous Bethe–Salpeter equation describes, in Poincaré-covariant manner, a bound state $|\mathrm{B}(P)\rangle$ of mass $\widehat{M}$ and momentum P, formed by two particles of relative momentum p, by its Bethe–Salpeter amplitude $\Phi(p, P)$. *One ingredient* are the constituents' full propagators, in the case of spin-$\frac{1}{2}$ fermions i given by mass $M_i(p^2)$ and wave-function renormalization $Z_i(p^2)$:

$$S_i(p) = \frac{\mathrm{i}\, Z_i(p^2)}{\not p - M_i(p^2) + \mathrm{i}\,\varepsilon}\,, \qquad \not p \equiv p^\mu\, \gamma_\mu\,, \qquad \varepsilon \downarrow 0\,, \qquad i = 1,2\,.$$

The *other ingredient* are the interactions responsible for bound-state formation. Ignoring their dependence on time components of momenta *and* the above propagators' dependence on zero momentum components *squared* led to a bound-state equation [1] for a Salpeter amplitude [6]

$$\phi(\boldsymbol{p}) \propto \int \mathrm{d}p_0\, \Phi(p, P)\,,$$

interpretable as equal-time bound-state wave function. An integral kernel $K(\boldsymbol{p}, \boldsymbol{q})$ captures the entirety of, by assumption instantaneous, effective interactions of all bound-state constituents.

*e-mail: Wolfgang.Lucha@oeaw.ac.at

© The Authors, published by EDP Sciences. This is an open access article distributed under the terms of the Creative Commons Attribution License 4.0 (http://creativecommons.org/licenses/by/4.0/).

The bound-state equation proposed in Ref. [1], formulated in terms of the kinetic energies,

$$E_i(\boldsymbol{p}) \equiv \sqrt{\boldsymbol{p}^2 + M_i^2(\boldsymbol{p}^2)}\,,$$

and the energy projectors onto positive and negative energies of the bound-state constituents i,

$$\Lambda_i^{\pm}(\boldsymbol{p}) \equiv \frac{E_i(\boldsymbol{p}) \pm \gamma_0\,[\boldsymbol{\gamma}\cdot\boldsymbol{p} + M_i(\boldsymbol{p}^2)]}{2\,E_i(\boldsymbol{p})}\,,$$

reads, for the case of fermion–antifermion bound states in the center-of-momentum frame [1],

$$\begin{aligned}\phi(\boldsymbol{p}) = Z_1(\boldsymbol{p}^2)\,Z_2(\boldsymbol{p}^2) \int \frac{\mathrm{d}^3 q}{(2\pi)^3} \Bigg(&\frac{\Lambda_1^+(\boldsymbol{p})\,\gamma_0\,[K(\boldsymbol{p},\boldsymbol{q})\,\phi(\boldsymbol{q})]\,\Lambda_2^-(\boldsymbol{p})\,\gamma_0}{\widehat{M} - E_1(\boldsymbol{p}) - E_2(\boldsymbol{p})} \\ &- \frac{\Lambda_1^-(\boldsymbol{p})\,\gamma_0\,[K(\boldsymbol{p},\boldsymbol{q})\,\phi(\boldsymbol{q})]\,\Lambda_2^+(\boldsymbol{p})\,\gamma_0}{\widehat{M} + E_1(\boldsymbol{p}) + E_2(\boldsymbol{p})} \Bigg).\end{aligned}$$

For any one-particle states $|B(P)\rangle$ normalized according to the Lorentz-invariant condition

$$\langle B(P)|B(P')\rangle = (2\pi)^3\,2\,P_0\,\delta^{(3)}(\boldsymbol{P} - \boldsymbol{P}')\,,$$

the normalization condition for the corresponding Salpeter amplitudes $\phi(\boldsymbol{p})$ is given by [9–11]

$$\int \frac{\mathrm{d}^3 p}{(2\pi)^3}\,\mathrm{Tr}\left[\phi^\dagger(\boldsymbol{p})\,\frac{\gamma_0\,[\boldsymbol{\gamma}\cdot\boldsymbol{p} + M_1(\boldsymbol{p}^2)]}{E_1(\boldsymbol{p})}\,\phi(\boldsymbol{p})\right] = 2\,P_0\,.$$

3 Assuming Generalized Flavour Symmetry

Things simplify considerably if the propagator functions of involved fermion and antifermion happen to be identical, $M_1(\boldsymbol{p}^2) = M_2(\boldsymbol{p}^2)$ and $Z_1(\boldsymbol{p}^2) = Z_2(\boldsymbol{p}^2)$; the Salpeter amplitude of every pseudoscalar bound state is then fully defined by just two Lorentz-scalar components, $\varphi_{1,2}(\boldsymbol{p})$:

$$\phi(\boldsymbol{p}) = \frac{1}{\sqrt{3}}\left[\varphi_1(\boldsymbol{p})\,\frac{\gamma_0\,[\boldsymbol{\gamma}\cdot\boldsymbol{p} + M(\boldsymbol{p}^2)]}{E(\boldsymbol{p})} + \varphi_2(\boldsymbol{p})\right]\gamma_5\,.$$

For $K(\boldsymbol{p},\boldsymbol{q})$ compatible with spherical and rather specific Fierz symmetries of the effective interactions, our bound-state equation governing $\phi(\boldsymbol{p})$ collapses to an eigenvalue problem [12] fixing the radial parts $\varphi_{1,2}(p)$, $p \equiv |\boldsymbol{p}|$, of $\varphi_{1,2}(\boldsymbol{p})$, with a single, spherically symmetric potential $V(r)$, $r \equiv |\boldsymbol{x}|$, encoding the configuration-space interactions between bound-state constituents:

$$\begin{aligned}E(p)\,\varphi_2(p) + \frac{2\,Z^2(p^2)}{\pi\,p} \int_0^\infty \mathrm{d}q\,q\,\mathrm{d}r \sin(p\,r)\sin(q\,r)\,V(r)\,\varphi_2(q) &= \frac{\widehat{M}}{2}\,\varphi_1(p)\,,\\ E(p)\,\varphi_1(p) &= \frac{\widehat{M}}{2}\,\varphi_2(p)\,.\end{aligned} \tag{1}$$

For the actual case of interest, mesonic bound states of quarks and antiquarks, the effective *interaction potential* $V(r)$ was extracted pointwise [2–4] by straightforward inversion [13, 14] of our Bethe–Salpeter-inspired bound-state equation [1], starting from that Salpeter amplitude $\phi(\boldsymbol{p})$ that represents massless pseudoscalar mesons. The latter, in turn, is connected [15–17] to the chiral-quark propagator [18, 19] by a Ward–Takahashi identity of QCD [7]. The emerging confining $V(r)$ rises, from its slightly negative value at $r = 0$, rather steeply to infinity (Fig. 1).

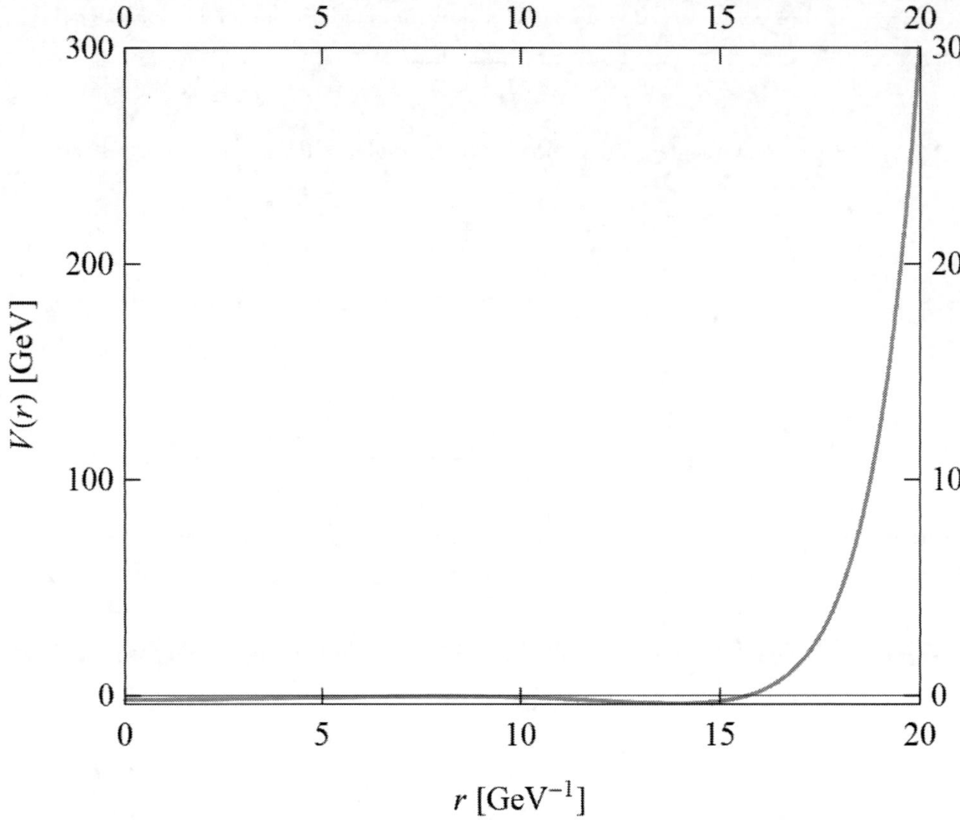

Figure 1. Effective interquark central potential $V(r)$ pinned down by inverting the radial Bethe–Salpeter problem (1) for a Salpeter-amplitude input derived from some quark-propagator model solution [18, 19]. The almost flatness of the potential $V(r)$ near the origin in combination with its extraordinarily steep rise to infinity renders the form of $V(r)$ pretty close but clearly not identical to that of a square-well potential.

4 Pseudoscalar-Meson Properties Revisited

In order to scrutinize pseudoscalar quark–antiquark bound states for physical (*i.e.*, non-chiral) quark masses, we ought to find the corresponding solutions to our set (1) of coupled equations for the two independent radial Salpeter components $\varphi_{1,2}(p)$. Our task is greatly facilitated by a definitely obvious move enabled by the purely algebraic nature of one of the two relations (1): inserting any of the two relations (1) into the other leads to single explicit eigenvalue problems for either $\varphi_1(p)$ or $\varphi_2(p)$ with mass-squared eigenvalue $\widehat{M}^2$ [2–4, 12]; conversion to equivalent matrix eigenvalue problems by expansion over a convenient function-space basis is among the standard solution procedures [20–24]. With these solutions at hand, we may create trust in the reliability of our approach by assessing and exploring its predictions for hadronic observables.

The spatial extension of the pion deduced, from the numerical ground-state solution to our bound-state formalism inferred along the above lines, in form of the pion's average interquark distance $\langle r\rangle = 0.478$ fm or root-mean-square radius $\sqrt{\langle r^2\rangle} = 0.529$ fm fits nicely to the pion's electromagnetic charge radius, $\sqrt{\langle r_\pi^2\rangle} = (0.672\pm0.008)$ fm [25]. Nevertheless, this agreement won't qualify as confirmation of the credibility of our framework since the Salpeter amplitude for chiral quarks served already as input to the inversion procedure that generated the shape of our potential $V(r)$, and the experimental u/d quark masses are pretty close to their chiral limit.

Consequently, we have to look for different and more significant criteria which allow us to appraise the reasonableness of any inversion considerations. Fortunately, this proves to be not too difficult: Equating the residues of pseudoscalar-meson pole terms in both axial-vector and pseudoscalar vertex functions entering into the axial-vector Ward–Takahashi identity of QCD expressing the invariance of QCD under chiral transformations leads to a generalization [7] of the Gell-Mann–Oakes–Renner relation [26]; this innovation relates, for a pseudoscalar bound state $|B(P)\rangle$, its (weak) *decay constant* f_B, defined in terms of the axial-vector quark current by

$$\langle 0|:\bar{\psi}_1(0)\,\gamma_\mu\,\gamma_5\,\psi_2(0):|B(P)\rangle = \mathrm{i}\,f_B\,P_\mu\ ,$$

— which, consequently, may be found by projection of $\phi(\boldsymbol{p})$ onto the axial-vector current, *i.e.*,

$$f_B \propto \int \mathrm{d}^3p\ \mathrm{Tr}[\gamma_0\,\gamma_5\,\phi(\boldsymbol{p})]$$

— and its *in-hadron condensate* [7] (universalizing the notion of quark vacuum condensates)[1]

$$\mathbb{C}_B \equiv \langle 0|:\bar{\psi}_1(0)\,\gamma_5\,\psi_2(0):|B(P)\rangle \propto \int \mathrm{d}^3p\ \mathrm{Tr}[\gamma_5\,\phi(\boldsymbol{p})]$$

to the mass $\widehat{M}_B$ of this pseudoscalar bound state and the two relevant quark mass parameters in the QCD Lagrangian [7]. For the simpler case of equal quark masses m, this relationship reads

$$f_B\,\widehat{M}_B^2 = 2\,m\,\mathbb{C}_B\ . \qquad (2)$$

Compatibility with Eq. (2) may be inspected by solving our formalism with the previously established potential $V(r)$ for bound states of chiral, u/d, and s quarks, taking advantage of the appropriate model propagator functions [18, 19]. Comparison of the quark masses m, fixed by the thus determined values of $\widehat{M}_B$, f_B, and $\mathbb{C}_B$, with the current-quark masses $\overline{m}(\mu)$ in modified minimal subtraction at scale μ proves that our findings for m are in the right ballpark (Table 1).

Table 1. Predictions of the Bethe–Salpeter-inspired bound-state equation (1), with effective interaction potential $V(r)$ as depicted in Fig. 1, for masses $\widehat{M}_B$, decay constants f_B and in-meson condensates $\mathbb{C}_B$ of the lightest pseudoscalar mesons, and confrontation of the quark-mass parameters m resulting from the more general Gell-Mann–Oakes–Renner relation (2) with their minimal-subtraction counterparts $\overline{m}(\mu)$.

Constituents	$\widehat{M}_B$ [MeV]	f_B [MeV]	$\mathbb{C}_B$ [GeV^2]	m [MeV]	$\overline{m}$(2 GeV) [MeV] [25]
chiral quarks	6.8	151	0.585	0.0059	—
u/d quarks	148.6	155	0.598	2.85	$3.5^{+0.7}_{-0.3}$
s quarks	620.7	211	0.799	51.0	96^{+8}_{-4}

5 Summary of Findings and Conclusions

In order to establish whether or not it is justified to lend trust to the outcomes of an approach to bound states proposed some time ago [1] and residing, as far as its compatibility with Poincaré covariance is concerned, somewhere in the vast range between bound-state descriptions along the ideas of Salpeter and Bethe, on the one hand, and static approximations, on the other hand, we evaluated the performance of the investigated framework's predictions for those properties of the lightest pseudoscalar mesons that happen to be related by an advancement of the insight gained by Gell-Mann, Oakes, and Renner [26]: In brief, the formalism of Ref. [1] is still alive.

[1] A recent, comprehensive, Bethe–Salpeter-rooted evaluation of in-hadron condensates may be found in Ref. [27].

References

[1] W. Lucha and F. F. Schöberl, J. Phys. G **31** (2005) 1133, arXiv:hep-th/0507281.
[2] W. Lucha and F. F. Schöberl, Int. J. Mod. Phys. A **31** (2016) 1650202, arXiv:1606.04781 [hep-ph].
[3] W. Lucha, EPJ Web Conf. **129** (2016) 00047, arXiv:1607.02426 [hep-ph].
[4] W. Lucha, EPJ Web Conf. **137** (2017) 13009, arXiv:1609.01474 [hep-ph].
[5] E. E. Salpeter and H. A. Bethe, Phys. Rev. **84** (1951) 1232.
[6] E. E. Salpeter, Phys. Rev. **87** (1952) 328.
[7] P. Maris, C. D. Roberts, and P. C. Tandy, Phys. Lett. B **420** (1998) 267, arXiv:nucl-th/9707003.
[8] W. Lucha and F. F. Schöberl, Int. J. Mod. Phys. A **33** (2018) 1850047, arXiv:1801.00264 [hep-ph].
[9] J.-F. Lagaë, Phys. Rev. D **45** (1992) 305.
[10] J. Resag, C. R. Münz, B. C. Metsch, and H. R. Petry, Nucl. Phys. A **578** (1994) 397, arXiv:nucl-th/9307026.
[11] M. G. Olsson, S. Veseli, and K. Williams, Phys. Rev. D **52** (1995) 5141, arXiv:hep-ph/9503477.
[12] Z.-F. Li, W. Lucha, and F. F. Schöberl, Phys. Rev. D **76** (2007) 125028, arXiv:0707.3202 [hep-ph].
[13] W. Lucha and F. F. Schöberl, Phys. Rev. D **87** (2013) 016009, arXiv:1211.4716 [hep-ph].
[14] W. Lucha, Proc. Sci., EPS-HEP 2013 (2013) 007, arXiv:1308.3130 [hep-ph].
[15] W. Lucha and F. F. Schöberl, Phys. Rev. D **92** (2015) 076005, arXiv:1508.02951 [hep-ph].
[16] W. Lucha and F. F. Schöberl, Phys. Rev. D **93** (2016) 056006, arXiv:1602.02356 [hep-ph].
[17] W. Lucha and F. F. Schöberl, Phys. Rev. D **93** (2016) 096005, arXiv:1603.08745 [hep-ph].
[18] P. Maris and P. C. Tandy, Phys. Rev. C **60** (1999) 055214, arXiv:nucl-th/9905056.
[19] P. Maris, in *Proceedings of the International Conference on Quark Confinement and the Hadron Spectrum IV*, editors W. Lucha and K. Maung Maung (World Scientific, Singapore, 2002), p. 163, arXiv:nucl-th/0009064.
[20] W. Lucha and F. F. Schöberl, Phys. Rev. A **56** (1997) 139, arXiv:hep-ph/9609322.
[21] W. Lucha and F. F. Schöberl, Int. J. Mod. Phys. A **14** (1999) 2309, arXiv:hep-ph/9812368.
[22] W. Lucha, K. Maung Maung, and F. F. Schöberl, Phys. Rev. D **63** (2001) 056002, arXiv:hep-ph/0009185.
[23] W. Lucha, K. Maung Maung, and F. F. Schöberl, Phys. Rev. D **64** (2001) 036007, arXiv:hep-ph/0011235.
[24] W. Lucha and F. F. Schöberl, Recent Res. Dev. Phys. **5** (2004) 1423, arXiv:hep-ph/0408184.
[25] C. Patrignani *et al.* (Particle Data Group), Chin. Phys. C **40** (2016) 100001.
[26] M. Gell-Mann, R. J. Oakes, and B. Renner, Phys. Rev. **175** (1968) 2195.
[27] T. Hilger, M. Gómez-Rocha, A. Krassnigg, and W. Lucha, Eur. Phys. J. A **53** (2017) 213, arXiv:1702.06262 [hep-ph].

Strong Isospin Breaking in Heavy-Meson Decay Constants: Employing Borelized QCD Sum Rules in Local-Duality Limit

Wolfgang Lucha[1,*], Dmitri Melikhov[1,2,3,**], and Silvano Simula[4,***]

[1]Institute for High Energy Physics, Austrian Academy of Sciences, Nikolsdorfergasse 18, A-1050 Vienna, Austria
[2]D. V. Skobeltsyn Institute of Nuclear Physics, M. V. Lomonosov Moscow State University, 119991 Moscow, Russia
[3]Faculty of Physics, University of Vienna, Boltzmanngasse 5, A-1090 Vienna, Austria
[4]INFN, Sezione di Roma Tre, Via della Vasca Navale 84, I-00146 Roma, Italy

Abstract. Applying the QCD sum-rule machinery in the so-called local-duality shape to heavy–light mesons reveals that, as a consequence of the non-zero mass gap between up and down quarks, the leptonic decay constants of the neutral and charged versions of the D, D^*, B and B^* mesons differ by approximately 1 MeV.

1 Target: Heavy-Pseudoscalar- and -Vector-Meson Decay Constants

The discrepancy $(m_d - m_u)(2\ \text{GeV}) \approx 2.5$ MeV [1] of the masses $m_{u,d}$ of up and down quarks is rather small compared to even the lightest meson masses but causes a strong-isospin-breaking disparity of the *leptonic decay constants* of mesons comprising heavy quark plus u or d quark.

We study pseudoscalar mesons P_q and vector mesons V_q, subsumed by H_q, as heavy–light mesons formed by a heavy quark $Q = c, b$ of mass m_Q and a light quark $q = u, d, s$ of mass m_q, characterized by their masses $M_{H_q} = M_{P_q}, M_{V_q}$, four-momenta p, polarization vectors $\varepsilon_\mu(p)$ in the case of vector mesons V_q, and leptonic decay constants $f_{H_q} = f_{P_q}, f_{V_q}$, defined according to

$$\langle 0|\,\bar{q}(0)\,\gamma_\mu\,\gamma_5\,Q(0)\,|P_q(p)\rangle = \mathrm{i}\, f_{P_q}\, p_\mu\,, \qquad \langle 0|\,\bar{q}(0)\,\gamma_\mu\,Q(0)\,|V_q(p)\rangle = f_{V_q}\, M_{V_q}\, \varepsilon_\mu(p)\,.$$

The formalism of QCD sum rules [2] provides an analytical approach to systems governed by the strong interactions. For our goals, it seems very promising to follow a path opened by a limiting case allowing us to concentrate all nonperturbative aspects within a single ingredient.

2 Tool: (Borel-Transformed) QCD Sum Rule in its Local-Duality Limit

A typical QCD sum rule [2] extracted from the two-point correlation function of appropriately chosen interpolating local operators relates basic properties of the hadron under consideration — in our case, of any heavy–light meson H_q — to the parameters of QCD. The latter are either of fundamental nature, such as its strong fine-structure coupling α_s and all quark masses, or of effective type, such as all the vacuum condensates $\langle \bar{q}\, q\rangle, \ldots$ encoding nonperturbative aspects.

*e-mail: Wolfgang.Lucha@oeaw.ac.at
**e-mail: dmitri_melikhov@gmx.de
***e-mail: simula@roma3.infn.it

© The Authors, published by EDP Sciences. This is an open access article distributed under the terms of the Creative Commons Attribution License 4.0 (http://creativecommons.org/licenses/by/4.0/).

For convenience, in order to remove necessary subtraction terms and to suppress contributions of hadronic excitations and continuum, usually a Borel transformation from one's momentum variable to some Borel parameter, called τ, is performed. The actual shape of such a QCD sum rule depends on its detailed formulation, reflected by a positive integer exponent $N = 0, 1, \ldots$:

$$f_{H_q}^2 \left(M_{H_q}^2\right)^N \exp(-M_{H_q}^2\,\tau) = \int\limits_{(m_Q+m_q)^2}^{s_{\rm eff}^{(N)}(\tau,m_Q,m_q,\alpha_s)} {\rm d}s \exp(-s\,\tau)\, s^N \rho(s, m_Q, m_q, \alpha_s\,|\,m_{\rm sea}) + \Pi_{\rm power}^{(N)}(\tau, m_Q, m_q, \alpha_s, \langle\bar{q}q\rangle, \ldots)\,. \quad (1)$$

On its QCD side, the three important ingredients are the τ-dependent [3–7] *effective threshold* $s_{\rm eff}^{(N)}(\tau, m_Q, m_q, \alpha_s)$, constructed such as to guarantee mutual cancellation of perturbative-QCD and hadron contributions above the threshold to the utmost achievable degree; the perturbative *spectral density* $\rho(s, m_Q, m_q, \alpha_s\,|\,m_{\rm sea})$, experiencing also the masses $m_{\rm sea}$ of the sea quarks and expressible as an expansion (Fig. 1) in powers of α_s depending on the renormalization scale μ,

$$\rho(s, m_Q, m_q, \alpha_s\,|\,m_{\rm sea}) = \rho_0(s, m_Q, m_q) + \frac{\alpha_s(\mu)}{\pi}\,\rho_1(s, m_Q, m_q, \mu) + \frac{\alpha_s^2(\mu)}{\pi^2}\,\rho_2(s, m_Q, m_q, \mu\,|\,m_{\rm sea}) + O(\alpha_s^3)\,; \quad (2)$$

and the τ-dependent *power corrections* $\Pi_{\rm power}^{(N)}(\tau, m_Q, m_q, \alpha_s, \langle\bar{q}q\rangle, \ldots)$ that subsume, together with the effective threshold $s_{\rm eff}^{(N)}(\tau, m_Q, m_q, \alpha_s)$, all the nonperturbative manifestations of QCD.

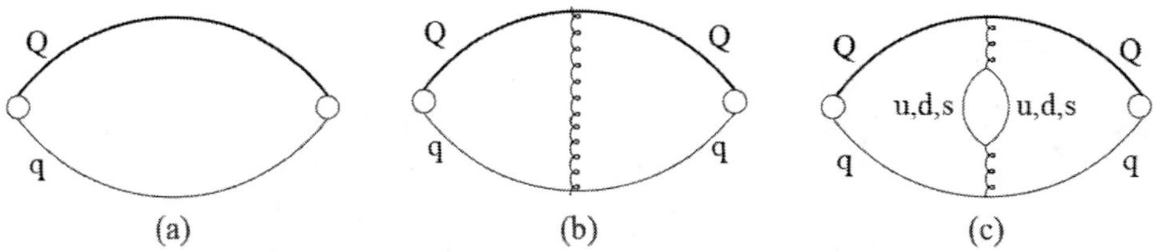

Figure 1. Perturbative expansion (2) of the spectral density $\rho(s, m_Q, m_q, \alpha_s\,|\,m_{\rm sea})$ in powers of the strong coupling α_s: (a) $O(1)$, (b) $O(\alpha_s)$, (c) $O(\alpha_s^2)$; sea quarks, and thus their masses, start to contribute at $O(\alpha_s^2)$.

The partitioning of nonperturbative contributions between power corrections and effective threshold reflects, in general, the details of the problem to which the QCD sum rule is applied. By considering, however, the local-duality limit $\tau \to 0$ of Eq. (1) for the handpicked exponent $N = 0$ such that all power corrections vanish we may achieve to shift all of the nonperturbative burden to the effective threshold. With all power corrections gone, Eq. (1) simplifies to [8–10]

$$f_{H_q}^2 = \int\limits_{(m_Q+m_q)^2}^{s_{\rm eff}(m_Q,m_q,\alpha_s)} {\rm d}s\, \rho(s, m_Q, m_q, \alpha_s\,|\,m_{\rm sea}) \equiv F(s_{\rm eff}(m_q), m_q\,|\,m_{\rm sea})\,. \quad (3)$$

All spectral densities required as input are known up to order $O(\alpha_s\, m_q)$ and $O(\alpha_s^2\, m_q^0)$ [11–14]. However, we first must ensure the well-definiteness of the local-duality limit $\tau \to 0$ of Eq. (1).

The dependence on the Borel parameter τ of our effective threshold $s_{\rm eff}^{(N)}(\tau, m_Q, m_q, \alpha_s)$ and of the power corrections $\Pi_{\rm power}^{(N)}(\tau, m_Q, m_q, \alpha_s, \langle\bar{q}q\rangle, \ldots)$ under the "disguise" of some quantity

$$\overline{\Pi}_{\rm power}^{(N)}(\tau) \equiv \Pi_{\rm power}^{(N)}(\tau)\, \frac{\exp(M_{H_q}^2\,\tau)}{M_{H_q}^{2\,N}} \quad (4)$$

https://doi.org/10.1051/epjconf/201819200041

is explored, for the three most interesting values $N = 0, 1, 2$ of the exponent N, in Fig. 2 for the pseudoscalar mesons and in Fig. 3 for the vector mesons. For the cases $N = 0$ and $N = 1$, the τ behaviour of the power corrections at $\tau = 0$ is regular: for $N = 0$, all power corrections vanish,

$$\lim_{\tau\to 0} \Pi^{(N)}_{\rm power}(\tau, m_Q, m_q, \alpha_s, \langle\bar{q}q\rangle, \dots) = 0 \qquad \text{for} \quad N = 0\,;$$

for $N = 1$, the power corrections approach a finite value. In contrast to this, for the case $N = 2$, due to an involvement of also $\log(\tau)$ terms the power corrections exhibit singularities at $\tau = 0$. Hence, in our pursuit of local-duality QCD sum rules the exponent of choice must read $N = 0$.

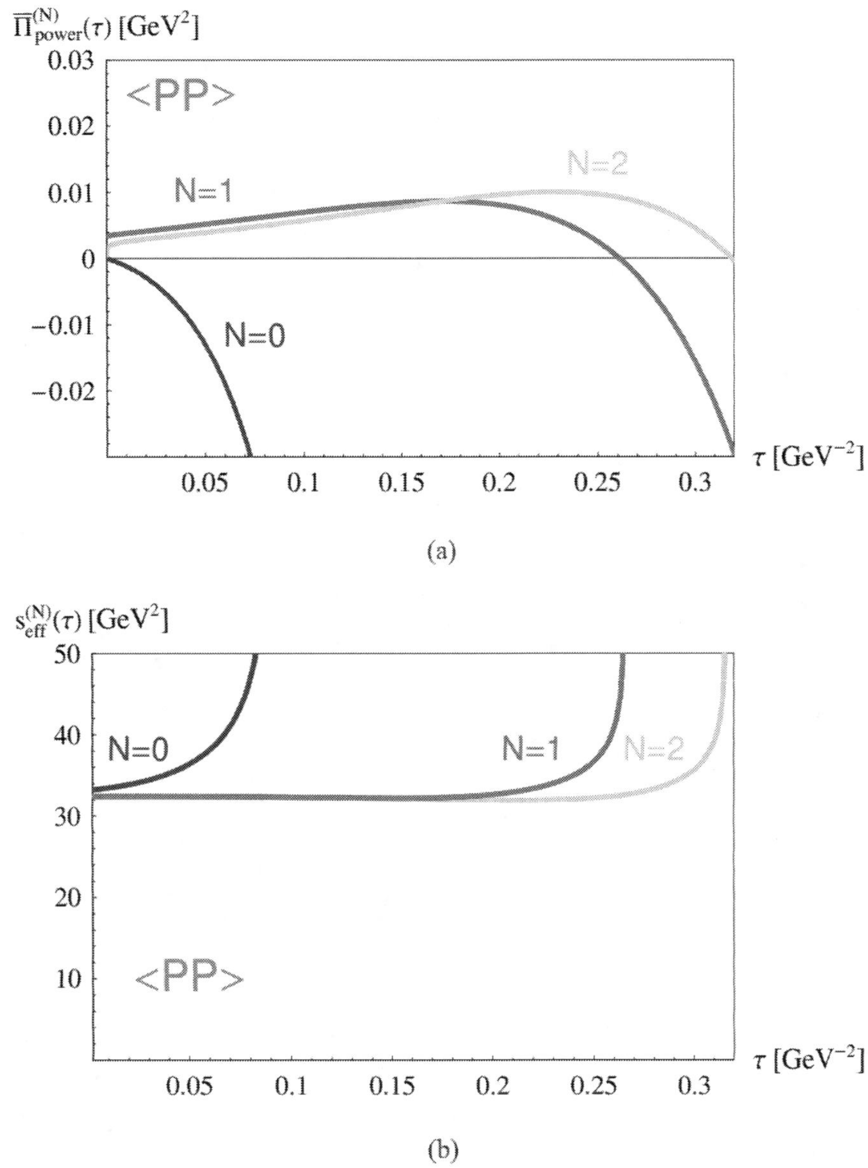

Figure 2. Dependence of (a) the power corrections $\Pi^{(N)}_{\rm power}(\tau)$, rescaled according to Eq. (4), as well as (b) our effective threshold $s^{(N)}_{\rm eff}(\tau)$ in the QCD sum rule for two pseudoscalar currents on the Borel variable τ.

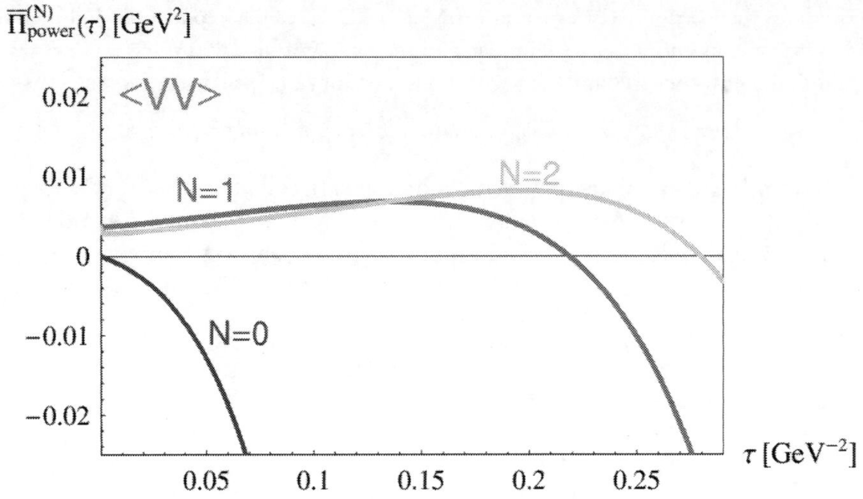

(a)

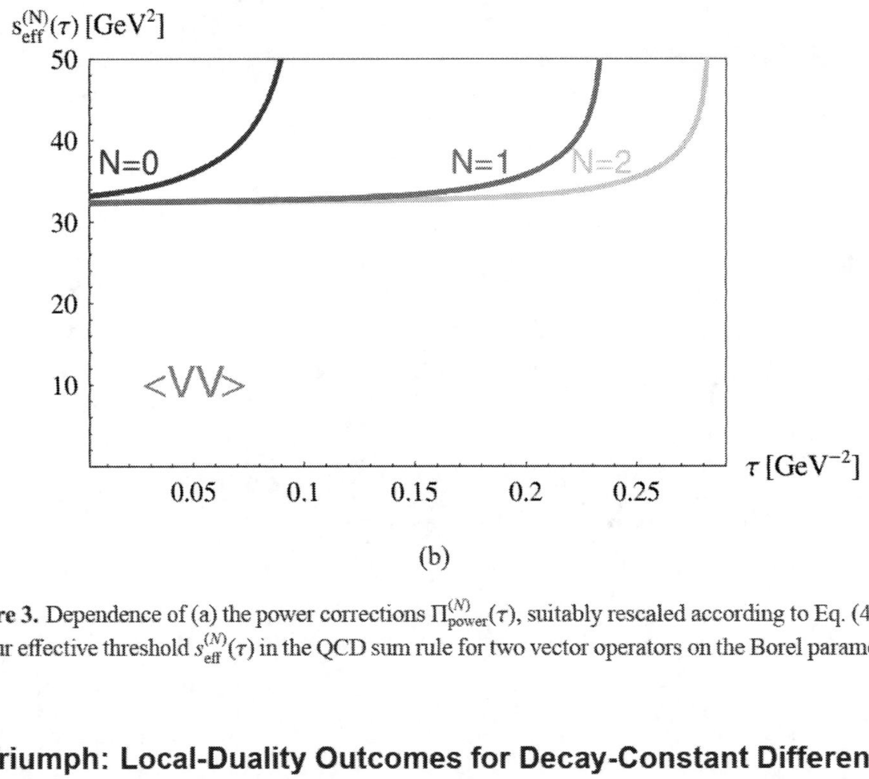

(b)

Figure 3. Dependence of (a) the power corrections $\Pi^{(N)}_{\rm power}(\tau)$, suitably rescaled according to Eq. (4), and (b) our effective threshold $s^{(N)}_{\rm eff}(\tau)$ in the QCD sum rule for two vector operators on the Borel parameter τ.

3 Triumph: Local-Duality Outcomes for Decay-Constant Differences

By collecting the available pieces of information about the m_q dependence of all the quantities entering in our local-duality limit (3) of the QCD sum rule (1), we extract the isospin-breaking difference $f_{H_d} - f_{H_u}$ from the difference of an associated function $F(s_{\rm eff}(m_q), m_q \,|\, m_{\rm sea})$, Eq. (3):

$$F(s_{\rm eff}(m_d), m_d \,|\, m_{\rm sea}) - F(s_{\rm eff}(m_u), m_u \,|\, m_{\rm sea})$$
$$= F(s_{\rm eff}(m_d), m_d \,|\, 0) - F(s_{\rm eff}(m_u), m_u \,|\, 0) + O(\alpha_s^2\,(m_d - m_u))\,. \qquad (5)$$

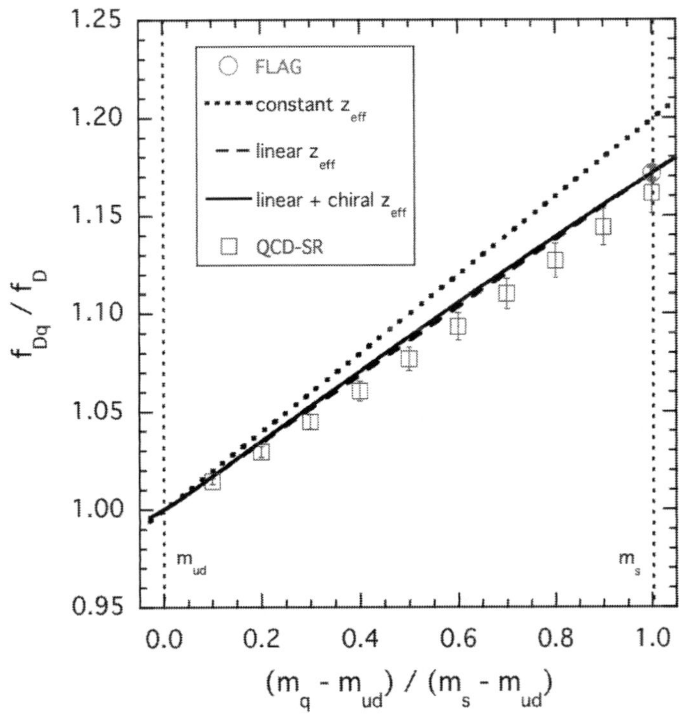

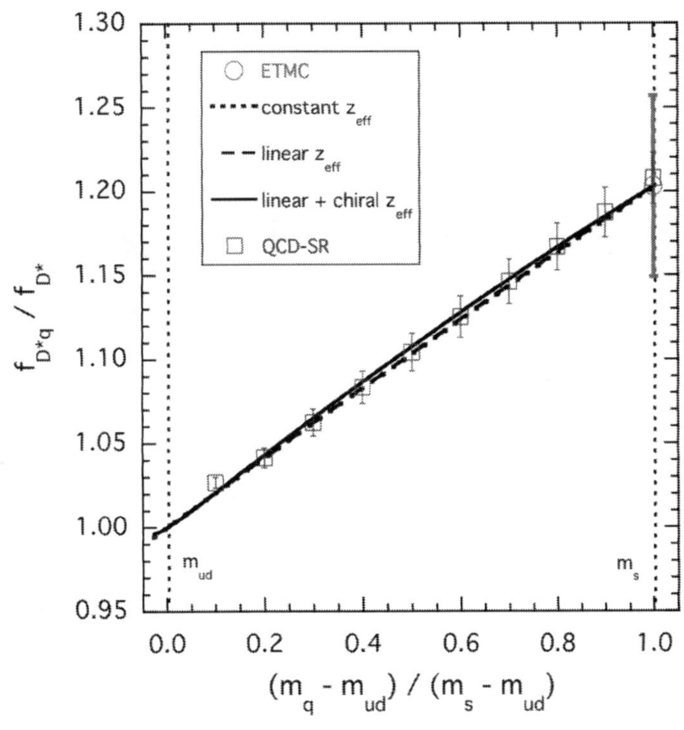

Figure 4. Dependence of the heavy–light charmed-meson decay constants, $f_{D_q^{(*)}} \equiv f_{D_q^{(*)}}(m_q)$, normalized to their values $f_{D^{(*)}} \equiv f_{D_q^{(*)}}(m_{ud})$ at the average light-quark mass, $m_{ud} \equiv (m_u + m_d)/2$, on the (conveniently shifted and rescaled) light-quark mass $(m_q - m_{ud})/(m_s - m_{ud})$, as inferred from three different assumptions about the behaviour of our effective threshold $z_{\rm eff} \equiv \sqrt{s_{\rm eff}} - m_c - m_q$ [8–10], vs. previous findings [15] (□).

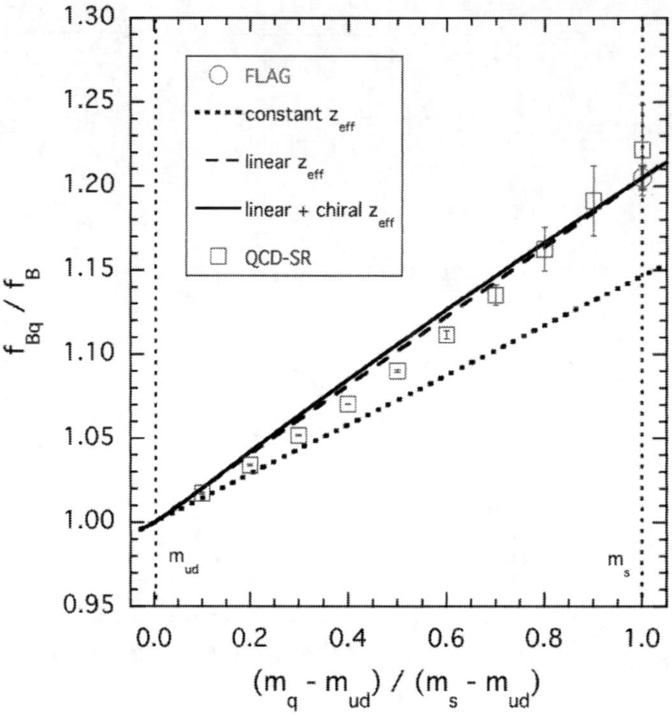

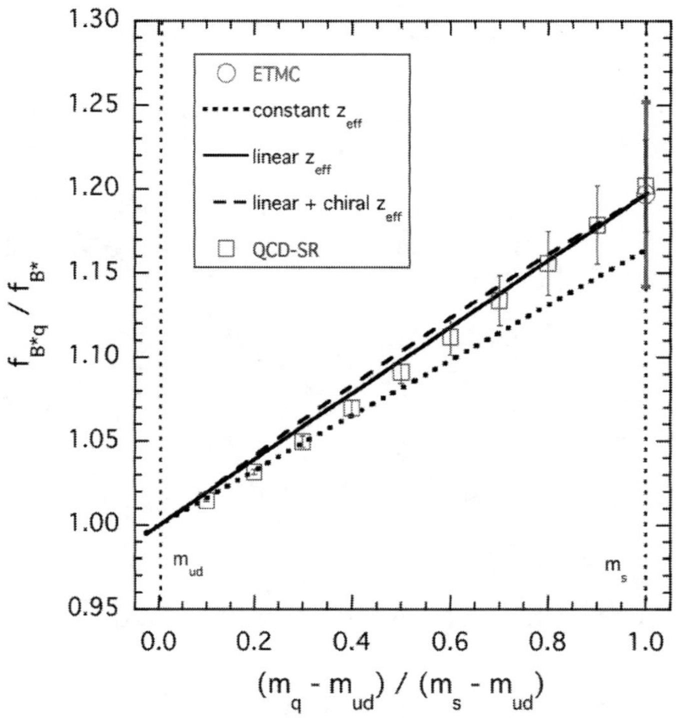

Figure 5. Dependence of the heavy–light bottom-meson decay constants $f_{B_q^{(*)}} \equiv f_{B_q^{(*)}}(m_q)$, normalized to their values $f_{B^{(*)}} \equiv f_{B_q^{(*)}}(m_{ud})$ at the average light-quark mass, $m_{ud} \equiv (m_u + m_d)/2$, on the — still suitably shifted and rescaled — light-quark mass $(m_q - m_{ud})/(m_s - m_{ud})$, inferred from three different assumptions about the behaviour of our effective threshold $z_{\text{eff}} \equiv \sqrt{s_{\text{eff}}} - m_b - m_q$ [8–10], vs. previous findings [15] (□).

All dependence on the light-quark mass m_q of our predictions for leptonic decay constants f_{H_q} by the local-duality QCD sum rule (3) originates in precisely two sources of different type:

1. The impact of the spectral density can be trivially inferred from existing knowledge due to perturbation theory. Equation (5) already takes into account that in this difference the contributions of *strange* sea quarks cancel. This reduces the uncertainties considerably.

2. The situation with the effective threshold is more delicate: we model it by allowing for a continuous variation of m_q within $[0, m_s]$ and matching all results for f_{H_q} to lattice QCD in three neighbouring ways, see Fig. 4 for charmed quarks and Fig. 5 for bottom quarks.

In proper naming, our results for the heavy–light meson decay-constant differences are [8–10]

$$f_{D^\pm} - f_{D^0} = (0.96 \pm 0.09)\ \text{MeV}\,, \qquad f_{D^{*\pm}} - f_{D^{*0}} = (1.18 \pm 0.35)\ \text{MeV}\,,$$
$$f_{B^0} - f_{B^\pm} = (1.01 \pm 0.10)\ \text{MeV}\,, \qquad f_{B^{*0}} - f_{B^{*\pm}} = (0.89 \pm 0.30)\ \text{MeV}\,.$$

Strong isospin breaking in these heavy–light-meson decay constants is pretty close to 1 MeV.

Acknowledgements

D. M. is grateful for support by the Austrian Science Fund (FWF) under project P29028-N27.

References

[1] Particle Data Group (M. Tanabashi *et al.*), Phys. Rev. D **98** (2018) 030001.
[2] M. A. Shifman, A. I. Vainshtein, and V. I. Zakharov, Nucl. Phys. B **147** (1979) 385.
[3] W. Lucha, D. Melikhov, and S. Simula, Phys. Rev. D **79** (2009) 096011, arXiv:0902.4202 [hep-ph].
[4] W. Lucha, D. Melikhov, and S. Simula, J. Phys. G **37** (2010) 035003, arXiv:0905.0963 [hep-ph].
[5] W. Lucha, D. Melikhov, H. Sazdjian, and S. Simula, Phys. Rev. D **80** (2009) 114028, arXiv:0910.3164 [hep-ph].
[6] W. Lucha, D. Melikhov, and S. Simula, Phys. Lett. B **687** (2010) 48, arXiv:0912.5017 [hep-ph].
[7] W. Lucha, D. Melikhov, and S. Simula, Phys. Atom. Nucl. **73** (2010) 1770, arXiv:1003.1463 [hep-ph].
[8] W. Lucha, D. Melikhov, and S. Simula, Eur. Phys. J. C **78** (2018) 168, arXiv:1702.07537 [hep-ph].
[9] W. Lucha, D. Melikhov, and S. Simula, PoS (EPS-HEP 2017) 669, arXiv:1709.02131 [hep-ph].
[10] W. Lucha, D. Melikhov, and S. Simula, PoS (Hadron2017) 175, arXiv:1711.07899 [hep-ph].
[11] K. G. Chetyrkin and M. Steinhauser, Phys. Lett. B **502** (2001) 104, arXiv:hep-ph/0012002.
[12] K. G. Chetyrkin and M. Steinhauser, Eur. Phys. J. C **21** (2001) 319, arXiv:hep-ph/0108017.
[13] M. Jamin and B. O. Lange, Phys. Rev. D **65** (2002) 056005, arXiv:hep-ph/0108135.
[14] P. Gelhausen, A. Khodjamirian, A. A. Pivovarov, and D. Rosenthal, Phys. Rev. D **88** (2013) 014015, arXiv:1305.5432 [hep-ph]; Phys. Rev. D **89** (2014) 099901(E); **91** (2015) 099901(E).
[15] W. Lucha, D. Melikhov, and S. Simula, Phys. Lett. B **765** (2017) 365, arXiv:1609.05050 [hep-ph].

Dynamical picture for the exotic XYZ states

Mikhail Ivanov[1,*]

[1]Bogoliubov Laboratory of Theoretical Physics, Joint Institute for Nuclear Research, 141980 Dubna, Russia

Abstract. We present a dynamical approach for description of the multi-quark states that is based on an effective interaction Lagrangian describing the coupling of hadrons to their constituent quarks. First, we explore the consequences of treating the $X(3872)$ meson as a tetraquark bound state. We calculate the decay widths of the observed channels and conclude that for reasonable values of the size parameter of the $X(3872)$ one finds consistency with the available experimental data. Then we have critically checked the tetraquark picture for the $Z_c(3900)$ state by analyzing its strong decays. We found that $Z_c(3900)$ has a much more stronger coupling to DD^* than to $J/\psi\pi$ that is in discord with experiment. As an alternative we have employed a molecular-type four-quark current to describe the decays of the $Z_c(3900)$ state as the charged particle in the isotriplet. We found that a molecular-type current gives the values of the above decays in accordance with the experimental observation. By using molecular-type four-quark currents for the recently observed resonances $Z_b(10610)$ and $Z_b(10650)$, we have calculated their two-body decay rates into a bottomonium state plus a light meson as well as into B-meson pairs.

1 Introduction

The most reliable prediction of the quark model is the spectrum of conventional hadrons composed either from quark-antiquark (mesons) or from three quarks (baryons). These states have been observed in experiments and their properties have been carefully studied by experimentalists as well as theorists for a long time. However, this situation has been changed since 2003 with the discovery of many charmonium- and bottomonium-like XYZ states that do not fit the simple quark-antiquark interpretation. The specific feature of these states is that their masses are close to meson-meson thresholds. There are several theoretical interpretations of them, e.g., as hadronic molecules, tetraquarks, threshold cusps, etc. (for review, see Refs. [1–3]).

The Y states are neutral with quantum numbers $J^{PC} = 1^{--}$ of charmonium but not the simple $c\bar{c}$ charmonium. Among the observed Y states are $Y(4005)$, $Y(4260)$, $Y(4360)$, etc. The Z states (Z_c and Z_b) are basically charged, for example, Z_c^+ has quark content $c\bar{c}u\bar{d}$ so it is certainly exotic state. The Z_b^+ has quark content $b\bar{b}u\bar{d}$. The X states are the non-$q\bar{q}$ mesons but other than the $Y's$ and $Z's$. The most famous is the $X(3872)$ with quantum numbers $J^{PC} = 1^{++}$.

We present a dynamical approach for description of the multi-quark states that is based on an effective interaction Lagrangian describing the coupling of hadrons to their constituent

*e-mail: ivanovm@theor.jinr.ru

© The Authors, published by EDP Sciences. This is an open access article distributed under the terms of the Creative Commons Attribution License 4.0 (http://creativecommons.org/licenses/by/4.0/).

quarks. We give a brief sketch of our results on the decay widths of some exotic four-quark states obtained within this framework and published in series of our papers of Refs. [4–9].

First, we explore the consequences of treating the $X(3872)$ meson as a tetraquark bound state. We calculate the decay widths of the observed channels and conclude that for reasonable values of the size parameter of the $X(3872)$ one finds consistency with the available experimental data. Then we have critically checked the tetraquark picture for the $Z_c(3900)$ state by analyzing its strong decays. We found that $Z_c(3900)$ has a much more stronger coupling to DD^* than to $J/\psi\pi$ that is in discord with experiment. As an alternative we have employed a molecular-type four-quark current to describe the decays of the $Z_c(3900)$ state as the charged particle in the isotriplet. We found that a molecular-type current gives the values of the above decays in accordance with the experimental observation. By using molecular-type four-quark currents for the recently observed resonances $Z_b(10610)$ and $Z_b(10650)$, we have calculated their two-body decay rates into a bottomonium state plus a light meson as well as into B-meson pairs.

2 Dynamical picture for multiquark states: covariant confined quark model

The main assumption of approach is that hadrons interact via quark exchange only. The interaction Lagrangian is written as

$$\mathcal{L}_{\rm int} = g_H \cdot H(x) \cdot J_H(x), \qquad \text{where the quark currents are given by}$$

$$J_M(x) = \int dx_1 \int dx_2 \, F_M(x; x_1, x_2) \cdot \bar{q}_1^a(x_1) \Gamma_M \, q_2^a(x_2) \qquad \text{Meson}$$

$$J_B(x) = \int dx_1 \int dx_2 \int dx_3 \, F_B(x; x_1, x_2, x_3) \qquad \text{Baryon}$$
$$\times \Gamma_1 \, q_1^{a_1}(x_1) \left(q_2^{a_2}(x_2) C \, \Gamma_2 \, q_3^{a_3}(x_3) \right) \cdot \varepsilon^{a_1 a_2 a_3}$$

$$J_T^\mu(x) = \int dx_1 \ldots \int dx_4 \, F_T(x; x_1, \ldots, x_4) \qquad \text{Tetraquark}$$
$$\times \left(q_1^{a_1}(x_1) C \Gamma_1 \, q_2^{a_2}(x_2) \right) \cdot \left(\bar{q}_3^{a_3}(x_3) \Gamma_2 C \, \bar{q}_4^{a_4}(x_4) \right) \cdot \varepsilon^{a_1 a_2 c} \varepsilon^{a_3 a_4 c}.$$

Here, the $F(x; x_1, \ldots, x_n)$ are the vertex functions characterizing the distribution of quarks inside the hadron. The matrix elements of the physical processes are described by the Feynman diagrams which are convolution of the quark propagators and vertex functions. The details of all calculations may be found in our published papers of Refs. [4–9]. For illustration, we show in Fig. 1 the tetraquark self-energy diagram which is needed for the determination of coupling constant g_T in the interaction Lagrangian.

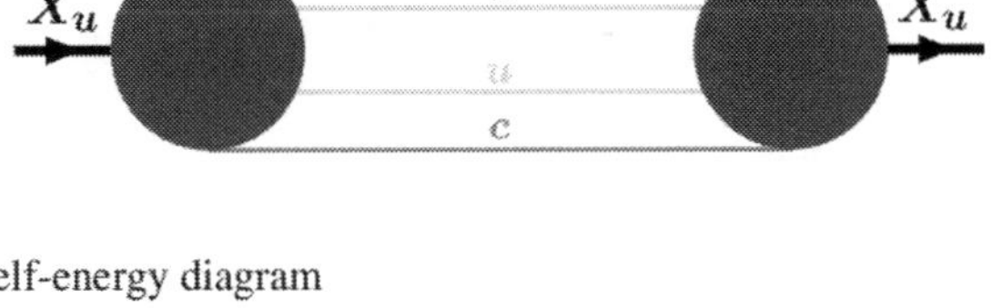

Figure 1. Tetraquark self-energy diagram

3 $X(3872)$

A narrow charmonium–like state $X(3872)$ was observed in 2003 in the exclusive decay process $B^\pm \to K^\pm\pi^+\pi^- J/\psi$ [10]. A mass was found to be very close to the $D^0 D^{*0}$ threshold and width less than 2.3 MeV. The state was confirmed in B-decays by the BaBar experiment [11] and in $p\bar{p}$ production by the Tevatron experiments CDF [12] and DØ[13]. From the angular analysis performed by several collaborations it was shown that the $X(3872)$ state has the quantum number $J^{PC} = 1^{++}$. Then it was found that the branching ratios of the modes $X \to \pi^+\pi^- J/\psi$ and $X \to \pi^+\pi^-\pi^0 J/\psi$ are almost the same that imply strong isospin violation. There are several different interpretations of the $X(3872)$ in the literature: a molecule bound state $(D^0\overline{D}^{*0})$, threshold cusps, hybrids and glueballs.

We employed the diquark-antidiquark interpretation suggested in Ref. [14] and performed an independent analysis of the properties of the $X(3872)$ within our covariant confined quark model in the papers Refs. [4, 5]. The quark content was chosen as $[cq]_{S=0}[\bar{c}\bar{q}]_{S=1} + [cq]_{S=1}[\bar{c}\bar{q}]_{S=0}$, $(q = u, d)$ [14] . The physical states are suppossed to be a linear superposition of the X_u and X_d states according to

$$X_l \equiv X_{\rm low} = X_u \cos\theta + X_d \sin\theta, \qquad X_h \equiv X_{\rm high} = -X_u \sin\theta + X_d \cos\theta.$$

The mixing angle θ is determined from fitting the ratio of branching ratios of the $X-$decays into $J/\psi + 2\pi$ and $J/\psi + 3\pi$.

The diagrams describing the strong decays $\Gamma(X \to J/\psi + n\pi)$ (n=2,3) and $X \to D^0\bar{D}^0\pi^0$ are shown in Fig. 2. The diagrams describing the radiative decays $\Gamma(X \to J/\psi + \gamma)$ are shown in Fig. 3.

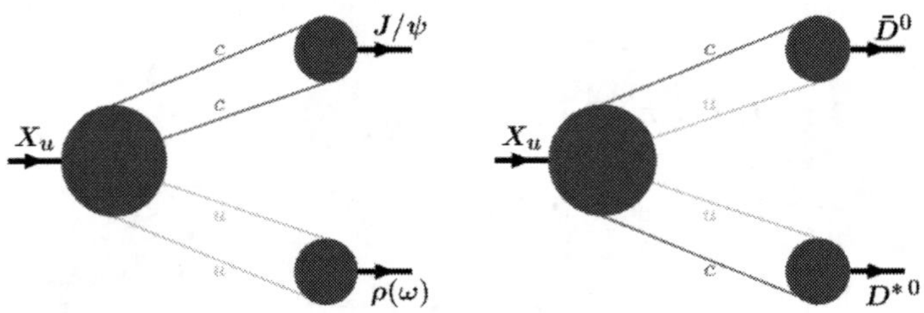

Figure 2. Diagrams describing the strong X-decays

The only free parameter is the so-called size parameter Λ_X which appears in the vertex function $F_T(x; x_1, \ldots, x_4)$. We plot the dependence of the calculated decay widths on this parameter in Fig. 4 in quite large range of $\Lambda_X \in [2.5 - 4]$ GeV.

Finally, we compare the obtained results for the ratios of decay widths with the available experimental data. One can see that there is reasonable agreement in the case of the strong decays and very good agreement in the case of the radiative decays.

$$\frac{\Gamma(X \to D^0\bar{D}^0\pi^0)}{\Gamma(X \to J/\psi\pi^+\pi^-)} = \left\{ \begin{array}{ll} 4.5 \pm 0.2 & \text{theor} \\ 10.5 \pm 4.7 & \text{expt} \end{array} \right|, \qquad \frac{\Gamma(X \to J/\psi + \gamma)}{\Gamma(X \to J/\psi + 2\pi)} = \left\{ \begin{array}{ll} 0.15 \pm 0.03 & \text{theor} \\ 0.14 \pm 0.05 & \text{expt} \end{array} \right|.$$

4 $Z_c(3900)$

The BESIII Collaboration [15] observed a structure in the $\pi^\pm J/\psi$ mass spectrum of the process $e^+e^- \to \pi^+ \underbrace{\pi^- J/\psi}_{Z_c^-}$. It was named as the $Z_c(3900)$ state with a mass near 3.9 GeV and a

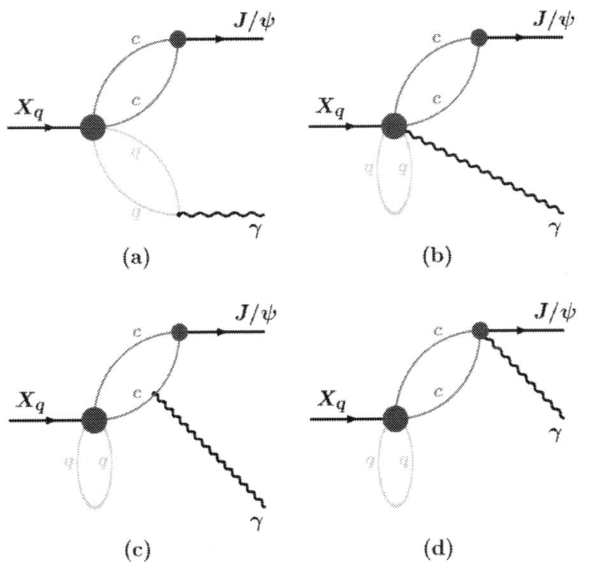

Figure 3. Diagrams describing the radiative X-decays

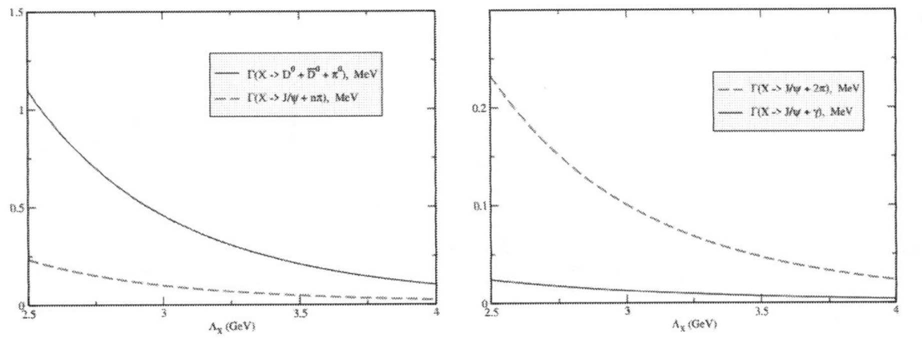

Figure 4. Dependence of the strong (left panel) and radiative (right panel) decay widths onthe size parameter

width around of 50 MeV. This state was confirmed by the Belle Collaboration [16]. Recently, the D0 Collaboration presented evidence for the exotic charged $Z_c(3900)$ in semi-inclusive weak decays of b-flavored hadrons [17]. This structure can be interpreted as a new charged charmonium-like state. Mode with $D\bar{D}^*$ in the final state was studied by the BESIII Collaboration [18]. A distinct charged structure was observed in the $(D\bar{D}^*)^{\mp}$ invariant mass distribution of the process $e^+e^- \to \pi^{\pm} \underbrace{(D\bar{D}^*)^{\mp}}_{Z_c^{\mp}}$. The angular distribution of the $\pi Z_c(3885)$ system favors a $J^P = 1^+$ quantum number assignment for the new structure. The ratio of partial widths was determined as

$$\frac{\Gamma(Z_c(3885) \to D\bar{D}^*)}{\Gamma(Z_c(3900) \to \pi J/\psi)} = 6.2 \pm 1.1 \pm 2.7 \tag{1}$$

Assuming that the Z_c the charged partner of the $X(3872)$ state one can write down a tetraquark-type current:

$$J^\mu = \frac{i}{\sqrt{2}}\varepsilon_{abc}\varepsilon_{dec}\left[(u_a^T C\gamma_5 c_b)(\bar{d}_d\gamma^\mu C\bar{c}_e^T) - (u_a^T C\gamma^\mu c_b)(\bar{d}_d\gamma_5 C\bar{c}_e^T)\right] \tag{2}$$

In Ref. [6] we have used the nonlocal generalization of this tetraquark current to calculate a number of the $Z_c(399)$ two-body decay widths of the process $1^+(p,\mu) \to 1^-(q_1,\nu) + 0^-(q_2)$. It was found that, in our model, the leading Lorentz metric structure in the matrix elements describing the decays $Z_c^+(3900) \to \bar{D}D^*$ vanishes analytically. This results in a significant suppression of these decay widths by the smallness of the relevant phase space factor $|\mathbf{q}|^5$. Since the experimental data [18] show that the $Z_c(3900)$ has a much more stronger coupling to DD^* than to $J/\psi\pi$, one has to conclude that the tetraquark-type current for the $Z_c(3900)$ is in discord with experiment.

As an alternative we have employed a molecular-type four-quark current to describe the decays of the $Z_c(3900)$ state:

$$J^\mu = \frac{1}{\sqrt{2}}\left[(\bar{d}\gamma_5 c)(\bar{c}\gamma^\mu u) + (\bar{d}\gamma^\mu c)(\bar{c}\gamma_5 u)\right]. \tag{3}$$

In this case we found that for a relatively large size parameter $\Lambda_{Z_c} \sim 3.3$ GeV one can obtain the partial widths of the decays $Z_c^+(3900) \to \bar{D}D^*$ at the order ~ 15 MeV for each mode. At the same time the partial widths for decays $Z_c^+(3900) \to J/\psi\pi^+, \eta_c\rho^+$ are suppressed by a factor of $6-7$ in accordance with experimental data [18]. If the Λ_{Z_c} is varied in the limits $\Lambda_{Z_c} = 3.3 \pm 1.1$ GeV then one has

$$\begin{aligned}
\Gamma(Z_c^+ \to J/\psi + \pi^+) &= (1.8 \pm 0.3)\,\text{MeV}, \qquad \Gamma(Z_c^+ \to \bar{D}^0 + D^{*+}) = (10.0^{+1.7}_{-1.4})\,\text{MeV},\\
\Gamma(Z_c^+ \to \eta_c + \rho^+) &= (3.2^{+0.5}_{-0.4})\,\text{MeV}, \qquad \Gamma(Z_c^+ \to \bar{D}^{*0} + D^+) = (9.0^{+1.6}_{-1.3})\,\text{MeV}.
\end{aligned}$$

Preliminary data from BESIII cited in [19] were reported for the ratio

$$R(Z) = \frac{\mathcal{B}(Z_c(3900) \to \rho\eta_c)}{\mathcal{B}(Z_c(3900) \to \pi J/\psi)} = 2.1 \pm 0.8. \tag{4}$$

They agree very well with our finding $R(Z) = 1.8 \pm 0.4$.

5 $Z_b(10610)$ and $Z'_b(10610)$

A few years ago the Belle Collaboration [20] reported on the observation of two charged bottomoniumlike resonances in the mass spectra of $\pi^\pm\Upsilon(nS)$ $(n = 1,2,3)$ and $\pi^\pm h_b(mP)$ $(m = 1,2)$ in the decays $\Upsilon(5S) \to \Upsilon(nS)\pi^+\pi^-, h_b(mP)\pi^+\pi^-$. The measured masses and widths were given by

$$\begin{aligned}
M_{Z_b} &= (10607.2 \pm 2.0)\,\text{MeV}, \qquad \Gamma_{Z_b} = (18.4 \pm 2.4)\,\text{MeV},\\
M_{Z'_b} &= (10652.2 \pm 1.5)\,\text{MeV}, \qquad \Gamma_{Z'_b} = (11.5 \pm 2.2)\,\text{MeV}.
\end{aligned}$$

and the quantum numbers are $I^G(J^P) = 1^+(1^+)$. The existence of these two states was later confirmed by the same collaboration [21, 22] in differing decay channels. In the paper [22] the Belle Collaboration reported on the results of an analysis of the three-body processes $e^+e^- \to B\bar{B}\pi^\pm, B\bar{B}^*\pi^\pm$, and $B^*\bar{B}^*\pi^\pm$. It was found that the transitions $Z_b^\pm(10610) \to [B\bar{B}^* + c.c.]^\pm$ and $Z_b^\pm(10650) \to [B^*\bar{B}^*]^\pm$ dominate among the corresponding final states.

Table 1. $Z_b(10610)$ and $Z_b'(10610)$: numerical results

Channel	Widths, MeV	
	$Z_b(10610)$	$Z_b'(10650)$
$\Upsilon(1S)\pi^+$	5.9 ± 0.4	$9.5^{+0.7}_{-0.6}$
$h_b(1P)\pi^+$	$(0.14 \pm 0.01) \cdot 10^{-1}$	$0.74^{+0.05}_{-0.04} \cdot 10^{-3}$
$\eta_b\rho^+$	4.4 ± 0.3	$7.5^{+0.6}_{-0.5}$
$B^+\bar{B}^{*0} + \bar{B}^0 B^{*+}$	$20.7^{+1.6}_{-1.5}$	–
$B^{*+}\bar{B}^{*0}$	–	$17.1^{+1.5}_{-1.4}$

Table 2. $Z_b(10610)$ and $Z_b'(10610)$: Total widths, MeV

	Theory	Belle Expt.
$Z_b(10610)$	$30.9^{+2.3}_{-2.1}$	25 ± 7
$Z_b'(10650)$	$34.1^{+2.8}_{-2.5}$	23 ± 8

Since the masses of the $Z_b^+(10610)$ and $Z_b'(10650)$ are very close to the respective $B^*\bar{B}$ (10604 MeV) and $B^*\bar{B}^*$ (10649 MeV) thresholds, it was suggested in Ref. [23] that they have molecular-type binding structures.

$$J^{\mu}_{Z_b^+} = \frac{1}{\sqrt{2}}\left[(\bar{d}\gamma_5 b)(\bar{b}\gamma^{\mu}u) + (\bar{d}\gamma^{\mu}b)(\bar{b}\gamma_5 u)\right], \qquad J^{\mu\nu}_{Z_b'^+} = \varepsilon^{\mu\nu\alpha\beta}(\bar{d}\gamma_\alpha b)(\bar{b}\gamma_\beta u). \tag{5}$$

Such a choice guarantees that the Z_b-state can only decay to the $[\bar{B}^*B + c.c.]$ pair whereas the Z_b'-state can decay only to a $\bar{B}^*B^*$ pair. Decays into the BB-channels are forbidden. Due to G-parity conservation $Z_b \to \Upsilon + \rho$, $Z_b \to \eta_b + \pi$, $Z_b \to \chi_{b1} + \pi$, $Z_b \to h_b + \rho$. The decay $Z_b \to \chi_{b1} + \rho$ is not allowed kinematically.

There are therefore only the three allowed decays: $Z_b^+ \to \Upsilon + \pi^+$, $Z_b^+ \to h_b + \pi^+$ and $Z_b^+ \to \eta_b + \rho^+$. The only two new parameters are the size parameters of the two exotic $Z_b(Z_b')$ states. As a guide to adjust them we take the experimental values of the largest branching fractions presented by Belle:

$$\mathcal{B}(Z_b^+ \to [B^+\bar{B}^{*0} + \bar{B}^0 B^{*+}]) = 85.6^{+1.5+1.5}_{-2.0-2.1}\,\%, \qquad \mathcal{B}(Z_b'^+ \to \bar{B}^{*+}B^{*0}) = 73.7^{+3.4+2.7}_{-4.4-3.5}\,\%. \tag{6}$$

By using the central values of these branching rates and total decay widths we find the central values of our size parameters $\Lambda_{Z_b} = 3.45$ GeV and $\Lambda_{Z_b'} = 3.00$ GeV. Allowing them to vary in the interval $\Lambda_{Z_b} = 3.45 \pm 0.05$ GeV and $\Lambda_{Z_b'} = 3.00 \pm 0.05$ GeV we obtain the values of various decay widths.

Acknowledgments

I am grateful to the organizers of International Workshop on QCD, theory and experiment (QCD@Work) for invitation and hospitality in Matera. This work was supported in part by the Joint Research Project of Institute of Physics, SAS and Bogoliubov Laboratory of Theoretical Physics, JINR, No. 01-3-1114.

References

[1] R. F. Lebed, R. E. Mitchell and E. S. Swanson, Prog. Part. Nucl. Phys. **93**, 143 (2017).

[2] M. Nielsen, F. S. Navarra and S. H. Lee, Phys. Rept. **497**, 41 (2010).

[3] E. Klempt and A. Zaitsev, Phys. Rept. **454**, 1 (2007).

[4] S. Dubnicka, A. Z. Dubnickova, M. A. Ivanov and J. G. Körner, Phys. Rev. D **81**, 114007 (2010).

[5] S. Dubnicka, A. Z. Dubnickova, M. A. Ivanov, J. G. Körner, P. Santorelli and G. G. Saidullaeva, Phys. Rev. D **84**, 014006 (2011).

[6] F. Goerke, T. Gutsche, M. A. Ivanov, J. G. Körner, V. E. Lyubovitskij and P. Santorelli, Phys. Rev. D **94**, no. 9, 094017 (2016).

[7] F. Goerke, T. Gutsche, M. A. Ivanov, J. G. Körner and V. E. Lyubovitskij, Phys. Rev. D **96**, no. 5, 054028 (2017).

[8] T. Gutsche, M. A. Ivanov, J. G. Körner, V. E. Lyubovitskij and K. Xu, Phys. Rev. D **96**, no. 11, 114004 (2017).

[9] T. Gutsche, M. A. Ivanov, J. G. Körner and V. E. Lyubovitskij, Phys. Rev. D **94**, no. 9, 094012 (2016).

[10] S. K. Choi *et al.* [Belle Collaboration], Phys. Rev. Lett. **91**, 262001 (2003).

[11] B. Aubert *et al.* [BaBar Collaboration], Phys. Rev. Lett. **93**, 041801 (2004).

[12] D. E. Acosta *et al.* [CDF Collaboration], Phys. Rev. Lett. **93**, 072001 (2004).

[13] V. M. Abazov *et al.* [D0 Collaboration], Phys. Rev. Lett. **93**, 162002 (2004).

[14] L. Maiani, F. Piccinini, A. D. Polosa and V. Riquer, Phys. Rev. D **71**, 014028 (2005).

[15] M. Ablikim *et al.* [BESIII Collaboration], Phys. Rev. Lett. **110**, 252001 (2013).

[16] Z. Q. Liu *et al.* [Belle Collaboration], Phys. Rev. Lett. **110**, 252002 (2013).

[17] V. M. Abazov *et al.* [D0 Collaboration], "Evidence for $Z_c^{\pm}(3900)$ in semi-inclusive decays of b-flavored hadrons," arXiv:1807.00183 [hep-ex] (accepted by PRD).

[18] M. Ablikim *et al.* [BESIII Collaboration], Phys. Rev. Lett. **112**, 022001 (2014).

[19] C. Z. Yuan, Int. J. Mod. Phys. A **33**, no. 21, 1830018 (2018).

[20] A. Bondar *et al.* (Belle Collaboration), Phys. Rev. Lett. **108**, 122001 (2012).

[21] A. Garmash *et al.* (Belle Collaboration), Phys. Rev. D **91**, 072003 (2015).

[22] A. Garmash *et al.* (Belle Collaboration), Phys. Rev. Lett. **116**, no. 21, 212001 (2016).

[23] A. E. Bondar, A. Garmash, A. I. Milstein, R. Mizuk and M. B. Voloshin, Phys. Rev. D **84**, 054010 (2011).

Compact pentaquark states

A. Giachino[1,*] and *E.* Santopinto[1,**]

[1]INFN Sezione di Genova, via Dodecaneso 33, 16164 Genova, Italy

Abstract. We investigated the lightest pentaquark state observed by the LHCb collaboration as a compact five-quark ground- state, with spin $S = 3/2$, and we showed that this state belongs to a $SU_f(3)$ flavor octet by means of a group theory approach. Moreover, we predicted in a parameter-free way the pentaquark state mass spectra for all the octet-pentaquark states. Finally, by means of an effective Lagrangian approach, we computed the octet-pentaquark partial decay widths in $J/\Psi\Lambda$, $J/\Psi\Sigma$ and $J/\Psi\Xi$ channels.

1 Introduction

In 2015, the Large Hadron Collider beauty experiment (LHCb) collaboration observed two hidden-charm pentaquarks, $P_c^+(4380)$ and $P_c^+(4450)$ in $\Lambda_b \to J/\Psi K^- p$ decay [1]. This decay can proceed both by intermediate $\Lambda^* \to K^- p$ resonances, expected to be dominant, and by intermediate exotic structures, which were referred to as charmonium pentaquark states by LHCb [1]. Fig. 1 shows the number of events observed in the $J/\Psi p$ invariant mass spectrum. The data show the presence of a wider structure with an average mass value of about 4380

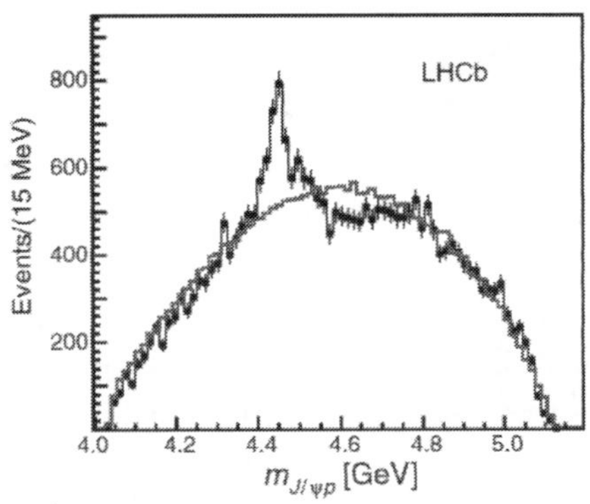

Figure 1. Number of events vs. the $J/\Psi p$ invariant mass spectrum. The red curve is the expectation from phase space. Figure adapted from ref. [1].

MeV, called $P_c^+(4380)$ and a narrower one, with an average mass value of about 4450 MeV, called $P_c^+(4450)$. These two pentaquark states are found to have masses of 4380 ± 8 ± 28

*e-mail: agiachino@ge.infn.it
**e-mail: Elena.Santopinto@ge.infn.it

© The Authors, published by EDP Sciences. This is an open access article distributed under the terms of the Creative Commons Attribution License 4.0 (http://creativecommons.org/licenses/by/4.0/).

MeV and 4449.8 ± 1.7 ± 2.5 MeV, with corresponding widths of 205 ± 18 ± 86 MeV and 39 ± 5 ± 19 MeV. The parities of these states are preferred to be opposite, and one state has $J = 3/2$ and the other $J = 5/2$.

The earliest prediction of a charmonium pentaquark state with $J^P = \frac{3}{2}^-$ was provided by J. J. Wu et al. [2]. Recently, Karliner e Rosner in Ref. [3] explained the heaviest pentaquark state, $P_c^+(4450)$ as a meson baryon molecular state, however the $P_c^+(4380)$ is missing in their description. For this reason, we focused on the lightest pentaquark structure by investigating it as a compact five quark state [4]. From the measured quantum numbers of the lightest resonant state, we show that it belongs to a $SU_f(3)$ octet with spin $S = \frac{3}{2}$. We also studied all the charmonium pentaquark states which belong to the octet, predicted their masses, and suggested possible decay channels in which the experimentalists can observe them. Recently, Wang et al. [5] calculated the pentaquark states cross section in J/Ψ photoproduction and compared it with the present experimental data [6–8]. As a result, to be consistent also with J/Ψ photoproduction data, they found an upper limit for branching ratio $\mathcal{B}(P_c^+ \to J/\Psi p)$. By using this upper limit and our predicted masses, we calculated the partial decay widths for the predicted pentaquark resonances [4].

2 Pentaquark mass spectra

We think the charmonium pentaquark wave function as $qqqc\bar{c}$ where $q = u, d, s$ is a light quark and c is the heavy charm quark. We adopt for each representation the notation $[f]_d = [f_1, \ldots, f_n]_d$, where f_i denotes the number of boxes in the i-th row of the Young tableau, and d is the dimension of the representation. From a group theory based approach we found that the allowed $SU_f(3)$ flavour representations to which the charmonium pentaquark states can belong are an octet and a decuplet [4]:

$$[21]_8\ ,\ \ [3]_{10} \tag{1}$$

In order to determine what is the the lowest mass multiplet, we performed the following extension of the Gürsey-Radicati (GR) mass formula [9]:

$$M_{GR} = M_0 + AS(S+1) + DY + E\left[I(I+1) - \frac{1}{4}Y^2\right] + GC_2(SU(3)) + FN_C \tag{2}$$

where M_0 is a scale parameter, I and Y are the isospin and hypercharge, respectively, while $C_2(SU(3))$ is the eigenvalue of the $SU_f(3)$ Casimir operator. Finally, N_C is a counter that takes into account the mass difference between a c quark (or a $\bar{c}$ antiquark) and the light quarks (u, d). The coefficients in the Gürsey-Radicati (GR) mass formula extension, A, D, E, G, F and the scale parameter M_0 were fitted to reproduce the spectrum of all the ground-state charmed baryons, the ground-state hyperons, and the ground-state non-strange baryons. We observe that this implies assuming that the coefficients in the GR formula are the same for different quark systems. The fitted parameters with their corresponding uncertainties are reported in Tab. 1.

According to the mass formula of Eq. 2 the lowest mass charmonium pentaquark state is the $[21]_8$ $SU_f(3)$ octet. Therefore, in this octet, we expect to find the pentaquark state observed. As one can see form Fig. 2, the charged state $P^{0+}(4377)$ has the same quantum numbers as the lightest resonance (charge, spin, parity) reported by the LHCb. Its predicted mass is M = 4377 ± 49 MeV, in agreement with the experimental mass, 4380 ± 8 ± 28 MeV [1].

Table 1. values of the parameters in the GR mass formula extension (Eq. 2) with the corresponding uncertainties.

	M_0	A	D	E	F	G
values (MeV)	940,0	23,0	-158,3	32,0	1354.6	52,5
uncertainties (MeV)	1.5	1.2	1.3	1.3	18.2	1.3

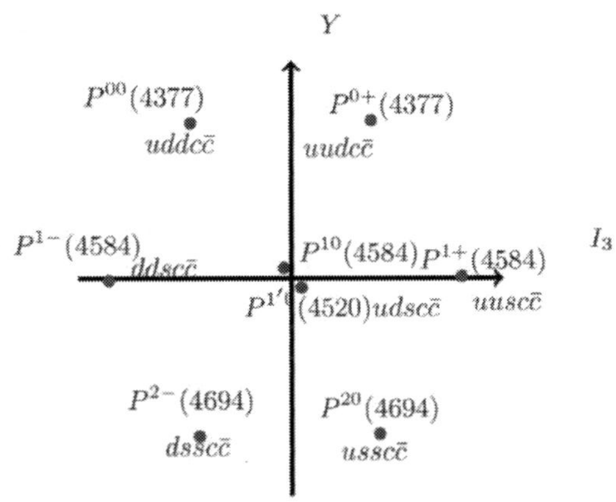

Figure 2. Octet of the charmonium pentaquark states. A pentaquark state is denoted with $P^{ij}(M)$, where $i = 0, 1, 2$ is the number of strange quarks, $j = -, 0, +$ is the pentaquark's electric charge, and M the predicted mass obtained with Eq. 2.

3 Pentaquark partial decay widths

In calculating the decay widths of the predicted pentaquark states, reported in Fig. 2, we adopted an effective Lagrangian proposed in Ref. [10] for pNJ/ψ coupling. Wang et al. [5] calculated the cross-section of the pentaquark states in J/Ψ photoproduction and compared it with the available experimental data ([6–8]). He noticed also that, since the pentaquark states have been observed in $J/\Psi p$ channel, it should be produced also in $J/\Psi p$ photoproduction [5]. This gives a constraint for the coupling strength between $J/\Psi p$ and the two pentaquark states. In fact, in order to have consistency with the available photoproduction data the branching ratio upper limit for both the pentaquark states is $\mathcal{B}(P^+ \to J/\Psi p) \leq 0.05$. From this constraint and the effective lagrangian proposed by S. H. Kim in [10] we calculated the $J/\Psi p$ coupling constant, $\bar{g}_{pJ/\Psi} = \frac{g_1}{2M_N}$. Moreover, from our pentaquark mass predictions we were also able to calculate the partial decay widths for the pentaquark states with open strangeness. For example, the P^{1+} partial decay width in the $\Sigma^+ J/\Psi$ channel, $\Gamma(P^{1+} \to \Sigma^+ J/\psi)$, is given by:

$$\Gamma(P^{1+} \to \Sigma^+ J/\psi) = \frac{\bar{g}^2_{\Sigma^+ J/\Psi}}{12\pi} \frac{p_{\Sigma^+}}{M_{P^{1'}}} (E_{\Sigma^+} + M_{\Sigma^+})[2E_{\Sigma^+}(E_{\Sigma^+} - M_{\Sigma^+}) + (M_{P^{1+}} - M_{\Sigma^+})^2 + 2M^2_{J/\psi}], \quad (3)$$

where kinematic variables are related by the relativistic equation $p_{\Sigma^+} = \sqrt{E_{\Sigma^+}^2 - M_{\Sigma^+}^2}$, and the coupling constant $\bar{g}_{\Sigma^+ J/\Psi}$, is related to the $J/\Psi p$ coupling constant by the following rescaling:

$$\bar{g}_{\Sigma^+ J/\Psi} = \frac{g_1}{2M_{\Sigma^+}} \; . \tag{4}$$

If we use the upper branching ratio limit extracted by Wang [5], $\mathcal{B}(P^+ \to J/\Psi p) = 0.05$, and the total width reported by LHCb, Γ_{tot} = 205 MeV, we obtain for the $P_c^+(4380)$ partial decay width in $J/\psi p$ channel:

$$\Gamma_{NJ/\Psi} = \mathcal{B}(P^+ \to J/\Psi p)\Gamma_{tot} = 10.25 \text{ MeV}. \tag{5}$$

The numerical results for the other channels are listed in Table 2.

Table 2. Partial decay widths for $\Lambda J/\Psi$, $\Sigma J/\Psi$ and $\Xi J/\Psi$ channels, calculated from the constraint that $J/\Psi p$ channel accounts for the 5% of the total pentaquark width, according to Ref. [5]. The notation used to denote a pentaquark state is the same as the one introduced in 2.

initial state	channel	partial width (MeV)
$P^{1'0}$	$\Lambda J/\Psi$	7.94
P^{1-}, P^{10}, P^{1+}	$\Sigma J/\Psi$	7.21
P^{2-}, P^{20}	$\Xi J/\Psi$	6.35

4 Conclusions

We have studied all the charmonium pentaquark states which belong to the flavour octet, predicted their masses, and suggested possible bottom baryon decay channels which involve the predicted resonances as intermediate states. Finally, we computed the partial decay widths for all the suggested octet-pentaquark decay channels.

References

[1] R. Aaij et al. [LHCb Collaboration], Phys. Rev. Lett. **115**, 072001 (2015).
[2] J. J. Wu et al., Phys. Rev. Lett. **105**, 232001 (2010)
[3] M. Karliner and J. L. Rosner, Phys. Rev. Lett. **115**, 122001 (2015).
[4] E.Santopinto and A Giachino, Phys.Rev. D **96**, 014014 (2017).
[5] Q. Wang, X. H. Liu, and Q. Zhao, Phys. Rev. D **92**, 034022 (2015).
[6] U. Camerini et al., Phys. Rev. Lett. **35**, 483 (1975).
[7] R. L. Anderson, Report No. SLAC-PUB-1417.
[8] B. Gittelman et al., Phys. Rev. Lett. **35**, 1616 (1975).
[9] F. Gursey and L.A. Radicati , Phys. Rev. Lett. **13**, 173 (1964)
[10] S. H. Kim, S. i. Nam, Y. Oh, and H.-Ch. Kim, Phys. Rev. D **84**, 114023 (2011).

Exotic Tetraquark Mesons in Large-N_c Limit: an Unexpected Great Surprise

Wolfgang Lucha[1,*], Dmitri Melikhov[1,2,3,**], and Hagop Sazdjian[4,***]

[1]Institute for High Energy Physics, Austrian Academy of Sciences, Nikolsdorfergasse 18, A-1050 Vienna, Austria
[2]D. V. Skobeltsyn Institute of Nuclear Physics, M. V. Lomonosov Moscow State University, 119991 Moscow, Russia
[3]Faculty of Physics, University of Vienna, Boltzmanngasse 5, A-1090 Vienna, Austria
[4]Institut de Physique Nucléaire, CNRS-IN2P3, Université Paris-Sud, Université Paris-Saclay, 91405 Orsay Cedex, France

Abstract. Two-ordinary-meson scattering in large-N_c QCD implies consistency criteria for intermediate-tetraquark contributions. Their fulfilment at N_c-leading order constrains the nature of the spectrum of genuinely exotic tetraquark states.

1 Introduction: N_c-Leading First-Principles Approach to Tetraquarks

Qualitative information on systems controlled by quantum chromodynamics may be collected by considering large-N_c QCD [1, 2], a quantum field theory generalizing QCD by enabling the number N_c of colour degrees of freedom to differ from $N_c = 3$ and grow beyond bounds while, simultaneously, demanding that the product $N_c\,\alpha_s$ of N_c and the strong fine-structure coupling

$$\alpha_s \equiv \frac{g_s^2}{4\pi}$$

approaches a finite value in the large-N_c limit. For the N_c behaviour of α_s, this demand implies

$$\alpha_s \propto \frac{1}{N_c} \qquad \text{for} \quad N_c \to \infty\,.$$

In an attempt to stay as close as possible to intuition, we adhere to the presumably very natural (but clearly not compulsory) assumption that the fermionic dynamical degrees of freedom, the quarks, continue to transform according to the N_c-dimensional, fundamental representation of the gauge group SU(N_c). By utilizing QCD's large-N_c limit ($N_c \to \infty$) and $1/N_c$ expansion (in powers of $1/N_c$) about this limit, we extract constraints on crucial features (*e.g.*, decay widths) [3, 4] of tetraquarks, meson bound states of two quarks and two antiquarks predicted by QCD.

With respect to their *flavour* degrees of freedom, tetraquarks can be classified (Table 1) by specifying — for the two quarks and two antiquarks constituting the tetraquark bound state —

- the number of different quark flavours encountered in such bound state, in combination with
- the total number of *open* quark flavours, defined as the number of quark flavours that are not counterbalanced by an antiquark of same flavour and hence carried by the observed mesons.

*e-mail: Wolfgang.Lucha@oeaw.ac.at
**e-mail: dmitri_melikhov@gmx.de
***e-mail: sazdjian@ipno.in2p3.fr

© The Authors, published by EDP Sciences. This is an open access article distributed under the terms of the Creative Commons Attribution License 4.0 (http://creativecommons.org/licenses/by/4.0/).

Table 1. Classification of tetraquark mesons by content of *different* and *open* quark flavour. The notion *open-flavour number* relates to the net sum of flavours not compensated by a corresponding antiflavour.

number of different quark flavours involved	tentative quark configuration $\bar{q}_\Box\, q_\Box\, \bar{q}_\Box\, q_\Box$	number of open quark flavours involved
4	$\bar{q}_1\, q_2\, \bar{q}_3\, q_4$	4
3	$\bar{q}_1\, q_2\, \bar{q}_3\, q_2$	4
	$\bar{q}_1\, q_2\, \bar{q}_1\, q_3$	4
	$\bar{q}_1\, q_2\, \bar{q}_2\, q_3$	2
	$\bar{q}_1\, q_2\, \bar{q}_3\, q_3$	2
2	$\bar{q}_1\, q_2\, \bar{q}_1\, q_2$	4
	$\bar{q}_1\, q_1\, \bar{q}_1\, q_2$	2
	$\bar{q}_1\, q_1\, \bar{q}_2\, q_1$	2
	$\bar{q}_1\, q_2\, \bar{q}_2\, q_1$	0
	$\bar{q}_1\, q_1\, \bar{q}_2\, q_2$	0
1	$\bar{q}_1\, q_1\, \bar{q}_1\, q_1$	0

Whereas, by definition, for ordinary mesons a count of open flavour should yield zero or two, the case of four open flavours signals the tetraquark nature of the respective quark bound state.

We focus to two variants of tetraquarks with particularly interesting quark flavour content:

1. *flavour-exotic* tetraquarks $T = (\bar{q}_1\, q_2\, \bar{q}_3\, q_4)$ with all four (anti-) quark flavours different;

2. *flavour-cryptoexotic* tetraquarks $T = (\bar{q}_1\, q_2\, \bar{q}_2\, q_3)$ involving three different (anti-) quark flavours, by containing a quark–antiquark pair of same flavour differing from the others.

In order to work out — at least, at a qualitative level — some basic features of such kind of tetraquarks, we investigate, for two ordinary mesons of appropriate flavour quantum numbers, their possible scattering reactions into two ordinary mesons with regard to potential s-channel contributions of intermediate poles interpretable as a manifestation of tetraquarks with narrow decay width. For both sets of tetraquark in our focus of interest, as well as for the one with two different flavours but no open flavour, we have to analyze two variants of scattering processes:

- flavour-preserving ones, with identical flavour content of the initial- and final-state mesons;
- flavour-rearranging ones, with unequal flavour content of the initial- and final-state mesons.

We identify all the contributions to four-point correlation functions of quark bilinear operators $j_{ij} \equiv \bar{q}_i\, q_j$ interpolating ordinary mesons M_{ij} (notationally exploiting the actual irrelevance of parity and spin therein) that are capable of supporting a pole related to a tetraquark built up by four (anti-) quarks of masses m_i, $i = 1, \ldots, 4$, by imposing an unambiguous selection criterion [3, 4]: for the scattering of two mesons of momenta p_1 and p_2, an allowable Feynman diagram not only has to depend on the Mandelstam variable $s \equiv (p_1+p_2)^2$ in a non-polynomial way but has to admit a four-quark intermediate state with related branch cut starting at the branch point

$$s = (m_1 + m_2 + m_3 + m_4)^2 \ .$$

2 Four different quark flavours $\Longleftrightarrow$ flavour-exotic tetraquark meson

Surprisingly or not, for truly flavour-exotic tetraquark states, $T = (\bar{q}_1\, q_2\, \bar{q}_3\, q_4)$, the leading-$N_c$ dependence of all contributions to the correlation functions of four quark bilinear operators j_{ij}

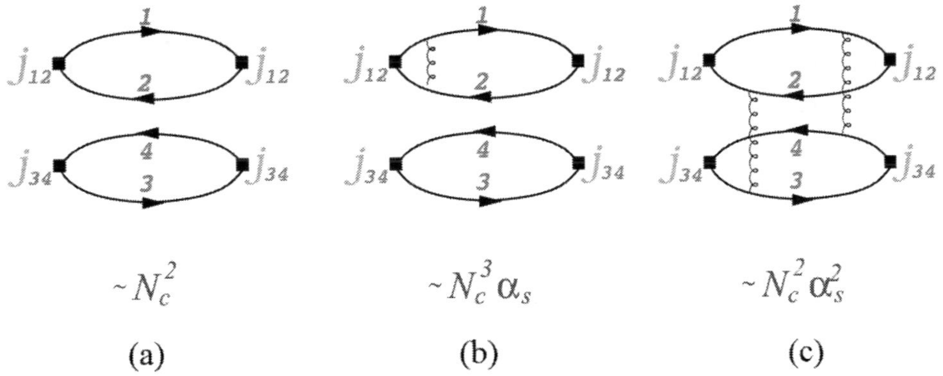

Figure 1. Flavour-preserving four-point correlation function of quark bilinear operators j_{ij}: examples of Feynman diagrams contributing at low order to this correlator's $1/N_c$ expansion, *viz.*, at order N_c^2 (a,b) or N_c^0 (c) [3, Fig. 1]. Tetraquark-friendly contributions of lowest order in $1/N_c$ turn out to be of order N_c^0 (c).

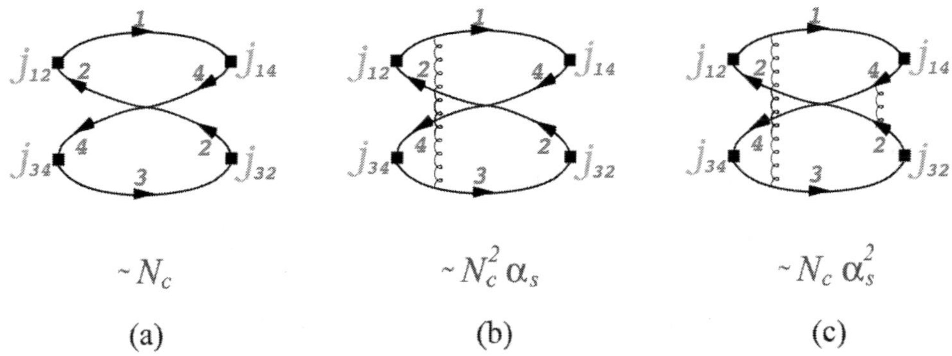

Figure 2. Flavour-rearranging four-point correlation function of quark bilinear currents j_{ij}: examples of Feynman diagrams contributing at low order to this correlator's $1/N_c$ expansion, that is, at order N_c (a,b) or N_c^{-1} (c) [3, Fig. 2]. Tetraquark-phile contributions of lowest order in $1/N_c$ prove to be of order N_c^{-1} (c).

(exemplified in Figs. 1 and 2) potentially capable of developing this tetraquark pole *differs*, for the flavour-preserving (Fig. 1(c)) and flavour-rearranging (Fig. 2(c)) cases, by one order of N_c:

$$\langle j_{12}^\dagger\, j_{34}^\dagger\, j_{12}\, j_{34}\rangle_T = O(N_c^0)\,, \qquad \langle j_{14}^\dagger\, j_{32}^\dagger\, j_{14}\, j_{32}\rangle_T = O(N_c^0)\,, \qquad \langle j_{14}^\dagger\, j_{32}^\dagger\, j_{12}\, j_{34}\rangle_T = O(N_c^{-1})\,.$$

The resulting N_c-leading tetraquark–two-ordinary-meson amplitudes A imply the presence of, at least, two[1] tetraquarks $T_{A,B}$ with, however, decay rates $\Gamma(T_{A,B})$ of similar large-N_c decrease:

$$\underbrace{A(T_A \longleftrightarrow M_{12}\, M_{34}) = O(N_c^{-1})}_{\Longrightarrow \quad \Gamma(T_A) = O(N_c^{-2})} \quad \stackrel{N_c}{>} \quad A(T_A \longleftrightarrow M_{14}\, M_{32}) = O(N_c^{-2})\,,$$

$$A(T_B \longleftrightarrow M_{12}\, M_{34}) = O(N_c^{-2}) \quad \stackrel{N_c}{<} \quad \underbrace{A(T_B \longleftrightarrow M_{14}\, M_{32}) = O(N_c^{-1})}_{\Longrightarrow \quad \Gamma(T_B) = O(N_c^{-2})}\,.$$

[1] In such a situation, their conclusions some people in the form may phrase "always two there are, . . . , no less" [5].

3 Two open quark flavours ≡ flavour-cryptoexotic tetraquark meson

In the case of flavour-cryptoexotic tetraquark mesons built from three different quark flavours, $T = (\bar{q}_1\, q_2\, \bar{q}_2\, q_3)$, the large-$N_c$ behaviour of flavour-preserving (Fig. 3) and flavour-reshuffling (Fig. 4) subcategories of those contributions to the correlation functions of four quark bilinear currents j_{ij} which might support the development of a tetraquark pole turns out to be identical:

$$\langle j_{12}^\dagger\, j_{23}^\dagger\, j_{12}\, j_{23}\rangle_{\rm T} = O(N_c^0)\,, \qquad \langle j_{13}^\dagger\, j_{22}^\dagger\, j_{13}\, j_{22}\rangle_{\rm T} = O(N_c^0)\,, \qquad \langle j_{13}^\dagger\, j_{22}^\dagger\, j_{12}\, j_{23}\rangle_{\rm T} = O(N_c^0)\,.$$

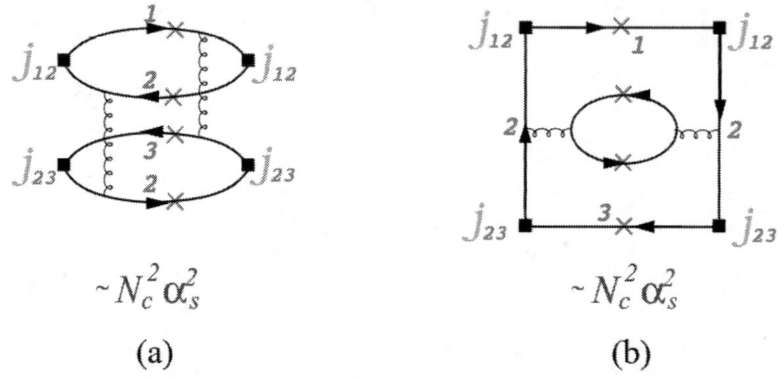

Figure 3. Flavour-preserving four-point correlation function of quark bilinear operators j_{ij}: examples of tetraquark-phile Feynman diagrams contributing at the N_c-leading order N_c^0 to this correlation function's $1/N_c$ expansion [3, Fig. 3]. (Purple crosses indicate quarks potentially contributing to a tetraquark pole.)

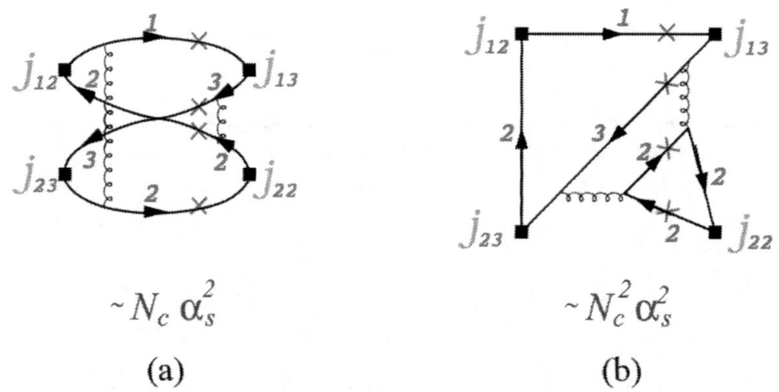

Figure 4. Flavour-rearranging four-point correlation function of quark bilinear currents j_{ij}: examples of tetraquark-phile Feynman diagrams contributing at low order to such correlator's $1/N_c$ expansion, *i.e.*, at order N_c^{-1} (a) or N_c^0 (b) [3, Fig. 4]; among these, the N_c-leading contributions prove to be of order N_c^0 (b).

So, a single tetraquark T_C (which, due to its cryptoexotic nature, can mix with the meson M_{13}) satisfies all constraints induced by the N_c-leading tetraquark–two-ordinary-meson amplitudes

$$\underbrace{A(T_C \longleftrightarrow M_{12}\, M_{23}) = O(N_c^{-1}) \quad \stackrel{N_c}{=} \quad A(T_C \longleftrightarrow M_{13}\, M_{22}) = O(N_c^{-1})}_{\Longrightarrow \quad \Gamma(T_C) = O(N_c^{-2})}\,.$$

4 Insights: N_c-Leading Conclusions for (Crypto-) Exotic Tetraquarks

In summary, we find that self-consistency conditions arising from the inspection of tetraquark contributions to the scattering amplitudes of two ordinary mesons into two ordinary mesons in the $1/N_c$ expansion of large-N_c QCD provide rigorous constraints on the features of tetraquark states [3, 4]. Demanding these constraints to be satisfied at N_c-leading order implies [3, 4] that

- genuinely exotic tetraquarks need to appear in pairs (T_A, T_B) the members of which differ in the large-N_c behaviour of their dominant or preferred decay modes to two ordinary mesons;
- both genuinely exotic $(T_{A,B})$ and cryptoexotic (T_C) tetraquarks exhibit narrow decay widths

$$\Gamma(T) \propto 1/N_c^2 \xrightarrow[N_c\to\infty]{} 0 \qquad \text{for} \quad T = T_A, T_B, T_C \, .$$

Table 2 confronts these findings for the rates of the large-N_c decrease of the total decay widths of exotic and cryptoexotic tetraquarks with corresponding outcomes of earlier analyses [6–8]. Imposition of additional requirements clearly may strengthen the predicted large-N_c decrease. Differences to our results arise from misidentifying the actually N_c-leading contribution to the tetraquark pole or from consideration of merely a single (say, the flavour-reshuffling) channel.

Table 2. Comparison: predictions of *upper bounds* on the large-N_c behaviour of tetraquark decay rates.

Author Collective	Decay Width Γ		Reference
	Exotic Tetraquarks	Cryptoexotic Tetraquarks	
Lucha *et al.*	$O(1/N_c^2)$	$O(1/N_c^2)$	[3, 4]
Knecht and Peris	$O(1/N_c^2)$	$O(1/N_c)$	[6]
Cohen and Lebed	$O(1/N_c^2)$	—	[7]
Maiani *et al.*	$O(1/N_c^3)$	$O(1/N_c^3)$	[8]

Acknowledgements

D. M. is grateful for support by the Austrian Science Fund (FWF) under project P29028-N27.

References

[1] G. 't Hooft, Nucl. Phys. B **72** (1974) 461.
[2] E. Witten, Nucl. Phys. B **160** (1979) 57.
[3] W. Lucha, D. Melikhov, and H. Sazdjian, Phys. Rev. D **96** (2017) 014022, arXiv:1706.06003 [hep-ph].
[4] W. Lucha, D. Melikhov, and H. Sazdjian, Eur. Phys. J. C **77** (2017) 866, arXiv:1710.08316 [hep-ph].
[5] Yoda, JM, in *Star Wars. Episode I: The Phantom Menace* (1999, in a galaxy far, far away), sc. 197.
[6] M. Knecht and S. Peris, Phys. Rev. D **88** (2013) 036016, arXiv:1307.1273 [hep-ph].
[7] T. D. Cohen and R. F. Lebed, Phys. Rev. D **90** (2014) 016001, arXiv:1403.8090 [hep-ph].
[8] L. Maiani, A. D. Polosa, and V. Riquer, J. High Energy Phys. 06 (2016) 160, arXiv:1605.04839 [hep-ph].

Search for a Λnn bound state in Pb-Pb collisions with ALICE at the LHC

Annalisa Mastroserio[1,2,*] on behalf of the ALICE Collaboration

[1]Universitá Degli Studi di Foggia
[2]INFN Sezione di Bari

Abstract. The extreme energy densities reached at the LHC lead to the production of a significant amount of baryons and strangeness. Such a regime allows for an increased production of potentially existing exotic QCD bound states containing nuclei and strange hadrons. An interesting measurement for the phenomenology of the nuclear interaction is the presence of a neutral bound state constituted by one Λ and two neutrons. The excellent particle identification, tracking and vertexing performance of the ALICE experiment allow for the search of this exotic bound state in the decay channel Λnn $\rightarrow {}^3$H+π^- in Pb-Pb collisions. In order to improve detection of this state despite the presence of a huge combinatorial background, the extraction of the signal is performed by means of a multivariate approach with the TMVA (Toolkit for Multivariate Data Analysis with ROOT). So far the indication for the existence of the Λnn state was reported only by the HypHI Collaboration, so the observation by ALICE would crucially contribute to the study of such an exotic state.

1 Introduction

The nuclear state composed by the Λ hyperon and two neutrons, namely the Λnn state, can be considered as an isospin partner state of the ${}^3_\Lambda$H which has a separation energy $\approx$ 130 keV and this is already a hint that the neutral state can hardly be formed. Theoretical models predict that such a neutral state should not be bound [1, 2], nevertheless recent measurements [3] indicate its formation at a mass of 2.993 GeV/c^2 as shown in Figure 1. The production of a significant amount of baryons and strangeness at LHC energies allows for an increased production of potentially existing exotic QCD bound states. The Λnn state, if existing, can be eventually detected with ALICE via the decay channel Λnn $\rightarrow {}^3$H+π^- and its observation would crucially contribute to the understanding of exotic nuclear bound states.

2 Analysis strategy for Λnn detection

The main challenge of this analysis is that the signal is not only rare, but it may not even exist. The triton identification is the main issue here, because above 2 GeV/c the sample of candidates selected by the Time Projection Chamber (TPC) contain several low mass particles (e.g.: pions and electrons) which enlarge dramatically the combinatorial background. The usage of the Time-Of-Flight detector (TOF) information helps in rejecting the background.

*e-mail: annalisa.mastroserio@ba.infn.it

© The Authors, published by EDP Sciences. This is an open access article distributed under the terms of the Creative Commons Attribution License 4.0 (http://creativecommons.org/licenses/by/4.0/).

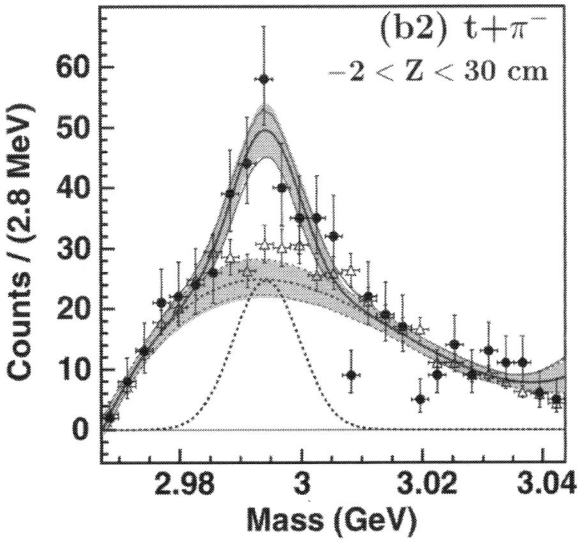

Figure 1. Invariant mass spectrum of the Λnn $\rightarrow^3$ H$+\pi^-$ as obtained from the reaction of ^{6}Li projectiles at 2 AGeV on a fixed graphite target at the GSI Helmholtz Centre for Heavy Ion Research (HypHI Collaboration [3]).

Nevertheless the signal efficiency is also reduced. Previous studies made use of rectangular cuts on several variables such as Distance of Closest Approach (DCA) of the mother particle and the daughter particles, momentum ranges, etc. and to improve the signal extraction more complicated cuts were introduced, such as linear combination of independent variables or triangular cuts. Given the difficulties in finding an optimal set of cuts, it was decided to

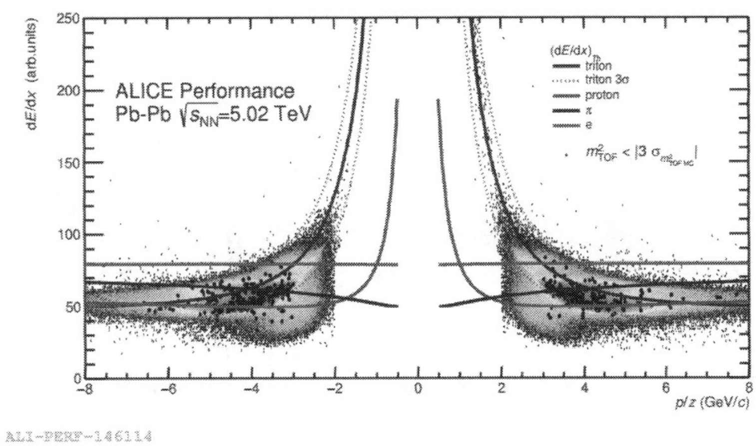

Figure 2. Energy loss in the TPC of the candidate V^0 daughters that were identified as tritons. The black dots refer to tritons which have a TOF mass in the triton expected range. The theoretical Bethe-Bloch curves of pions, protons and electrons are also shown. The two dotted lines refers to the expected trend of the energy loss a ± 3 σ, where σ is the expected triton energy loss width of the specific energy-loss signal in the TPC.

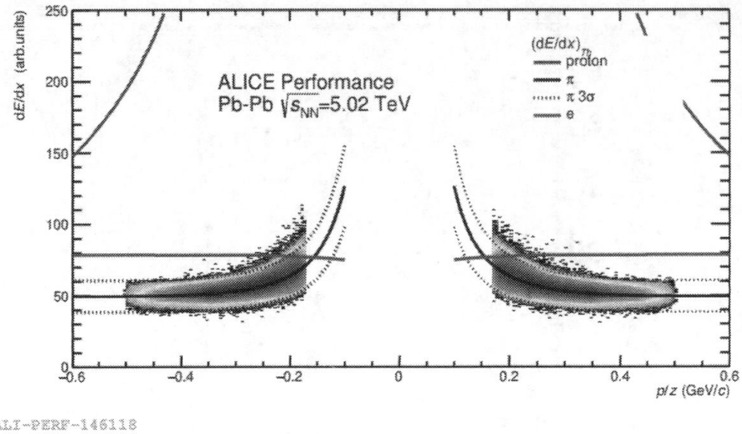

Figure 3. Energy loss in the TPC of the candidate V^0 daughters that were identified as pions. The theoretical Bethe-Bloch curves of pions, protons and electrons are also shown. The two dotted lines refer to the expected trend of the energy loss a $\pm 3\ \sigma$, where σ is the expected pion energy loss width of the specific energy-loss signal in TPC.)

use machine learning (ML) tools because such algorithms are trained to understand *all* the features of the signal and of the background. The main advantage of such a technique is that it is able to make n-dimensional cuts more efficiently than standard analysis techniques and they can improve both background rejection and signal yield extraction. In particular the Boost Decision Tree (BDT) algorithm within the TMVA package was used [4] since it already proved to improve the significance of rare hadron identification within a topological driven analysis [5] such as the Λ_C^+. Furthermore the doubtful existence of Λnn may result in a huge effort in signal extraction with a standard analysis. The Machine Learning algorithms, instead, with their supply tools (e.g.: variable ranking for the cuts, fake rate probability, etc.) are able to optimize the selection process and to speed up the analysis results, also in the missing signal case.

In the specific case of the Λnn identification in ALICE, the same procedure was applied at both Pb-Pb collision energies $\sqrt{s_{NN}}$=2.76 TeV and $\sqrt{s_{NN}}$ =5.02 TeV. The Particle IDentification (PID) for tritons was performed via TPC dE/dx and TOF information. The TOF signal was used in parallel to reject protons and pions below 4 GeV/c and then the hadron squared mass by the TOF β measurement was used above 3 GeV/c to have enough statistics. The energy loss in TPC of the selected pions and tritons after such PID cuts is shown in Figures 3 and 2 respectively. After some further checks, only candidates containing antitritons were considered because it improved the performance of the machine learning algorithm.

3 Λnn signal extraction with a Machine Learning approach

The machine learning procedure adopted in this analysis was a supervised one, which means that the signal is extracted from the real data after a *learning* phase of the BDT algorithm. The full procedure foresees an initial stage where the algorithm is trained to identify signal candidates and background candidates, and a final stage which consists in applying the updated algorithm to identify the signal-like candidates in the data. The first stage must consider candidates which should belong only to the signal type and to the background type and this is typically achieved making use of Monte Carlo simulations. The last stage is done always on

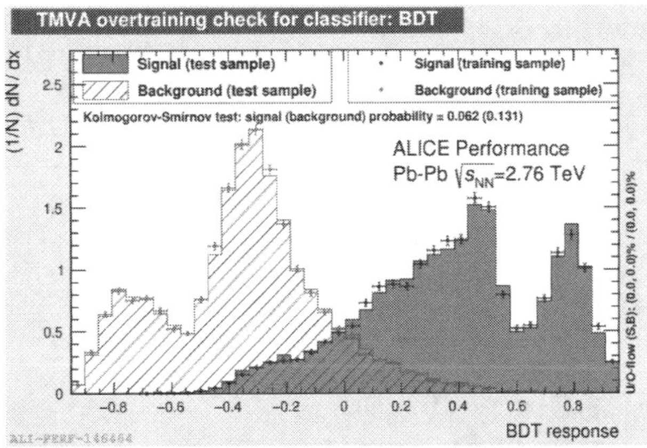

Figure 4. BDT classifier for signal and background Λnn candidates in Pb-Pb collisions at $\sqrt{s_{\rm NN}}$=2.76 TeV.

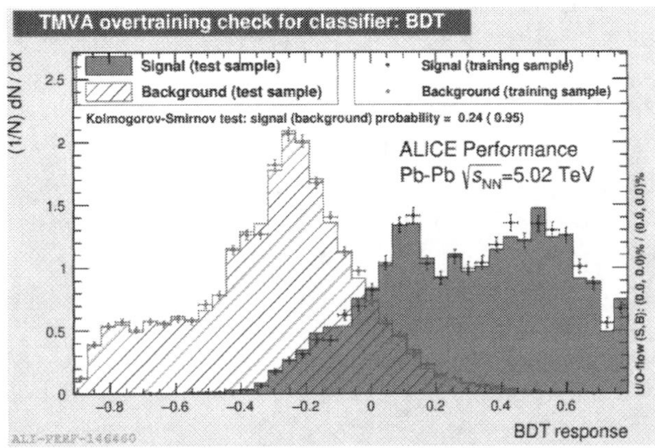

Figure 5. BDT classifier for signal and background Λnn candidates in Pb-Pb collisions at $\sqrt{s_{\rm NN}}$=5.02 TeV.

real data. In this analysis the signal candidates were indeed extracted from Monte Carlo simulations based on the HIJING generator [6] at both energies with injected Λnn signal (flat in p_T) that had a mass value of 2.982 GeV/c^2 that is a bit lower than the measured one to comply with the expected value $p_{\Lambda\to\pi p}$ = 101 MeV/c [1]. The experimental resolution from simulations was found to be $\sigma_{\rm MC}(\Lambda{\rm nn})$ =2 MeV/c^2. The choice of the background candidates, instead, needed some further study. Due to the missing nuclei simulation in HIJING, the background combinatorics cound not be extracted from the Monte Carlo. It was decided, then, to use the candidates outside the signal region in the invariant mass distribution as background candidates.

In particular the signal region was chosen to be the one between 2.982 GeV/c^2-$3\sigma_{\rm MC}(\Lambda{\rm nn})$ < $M(t\pi)$ <2.993 GeV/c^2+$3\sigma_{\rm MC}(\Lambda{\rm nn})$. The value 2.982 GeV/$c^2$ is considered

[1]In the two body decay Λnn→^{3}Hπ, if assuming a mass of 2.993 GeV/c^2 then the momentum of the decay products in the rest frame of the mother is higher then the value 101 MeV/c^2, that is the value of the corresponding momentum in the free lambda case $\Lambda\to\pi p$. In order to have the same value of that momentum, the Λnn state shold have a lower mass value 2.982 GeV/c^2

in the lower limit case because it is expected mass value at the p_Λ = 101 MeV/c of the free Λ, whereas 2.993 GeV/c^2 is the available measured value. The PID selection was applied to tritons to have 10^4 candidates in both the signal sample and the background sample. Four variables were used and they were selected to avoid any correlation with the invariant mass in the background data sample. The results of the machine learning training stage are shown in Figure 4 and 5 for Pb-Pb collisions at $\sqrt{s_{\mathrm{NN}}}$=2.76 TeV and $\sqrt{s_{\mathrm{NN}}}$=5.02 TeV respectively. The classifier distributions for signal and background are well separated providing a low bkg contamination for the selected signal.
Applying the trained BDT algorithm to real data the background rejection was improved and in the end a sample of Λnn was selected. The retrieved number was compatible with the number of expected false positive candidates foreseen by the algorithm (and provided during the training stage). Further work is necessary and studies are ongoing to include positive tritons. The strategy is to improve the triton PID and include the Transition Radiation Detector (TRD) to further reject low mass particles from the selected sample of nuclei.

References

[1] R. H. Dalitz and B. W. Downs in Phys. Rev. **110**, 958 (1958); **111**, 967 (1958);**114**,593 (1959).
[2] H. Garcilazo and A. Valcarce , Phys. Rev. C **89**, 057001 (2014).
[3] C. Rappold, et al. (HypHI Collaboration), Phys. Rev. C **88**, 041001 (2013).
[4] https://root.cern.ch/tmva
[5] S. Acharya, et al. (ALICE Collaboration), JHEP 04, **108** (2014).
[6] X.-N. Wang, M. Gyulassy, Phys. Rev. D **44**, 3501 (1991).

Fractal structure of hadrons and non-extensive statistics *

Eugenio Megías[1,2,**], *Airton* Deppman[3,***], *Tobias* Frederico[4,****], and *Débora P.* Menezes[5,†]

[1]Departamento de Física Atómica, Molecular y Nuclear and Instituto Carlos I de Física Teórica y Computacional, Universidad de Granada, Avenida de Fuente Nueva s/n, 18071 Granada, Spain
[2]Departamento de Física Teórica, Universidad del País Vasco UPV/EHU, Apartado 644, 48080 Bilbao, Spain
[3]Instituto de Física, Universidade de São Paulo, Rua do Matão Travessa R Nr. 187, Cidade Universitária, CEP 05508-090 São Paulo, Brazil
[4]Instituto Tecnológico da Aeronáutica, 12228-900 São José dos Campos, Brazil
[5]Departamento de Física, CFM, Universidade Federal de Santa Catarina, CP 476, CEP 88040-900 Florianópolis, Brazil

Abstract. The role played by non-extensive thermodynamics in physical systems has been under intense debate for the last decades. Some possible mechanisms that could give rise to non-extensive statistics have been formulated along the last few years, in particular the existence of a fractal structure in thermodynamic functions for hadronic systems. We investigate the properties of such fractal thermodynamical systems, in particular the fractal scale invariance is discussed in terms of the Callan-Symanzik equation. Finally, we propose a diagrammatic method for calculations of relevant quantities.

1 Introduction

Entropy emerges as an important quantity in different areas that try to describe systems of increasing complexity. The formulation of new entropic forms that generalize the one proposed by Boltzmann constitutes an important research line. In particular, the non-additive entropy introduced by Tsallis [1] has found wide applicability in the last few years, see e.g. [2, 3]. However, the full understanding of the non-extensive statistics formulated by Tsallis has not been accomplished yet. Several connections between Boltzmann and Tsallis statistics have been proposed so far, see e.g. [4–6], but it seems that the physical meaning of the entropic index q is not understood in the general case. In the present work, we make a detailed analysis of the connection based on a system featuring fractal structure in its thermodynamic properties. Fractals are conceived as objects with an internal structure that can be considered as an ideal gas of a specific number of subsystems, which, in turn, are also fractals of the same kind. The self-similarity between fractals at different levels of the internal structure reveals the typical scale invariance. The results obtained in the present work offer a new perspective in the analysis of hadron structure.

*Presented by E. Megías at the QCD@Work: International Workshop on QCD, 25-28 June 2018, Matera, Italy.
**e-mail: emegias@ugr.es
***e-mail: deppman@if.usp.br
****e-mail: tobias@ita.br
†e-mail: debora.p.m.26@gmail.com

© The Authors, published by EDP Sciences. This is an open access article distributed under the terms of the Creative Commons Attribution License 4.0 (http://creativecommons.org/licenses/by/4.0/).

2 Tsallis statistics and QCD thermodynamics

Tsallis statistics constitutes a generalization of Boltzmann-Gibbs (BG) statistics, under the assumption that the entropy of the system is non-additive. For two independent systems A and B

$$S_{A+B} = S_A + S_B + (1-q)S_A S_B\,, \tag{1}$$

where the entropic index q measures the degree of non-extensivity [1]. Let us define the q-exponential $e_q^{(\pm)}(x) = [1 \pm (q-1)x]^{\pm 1/(q-1)}$, with $e_q^{(+)}(x)$ defined for $x \geq 0$ and $e_q^{(-)}(x)$ for $x < 0$, and the q-log function $\log_q^{(\pm)}(x) = \pm(x^{\pm(q-1)} - 1)/(q-1)$. Then the grand-canonical partition function for a non-extensive ideal quantum gas at finite chemical potential is [7]

$$\log \Xi_q(V,T,\mu) \;=\; -\xi V \int \frac{d^3p}{(2\pi)^3} \sum_{r=\pm} \Theta(rx) \log_q^{(-r)} \left(\frac{e_q^{(r)}(x) - \xi}{e_q^{(r)}(x)} \right), \tag{2}$$

where $x = (E_p - \mu)/(kT)$, the particle energy is $E_p = \sqrt{p^2+m^2}$ with m the mass and μ the chemical potential, $\xi = \pm 1$ for bosons and fermions respectively, and Θ is the step function. Eq. (2) reduces to the Bose-Einstein and Fermi-Dirac partition functions in the limit $q \to 1$.

The thermodynamics of Quantum Chromodynamics (QCD) in the confined phase can be studied within the Hadron Resonance Gas (HRG) approach, which is based on the assumption that physical observables in this phase admit a representation in terms of hadronic states which are treated as non-interacting and point-like particles [8]. These states are taken as the conventional hadrons listed in the review by the Particle Data Group. Within this approach the partition function is then given by [7, 9]

$$\log \Xi_q(V,T,\{\mu\}) = \sum_i \log \Xi_q(V,T,\mu_i)\,, \tag{3}$$

where μ_i refers to the chemical potential for the *i-th* hadron. The thermodynamic functions

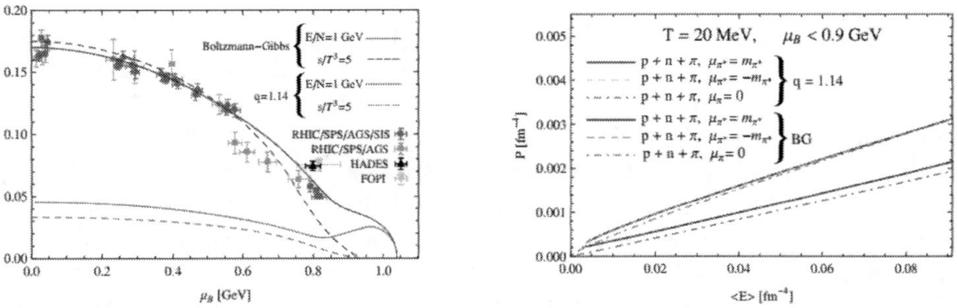

Figure 1. Left panel: Chemical freeze-out line $T = T(\mu_B)$. Right panel: EoS of hadronic matter.

can be obtained from Eq. (3) by using the standard thermodynamic relations. There are different proposals for the conditions determining the transition line between the confined and the deconfined regimes of QCD. Using the arguments of [10], the phase transition line in the $T \times \mu_B$ diagram can be determined by the condition $\langle E \rangle / \langle N \rangle = 1\,\mathrm{GeV}$. A different method based on the entropy density has been proposed in [11]. The result for the chemical freeze-out line is displayed in Fig. 1 (left). We also display in Fig. 1 (right) the Equation of State (EoS) for hadronic matter at finite baryonic chemical potential. It it remarkable that $P(E)$ becomes larger in Tsallis statistics as compared to BG statistics (see a discussion in [7, 9]).

3 Thermofractals and Tsallis statistics

The emergence of the non-extensive behavior has been attributed to different causes: long-range interactions, correlations and memory effects [4]; temperature fluctuations [5]; and finite size of the system. In this work we will study a natural derivation of non-extensive statistics in terms of thermofractals. These are systems in thermodynamical equilibrium presenting the following properties [6, 12]:

1. The total energy is given by
$$U = F + E\,, \tag{4}$$
where F is the kinetic energy of N constituent subsystems, and E is the internal energy of those subsystems, which behave as particles with an internal structure.

2. The constituent particles are thermofractals. The energy distribution $P_{\text{TF}}(E)$ is self-similar or self-affine, so that at some level n of the hierarchy of subsystems, $P_{\text{TF}(n)}(E)$ is equal to the distribution in the other levels.

3. At level n in the hierarchy of subsystems the phase space is so narrow that one can consider $P_{\text{TF}(n)}(E_n)dE_n = \rho dE_n$, with ρ being independent of the energy E_n.

The energy distribution of a thermodynamical system is given, according to BG statistics, by
$$P_{\text{BG}}(U)dU = A\exp(-U/kT)dU\,, \tag{5}$$
where A is a normalization constant. However, in the case of thermofractals the phase space must include the momentum degrees of freedom of free particles as well as their internal degrees of freedom. According to property 2 of self-similar thermofractals, the internal energy is given by [6]
$$dE = \frac{F}{kT}\,[P_{\text{TF}}(\varepsilon)]^{\nu}\,d\varepsilon\,, \tag{6}$$
where $\varepsilon/(kT) = E/F$ and ν is to be determined. Then, the total energy distribution is given by
$$P_{\text{TF}(0)}(U)dU = A'F^{\frac{3N}{2}-1}\exp\left(-\frac{\alpha F}{kT}\right)dF\,\left[P_{\text{TF}(1)}(\varepsilon)\right]^{\nu}d\varepsilon\,, \qquad \alpha = 1+\frac{\varepsilon}{kT}\,. \tag{7}$$
This expression relates the distributions at level 0 and level 1 of the subsystem hierarchy. After integration in F, one finds
$$\Omega = \int P_{\text{TF}(0)}(U)dU = \int_0^\infty A\cdot\left[1+\frac{\varepsilon}{kT}\right]^{-3N/2}\left[P_{\text{TF}(1)}(\varepsilon)\right]^{\nu}d\varepsilon\,, \qquad A = \Gamma\left[\frac{3}{2}N\right](kT)^{\frac{3}{2}N}A'\,, \tag{8}$$
where $\Gamma[z]$ is the Euler gamma function, and obviously $\Omega = 1$. At this point it is possible to impose the identity
$$P_{\text{TF}(0)}(U) \propto P_{\text{TF}(1)}(\varepsilon)\,, \tag{9}$$
corresponding to a self-similar solution for the thermofractal probability distribution. Then, the simultaneous solution of Eqs. (8) and (9) is obtained with [12, 13]
$$P_{\text{TF}(n)}(\varepsilon) = A_n\cdot\left[1+(q-1)\frac{\varepsilon}{k\tau}\right]^{-\frac{1}{q-1}}\,, \tag{10}$$
where $q-1 = 2(1-\nu)/(3N)$ and $\tau = N(q-1)T$. The energy distribution of thermofractals then obeys Tsallis statistics, and this constitutes one of the most important results of the present work. [1]

[1] In the following we will define $N' = N + 2/3$ as the effective number of particles taking into account their internal degrees of freedom (see [13] for details).

4 Callan-Symanzik equation and diagrammatic representation

Due to the evident similarities between hadron structure and thermofractal structure [12, 14–17], it is possible to show that the thermofractal description has close connections to quantum field theory as far as scaling properties are concerned. This will be done in a future work [18], but we will advance some aspects as follows.

4.1 Scale invariance and Callan-Symanzik equation

Thermofractals are scale invariant. Formally, this property should be accomplished with the scale invariance of the distribution of kinetic and internal energy. Using $F^{(0)}/T^{(0)} = F^{(n)}/T^{(n)}$ and $\varepsilon/(kT) = E^{(n)}/F^{(n)}$, one finds [13]

$$\lambda_n := \frac{E^{(n)}}{E^{(0)}} = \left(\frac{1}{N}\right)^{\frac{n}{1-D}}, \tag{11}$$

where D is the fractal dimension. Using thermofractals it can be obtained the fractal dimension of hadrons, resulting in $D = 0.69$ [6], a value close to that resulting from intermittence analysis [19]. Using these scaling properties, the vertex function of thermofractals writes [13]

$$\Gamma(E,\varepsilon,T) \propto (kT')^{-(1-D)} g \prod_{i=1}^{N'} \left(2\pi \frac{E_i}{kT'}\right)^{-3/2} [P_{\text{TF}}(\varepsilon_i)]^{\nu}, \tag{12}$$

where $T' = T^{(n)}$. Introducing $M = kT'$ as the scale parameter that determines the fractal structure of the subsystem at a certain level, then one can derive the Callan-Symanzik equation for thermofractals, which writes

$$\left[M\frac{\partial}{\partial M} + \sum_{i=1}^{N'} E_i \frac{\partial}{\partial E_i} + d\right] \Gamma = 0, \tag{13}$$

where $d = 1 - D$ is the anomalous dimension for thermofractals. This result is equivalent to the one obtained in [20].

4.2 Diagrammatic representation

Finally, let us mention that by using the formalism presented in Sec. 3, it is possible to have a diagrammatic representation of the probability densities of thermofractals that can facilitate the calculation of the partition function and other relevant quantities [13]. The basic diagrams are summarized in Fig. 2 (left). An example of diagram which gives the probability to find a constituent subsystem fractal f at the third level of the initial fractal i is displayed in Fig. 2 (right) (see [13] for further details).

5 Conclusions

We have introduced the non-extensive statistics in the form of Tsallis statistics of a quantum gas at finite temperature and chemical potential, and we have shown its application to study the EoS of QCD. In a second step we have investigated the structure of a thermodynamical system presenting fractal structure. It has been shown that the fractal structure in thermodynamics naturally leads to non-extensive statistics in the form of Tsallis. Using this formalism, we have introduced a diagrammatic formulation for practical calculations. Finally, based on the scale invariance of thermofractals, the Callan-Symanzik equation has been obtained. These results open the opportunity to develop a *'field theoretical approach'* for thermofractals, leading to a possible theoretical understanding of the non-extensive properties of hadronic systems. This study will be addressed in details in [18].

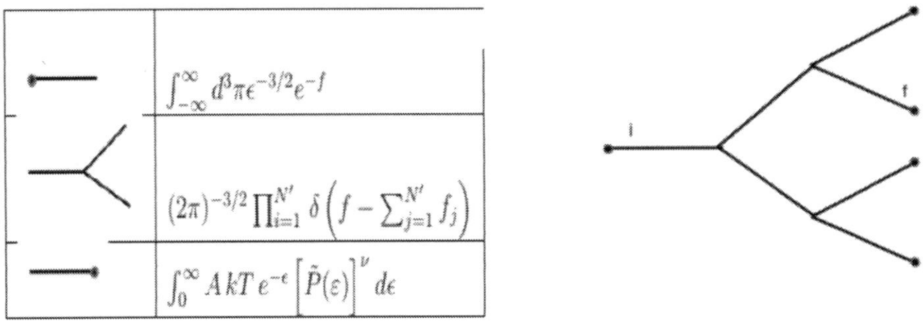

Figure 2. Left panel: Basic diagrams for thermofractals and their mathematical expressions. Right panel: Example of a tree graph representing different levels of a thermofractal.

Acknowledgement

A.D., D.P.M. and T.F. are partially supported by the Conselho Nacional de Desenvolvimento Científico e Tecnológico (CNPq-Brazil) and by Project INCT-FNA Proc. No. 464898/2014-5. T.F. thanks the partial support from Fundação de Amparo à Pesquisa do Estado de São Paulo, FAPESP Grant No. 17/05660-0. The work of E.M. is supported by the Spanish MINEICO under Grants FPA2015-64041-C2-1-P and FIS2017-85053-C2-1-P, the Junta de Andalucía under Grant FQM-225, the Basque Government under Grant IT979-16, and the Spanish Consolider Ingenio 2010 Programme CPAN (CSD2007-00042). The research of E.M. is also supported by the Ramón y Cajal Program of the Spanish MINEICO, and by the Universidad del País Vasco UPV/EHU, Bilbao, Spain, as a Visiting Professor.

References

[1] C. Tsallis, J. Statist. Phys. **52**, 479 (1988).
[2] P. Tempesta, Phys. Rev. **A84**, 021121 (2011).
[3] N. Kalogeropoulos, Int. J. Mod. Phys. **B28**, 1450162 (2014).
[4] L. Borland, Phys. Lett. **A245**, 67 (1998).
[5] G. Wilk, Z. Wlodarczyk, Phys. Rev. **C79**, 054903 (2009).
[6] A. Deppman, Phys. Rev. **D93**, 054001 (2016).
[7] E. Megias, D.P. Menezes, A. Deppman, Physica **A421**, 15 (2015).
[8] R. Hagedorn, Lect. Notes Phys. **221**, 53 (1985).
[9] D.P. Menezes, A. Deppman, E. Megias, L.B. Castro, Eur. Phys. J. **A51**, 155 (2015).
[10] J. Cleymans, K. Redlich, Phys. Rev. **C60**, 054908 (1999).
[11] A. Tawfik, Nucl. Phys. **A764**, 387 (2006).
[12] A. Deppman, E. Megias, EPJ Web Conf. **141**, 01011 (2017).
[13] A. Deppman, T. Frederico, E. Megias, D.P. Menezes, Entropy **20**, 633 (2018).
[14] A. Deppman, Physica **A391**, 6380 (2012).
[15] E. Megias, D.P. Menezes, A. Deppman, EPJ Web Conf. **80**, 00040 (2014).
[16] A. Deppman, E. Megias, D. Menezes, J. Phys. Conf. Ser. **607**, 012007 (2015).
[17] A. Deppman, Universe **3**, 62 (2017).
[18] A. Deppman, T. Frederiro, E. Megias, D.P. Menezes, in progress (2018).
[19] D. Ghosh, A. Deb, R. Chattopadhyay, S. Sarkar, A.K. Jafry, M. Lahiri, S. Das, K. Purkait, B. Biswas, J. Roychoudhury, Phys. Rev. **C58**, 3553 (1998).
[20] A. Deppman, Adv. High Energy Phys. **2018**, 9141249 (2018).

Conformal Ward Identities and the Coupling of QED and QCD to Gravity

Claudio Corianò[1,*], *Matteo Maria* Maglio[1,**]

[1] *Dipartimento di Matematica e Fisica "Ennio De Giorgi", Università del Salento and INFN-Lecce, Via Arnesano, 73100 Lecce, Italy*

Abstract. We present a general study of 3-point functions of conformal field theory (CFT) in momentum space, following a reconstruction method for tensor correlators, based on the solution of the conformal Ward identities (CWIs), introduced in recent works. We investigate and detail the structure of the CWIs and their non-perturbative solutions, and compare them to perturbation theory, taking QED and QCD as examples. Exact solutions of CFT's in the flat background limit in momentum space are matched by the perturbative realizations in free field theories, showing that the origin the conformal anomaly is related to effective scalar interactions, generated by the renormalization of the longitudinal components of the corresponding operators.

1 Introduction

Correlation functions of 3-point functions play a special role in conformal field theory (CFT) in $d = 4$ since they can be determined by solving the constraints imposed by the conformal group, which in $d = 4$ is SO(2, 4). The functional form of the solutions are determined more easily using methods of coordinate space rather than momentum space. Operator product expansion techniques, for instance, which allow to define a program for the bootstrap of higher point functions, as well as other analysis, have so far favoured coordinate space.
On the other hand, studies of CFT's based on the standard perturbative expansion in the gauge coupling, such as in $N = 4$ supersymmetric Yang-Mills theory, as far as the computation of the amplitudes is concerned, have been largely performed in momentum space. A Lagrangian realization of a CFT is, in any case, a theory which is often investigated perturbatively in terms of its Feynman diagrams. On the contrary, a direct solution of the conformal Ward identities - for a given correlation function - can be interpreted, in this respect, as being strictly non perturbative, in the sense that it is not related to any coupling but only to the underlying symmetry.
Establishing a link between the two approaches in momentum space, with exact CFT solutions on one side and Lagrangian realizations on the other, turns useful for various reasons. One of them is to provide a simplification of general CFT results, as we are going to show next. A second one is the possibility of offering a simple and physical interpretation of the origin of the conformal anomaly in 3-point functions, for correlators with one or more insertions of stress energy tensors.

* e-mail: claudio.coriano@le.infn.it
** e-mail: matteomaria.maglio@le.infn.it

© The Authors, published by EDP Sciences. This is an open access article distributed under the terms of the Creative Commons Attribution License 4.0 (http://creativecommons.org/licenses/by/4.0/).

2 The origin of the conformal anomaly

Anomalies are associated, in coordinate space, to configurations of a correlator in which all the points coalesce. For a 3-point function these are configurations of the form $\sim \delta(x_1 - x_2)\delta(x_2 - x_3)$. Configurations of partial coalescence are defined to be *semilocal* or simply *contact* terms. Two counterterms are needed in $d = 4$ in the renormalization of of correlators with multiple stress energy tensors, responsible for the generation of the scheme independent part of the anomaly: E and C^2. They correspond to the Euler-Poincare density and the Weyl tensor squared respectively, which find application in the TTT (3-graviton) vertex. For other correlators, such as the TJJ vertex the renormalization of the vector 2-point function JJ is sufficient to generate a finite correlator, which is reflected in the F^2 term of the anomaly functional, with F the field strength of the photon (see [1, 2] for related studies). The solutions of the CWIs for 3-point functions in coordinate space have been obtained in the past by superposition of a traceless contribution and of contact terms proportional to the two operators mentioned above. Obviously, the coordinate space approach provides limited information on the origin of the anomaly, except for telling us that its origin is a short distance effect.

3 Anomalies as light cone processes

Once a CFT is matched to a free field theory, one has the possibility of handling far more simplified expressions of such correlators and can proceed with the analysis of an ordinary Feynman expansion. As discussed in several previous works, the analysis in momentum space provides additional information on this phenomenon. This is described by the emergence of massless effective scalar degrees of freedom in 3-point functions containing insertions of stress energy tensors, which can be interpreted as light-cone interactions. As discussed in [3–5], this phenomenon clearly points towards an interpretation of the origin of the conformal anomaly as mediated by correlated pairs of fermions/scalars, as emerging from the spectral representation of a given perturbative correlator. More recently, it has been pointed out that such type of interactions play a role in the context of Weyl semimetals, with the paired electrons (representing the massless pole) interacting with the lattice of such materials [6]. Both chiral and conformal anomaly poles play a role in this phenomenon.

Establishing conclusively the presence of such interactions at nonperturbative level is possible for 3-point functions, by matching the perturbative result with the nonperturbative solutions derived from the CWI's. Using this approach we have conclusively shown in [7] that these interactions are associated to renormalization and are not related to specific parametrization of the tensor correlators. The proof can be illustrated more easily in the case of the TJJ correlator in QED, that we will sketch below. A similar analysis can be performed for QCD, though it is more involved.

4 Effective interactions

The key construct in this analysis is the anomaly action, which for 3-point functions can be completely fixed, modulo few constants, which are specific to a given CFT.

It has been shown that in an uncontracted anomaly vertex of either chiral, conformal [3–5] or superconformal type [8], the origin of an anomaly has to be attributed to the appearance of specific form factors in its tensor structure, which are proportional to $1/k^2$ in the massless limit. Such anomaly poles define massless exchanges in momentum space and are the direct signature of the anomaly. In all such cases k denotes the momentum of an axial-vector current in an AV V (axial-vector/vector/vector) correlator or that of a stress energy tensor (T) in a TJJ vertex.

The existence of chiral anomaly poles has been discussed in the literature in the 70's, while conformal anomaly poles have been shown to be part of the TJJ vertex in QED, QCD and the electroweak sector of the Standard Model [5, 9] only more recently. Studies of such interactions in the context of both chiral and conformal anomaly diagrams have always been performed at the perturbative level in the past, with the obvious limitations of the case. These studies show the presence of some universal features of these interactions. The chiral and conformal anomaly coefficients are then proportional to the residues of the corresponding correlators evaluated at the anomaly pole (times a tensor structure which is the anomaly functional). By combining nonperturbative information coming from the solution of the CWI's with that from free field theories it is possible to show rigorously the existence of such interpolating states.

5 Conformal Ward identities in momentum space

The CWIs in momentum space, in d dimensions, for a correlation function, in the simpler case of a scalar correlator

$$\Phi(x_1, x_2, \dots x_n) = \langle \phi_1(x_1)\phi_2(x_2)\dots\phi_3(x_n)\rangle \tag{1}$$

in momentum space take the form [10, 11]

$$\left[\sum_{j=1}^{n}\Delta_j - (n-1)d - \sum_{j=1}^{n-1} p_j^\alpha \frac{\partial}{\partial p_j^\alpha}\right]\Phi(p_1, p_2, \dots, \overline{p}_n) = 0. \tag{2}$$

(with $\bar{p}_n^\mu = -\sum_{j=1}^{n-1} p_j^\mu$) for the dilatation and

$$\sum_{j=1}^{n-1}\left(p_j^\kappa \frac{\partial^2}{\partial p_j^\alpha \partial p_j^\alpha} + 2(\Delta_j - d)\frac{\partial}{\partial p_j^\kappa} - 2p_j^\alpha \frac{\partial^2}{\partial p_j^\kappa \partial p_j^\alpha}\right)\Phi(p_1, \dots p_{n-1}, \bar{p}_n) = 0. \tag{3}$$

for the special CWIs. We will choose $n = 3$. The latter equation can be rewritten in the form

$$K^\kappa_{scalar}\Phi\,(p_1, p_2, \bar{p}_3) = 0, \tag{4}$$

in terms of the magnitudes of the momenta $p_i \equiv \sqrt{p_i^2}$ with

$$K^\kappa_{scalar} = \sum_{i=1}^{3} p_i^\kappa\, K_i \tag{5}$$

and the scalar operator

$$K_i \equiv \frac{\partial^2}{\partial p_i \partial p_i} + \frac{d+1-2\Delta_i}{p_i}\frac{\partial}{\partial p_i}. \tag{6}$$

Eq. (5) can be split into the two independent equations

$$K_{13} \equiv K_2 - K_3 = \frac{\partial^2\Phi}{\partial p_1 \partial p_1} + \frac{1}{p_1}\frac{\partial\Phi}{\partial p_1}(d+1-2\Delta_1) - \frac{\partial^2\Phi}{\partial p_3 \partial p_3} - \frac{1}{p_3}\frac{\partial\Phi}{\partial p_3}(d+1-2\Delta_3) = 0. \tag{7}$$

Following the approach presented in [12], the ansatz for the solution can be taken of the form

$$\Phi(p_1, p_2, p_3) = p_3^{\Delta-2d} x^a y^b F(x, y) \tag{8}$$

with $x = \frac{p_1^2}{p_3^2}$ and $y = \frac{p_2^2}{p_3^2}$. Here we are taking p_3 as "pivot" in the expansion, but we could equivalently choose any of the 3 momentum invariants. Φ is required to be homogeneous of degree $\Delta - 2d$, where

$\Delta = \Delta_1 + \Delta_2 + \Delta_3$ is the total scaling dimension of Φ. In the case of a scalar correlator the four fundamental solutions are expressed in terms of the Appell function

$$F_4(\alpha(a,b),\beta(a,b);\gamma(a),\gamma'(b);x,y) = \sum_{i=0}^{\infty}\sum_{j=0}^{\infty}\frac{(\alpha(a,b))_{i+j}\,(\beta(a,b))_{i+j}}{(\gamma(a))_i\,(\gamma'(b))_j}\frac{x^i}{i!}\frac{y^j}{j!} \tag{9}$$

where $(\alpha)_i = \Gamma(\alpha+i)/\Gamma(\alpha)$ is the Pochammer symbol. The general solution is expressed as linear combinations of the 4 independent special solutions of type (8) as [10, 11]

$$\Phi(p_1,p_2,p_3) = p_3^{\Delta-2d-2}\sum_{a,b} c(a,b,\vec{\Delta})\,x^a y^b\,F_4(\alpha(a,b),\beta(a,b);\gamma(a),\gamma'(b);x,y) \tag{10}$$

where the sum runs over the four values a_i, b_i $i = 0, 1$ which define the Fuchsian exponents, with constants $c(a,b,\vec{\Delta})$, with $\vec{\Delta} = (\Delta_1,\Delta_2,\Delta_3)$. They have been shown in [13] to be identical for all the type of correlators investigated so far. The CWIs correspond to hypergeometric systems of equations which can be solved in two different ways [7, 11].

6 Tensor Correlators

In the case of the TJJ vertex, the proof in CFT that such poles are not artificial exploits two different representation of the same vertex. We sketch the proof.
In a basis of 13 form factors introduced in [3] the correlator takes the form

$$\Gamma^{\mu_1\nu_1\mu_2\mu_3}(p_2,p_3) \equiv \langle T^{\mu_1\nu_1}(p_1)\,J^{\mu_2}(p_2)\,J^{\mu_3}(p_3)\rangle = \sum_{i=1}^{13} F_i(s;s_1,s_2,0)\,t_i^{\mu_1\nu_1\mu_2\mu_3}(p_2,p_3), \tag{11}$$

in terms of a suitable set of independent tensor structures $t_i^{\mu_1\nu_1\mu_2\mu_3}$ and form factors F_i. The basis is non minimal, since it is not built by imposing the conformal constraints but only the conservation and symmetry properties of the correlator. It contains only two form factors whose tensor structures are tracefull (t_1, t_2) and only one form factor which needs to be renormalized as $d \to 4$, denoted as F_{13}. The form factors F_i are functions of the kinematic invariants $s = p_1^2 = (p_2+p_3)^2$, $s_1 = p_2^2$, $s_2 = p_3^2$. This representation of the TJJ has been studied at 1-loop in QED, with an extension to QCD.
An alternative and minimal basis has been introduced in [11] using a decomposition of the same correlator in terms of transverse traceless (tjj) and longitudinal components $(t_{loc}JJ)$ in the form

$$\langle T^{\mu_1\nu_1}(p_1)\,J^{\mu_2}(p_2)\,J^{\mu_3}(p_3)\rangle = \langle t^{\mu_1\nu_1}(p_1)\,j^{\mu_2}(p_2)\,j^{\mu_3}(p_3)\rangle + \langle t_{loc}^{\mu_1\nu_1}(p_1)\,J^{\mu_2}(p_2)\,J^{\mu_3}(p_3)\rangle \tag{12}$$

with the transverse traceless sector expanded in a set of 4 form factors A_l [7, 14, 15]

$$\begin{aligned}\langle t^{\mu_1\nu_1}(p_1)j^{\mu_2}(p_2)j^{\mu_3}(p_3)\rangle =& \Pi_1{}^{\mu_1\nu_1}_{\alpha_1\beta_1}\pi_2{}^{\mu_2}_{\alpha_2}\pi_3{}^{\mu_3}_{\alpha_3}\Big(A_1\,p_2^{\alpha_1}p_2^{\beta_1}p_3^{\alpha_2}p_1^{\alpha_3} + A_2\,\delta^{\alpha_2\alpha_3}p_2^{\alpha_1}p_2^{\beta_1}\\ &+A_3\,\delta^{\alpha_1\alpha_2}p_2^{\beta_1}p_1^{\alpha_3} + A_3(p_2\leftrightarrow p_3)\delta^{\alpha_1\alpha_3}p_2^{\beta_1}p_3^{\alpha_2} + A_4\,\delta^{\alpha_1\alpha_3}\delta^{\alpha_2\beta_1}\Big).\end{aligned} \tag{13}$$

The two basis can be related and the exact solution of the CWI's fixed up to 2 constants in $d = 4$. The analysis of the exact solutions shows that the singularities in the A_i basis are in exact correspondence with the presence of F_{13}, setting a direct link between the two basis. By imposing the traceless condition in d dimensions one gets the constraints

$$\begin{aligned}F_1 &= \frac{(d-4)}{p_1^2(d-1)}\Big[F_{13} - p_2^2F_3 - p_3^2F_5 - p_2\cdot p_3F_7\Big]\\ F_2 &= \frac{(d-4)}{p_1^2(d-1)}\Big[p_2^2F_4 + p_3^2F_6 + p_2\cdot p_3F_8\Big],\end{aligned} \tag{14}$$

which as $d \to 4$ show that $F_2 \to 0$ and $F_1 \sim 1/p_1^2$ if F_{13} develops a single pole in $1/\epsilon$, as is the case in a CFT. Indeed, in the $d \to 4$ limit

$$\lim_{d\to 4} F_1 = -\frac{4}{9}\frac{\pi^2}{p_1^2}, \tag{15}$$

showing the appearance of an anomaly pole in the single form factor which is responsible for the trace anomaly. It is then clear that the emergence of an anomaly pole in the TJJ is not limited to perturbation theory but is a specific feature of the nonperturbative solution as well, if we can show that the general expression for the TJJ can be exactly matched in free field theory [7, 13] this is indeed the case. In the dispersive representation of the unique form factor which is responsible for the appearance of the anomaly, it is related to the exchange of a collinear fermion/antifermion pair in the s variable ($\rho(s) \sim \delta(s)$), where $\rho(s)$ is the spectral density of the single form factor F_1 which contributes to the trace. The study can be performed in the non conformal limit, with a nonzero mass fermion in the loop, taking the massless limit at a second stage. This configuration provides a contribution to the anomaly action of the form

$$\mathcal{S}_{pole} = -\frac{e^2}{36\pi^2}\int d^4x d^4y \left(\Box h(x) - \partial_\mu\partial_\nu h^{\mu\nu}(x)\right)\Box^{-1}_{x\,y} F_{\alpha\beta}(x) F^{\alpha\beta}(y) \tag{16}$$

with the appearance of a massless nonlocal interaction. In view of the equivalence between the perturbative and the nonperturbative solution for the A_i (and henceforth for the F- basis), this behaviour is clearly present in the nonperturbative solution, due to the complete matching between the two [7, 13].

7 Conclusions

The breaking of conformal symmetry by the conformal anomaly is an aspect of QED and QCD which is of general significance. It is associated to effective massless interactions. This provides a new perspective on the analysis of the conformal phases of realistic field theories and their renormalization group flows in $D = 4$.

References

[1] F. Bastianelli, O. Corradini, J. M. Davila, and C. Schubert, Phys. Lett. **B716**, 345 (2012)
[2] F. Bastianelli, O. Corradini, J. M. Davila, and C. Schubert, Phys. Part. Nucl. **43**, 630 (2012)
[3] M. Giannotti and E. Mottola, Phys. Rev. **D79**, 045014 (2009), arXiv:0812.0351.
[4] R. Armillis, C. Corianò, and L. Delle Rose, Phys. Rev. **D81**, 085001 (2010), arXiv:0910.3381.
[5] R. Armillis, C. Corianò, and L. Delle Rose, Phys. Rev. **D82**, 064023 (2010), arXiv:1005.4173.
[6] M. N. Chernodub, A. Cortijo, and M. A. H. Vozmediano, Phys. Rev. Lett. **120**, 20 (2018).
[7] C. Corianò and M. M. Maglio, (2018), arXiv:1802.07675.
[8] C. Corianò, A. Costantini, L. Delle Rose, and M. Serino. JHEP, **06**:136, 2014. .
[9] C. Corianò, L. Delle Rose, and M. Serino, Phys. Rev. **D83**, 2011,125028.
[10] C. Corianò, L. Delle Rose, E. Mottola, and M. Serino, JHEP **08**, 147 (2012), arXiv:1203.1339.
[11] A. Bzowski, P. McFadden, and K. Skenderis, JHEP **03**, 111 (2014)
[12] C. Corianò, L. Delle Rose, E. Mottola and M. Serino, JHEP **07**, 011 (2013)
[13] C. Corianò and M. M. Maglio, (2018), arXiv:1802.01501.
[14] A. Bzowski, P. McFadden, and K. Skenderis, (2018)
[15] A. Bzowski, P. McFadden, and K. Skenderis, JHEP **02**, 068 (2016)

The Dual QCD @ Work: 2018

Andrzej J. Buras*
[1]TUM Institute for Advanced Study, Lichtenbergstr. 2a, D-85748 Garching, Germany

Abstract. The Dual QCD (DQCD) framework, based on the ideas of 't Hooft and Witten, and developed by Bill Bardeen, Jean-Marc Gérard and myself in the 1980s is not QCD, a theory of quarks and gluons, but a successful low energy approximation of it when applied to $K \to \pi\pi$ decays and $K^0 - \bar{K}^0$ mixing. After years of silence, starting with 2014, this framework has been further developed in order to improve the SM prediction for the ratio ε'/ε, the $\Delta I = 1/2$ rule and $\hat{B}_K$. Most importantly, this year it has been used for the calculation of *all* $K \to \pi\pi$ hadronic matrix elements of BSM operators which opened the road for the general study of ε'/ε in the context of the SM effective theory (SMEFT). This talk summarizes briefly the past successes of this framework and discusses recent developments which lead to a master formula for ε'/ε valid in *any* extension of the SM. This formula should facilitate the search for new physics responsible for the ε'/ε anomaly hinted by 2015 results from lattice QCD and DQCD.

1 Introduction

Among the main stars of kaon flavour physics is the ratio ε'/ε that describes the amount of direct CP violation in $K_L \to \pi\pi$ decays relative to the indirect one. The latter one is related to $K^0 - \bar{K}^0$ mixing, where also the $K_L - K_S$ mass difference, ΔM_K, is an important quantity. Among rare kaon decays the most important roles these days are played by rare decays $K^+ \to \pi^+\nu\bar{\nu}$ and $K_L \to \pi^0\nu\bar{\nu}$, soon to be measured by NA62 and KOTO experiments, respectively.

Various aspects of these decays and observables are reviewed at length in [1, 2]. Here we will concentrate our discussion on ε'/ε that has been measured already many years ago with the experimental world average from NA48 [3] and KTeV [4, 5] collaborations given by

$$(\varepsilon'/\varepsilon)_{\rm exp} = (16.6 \pm 2.3) \times 10^{-4}\,. \qquad (1)$$

There are at least two reasons for concentrating on this ratio. First the situation of ε'/ε in the SM can be briefly summarized as follows:

- The analysis of ε'/ε by the RBC-UKQCD collaboration based on their lattice QCD calculation of $K \to \pi\pi$ matrix elements [6, 7], as well as the analyses performed in [8, 9] that are based on the same matrix elements but also include isospin breaking effects, find ε'/ε in the ballpark of $(1-2)\times 10^{-4}$. This is by one order of magnitude below the data, but with an error in the ballpark of 5×10^{-4}. Consequently, based on these analyses, one can talk about an ε'/ε anomaly of at most 3σ.

*e-mail: aburas@ph.tum.de

© The Authors, published by EDP Sciences. This is an open access article distributed under the terms of the Creative Commons Attribution License 4.0 (http://creativecommons.org/licenses/by/4.0/).

- An independent analysis based on hadronic matrix elements from the Dual QCD (DQCD) approach [10, 11] gives a strong support to these values and moreover provides an *upper bound* on ε'/ε in the ballpark of 6×10^{-4}.
- A different view has been expressed in [12] where, using ideas from chiral perturbation theory, the authors find $\varepsilon'/\varepsilon = (15 \pm 7) \times 10^{-4}$. While in agreement with the measurement, the large uncertainty, that expresses the difficulties in matching long distance and short distance contributions in this framework, does not allow for clear-cut conclusions. Consequently, values above 2×10^{-3}, that are rather unrealistic from the point of view of lattice QCD and DQCD, are not excluded in this approach.

Based on the results from RBC-UKQCD and the DQCD approach of 2015 a number of analyses have been performed in specific models beyond the SM (BSM) with the goal to obtain a sufficient upward shift in ε'/ε and thereby its experimental value. They are collected in Table 1 of [2]. An important limitation of these analyses is that they address the ε'/ε anomaly only in models in which new physics (NP) enters exclusively through modifications of the Wilson coefficients of SM operators. However, generally, BSM operators with different Dirac structures – like the ones resulting from tree-level scalar exchanges and leading to scalar and tensor operators – or chromo-magnetic dipole operators could play a significant role in ε'/ε. Until recently, no quantitative judgment of the importance of such operators was possible because of the absence of even approximate calculations of the relevant hadronic matrix elements in QCD.

This brings me to the second reason for concentrating here on ε'/ε. Namely, recently a significant theoretical progress has been made towards the general search for NP responsible for the hinted ε'/ε anomaly. This progress is based on the following analyses:

- the first to date calculations of the $K \to \pi$ and $K \to \pi\pi$ matrix elements of the chromo-magnetic dipole operators by lattice QCD [13] and DQCD [14], respectively and in particular the calculation of $K \to \pi\pi$ matrix elements of *all* four-quark BSM operators, including scalar and tensor operators, by DQCD [15].
- derivation of a master formula for ε'/ε [16], which can be applied to any theory beyond the SM in which the Wilson coefficients of all contributing operators have been calculated at the electroweak scale. The relevant hadronic matrix elements of BSM operators used in this formula are from the DQCD, as lattice QCD did not calculate them yet, and the SM ones from lattice QCD.
- the first to date model-independent anatomy of the ratio ε'/ε in the context of the $\Delta S = 1$ effective theory with operators invariant under QCD and QED and in the context of the Standard Model Effective Field Theory (SMEFT) with the operators invariant under the full SM gauge group [17].

The main goal of this talk is to review briefly these advances and to stress that they would not be possible in this decade without the existence of DQCD approach [10, 18–20].

We begin our presentation in Section 2 by recalling the general structure of the effective Hamiltonian valid in any BSM theory and summarizing the number of BSM operators one has to consider. In Section 3 we will briefly recall the main virtues of the DQCD approach. In this section we will stress the importance of the so-called *meson evolution*, an important ingredient of DQCD, which describes *analytically* very important dynamics at scales lower than 1 GeV. In Section 4 we will briefly review the present status of ε'/ε anomaly and in Section 5 we will present a master formula for ε'/ε that is valid in any theory beyond the SM [16]. In Section 6 we summarize the main lessons from the EFT and SMEFT analyses of ε'/ε in [17]. We conclude in Section 7 with an outlook for coming years.

2 Effective Hamiltonian for $K \to \pi\pi$ Beyond the SM

The technology of operator product expansion and renormalization group can easily be extended to any model containing heavy particles and new interactions. Starting with a NP scale $\Lambda_{\rm NP}$ which could still be of order of a few TeV but also much larger, like 100 TeV, one can consecutively integrate out heavy particles and construct a series of effective theories until one reaches the electroweak scale. Below it only gluons, photon and SM quarks and leptons except for the top quark appear as dynamical degrees of freedom. Subsequently the known RG evolution down to scales relevant for a given decay is performed. The resulting effective Hamiltonian at these low energy scales has the following general structure:

$$\mathcal{H}_{\rm eff} = \sum_i C_i O_i^{\rm SM} + \sum_j C_j^{\rm NP} O_j^{\rm NP} . \tag{2}$$

The four objects appearing in this formula are as follows:

- $O_i^{\rm SM}$ are local operators present in the SM and $O_j^{\rm NP}$ are new operators having typically new Dirac structures, in particular scalar-scalar and tensor-tensor operators. They will play an important role soon.
- C_i and $C_j^{\rm NP}$ are the Wilson coefficients of these operators. But whereas $C_j^{\rm NP}$ are only non-vanishing in the presence of NP,

$$C_i = C_i^{\rm SM} + \Delta_i^{\rm NP}. \tag{3}$$

This means that in the presence of NP the Wilson coefficients of SM operators are generally modified.

The amplitudes for $K \to \pi\pi$ decays can now be written as follows

$$\mathcal{A}(K \to \pi\pi) = \sum_i C_i(\mu)\langle\pi\pi|O_i^{\rm SM}(\mu)|K\rangle + \sum_j C_j^{\rm NP}(\mu)\langle\pi\pi|O_j^{\rm NP}(\mu)|K\rangle . \tag{4}$$

The coefficients C_i and $C_j^{\rm NP}$ can be calculated in the RG improved perturbation theory, although generally the analysis is rather complicated because of the presence of many operators that mix under renormalization. As far as QCD and QED corrections are concerned, they have been known already for 25 years at NLO [21–26] and for the BSM operators two-loop anomalous dimensions have been known [27, 28] for almost two decades. First steps towards NNLO predictions for ε'/ε have been made in [29–32] and further progress towards a complete NNLO result is under way [33].

In the context of the so-called SM effective theory not only QCD and QED effects below the electroweak scale but also electroweak effects above this scale and in particular the top quark Yukawa coupling have an important impact on the values of the coefficients C_i and $C_j^{\rm NP}$ at the electroweak scale. But fortunately all these effects are already summarized for EFT in [34, 35] and for SMEFT in [36–38].

Until recently only the first sum in (4) could be evaluated because the matrix elements of the new operators were unknown. The new development is the calculation of the matrix elements of all BSM four-quark operators using DQCD in the SU(3) chiral limit [15]. As it will still take some time before corresponding results in lattice QCD will be available, already these approximate results from DQCD can teach us a lot about the relevance of various operators. Putting aside the chromomagnetic dipole operator, for a given chirality there are 13 BSM four-quark operators in the EFT and 7 BSM four-quark operators in the SMEFT for which hadronic matrix elements have to be calculated. This counting is demonstrated in [15] for the case of (u, d, s) quarks. The full list of operators including those with (c, b) quarks can be found in Table 1 of [16].

3 Basics of Dual QCD Approach

3.1 Basic framework

This analytic approach to $K \to \pi\pi$ decays and $K^0 - \bar{K}^0$ mixing originated in the ideas of 't Hooft and Witten [39–42] based on large number N of colours. In this limit QCD at very low momenta becomes a theory of weakly interacting mesons with the coupling $O(1/N)$ and in particular in the strict large N limit becomes a free theory of mesons simplifying the calculations significantly. With non-interacting mesons the factorization of matrix elements of four-quark operators into matrix elements of quark currents and quark densities used adhoc in the 1970s and early 1980s is automatic and can be considered as a property of QCD in this limit. But factorization cannot be the whole story as the most important QCD effects related to asymptotic freedom are related to non-factorizable contributions generated by exchanges of gluons. In DQCD this role is played by meson loops that represent dominant non-factorizable contributions at the very low energy scale. Calculating these loops one discovers then that factorization in question does not take place at values of $\mu \geq 1\,\text{GeV}$ at which Wilson coefficients are calculated but rather at very low momentum transfer between colour-singlet currents or densities.

The explicit calculation of the contributions of pseudoscalars to hadronic matrix elements of local operators is based on a truncated chiral Lagrangian describing the low energy interactions of the lightest mesons [43–45]

$$L_{tr} = \frac{F^2}{8}\left[\mathrm{Tr}(D^\mu U D_\mu U^\dagger) + r\mathrm{Tr}(mU^\dagger + \text{h.c.}) - \frac{r}{\Lambda_\chi^2}\mathrm{Tr}(mD^2U^\dagger + \text{h.c.})\right] \tag{5}$$

where

$$U = \exp(i\sqrt{2}\frac{\Pi}{F}), \qquad \Pi = \sum_{\alpha=1}^{8} \lambda_\alpha \pi^\alpha \tag{6}$$

is the unitary chiral matrix describing the octet of light pseudoscalars. The parameter F is related to the weak decay constants $F_\pi \approx 130\,\text{MeV}$ and $F_K \approx 156\,\text{MeV}$ through

$$F_\pi = F\left(1 + \frac{m_\pi^2}{\Lambda_\chi^2}\right), \qquad F_K = F\left(1 + \frac{m_K^2}{\Lambda_\chi^2}\right), \tag{7}$$

so that $\Lambda_\chi \approx 1.1\,\text{GeV}$. The diagonal mass matrix m involving m_u, m_d and m_s is such that

$$r(\mu) = \frac{2m_K^2}{m_s(\mu) + m_d(\mu)}, \tag{8}$$

with $r(1\,\text{GeV}) \approx 3.75\,\text{GeV}$ for $(m_s + m_d)(1\,\text{GeV}) \approx 132\,\text{MeV}$.

In order to calculate the matrix elements of the local operators in question we need meson representations of colour-singlet quark currents and densities. They are directly obtained from the effective Lagrangian in (5) and are given respectively as follows

$$\bar{q}_L^b \gamma_\mu q_L^a = i\frac{F^2}{8}\left\{(\partial_\mu U)U^\dagger - U(\partial_\mu U^\dagger) + \frac{r}{\Lambda_\chi^2}\left[(\partial_\mu U)m^\dagger - m(\partial_\mu U^\dagger)\right]\right\}^{ab}, \tag{9}$$

$$\bar{q}_R^b q_L^a = -\frac{F^2}{8} r\left[U - \frac{1}{\Lambda_\chi^2}\partial^2 U\right]^{ab}, \tag{10}$$

with U turned into $U^\dagger$ under parity.

At the tree level, corresponding to leading order in $1/N$, one uses these representations to simply express the operators in terms of the meson fields and expands the matrix U in powers of $1/F$. For $K \to \pi\pi$ decays the relevant contribution to hadronic matrix elements is read off from terms involving only one kaon and two pions. However, in the large N limit the hadronic matrix elements factorize and in order to combine them with the Wilson coefficients loops in the meson theory have to be calculated. In contrast to chiral perturbation theory, in DQCD a physical cut-off Λ is used in the integration over loop momenta. As discussed in detail in [18, 20] this allows to achieve much better matching with short distance contributions than it is possible in chiral perturbation theory which uses dimensional regularization. The cut-off Λ is typically chosen around 0.7 GeV when only pseudoscalar mesons are exchanged in the loops [18] and can be increased to 0.9 GeV when contributions from lowest lying vector mesons are taken into account as done in [20]. These calculations are done in a momentum scheme but as demonstrated in [20] they can be matched to the commonly used NDR scheme. Once this is done it is justified to set $\Lambda \approx \mu$. We ask sceptical readers to study a detailed exposition of DQCD in [20] where also the differences from the usual chiral perturbation calculations are emphasized.

3.2 Grand view of DQCD

The application of DQCD to weak decays consists in any model of the following steps:

Step 1: At $\Lambda_{\rm NP}$ one integrates out the heavy degrees of freedom and performs the RG evolution including Yukawa couplings and all gauge interactions present in the SM down to the electroweak scale. This evolution involves in addition to SM operators also BSM operators. This is SMEFT.

Step 2: At the electroweak scale W, Z, top quark and the Higgs are integrated out and the SMEFT is matched on to EFT with only SM quarks but top-quark, the photon and the gluons. Subsequently QCD and QED evolution is performed down to scales $O(1\,\text{GeV})$.

Step 3: Around scales $O(1\,\text{GeV})$ the matching to the theory of mesons is performed and meson evolution to the factorization scale is performed.

Step 4: The matrix elements of all operators are calculated in the large N limit, that is using factorization of matrix elements into products of currents or densities.

We do not claim that these are all QCD effects responsible for non-leptonic transitions but these evolutions based entirely on non-factorizable QCD effects both at short distance and long distance scales appear to be the main bulk of QCD dynamics responsible for the $\Delta I = 1/2$ rule, ε'/ε and $K^0 - \bar{K}^0$ mixing.

3.3 Past successes of DQCD

Past successes of this approach have been recently reviewed in [2]. In particular the non-perturbative parameter $\hat{B}_K$ has been predicted already in 1987 to be close to its large N value 0.75 and found in 2014 with higher precision to be $\hat{B}_K = 0.73 \pm 0.02$ [20] in a very good agreement with the present FLAG average $\hat{B}_K = 0.762 \pm 0.010$ [46].

DQCD also allowed for the first time to identify already in 1986 the dominant mechanism behind the $\Delta I = 1/2$ rule [18]. In the framework of DQCD the current-current operators and not QCD penguins are responsible dominantly for this rule. In fact this dynamics is rather simple and follows the general pattern outlined above. It is just short distance (quark-gluon) evolution of current-current operators down to scales $O(1\,\text{GeV})$ followed by meson evolution down to scales $O(m_\pi)$ at which the hadronic matrix elements factorize and can easily be calculated. Improved calculations in 2014 [20] resulted in

$$\left(\frac{\mathrm{Re}A_0}{\mathrm{Re}A_2}\right)_{\mathrm{DQCD}} = 16.0 \pm 1.5, \qquad \left(\frac{\mathrm{Re}A_0}{\mathrm{Re}A_2}\right)_{\mathrm{lattice\ QCD}} = 31.0 \pm 11.1\,, \tag{11}$$

where we also give the result from the RBC-UKQCD collaboration [6]. Also lattice result is governed by current-current operators. But the uncertainty is still very large and it will be interesting to see whether the lattice QCD will be able to come closer to the data

$$\left(\frac{\mathrm{Re}A_0}{\mathrm{Re}A_2}\right)_{\mathrm{exp}} = 22.35\,, \tag{12}$$

than it is possible using DQCD.

I doubt that the remaining piece can be fully explained by NP as this would lead to a large fine-tuning in ΔM_K as demonstrated in [47]. But as analysed in that paper a colour octet of heavy gluons (G') could bring the theory closer to the data. It is however likely that also final state interactions (FSI) as stressed by Pich and collaborators and additional subleading corrections not included in the DQCD result in (11) could be responsible for the missing piece. Yet, I do not think that the present analytic methods like DQCD or the methods advocated by Pich and collaborators, as reviewed recently in [12], are sufficiently powerful to answer the question at which level NP enters the amplitudes $\mathrm{Re}A_0$ and $\mathrm{Re}A_2$. Here lattice QCD should provide valuable answers and I am looking forward to improved results on these two amplitudes from RBC-UKQCD collaboration and other lattice groups. This would provide two additional important constraints on NP models.

DQCD is really not QCD, the theory of quarks and gluons, but as we have seen above and we will see below a successful low energy approximation of QCD. Moreover, it has several virtues:

- It is an efficient approximate method for obtaining results for non-leptonic decays, years and even decades before useful results from numerically sophisticated and demanding lattice calculations could be obtained.
- It is the only existing method allowing to study analytically the dominant dynamics below 1 GeV scale, represented by the meson evolution, which turns out to have the pattern of operator mixing, both for SM and BSM operators, to agree with the one found perturbatively at short distance scales. This allows for satisfactory, even if approximate, matching between Wilson coefficients and hadronic matrix elements.
- It provides insight in the purely numerical results obtained by lattice QCD for both $K^0-\bar{K}^0$ hadronic matrix elements of BSM operators and of SM QCD and electroweak penguin operators relevant for ε'/ε. See below.
- Most importantly it allowed already now to evaluate $K \to \pi\pi$ matrix elements of all BSM operators providing in this manner for the first time a global view on NP contributions to ε'/ε.

4 More on ε'/ε Anomaly

While the results based on the hadronic matrix elements from RBC-UKQCD lattice collaboration, suggest some evidence for the presence of NP in hadronic K decays, the large uncertainties in the hadronic matrix elements in question do not yet preclude that eventually the SM will agree with data. In this context the upper bounds on the matrix elements of the dominant penguin operators from DQCD [10] are important as they give presently the strongest support to the anomaly in question, certainly stronger than present lattice results. To see this in explicit terms let us look at the parameters $B_6^{(1/2)}$ and $B_8^{(3/2)}$ that represent the relevant hadronic matrix elements of the QCD penguin operator Q_6 and the electroweak penguin operator Q_8, respectively.

In the strict large N limit [19, 44, 48] one simply has $B_6^{(1/2)} = B_8^{(3/2)} = 1$, but RBC-UKQCD results [6, 7] imply [8, 49]

$$B_6^{(1/2)} = 0.57 \pm 0.19\,, \qquad B_8^{(3/2)} = 0.76 \pm 0.05\,, \qquad \text{(RBC-UKQCD)}, \qquad (13)$$

and this suppression of both parameters below unity, in particular of $B_6^{(1/2)}$, is the main origin of the strong suppression of ε'/ε within the SM below the data. Yet in view of the large error in $B_6^{(1/2)}$ one could be sceptical about claims made by me and my collaborators that there is NP in ε'/ε. Future lattice results could in principle raise $B_6^{(1/2)}$ towards its large N value and above $B_8^{(3/2)}$ bringing the SM result for ε'/ε close to its experimental value.

However, the analyses of $B_6^{(1/2)}$ and $B_8^{(3/2)}$ within DQCD in [10, 11] show that such a situation is rather unlikely. Indeed, in this approach going beyond the strict large N limit one can understand the suppression of $B_6^{(1/2)}$ and $B_8^{(3/2)}$ below the unity as the effect of the meson evolution from scales $\mu = \mathcal{O}(m_\pi)$ at which $B_6^{(1/2)} = B_8^{(3/2)} = 1$ is valid to $\mu = \mathcal{O}(1\,\text{GeV})$ at which Wilson coefficients of Q_6 and Q_8 are evaluated [10]. This evolution has to be matched to the usual perturbative quark evolution for scales higher than 1 GeV and in fact the supressions in question and the property that $B_6^{(1/2)}$ is more strongly suppressed than $B_8^{(3/2)}$ are consistent with the perturbative evolution of these parameters above $\mu = \mathcal{O}(1\,\text{GeV})$. Thus we are rather confident that [10]

$$B_6^{(1/2)} < B_8^{(3/2)} < 1 \qquad \text{(dual QCD)}. \qquad (14)$$

Explicit calculation in this approach gives $B_8^{(3/2)}(m_c) = 0.80 \pm 0.10$. The result for $B_6^{(1/2)}$ is less precise but in agreement with (13). For further details, see [10].

It should be recalled that in the past values $B_6^{(1/2)} = B_8^{(3/2)} = 1.0$ have been combined in phenomenological applications with the Wilson coefficients evaluated at scales $\mu = \mathcal{O}(1\,\text{GeV})$. The discussion above shows that this is incorrect. The meson evolution from $\mu = \mathcal{O}(m_\pi)$ to $\mu = \mathcal{O}(1\,\text{GeV})$ has to be performed and this effect turns out to be stronger than the scale dependence of $B_6^{(1/2)}$ and $B_8^{(3/2)}$ in the perturbative regime, where it is very weak.

Additional support for the small value of ε'/ε in the SM comes from the reconsideration of the role of final state interactions (FSI) in ε'/ε in [11]. In this paper using the DQCD approach we have demonstrated that in contrast to claims made by chiral perturbation theory practitioners, in particular Pich and collaborators as summarised recently in [12], FSI are not expected to be important for ε'/ε. But we agree with these authors that FSI are likely to be important for the $\Delta I = 1/2$ rule. I should remark that in private conversations the authors of [12] disagree with our arguments and are convinced that FSI enhance the QCD penguin contribution to ε'/ε bringing it within the SM to agree with the data. Unfortunately no public response in arxiv to our claim in [11] has been made by these authors and they were unable to demonstrate in private conversations that we are wrong. While lattice QCD calculations are likely to resolve this controversy in due time, it would be desirable if other experts had a look at [11] in order to confirm our claim or invalidate it explicitly in public.

But the controversy on FSI is not the whole story. Also the existence of meson evolution, crucial in the DQCD approach, has been questioned over the last 30 years by some chiral and lattice experts. We have demonstrated in [50] that meson evolution is crucial in reproducing the results for four BSM $K^0 - \bar{K}^0$ matrix elements from lattice QCD (ETM, SWME and

RBC-UKQCD) collaborations [51–53]. The important virtue of this exercise is the absence of FSI so that the meson evolution can be better exposed than in $K \to \pi\pi$ decays. In this analysis it is also demonstrated that in contrast to $\hat{B}_K$, the strict large N limit in the case of the BSM parameters B_{2-5} is a very bad approximation missing lattice results by factors of $2-3$. Including non-factorizable contributions represented by meson evolution allows to understand the lattice data.

The latter analysis demonstrates the importance of the QCD dynamics at scales below 1 GeV and gives additional support to our claim that meson evolution is the dominant QCD dynamics responsible for the $\Delta I = 1/2$ rule and also the ε'/ε anomaly. We are not aware of any analytical approach that could provide such insight in lattice QCD results in question. We challenge the chiral perturbation theory experts to provide an insight into the values of B_i parameters from Lattice QCD in their framework, in particular without using lower energy constants obtained from lattice QCD. Motivated by this result we now turn our attention to NP.

5 Master Formula for ε'/ε beyond the SM

Having both the RG evolution and all matrix elements at the low-energy scale μ for the first time at hand allowed us recently [16] to present a master formula for $(\varepsilon'/\varepsilon)_{\rm BSM}$ that exhibits its dependence on each Wilson coefficient at the scale $\mu_{\rm ew}$ and consequently is valid in *any* theory beyond the SM that is free from non-standard light degrees of freedom below the electroweak scale. We will now present this formula. Technical details which led to this formula can be found in [16] and in particular in [17].

Writing ε'/ε as a sum of the SM and BSM contributions,

$$\frac{\varepsilon'}{\varepsilon} = \left(\frac{\varepsilon'}{\varepsilon}\right)_{\rm SM} + \left(\frac{\varepsilon'}{\varepsilon}\right)_{\rm BSM}, \tag{15}$$

the master formula of [16] for the BSM part then reads

$$\left(\frac{\varepsilon'}{\varepsilon}\right)_{\rm BSM} = \sum_i P_i(\mu_{\rm ew})\,{\rm Im}\left[C_i(\mu_{\rm ew}) - C_i'(\mu_{\rm ew})\right] \times (1\,{\rm TeV})^2, \tag{16}$$

where

$$P_i(\mu_{\rm ew}) = \sum_j \sum_{I=0,2} p_{ij}^{(I)}(\mu_{\rm ew},\mu)\left[\frac{\langle O_j(\mu)\rangle_I}{\rm GeV^3}\right], \tag{17}$$

with the sum over i extending over the Wilson coefficients C_i of all operators and their chirality-flipped counterparts, that is $36 + 36'$ linearly independent four-quark operators and $1 + 1'$ chromo-magnetic dipole operators. The C_i' are the Wilson coefficients of the corresponding chirality-flipped operators obtained by interchanging $P_L \leftrightarrow P_R$. The relative minus sign accounts for the fact that their $K \to \pi\pi$ matrix elements differ by a sign. Among the contributing operators are also operators present already in the SM but their Wilson coefficients in (16) include only BSM contributions. In view of space limitations we cannot list all these operators here and the readers are invited to look up our last three papers [15–17].

The dimensionless coefficients $p_{ij}^{(I)}(\mu_{\rm ew},\mu)$ include the QCD and QED RG evolution from $\mu_{\rm ew}$ to μ for each Wilson coefficient as well as the relative suppression of the contributions to the $I = 0$ amplitude due to ${\rm Re}A_2/{\rm Re}A_0 \ll 1$ for the matrix elements $\langle O_j(\mu)\rangle_I$ of all the operators O_j present at the low-energy scale. The index j includes also i so that the effect of self-mixing is included. The $P_i(\mu_{\rm ew})$ do not depend on μ to the considered order, because the

μ-dependence cancels between matrix elements and the RG evolution operator. Moreover, it should be emphasized that their values are *model-independent* and depend only on the SM dynamics below the electroweak scale, which includes short distance contributions down to μ and the long distance contributions represented by the hadronic matrix elements. The BSM dependence enters our master formula in (16) *only* through the Wilson coefficients $C_i(\mu_{\rm ew})$ and $C'_i(\mu_{\rm ew})$. That is, even if a given P_i is non-zero, the fate of its contribution depends on the difference of these two coefficients. In particular, in models with exact left-right symmetry this contribution vanishes as first pointed out in [54].

The numerical values of the $P_i(\mu_{\rm ew})$ are collected in the tables presented in [16, 17]. As seen in (17), the P_i depend on the hadronic matrix elements $\langle O_j(\mu)\rangle_I$ and the RG evolution factors $p_{ij}^{(I)}(\mu_{\rm ew},\mu)$. The numerical values of the hadronic matrix elements rely on lattice QCD in the case of SM operators and DQCD in the case of BSM operators as summarized above.

Inspecting the results in the tables in [17] the following observations can be made, here given in a simplified form in terms of three islands of hadronic matrix elements: SM-like island governed by SM operators $Q_{7,8}$, chromo-magnetic one governed by chromo-magnetic penguin operator O_{8g} and the BSM-island conquered in [15]:

- The largest P_i values on SM-like island can be traced back to the large values of the matrix elements $\langle Q_{7,8}\rangle_2$, the dominant electroweak penguin operators in the SM, and the enhancement by 22 of the $I = 2$ contributions relative to $I = 0$ ones.
- The small P_i values on the chromo-magnetic island are the consequence of the fact that each one is proportional to $\langle O_{8g}\rangle_0$, which has recently been found to be much smaller than previously expected [13, 14]. Moreover, as $\langle O_{8g}\rangle_2 = 0$, all contributions in this class are suppressed by the factor 22 relative to contributions from other classes.
- The large P_i values on the BSM-island can be traced back to the large hadronic matrix elements of scalar and tensor operators calculated recently in [15].

The question then arises which creatures are living on these islands. Here comes the answer.

6 Lessons from the SMEFT Analysis of ε'/ε

In [17] we have presented for the first time a model-independent anatomy of the ratio ε'/ε in the context of the $\Delta S = 1$ EFT with operators invariant under QCD and QED and in the context of the Standard Model Effective Field Theory (SMEFT) with the operators invariant under the full SM gauge group. This was only possible thanks to the very recent calculations of the $K \to \pi\pi$ matrix elements of BSM operators, namely of the chromo-magnetic dipole operators by lattice QCD [13] and DQCD [14] and in particular through the calculation of matrix elements of all four-quark BSM operators, including scalar and tensor operators, by DQCD [15]. Even if the latter calculations have been performed in the chiral limit, they offer for the first time a look into the world of BSM operators contributing to ε'/ε.

Our main goal was to identify those NP scenarios which are probed by ε'/ε and which could help to explain the emerging anomaly in ε'/ε discussed above.

In the last three years a number of analyses, addressing the ε'/ε anomaly in concrete models, appeared in the literature. They are collected in Table 1 of [2] and also listed in [17]. But all these analyses concentrated on models in which NP entered exclusively through modifications of the Wilson coefficients of SM operators. In particular the Wilson coefficient of the dominant electroweak penguin operator Q_8. Thus these analyses have been performed on the SM-like island. This is a significant limitation if one wants to have a general view of possible BSM scenarios responsible for the ε'/ε anomaly. In particular, in the absence of even approximate values of hadronic matrix elements of BSM operators, no complete model-independent analysis was possible until recently.

The recent calculations of BSM $K \to \pi\pi$ matrix elements, in particular of those of scalar and tensor operators in [15] allowed to perform the analysis also on the BSM-island which combined with the EFT and in particular SMEFT technology in [17] widened significantly our view on BSM contributions to ε'/ε. The analysis in [17] has two main virtues:

- It opens the road to the analyses of ε'/ε in any theory beyond the SM and allows with the help of the master formula in (16) to search very efficiently for BSM scenarios behind the ε'/ε anomaly. In particular the values of P_i collected in these two papers indicate which routes could be more successful than others both in the context of the low-energy EFT and SMEFT. By implementing our results in the open source code `flavio` [55], testing specific BSM theories becomes particularly simple.
- Through our SMEFT analysis we were able to identify correlations between ε'/ε and various observables that depend sensitively on the operators involved. Here $\Delta S = 2$, $\Delta C = 2$ and electro-magnetic dipole moments (EDM) play a prominent role but also correlations with $\Delta S = 1$ and $\Delta C = 1$ provide valuable informations. We ask the interested readers to look at our papers.

Our main messages to take home are as follows:

- Tree-level vector exchanges, like Z', W' and G' contributions (SM-like island), discussed already by various authors and vector-like quarks [56] can be responsible for the observed anomaly but generally one has to face important constraints from $\Delta S = 2$ and $\Delta C = 2$ transitions as well as direct searches and often some fine tuning is required. Here the main role is played by the electroweak penguin operator Q_8 with its Wilson coefficient significantly modified by NP.
- Models with tree-level exchanges of heavy colourless or coloured scalars (BSM-island) are a new avenue, opened with the results for BSM operators from DQCD in [15]. In particular scalar and tensor operators, having chirally enhanced matrix elements and consequently large coefficients P_i, are candidates for the explanation of the anomaly in question. Moreover, some of these models, in contrast to models with tree-level Z' and G' exchanges, are free from both $\Delta S = 2$ and $\Delta C = 2$ constraints. The EDM of the neutron is an important constraint for these models, depending on the couplings, but does not preclude a sizable NP effect in ε'/ε.
- Models with modified $W^{\pm}$ or Z^0 couplings can induce sizable effects in ε'/ε without appreciable constraints from semi-leptonic decays such as $K^+ \to \pi^+\nu\bar{\nu}$ or $K_L \to \pi^0 l^+ l^-$. In the case of a SM singlet Z' mixing with the Z^0, sizable Z^0-mediated contributions are disfavoured by electroweak precision tests. Yet, as discussed in [2] and in references given there, also such models could play some role in understanding the role of NP in ε'/ε. This is in particular the case of model with vector-like quarks [56].

The future of ε'/ε in the SM and in the context of searches for NP will depend on how accurately it can be calculated. This requires improved lattice calculations not only of the matrix elements of SM operators but also of the BSM ones, which are known presently only from the DQCD approach in the chiral limit. Moreover, the impact of FSI on the values of P_i have to be investigated. As these values are dominated by $I = 2$ matrix elements we expect that these effects amount to at most $10 - 20\%$ of the present values. It is also hoped that lattice QCD will be able to take into account isospin breaking corrections and that other lattice collaborations will attempt to calculate hadronic matrix elements of all relevant operators. In this context we hope that the new analysis of the RBC-UKQCD collaboration with improved matrix elements to be expected this year will shed new light on the hinted anomaly. Such future updates can be easily accounted for by the supplementary details on the master formula in the appendices of [17].

On the short-distance side the NNLO results for QCD penguins should be available soon [33]. The dominant NNLO corrections to electroweak penguins have been calculated almost 20 years ago [30]. As pointed out in [17] all the existing estimates of ε'/ε at NLO suffer from short-distance renormalization scheme uncertainties in the electroweak penguin contributions and also scale uncertainties in $m_t(\mu)$ that are removed only in the NNLO matching at the electroweak scale [30]. In the naive dimensional regularization (NDR) scheme, used in all recent analyses, these corrections enhance the electroweak penguin contribution by roughly 16%, thereby leading to a *negative* shift of -1.3×10^{-4} decreasing further the value of ε'/ε, similarly to isospin breaking effects. This could appear small in view of other uncertainties. However, on the one hand, potential scale and renormalization scheme uncertainties have been removed in this manner and on the other hand, one day such corrections could turn out to be relevant. Finally, the fact that this correction further decreases ε'/ε within the SM gives another motivation for the search for new physics responsible for it, and thus for our recent analyses.

As far as BSM operators are concerned, a NLO analysis of their Wilson coefficients is in progress, but its importance is not as high as of hadronic matrix elements due to significant additional parametric uncertainties residing in any NP model. In any case, in the coming years the ratio ε'/ε is expected to play a significant role in the search for NP. In this respect, the results presented in [16, 17] will be helpful in disentangling potential models of new CP violating sources beyond the SM as well as constraining the magnitude of their effects.

7 Outlook

Our outlook is very short. The future of ε'/ε, in particular in correlation with rare decays $K^+ \to \pi^+ \nu\bar{\nu}$ and $K_L \to \pi^0 \nu\bar{\nu}$, looks great and the coming years should be very exciting. I am looking forward to QCD@Work 2020 when the landscape of NP will be more transparent than it is now.

Acknowledgements

It is a pleasure to thank the organizers of this workshop for inviting me to this interesting event and for an impressive hospitality. Particular thanks go to my four great collaborators: Jason Aebischer, Christoph Bobeth, Jean-Marc Gérard and David Straub for fantastic time we spent together in the last six months. This research was supported by the DFG cluster of excellence "Origin and Structure of the Universe".

References

[1] A.J. Buras, J. Girrbach, Rept. Prog. Phys. **77**, 086201 (2014), 1306.3775
[2] A.J. Buras, Acta Phys. Polon. **B49**, 1043 (2018), 1805.11096
[3] J. Batley et al. (NA48), Phys. Lett. **B544**, 97 (2002), hep-ex/0208009
[4] A. Alavi-Harati et al. (KTeV), Phys. Rev. **D67**, 012005 (2003), hep-ex/0208007
[5] E. Abouzaid et al. (KTeV), Phys. Rev. **D83**, 092001 (2011), 1011.0127
[6] Z. Bai et al. (RBC, UKQCD), Phys. Rev. Lett. **115**, 212001 (2015), 1505.07863
[7] T. Blum et al., Phys. Rev. **D91**, 074502 (2015), 1502.00263

[8] A.J. Buras, M. Gorbahn, S. Jäger, M. Jamin, JHEP **11**, 202 (2015), 1507.06345
[9] T. Kitahara, U. Nierste, P. Tremper (2016), 1607.06727
[10] A.J. Buras, J.M. Gerard, JHEP **12**, 008 (2015), 1507.06326
[11] A.J. Buras, J.M. Gerard, Eur. Phys. J. **C77**, 10 (2017), 1603.05686
[12] H. Gisbert, A. Pich, Rept. Prog. Phys. **81**, 076201 (2018), 1712.06147
[13] M. Constantinou, M. Costa, R. Frezzotti, V. Lubicz, G. Martinelli, D. Meloni, H. Panagopoulos, S. Simula (ETM), Phys. Rev. **D97**, 074501 (2018), 1712.09824
[14] A.J. Buras, J.M. Gérard, JHEP **07**, 126 (2018), 1803.08052
[15] J. Aebischer, A.J. Buras, J.M. Gérard (2018), 1807.01709
[16] J. Aebischer, C. Bobeth, A.J. Buras, J.M. Gérard, D.M. Straub (2018), 1807.02520
[17] J. Aebischer, C. Bobeth, A.J. Buras, D.M. Straub (2018), 1808.00466
[18] W.A. Bardeen, A.J. Buras, J.M. Gérard, Phys. Lett. **B192**, 138 (1987)
[19] A.J. Buras, J.M. Gérard, Nucl. Phys. **B264**, 371 (1986)
[20] A.J. Buras, J.M. Gérard, W.A. Bardeen, Eur. Phys. J. **C74**, 2871 (2014), 1401.1385
[21] A.J. Buras, M. Jamin, M.E. Lautenbacher, P.H. Weisz, Nucl. Phys. **B370**, 69 (1992)
[22] A.J. Buras, M. Jamin, M.E. Lautenbacher, P.H. Weisz, Nucl. Phys. **B400**, 37 (1993), hep-ph/9211304
[23] A.J. Buras, M. Jamin, M.E. Lautenbacher, Nucl. Phys. **B400**, 75 (1993), hep-ph/9211321
[24] M. Ciuchini, E. Franco, G. Martinelli, L. Reina, Phys. Lett. **B301**, 263 (1993), hep-ph/9212203
[25] A.J. Buras, M. Jamin, M.E. Lautenbacher, Nucl. Phys. **B408**, 209 (1993), hep-ph/9303284
[26] M. Ciuchini, E. Franco, G. Martinelli, L. Reina, Nucl. Phys. **B415**, 403 (1994), hep-ph/9304257
[27] M. Ciuchini, E. Franco, V. Lubicz, G. Martinelli, I. Scimemi et al., Nucl. Phys. **B523**, 501 (1998), hep-ph/9711402
[28] A.J. Buras, M. Misiak, J. Urban, Nucl. Phys. **B586**, 397 (2000), hep-ph/0005183
[29] C. Bobeth, M. Misiak, J. Urban, Nucl. Phys. **B574**, 291 (2000), hep-ph/9910220
[30] A.J. Buras, P. Gambino, U.A. Haisch, Nucl. Phys. **B570**, 117 (2000), hep-ph/9911250
[31] M. Gorbahn, U. Haisch, Nucl. Phys. **B713**, 291 (2005), hep-ph/0411071
[32] J. Brod, M. Gorbahn, Phys. Rev. **D82**, 094026 (2010), 1007.0684
[33] M. Cérdà-Sevilla, M. Gorbahn, S. Jäger, A. Kokulu, J. Phys. Conf. Ser. **800**, 012008 (2017), 1611.08276
[34] J. Aebischer, A. Crivellin, M. Fael, C. Greub, JHEP **05**, 037 (2016), 1512.02830
[35] E.E. Jenkins, A.V. Manohar, P. Stoffer, JHEP **03**, 016 (2018), 1709.04486
[36] E.E. Jenkins, A.V. Manohar, M. Trott, JHEP **10**, 087 (2013), 1308.2627
[37] E.E. Jenkins, A.V. Manohar, M. Trott, JHEP **01**, 035 (2014), 1310.4838
[38] R. Alonso, E.E. Jenkins, A.V. Manohar, M. Trott, JHEP **04**, 159 (2014), 1312.2014
[39] G. 't Hooft, Nucl. Phys. **B72**, 461 (1974)
[40] G. 't Hooft, Nucl. Phys. **B75**, 461 (1974)
[41] E. Witten, Nucl. Phys. **B160**, 57 (1979)
[42] S.B. Treiman, E. Witten, R. Jackiw, B. Zumino, *Current Algebra and Anomalies* (Singapore, Singapore: World Scientific (1985) 537p, 1986)
[43] R.S. Chivukula, J. Flynn, H. Georgi, Phys.Lett. **B171**, 453 (1986)
[44] W.A. Bardeen, A.J. Buras, J.M. Gérard, Phys. Lett. **B180**, 133 (1986)

[45] W.A. Bardeen, A.J. Buras, J.M. Gérard, Nucl. Phys. **B293**, 787 (1987)
[46] S. Aoki et al., Eur. Phys. J. **C77**, 112 (2017), 1607.00299
[47] A.J. Buras, F. De Fazio, J. Girrbach, Eur. Phys. J. **C74**, 2950 (2014), 1404.3824
[48] A.J. Buras, J.M. Gérard, Phys. Lett. **B192**, 156 (1987)
[49] A.J. Buras, D. Buttazzo, J. Girrbach-Noe, R. Knegjens, JHEP **11**, 033 (2015), 1503.02693
[50] A.J. Buras, J.M. Gerard (2018), 1804.02401
[51] N. Carrasco, P. Dimopoulos, R. Frezzotti, V. Lubicz, G.C. Rossi, S. Simula, C. Tarantino (ETM), Phys. Rev. **D92**, 034516 (2015), 1505.06639
[52] B.J. Choi et al. (SWME), Phys. Rev. **D93**, 014511 (2016), 1509.00592
[53] P. Boyle, N. Garron, J. Kettle, A. Khamseh, J.T. Tsang, EPJ Web Conf. **175**, 13010 (2018), 1710.09176
[54] G.C. Branco, J.M. Frere, J.M. Gerard, Nucl. Phys. **B221**, 317 (1983)
[55] D.M. Straub et al., *flavio – flavour phenomenology in the standard model and beyond*, https://flav-io.github.io
[56] C. Bobeth, A.J. Buras, A. Celis, M. Jung, JHEP **04**, 079 (2017), 1609.04783

AUTHOR INDEX

EDP Sciences – Web of Conferences
17, Avenue du Hoggar
Parc d'Activité de Courtabouf
BP 112
F-91944 Les Ulis Cedex A
France

ISBN 978-1-5108-7478-7

9 781510 874787

CURRAN ASSOCIATES INC.
proceedings.com

International Journal on Advances in Intelligent Systems

2017 Volume 10 Number 3&4

ISSN: 1942-2679

IARIA